METHODS IN MOLECULAR BIOLOGY™

AF180905

Series Editor
John M. Walker
School of Life Sciences
University of Hertfordshire
Hatfield, Hertfordshire, AL10 9AB, UK

For other titles published in this series, go to
www.springer.com/series/7651

MAP Kinase Signaling Protocols

Second Edition

Edited by

Rony Seger

Department of Biological Regulation, The Weizmann Institute of Science, Rehovot, Israel

Humana Press

Editor
Rony Seger, Ph.D.
Department of Biological Regulation
The Weizmann Institute of Science
Rehovot, Israel
rony.seger@weizmann.ac.il

ISSN 1064-3745 e-ISSN 1940-6029
ISBN 978-1-4939-6105-4 ISBN 978-1-60761-795-2 (eBook)
DOI 10.1007/978-1-60761-795-2
Springer New York Dordrecht Heidelberg London

Cover illustration: Figure 3 from Chapter 26

Printed on acid-free paper

Humana Press is part of Springer Science+Business Media (www.springer.com)

Preface

Mitogen-activated protein (MAP) kinase (MAPK) cascades are parallel signaling pathways that mediate the intracellular transmission of many extracellular cues. As of today, four such cascades have been identified, and these are characterized by a sequential phosphorylation and activation of several protein kinases, organized in 3–5 tiers. These cascades cooperate with each other and with additional intracellular signaling pathways to form a complex network that is able to direct signals to their proper destination in a timely fashion. Thus, these cascades operate as major lines of communication that are responsive to most membranal receptors and regulate many cellular processes including proliferation, differentiation, development, oncogenic transformation, stress response, and apoptosis. Not less than 55,000 papers on MAPKs have been published to date, and the number of new publications is still very high, despite 20 years of research in the field. This reflects the importance of the cascades in a wide array of distinct systems and particularly in the induction or cure of severe diseases, such as diabetes, inflammation, and cancer.

Although much information has been gathered on the kinetics and the role of the MAPK cascade, there are still many challenges in the field that need to be met. Since each of the cascades is responsive to a large number of distinct and even opposing signals, one of the main questions being studied today is the specificity determination of the MAPK cascades. Indeed, several specificity-directing mechanisms have been proposed, including spatio-temporal regulation, protein–protein interaction, and distinctly functioning alternatively spliced isoforms of components of the MAPK cascades. However, these mechanisms still require clarification in order for us to fully understand their function in determining the MAPK-dependent physiological functions. Another challenge is to utilize the MAPK cascades for the cure of severe diseases, such as cancer and diabetes. In this sense, the ERK cascade was shown to act downstream of most oncogenes in more than 85% of cancer types. Therefore, inhibition of the cascade could, in principle, be a general useful tool in combating cancer. However, this seems to be problematic, as specific inhibitors of the cascade that have recently been developed have shown very limited efficacy in only a few types of cancer. The main reason for this ineffectiveness is thought to be inhibition of negative feedback loops normally induced by the ERK cascade. This causes induction of parallel signaling cascades, normally suppressed by the ERK cascade, that overcome the partial ERK inhibition by the developed drugs, and thereby allow survival and proliferation. Therefore, reagents that inhibit the ERK signaling to proliferation without affecting the negative feedback loop, or other types of inhibitors with higher efficacy, should be developed to allow better treatment of cancer. Finally, it was recently shown that ERK1/2/5 can either act as transcription factors or transcription receptors or can activate nuclear enzymes by phosphorylation-independent mechanisms. These represent a new facet of MAPK signaling action and may indicate that more unexpected cellular mechanisms regulated by the cascades would be identified in the future. Such mechanisms, as well as the full scope of the currently known processes, should be identified using a systematic approach for the study of MAPK that is an important immediate challenge in the field.

The aim of this book is to provide updated information on the various techniques used in the study of MAPK signaling cascades in various cellular contexts. For this purpose, many of the leading investigators in the field contributed chapters that describe the common techniques used in their laboratories. The techniques covered in the book can be subdivided into six distinct sections which describe: (1) activation and function of components of the MAPK signaling cascades; (2) the study of MAPK cascades as transmitters of membranal receptor signals; (3) structure–function relationships of MAPKs; (4) studies on the regulation of MAPK cascades; (5) the use of lower organisms, animal models, and human genetics in the study of MAPKs; and (6) the study of MAPKs in specific systems and diseases. The 32 chapters of this book should provide enough information for both newcomers as well as veterans in the field on how to perform the necessary experiments. Users of this book would generally be biochemists and cell biologists from various fields of interest, who would like to study MAPK signaling in their experimental systems. However, the book will also interest physicians who would like to study the involvement of MAPK cascades in health and disease, as well as biotechnologists who are interested in the use of MAPK signaling as readout for the influence of drugs. We certainly hope that this cutting-edge and comprehensive protocol book will facilitate the study of MAPKs and allow quicker progress in our knowledge of many cellular processes as well as diseases.

Rehovot, Israel *Rony Seger*

Contents

PART VI STUDY OF MAP KINASES IN SPECIFIC PHYSIOLOGICAL SYSTEMS

Contributors

NATALIE G. AHN • *Department of Chemistry and Biochemistry, HHMI, University of Colorado, Boulder, CO, USA*
ADNAN AL-AYOUBI • *Department of Cell and Molecular Pharmacology and Experimental Therapeutics, Medical University of South Carolina, Charleston SC, USA*
STEPHEN P. ARMSTRONG • *Henry Wellcome Laboratories for Integrative Neuroscience and Endocrinology (LINE), Department of Clinical Sciences at South Bristol, University of Bristol, Bristol, UK*
MANUELA BACCARINI • *Max F. Perutz Laboratories, Center for Molecular Biology of the University of Vienna, Vienna, Austria*
EYAL BENGAL • *Department of Biochemistry, Faculty of Medicine, Rappaport Institute for Research in the Medical Sciences, Technion-Israel Institute of Technology, Haifa, Israel*
JOHN BLENIS • *Department of Cell Biology, Harvard Medical School, Boston, MA, USA*
RICCARDO BRAMBILLA • *Division of Neuroscience, Institute of Experimental Neurology, San Raffaele Foundation and University, Milano, Italy*
RAINER BREITLING • *Groningen Bioinformatics Centre, Groningen Biomolecular Sciences and Biotechnology Institute, University of Groningen, Haren, The Netherlands*
KIMBERLY BURKHARD • *Department of Pharmaceutical Sciences, School of Pharmacy, University of Maryland, Baltimore, MD, USA*
CHRISTOPHER J. CAUNT • *Henry Wellcome Laboratories for Integrative Neuroscience and Endocrinology (LINE), Department of Clinical Sciences at South Bristol, University of Bristol, Bristol, UK*
ROCÍO CEJUDO-MARÍN • *Centro de Investigación, Príncipe Felipe, Valencia, Spain*
DEBORAH N. CHADEE • *Department of Biological Sciences, University of Toledo, Toledo, OH, USA*
JIANMING CHEN • *School of Life Sciences, Xiamen University, Xiamen, Fujian, China*
HONGBO CHI • *Department of Immunology, St Jude Children's Research Hospital, Memphis, TN, USA*
MARIO CHIARIELLO • *Istituto Toscano Tumori and Consiglio, Nazionale delle Ricerche, Siena, Italy*
MELANIE H. COBB • *Department of Pharmacology, The University of Texas Southwestern Medical Center, Dallas, TX, USA*
MARIO COSTA • *CNR, Institute of Neuroscience, Pisa, Italy*
PIERO CRESPO • *Instituto de Biomedicina y Biotecnología de Cantabria (IBBTEC), Consejo Superior de Investigaciones Científicas (CSIC), IDICAN, Departamento de Biología Molecular, Facultad de Medicina, Universidad de Cantabria, Santander, Cantabria, Spain*

SCOTT T. EBLEN • *Department of Cell and Molecular Pharmacology and Experimental Therapeutics and Hollings Cancer Center, Medical University of South Carolina, Charleston, SC, USA*
ELAINE A. ELION • *Department of Biological Chemistry and Molecular Pharmacology, Harvard Medical School, Boston, MA, USA*
DAVID ENGELBERG • *Department of Biological Chemistry, Hebrew University, Jerusalem, Israel*
EKREM EMRAH ER • *Department of Cell Biology, Harvard Medical School, Boston, MA, USA*
KHADIJA ESSAFI-BENKHADIR • *Biochemistry and Experimental Pathology Unit, Pasteur Institute of Tunis, Tunis, Tunisia*
EVA M. EVES • *Ben May Department for Cancer Research, University of Chicago, Chicago, IL, USA*
RICHARD A. FLAVELL • *Howard Hughes Medical Institute and Department of Immunobiology, Yale University School of Medicine, New Haven, CT, USA*
GERGANA GALABOVA-KOVACS • *Max F. Perutz Laboratories, Center for Molecular Biology of the University of Vienna, Vienna, Austria*
DAVID GILBERT • *School of Information Science, Computing and Mathematics, Brunel University, Uxbridge, Middlesex, UK*
ELIZABETH J. GOLDSMITH • *Department of Biochemistry, The University of Texas Southwestern Medical Center at Dallas, Dallas, TX, USA*
J. SILVIO GUTKIND • *Oral and Pharyngeal Cancer Branch, National Institute of Dental and Craniofacial Research, National Institutes of Health, Bethesda, MD, USA*
JIAHUAI HAN • *School of Life Sciences, Xiamen University, Xiamen, Fujian, China*
HAIXIA HE • *Department of Biochemistry, The University of Texas Southwestern Medical Center at Dallas, Dallas, TX, USA*
MONIKA HEINER • *Department of Computer Science, Brandenburg University of Technology, Cottbus, Germany*
AHARON HELMAN • *Department of Developmental Biology and Cancer Research, IMRIC, Faculty of Medicine, The Hebrew University, Jerusalem, Israel*
MARZIA INDRIGO • *Division of Neuroscience, Institute of Experimental Neurology, San Raffaele Foundation and University, Milano, Italy*
ARIF JIVAN • *Department of Pharmacology, The University of Texas Southwestern Medical Center, Dallas, TX, USA*
GARY L. JOHNSON • *Department of Pharmacology and Lineberger Comprehensive Cancer Center, School of Medicine, University of North Carolina, Chapel Hill, NC, USA*
AVIAD KEREN • *Department of Biochemistry, Faculty of Medicine, Rappaport Institute for Research in the Medical Sciences, Technion-Israel Institute of Technology, Haifa, Israel*
YONAT KESHET • *Department of Biological Regulation, The Weizmann Institute of Science, Rehovot, Israel*
SARAH KRAUS • *Department of Biological Regulation, The Weizmann Institute of Science, Rehovot, Israel*

JOHN M. KYRIAKIS • *The Molecular Cardiology Research Institute at Tufts Medical Center and Department of Medicine, School of Medicine, Tufts University, Boston, MA, USA*
VERED LEVIN-SALOMON • *Department of Biological Chemistry, Hebrew University, Jerusalem, Israel*
ODED LIVNAH • *Department of Biological Chemistry, The Wolfson Center for Applied Structural Biology, Hebrew University, Jerusalem, Israel*
MATILDE MARCHI • *NEST/INFM and Scuola Normale Superiore, Pisa, Italy*
JORGE MARTÍN-PÉREZ • *Instituto de Investigaciones Biomédicas A. Sols (CSIC/UAM), Madrid, Spain*
CRAIG A. MCARDLE • *Henry Wellcome Laboratories for Integrative Neuroscience and Endocrinology (LINE), Department of Clinical Sciences at South Bristol, University of Bristol, Bristol, UK*
MELISSA M. MCKAY • *Laboratory of Cell and Developmental Signaling, Center for Cancer Research, National Cancer Institute-Frederick, Frederick, MD, USA*
MICHELLE C. MENDOZA • *Department of Cell Biology, Harvard Medical School, Boston, MA, USA*
XIAOSHAN MIN • *Department of Biochemistry, The University of Texas Southwestern Medical Center at Dallas, Dallas, TX, USA*
DEBORAH K. MORRISON • *Laboratory of Cell and Developmental Signaling, Center for Cancer Research, National Cancer Institute-Frederick, Frederick, MD, USA*
KAZUHIRO NAKAMURA • *Department of Pharmacology and Lineberger Comprehensive Cancer Center, School of Medicine, University of North Carolina, Chapel Hill, NC, USA*
CAROLINE NUNES-XAVIER • *Centro de Investigación Príncipe Felipe, Valencia, Spain*
DANIEL ORELLANA • *Division of Neuroscience, Institute of Experimental Neurology, San Raffaele Foundation and University, Milano, Italy*
RICHARD ORTON • *Institute of Comparative Medicine, Faculty of Veterinary Medicine, University of Glasgow, G611QH, Glasgow*
GILLES PAGÈS • *Institute of Developmental Biology and Cancer Research UMR, University of Nice Sophia Antipolis, UMR CNRS 6543, Nice, France*
ALESSANDRO PAPALE • *Division of Neuroscience, Institute of Experimental Neurology, San Raffaele Foundation and University, Milano, Italy*
ZE'EV PAROUSH • *Department of Developmental Biology and Cancer Research, IMRIC, Faculty of Medicine, The Hebrew University, Jerusalem, Israel*
RICCARDO PARRA • *NEST/INFM and Scuola Normale Superiore, Pisa, Italy*
ADÁN PINTO • *Instituto de Biomedicina y Biotecnología de Cantabria (IBBTEC), Consejo Superior de Investigaciones Científicas (CSIC), IDICAN, Departamento de Biología Molecular, Facultad de Medicina, Universidad de Cantabria, Santander, Cantabria, Spain*

JACQUES POUYSSÉGUR • *Institute of Developmental Biology and Cancer Research UMR CNRS, University of Nice Sophia Antipolis, UMR CNRS 6543, Nice, France*
SHIRI PROCACCIA • *Department of Biological Regulation, The Weizmann Institute of Science, Rehovot, Israel*
RAFAEL PULIDO • *Centro de Investigación Príncipe Felipe, Valencia, Spain*
AARATI RANGANATHAN • *Department of Pharmacology, The University of Texas Southwestern Medical Center, Dallas, TX, USA*
GIAN MICHELE RATTO • *NEST/INFM and Scuola Normale Superiore, Pisa, Italy*
KATHERINE A. RAUEN • *Department of Pediatrics, Division of Medical Genetics, UCSF Helen Diller Family Comprehensive Cancer Center, University of California San Francisco, San Francisco, CA, USA*
MARSHA RICH ROSNER • *Ben May Department for Cancer Research, University of Chicago, Chicago, IL, USA*
DAVID B. SACKS • *Brigham and Women's Hospital and Harvard Medical School, Boston, MA, USA*
RUPAM SAHOO • *Department of Biological Chemistry and Molecular Pharmacology, Harvard Medical School, Boston, MA, USA*
RONY SEGER • *Department of Biological Regulation, The Weizmann Institute of Science, Rehovot, Israel*
PAUL SHAPIRO • *Department of Pharmaceutical Sciences, School of Pharmacy, University of Maryland, Baltimore, MD, USA*
ANDREW D. SHARROCKS • *Faculty of Life Sciences, University of Manchester, Manchester, UK*
MARIA SIBILIA • *Department of Medicine I, Institute for Cancer Research, Medical University of Vienna, Vienna, Austria*
KEVIN M. SOURS • *Department of Chemistry and Biochemistry, University of Colorado, Boulder, CO, USA*
GABI TARCIC • *Department of Biological Regulation, The Weizmann Institute of Science, Rehovot, Israel*
CÉLINE TÁRREGA • *Centro de Investigación Príncipe Felipe, Valencia, Spain*
WILLIAM E. TIDYMAN • *Department of Pediatrics, Division of Medical Genetics, University of California San Francisco, San Francisco, CA, USA*
JOSE P. VAQUÉ • *Oral and Pharyngeal Cancer Branch, National Institute of Dental and Craniofacial Research, National Institutes of Health, Bethesda, MD, USA*
BETTINA WAGNER • *APEIRON Biologics AG, Vienna, Austria*
COLIN D. WHITE • *Brigham and Women's Hospital and Harvard Medical School, Boston, MA, USA*
SHEN-HSI YANG • *Faculty of Life Sciences, University of Manchester, Manchester, UK*
YOSEF YARDEN • *Department of Biological Regulation, The Weizmann Institute of Science, Rehovot, Israel*
HUI ZHENG • *Department of Cell and Molecular Pharmacology and Experimental Therapeutics, Medical University of South Carolina, Charleston, SC, USA*
HUAMIN ZHOU • *School of Life Sciences, Xiamen University, Xiamen, Fujian, China*
TIANJUN ZHOU • *Department of Biochemistry, The University of Texas Southwestern Medical Center at Dallas, Dallas, TX, USA*

Part I

Activation and Function of Components of the MAP Kinase Signaling Cascades

Chapter 1

The MAP Kinase Signaling Cascades: A System of Hundreds of Components Regulates a Diverse Array of Physiological Functions

Yonat Keshet and Rony Seger

Abstract

Sequential activation of kinases within the mitogen-activated protein (MAP) kinase (MAPK) cascades is a common, and evolutionary-conserved mechanism of signal transduction. Four MAPK cascades have been identified in the last 20 years and those are usually named according to the MAPK components that are the central building blocks of each of the cascades. These are the extracellular signal-regulated kinase 1/2 (ERK1/2), c-Jun N-Terminal kinase (JNK), p38, and ERK5 cascades. Each of these cascades consists of a core module of three tiers of protein kinases termed MAPK, MAPKK, and MAP3K, and often two additional tiers, the upstream MAP4K and the downstream MAPKAPK, which can complete five tiers of each cascade in certain cell lines or stimulations. The transmission of the signal via each cascade is mediated by sequential phosphorylation and activation of the components in the sequential tiers. These cascades cooperate in transmitting various extracellular signals and thus control a large number of distinct and even opposing cellular processes such as proliferation, differentiation, survival, development, stress response, and apoptosis. One way by which the specificity of each cascade is regulated is through the existence of several distinct components in each tier of the different cascades. About 70 genes, which are each translated to several alternatively spliced isoforms, encode the entire MAPK system, and allow the wide array of cascade's functions. These components, their regulation, as well as their involvement together with other mechanisms in the determination of signaling specificity by the MAPK cascade is described in this review. Mis-regulation of the MAPKs signals usually leads to diseases such as cancer and diabetes; therefore, studying the mechanisms of specificity-determination may lead to better understanding of these signaling-related diseases.

Key words: ERK, JNK p38, Signaling cascades, Phosphorylation

1. Overview

In order to perform their functions and to survive, cells need to respond to a large number of extracellular stimuli and environmental changes, including mitogens, hormones, stresses, as well

Rony Seger (ed.), *MAP Kinase Signaling Protocols: Second Edition*, Methods in Molecular Biology, vol. 661,
DOI 10.1007/978-1-60761-795-2_1, © Springer Science+Business Media, LLC 2010

as changes in temperature, osmotic pressure, and more. The response of cells to the plethora of extracellular signals is often mediated through the activation of transcription factors, which, in turn, induce the necessary cellular processes. However, most extracellular agents cannot cross the plasma membrane in order to activate their corresponding genes. Instead, these agents use intracellular "communication lines," which are known in fact as signaling pathways, to transmit their signals to various cytoplasmatic and nuclear targets (1). In many cases, these pathways operate through sequential phosphorylation events that are termed protein kinase cascades. This type of signaling mechanism is utilized by the mitogen-activated protein (MAP) kinase (MAPK) signaling cascades, which are evolutionary-conserved, central signal transduction pathways. Among others, these cascades respond to various extracellular factors and consequently regulate diverse cellular processes such as proliferation, differentiation, stress response, and apoptosis.

The MAPK signaling cascades have been extensively studied over the past two decades and showed to operate in a large number of cells and conditions (for recent reviews see (2–8)). Transmission of signals via these cascades is usually initiated by activation of a small G protein (e.g., Ras) or by activating interaction of upstream components of the cascade with adaptor proteins. Then, the signals are further transmitted downstream the cascade by cytosolic protein kinases that are organized in three to five tiers. The kinases in each tier phosphorylate and activate the kinases located in their downstream tier to allow a rapid and regulated transmission of the signals to various targets of the cascades. Three of the tiers, MAP3K, MAPKK, MAPK, are considered as core ones (Fig. 1), while the upstream (MAP4K) or the downstream (MAP Kinase-activated kinase; MAPKAPK) tiers are not always necessary for signaling through the cascades. Kinases downstream of MAPKAPKs do exist as well, but those are not considered as part of the cascades. Importantly, each of the tiers of the different cascades is composed of several components that are usually distinct gene products, and are often translated to several alternative spliced isoforms. About 70 genes are known today to encode for close to 200 distinct components that compose the entire MAPK system. This multiplicity of components allows the extended specificity and tight regulation, which are hallmarks of these cascades.

Four MAPK cascades have been fully elucidated by now, and named according to the components in their MAPK tier. These are extracellular signal-regulated kinase 1 and 2 (ERK1/2; (9)), c-Jun N-terminal kinase 1–3 (JNK1–3, (10, 11)), p38MAPK α, β, γ, δ (p38α–δ; (12–14)) and ERK5 ((15, 16); Fig. 1 and Table 1). Other MAPKs or MAPK-like components, have been identified as well (e.g., ERK3/4 and ERK7/8; (17)), but those

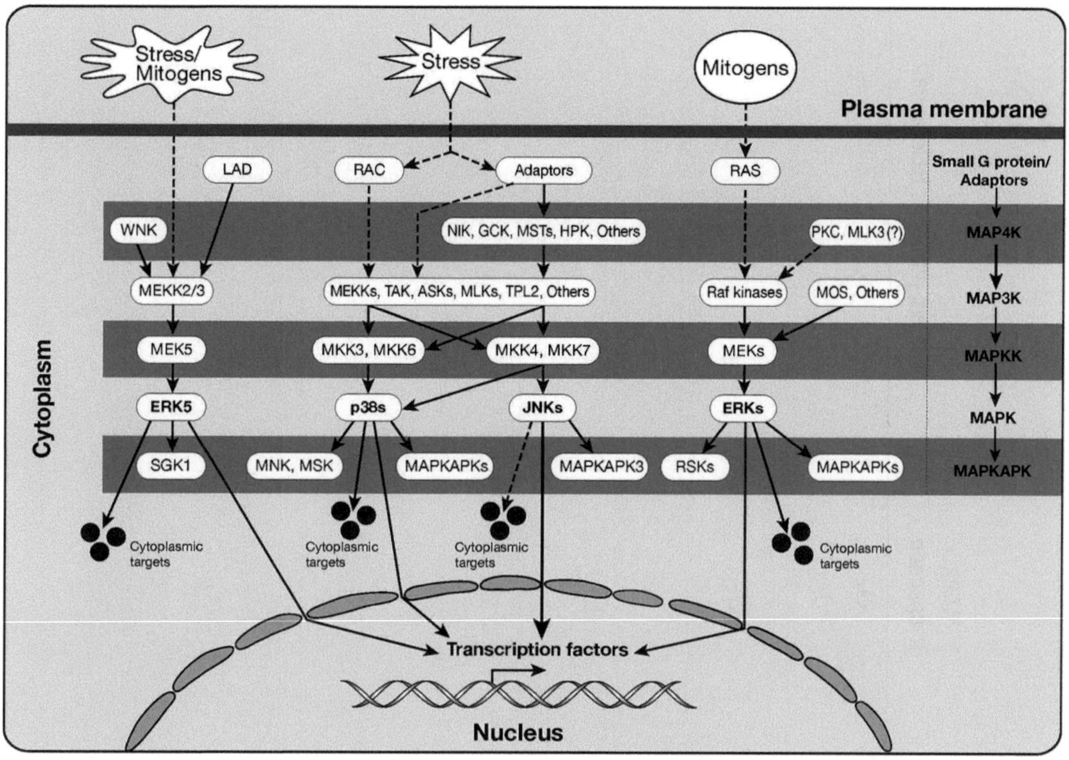

Fig. 1. Schematic representation of the four MAPK signaling cascades. More components are described in the text and in Table 1.

do not seem to operate within a kinase cascade, or are activated by distinct mechanisms than MAPKs, and therefore are not considered as genuine MAPKs. Each of the cascades can regulate several distinct, and sometime overlapping cellular processes, and they generally seem to differ in their main physiological activities. Usually, the ERK1/2 cascade plays a role in proliferation and differentiation, JNK and p38 cascades are activated mainly by cellular stresses, and therefore their MAPK components are termed stress-activated protein kinases (SAPKs), and the ERK5 cascade seems to respond equally to certain stresses as well as mitogenic signals. However, dependent on the cell lines and stimulation, the distinct cascades may regulate noncanonical and even opposing functions. Thus, under rare conditions, ERK1/2 may participate in the response to stress and apoptosis (18); while JNK can occasionally mediate proliferation (19). Taken together, the activation of MAPK cascades by a large variety of stimuli indicates that these cascades are key mediators of essentially all stimulation-induced cellular processes. Importantly, mis-regulation of the cascades often leads to diseases such as cancer, diabetes, immune response inflammation (for reviews see (20–22)). Therefore, better understanding of MAPK signaling and its regulation may

Table 1
Components of the MAPK system

Tier	Name	Other names	MAPK cascade	Mol. wt. (kDa)	Reference
MAP4K	GCK	Germinal center kinase (GCK), Rab8-interacting protein (RAB8IP1), MAP4K2, MEKKK2	JNK	91	(196)
	GCKR	GCK-related kinase (GCKR) Kinase homologous to SPS1/STE 20 (KHS) MAP4K5 MEKKK5	JNK	95	(197)
	GLK	GCK-like kinase (GLK) RAB8IP-like kinase1 (RAB8IPL1) MAP4K3 MEKKK3	JNK	104	(198)
	HPK	Hematopoietic progenitor kinase (HPK) MAP4K1 isoform 2 MEKKK1	JNK	97	(199)
	MINK	Misshapen/NIKs-related kinase (MINK) MAP4K6 MEKKK6	JNK, p38	150–160	(200)
	MST1	Mammalian Ste20-like kinase 1 (Mst1)	JNK, p38, ERK1/2	59	(201)
	MST4	Mammalian Ste20-like kinase 4 (Mst4) Mst3 and SOK1-related kinase (MASK)	ERK (?)	52	(202)
	NESK	NIK-related protein kinase NIK-like embryo-specific kinase (NESK)	JNK	175	(203)

	NIK	Nck Interacting Kinase (NIK) HPK/GCK-like kinase (HGK) MAP4K4 MEKKK4	JNK	100–130	(204)
	OSR1	Oxidative stress-responsive 1 (OSR1)	JNK	60	(205)
	PAK1	p21-activated kinase 1 (PAK1) p68-PAK Alpha-PAK MAPK upstream kinase (MUK)	JNK, p38 (?)	68	(206)
	PAK5	p21-activated kinase 5 (PAK5)	JNK	90	(207)
	RIP	Receptor (TNFR)-interacting serine-threonine kinase 1 (RIPK1)	JNK, p38	78	(208)
	SPAK	Ste20-related proline/alanine-rich kinase (SPAK) Proline-alanine-rich Ste20-related kinase (PASK) Serine/threonine kinase 39	p38	64	(209)
	SLK	Ste20-like kinase (SLK) Serine/threonine kinase 2 Long Ste20-like kinase (LOSK)	JNK	150 (210)	(210)
	TNIK	TRAF2 and NCK interacting kinase (TNIK)	JNK	160	(211)
	WNK1	With no lysine 1 (WNK1)	ERK5	230	(110)
MAP3K	ASK1	Apoptosis signal-regulating kinase 1 (ASK1) MAP3K5 MEKK5	JNK, p38	155	(212)
	ASK2	Apoptosis signal-regulating kinase 2 (ASK2) MAP3K6 MEKK6	JNK	140	(213)

(continued)

Table 1 (continued)

Tier	Name	Other names	MAPK cascade	Mol. wt. (kDa)	Reference
	DLK	Dual leucine-zipper-bearing kinase (DLK1) Leucine-zipper protein kinase (ZPK) Mixed-lineage kinase DLK MAP3K12	JNK, p38	110	(214)
	LZK	Leucine-zipper-bearing kinase 1 (LZK1) Mixed-lineage kinase LZK MAP3K13	JNK	140	(215)
	MAP3K15	MEKK15 ASK family kinase Not tested experimentally	JNK (?)	85	http://www.sanger.ac.uk
	MAP3K19	Not tested experimentally	JNK (?)	120	http://www.sanger.ac.uk
	MEKK1	MAP3K1	JNK, p38, ERK1/2	185	(90)
	MEKK2	MAP3K2	JNK, p38 ERK5	70	(216, 217)
	MEKK3	MAP3K3	JNK, p38, ERK5	71	(217)
	MEKK4	MAP3K4	JNK	180	(218)
	MLK1	Mixed-lineage kinase 1 (MLK1) PRKE1 MAP3K9	JNK	130	(115, 219)
	MLK2	Mixed-lineage kinase 2 (MLK2) Mammalian Ste20-like kinase (MST) MAP3K10	JNK, p38 (?)	115	(220)

MLK3	Mixed-lineage kinase 3 (MLK3) SH3 domain-containing proline-rich kinase (SPRK) MAP3K11	JNK, p38ERK (?)	93	(221)
MLK4	Mixed-lineage kinase 4 (MLK4) KIAA1804	JNK (?), p38 (?)	120	(222)
MLTK α/β	MLK-like MAPK triple kinase (MLT, MLTK) Sterile alpha motif- and leucine-zipper-containing kinase (AZK) Mixed-lineage kinase-related kinase (MRK) Mixed-lineage kinase 7 (MLK7) ZAK – sterile alpha motif and leucine-zipper-containing kinase	JNK, p38, ERK1/2 (?), ERK5 (?)	95/50	(115)
MOS	Moloney murine sarcoma virus (MOS) Oocyte maturation factor mos	ERK1/2	39	(223)
A-Raf	ARAF1 RAFA1 Proto-oncogene serine/threonine-protein kinase, (PSK)	ERK1/2	68	(224)
B-Raf	BRAF1RAFB1	ERK1/2	94	(225)
Raf-1	c-RafRAF proto-oncogene serine/threonine-protein kinase	ERK1/2	74	(30)
TAK1	TGFβ-activated kinase 1 (TAK1) MAP3K7	JNK, p38	82	(226)
TAO1	Thousand and one amino acid protein 1 (TAO1) Tao kinase 1 (TAOK1) Protein Ser/Thr Kinase2 (PSK2) MARK/PAR-1 kinase (MARKK) MAP3K16	JNK, p38	140	(227)

(continued)

Table 1 (continued)

Tier	Name	Other names	MAPK cascade	Mol. wt. (kDa)	Reference
	TAO2	Thousand and one amino acid protein 2 (TAO2) Protein Ser/Thr Kinase1 (PSK1) Tao kinase 2 (TAOK2) MAP3K17	JNK, p38	120	(228)
	TAO3	Thousand and one amino acid protein 3 (TAO3) JNK-inhibitory kinase (JIK) Dendritic cell-derived protein kinase (DPK) Tao kinase 3 (TAOK3) MAP3K18	JNK	105	(229)
	TPL2	Tumor progression locus 2 (TPL2) Cot proto-oncogene c-Cot	JNK, p38, ERK1/2, ERK5	60+	(230)
MAPKK	MEK1	MAPK/ERK kinase 1 (MEK1) ERK activator MAPKK1 MAP2K1 MKK1	ERK1/2	45	(35, 37, 38)
	MEK2	MAPK/ERK kinase 2 (MEK2) ERK activator kinase 2 MAPKK2 MAP2K2 MKK2	ERK1/2	46	(39)
	MEK5	MAPK/ERK kinase 5 (MEK5) MAPKK5 MKK5 MAP2K5	ERK5	45	(15, 116)

	MKK3	MAPKK3MAP2K3 MAPK/ERK kinase 3 (MEK3) SAPK kinase 3 (SKK3)	p38	39	(65)
	MKK4	MAPK/ERK kinase 4 (MEK4) JNK kinase 1 (JNKK1) SAPK/ERK1 (SEK1) MAPKK4 MAP2K4	JNK, p38-limitted	44	(89, 90)
	MKK6	MAPKK6MAP2K6 MAPK/ERK kinase 6 (MEK6) SKK3	p38	38	(66, 67)
	MKK7	JNK kinase 2 (JNKK2) MAPKK7 MAP2K7 SAPK kinase 7 (SKK7)	JNK	48	(91–93)
MAPK	ERK1	Extracellular signal-regulated kinase 1 (ERK1) p44MAPK MAPK3	ERK1/2	44 + other forms	(44, 231)
	ERK2	Extracellular signal-regulated kinase 2 (ERK2) p42MAPK MAPK1	ERK1/2	42 + other forms	(9, 44)
	ERK3	Extracellular signal-regulated kinase 3 (ERK3) MAPK6	Not a genuine MAPK	82, (63?)	(9)
	ERK4	Extracellular signal-regulated kinase 4 (ERK4) ERK3-related MAPK MAPK4	Not a genuine MAPK	97	(47)

(continued)

Table 1 (continued)

Tier	Name	Other names	MAPK cascade	Mol. wt. (kDa)	Reference
	ERK5	Extracellular signal-regulated kinase 5 (ERK5) Big MAPK 1 (BMK1) MAPK7	ERK5	110	(15, 16)
	ERK7/8	Extracellular signal-regulated kinase 7/8 (ERK7/8 – mouse/human) MAPK15	Not within a MAPK cascade	63/62	(127, 128, 232)
	JNK1	c-Jun-N-terminal kinase1 (JNK1) Stress-activated protein kinase 1 (SAPK1) MAPK8	JNK	46 (54)	(10, 233)
	JNK2	c-Jun-N-terminal kinase 2 (JNK2) Stress-activated protein kinase (SAPK) p54aSAPK MAPK9	JNK	54 (46)	(11, 234)
	JNK3	c-Jun-N-terminal kinase 3 (JNK3) SAPK p54bSAPK p493F12 MAPK10	JNK	52 (46)	(95)
	p38α	p38MAPKαCytokine suppressive anti-inflammatory drug-binding protein (CSBP) Stress-activated protein kinase (SAPK2a) Hog-like RK MAPK14	p38	43, 38 + other forms	(13, 14, 69)

	p38β	p38MAPKβ Stress-activated protein kinase (SAPK2b) MAPK11	p38	41 + others	(70)
	p38γ	p38MAPKγ Stress-activated protein kinase (SAPK3) Extracellular signal-regulated kinase 6 (ERK6) MAPK12	p38	45	(71, 72)
	p38δ	p38MAPKδ Stress-activated protein kinase (SAPK4) MAPK13	p38	47	(73)
MAPKAPK	MAPK-APK2	MAPK-activated protein kinase 2 (MAPKAPK2) MK2	p38	45 + others	(74)
	MAPK-APK3	MAPKAPK2 MK3 3PK	p38, ERK1/2JNK	43	(75)
	MK5	MAPK-activated protein kinase 5 (MAPKAPK5) p38-regulated/activated protein kinase (PRAK) p38-activated kinase	p38, ERK1/2 (?)	56	(76, 77)
	MNK1	MAP kinase-interacting ser/thr-protein kinase 1/MAP kinase signal-integrating kinase 1 (MNK1) MKNK1	p38, ERK1/2	52	(52, 53)
	MNK2	MAP kinase-interacting ser/thr-protein kinase 2/MAP kinase signal-integrating kinase 2 (MNK2) G-protein-coupled receptor kinase 7 (GPRK7) MKNK2	p38, ERK1/2		(53)

(continued)

Table 1 (continued)

Tier	Name	Other names	MAPK cascade	Mol. wt. (kDa)	Reference
	MSK1	Mitogen- and stress-activated protein kinase-1 (MSK1) MSPK RSK-related protein kinase (RLPK) Ribosomal protein S6 kinase, 90 kDa, 5 (RPS6KA5) Ribosomal protein S6 kinase 5 (RSK5)	p38, ERK1/2	90	(51)
	MSK2	Mitogen- and stress-activated protein kinase-2 (MSK2) Ribosomal protein S6 kinase, 90 kDa, 4 (RPS6KA4) Ribosomal protein S6 kinase 4 (RSK4)	p38, ERK1/2 (?)	90	(235, 236)
	RSK1	Ribosomal protein S6 kinase, 90 kDa, 1 (RPS6KA1) Ribosomal protein S6 kinase 1 (RSK1) MAPK-activated protein kinase 1A (MAPKAPK1A) S6 kinase (S6K) II S6 kinase (S6K) alpha1 P90RSK	ERK1/2	90	(50, 237)
	RSK2	Ribosomal protein S6 kinase, 90 kDa, 3 (RPS6KA3) Ribosomal protein S6 kinase 2 (RSK2) MAPK-activated protein kinase 1B (MAPKAPK1B) Insulin-stimulated protein kinase 1 (ISPK1) Coffin–Lowry syndrome-related kinase (CLS) S6 kinase (S6K)-alpha3	ERK1/2	90	(238, 239)
	RSK3	Ribosomal protein S6 kinase, 90 kDa, 2 (RPS6KA2) Ribosomal protein S6 kinase 3 (RSK3) MAPK-activated protein kinase 1C (MAPKAPK1C) S6 kinase (S6K)-alpha2	ERK1/2	90	(240)
	SGK1	Serum/glucocorticoid-regulated kinase 1 (SGK1)	ERK5	54	(120, 241)

The protein kinases of each tiers are presented in an alphabetical order (?) – indicates controversial or unclear effect result in the designation of better strategies to combat the signaling-related diseases. In this review we describe the large number of MAPK components in each cascade and present the way by which their different regulation contributes to specificity-determination of the distinct cascades.

2. The ERK Cascade

The ERK cascade is activated by a variety of extracellular agents, including growth factors, hormones and also cellular stresses to induce cellular processes that include mainly proliferation and differentiation, but under some conditions also stress response and others (for recent reviews see (3, 23–26)). The extracellular factors act via tyrosine kinase receptors (RTK; (27)), G-protein-coupled receptors (GPCR; (28)), ion channels (29), and others. Those membranal receptors further transmit the signals to the ERK cascade by a plethora of signaling processes, which in many cases involve recruitments of adaptor proteins such as Shc or Grb2 to the activated receptors or their effector proteins (e.g., Fak1). In turn, the adaptors direct guanine nucleotides exchange factors (GEFs) to membrane-bound small GTP-binding proteins (e.g., Ras, Rap), rendering them to their GTP-bound, active form. This further allows transmission of the signal to the components of the MAP3K tier of the ERK cascade, which are mostly Raf kinases (Raf-1, B-Raf, A-Raf; (30)), but possibly also TPL2 and the stress-activated MEKK1 and MLTK (see Fig. 1 and Table 1). MOS is another MAP3K of the ERK cascade, but it operates mainly in the reproductive system by a distinct mode of regulation (31). One well-studied example for the mechanism of signal transmission from extracellular agents to the ERK cascade is the one mediated by growth factor receptors. Upon binding of the factors, the receptors dimerize and undergo autophosphorylation on several tyrosine residues. Then, the adaptor protein, Grb2, is recruited to the receptor via its SH2 domain, and by this allows further recruitment of the GEF, SOS, to its proper location next to membrane-bound Ras proteins. The activation of Ras by SOS is followed by recruitment and activation of Raf1 and/or B-Raf to the plasma membrane where they are activated by a mechanism that is not fully understood yet (32). Under some conditions, the activation of the Raf kinases can be mediated or enhanced by other Ser/Thr kinases such as PKC (33) and MLK3 (34), and therefore those can be considered as MAP4Ks of this cascade.

From the MAP3K level, the signal is transmitted down the cascade through the MAPKK components (35, 36), termed MAPK/ERK kinases 1 and 2 (MEK1/2; (37–39)). These MAPKK components are activated through serine phosphorylation at the typical Ser-Xaa-Ala-Xaa-Ser/Thr motif in their activation loop (Ser 218, 222 in human MEK1; (40, 41)). The activated MEKs are dual specificity kinases, which demonstrate a unique selectivity toward ERKs in the MAPK level (42). An alternatively spliced isoform of MEK1, termed MEK1b, was identified as well (38), and it was recently shown that this isoform acts as a specific mediator of limited cellular processes downstream of Rafs (43). Two genes are known to encode the ERKs, and those are designated as ERK1 (MAPK3) and ERK2 (MAPK1). These two genes encode two main proteins, p44 and p42, respectively, (9, 44), as well as a few alternatively spliced isoforms such as the rodent ERK1b (45), the primate ERK1c (46) and probably ERK2b (47). The activation of ERKs is mediated by MEKs phosphorylation of both Tyr and Thr residues in the activation loop Thr-Glu-Tyr motif (Thr183, Tyr185 in human ERK2). These are ubiquitous Ser/Thr kinases that phosphorylate hundreds of substrates either in the cytosol (e.g., PLA2 RSK), or upon translocation, in the nucleus (e.g., Elk1 (48)). Most of the ERKs' substrates are regulatory proteins, including one or more MAPKAPKs (49). The MAPKAPK tier includes the ribosomal S6 kinase (RSK; (50)), the MAPK/SAPK-activated kinase (MSK; (51)), MAPK signal-interacting kinases 1 and 2 (MNK1/2; (52, 53)), and MAPKAPK3/5 (49), although the later ones can also be activated by p38. Finally, protein kinases such as GSK3 (54) and LKB1 (55) have been identified as immediate substrates for MAPKAPKs, but those are not usually considered as integral components of the cascade. The inactivation of ERKs, which is similar to the other MAPKs is described below under "Regulation and specificity-determination of MAPKs."

3. The p38 Cascade

The p38 MAPK cascade participates primarily in responses of cells to stress, but also other processes such as immune response and inflammation (for recent reviews see (56–60)). The activity of this cascade is induced by various stress factors and ligands that operate via different receptors including apoptosis-related receptors, GPCRs, and even RTKs. In addition, some of the physical stresses (e.g., heat, osmotic shock), which are among the strongest stimulators of the p38 cascade, are thought to operate in many cases via receptor-independent machinery that requires changes in membrane fluidity or other specialized signaling systems (61). Then, the signals are transmitted by induction of a complex network of signaling molecules that often results either in activation of small

GTPases such as Rac and CDC42 (62), or sometime via activatory interactions of adaptor proteins (63). These two processes then induce activation of protein kinases at either the MAP4K or directly the MAP3K tiers of the p38 cascade. Many kinases at these tiers have been identified (Fig. 1 and Table 1); however, their individual roles and specificities are not yet fully elucidated. Thus, about 20 distinct genes encode kinases implicated at the MAP3K tier of this cascade (reviewed in part in (64)), and many of them have more than one spliced isoform, significantly extending their number. These components are very similar to those of the JNK cascade, and require a specificity-determination mechanism for their action upstream the p38 cascade, as described below for the JNK cascade. In the case of the p38 cascade, the MAP3K tier kinases are activated mainly by small GTPases, but under some conditions, also by quite a few MAP4Ks or even directly by adaptor proteins. This large number of MAP3K tier kinases transmit their signals to a much smaller number of MAPKKs, phosphorylating them on Ser and Thr residues at the typical Ser-Xaa-Ala-Xaa-Ser/Thr motif in their activation loop. The main MAPKK of the p38 cascade are MKK6 and MKK3 (65–67), although MKK4 (65) and MKK7 (68) have been somewhat implicated in p38 activation as well.

The next tier of the cascade is composed of products of four MAPKs genes, including p38α (SAPK2a; (13, 14, 69)) p38β (SAPK2b; (70)), and also p38γ and δ (71–73). Importantly, p38 genes also express several alternatively spliced forms, bringing the number of isoforms of this group to ten, which are all activated by phosphorylation of the Tyr and Thr residues in the Thr-Gly-Tyr motif in their activation loop (Thr180 and Tyr182 of human p38MAPKα). Although all p38s share a similar mechanism of activation and substrate specificity (~60%), their distinct sequence identity and sensitivity to inhibitors indicate that they can be subdivided into two groups, p38α+p38β and p38γ+p38δ. However, it is not clear yet whether these subgroups also have distinct physiological function. Once these p38s are activated, they either transmit the signal to the MAPKAPK level components MAPKAPK2 (74), MAPKAPK3 (75), MNK1/2, MSK1/2, and MK5/PARK (76, 77), or phosphorylate regulatory molecules such as PLA_2 (78), heat shock proteins (14), the transcription factors ATF2, ELK1, CHOP, MEF2C, and more (49). Unlike ERK1/2, the p38s can be localized both in the nucleus and/or in the cytosol, and their translocation upon stimulation seems to be bi-directional (79). In addition, the MAPKAPKs can complete a plausible six-tiered p38 cascade by phosphorylating protein kinases such as LKB1 (55), but as for the ERK1/2 cascade, those kinases are not usually considered as integral components of the cascade.

4. The JNK Cascade

Another stress-activated MAPK cascade is that of the c-Jun N-terminal kinases (JNKs; for recent review see (80–84)). The JNK cascade plays an important role in the response to stress, in inducing apoptosis upon various stimulations, but may play a role in an array of other processes. As the p38 cascade, this cascade is responsive to stress/apoptosis-related receptors, receptor-independent physical stresses, GPCRs and even RTKs. Those receptors or receptor-independent stress-induced membranal changes further transmit the signals to adaptor proteins that can by themselves activate kinases in the MAP4K, and sometime, MAP3K tiers of the JNK cascade (reviewed in (63)). Alternatively, the membranal receptors/moieties can activate a network of interacting proteins that eventually induce either activatory changes in adaptor proteins (e.g., TRAF (85)) or activation of small GTPases (e.g., Rac, CDC42; (86)). These two processes then transmit the signal further by activating MAP4K, or sometime directly MAP3K tiers kinases (Fig. 1 and Table 1). The kinases at the MAP4K tiers, include GCK, GLK, GCKR, HPK, and other Ste20-like kinases (reviewed in part in (87)), which are mostly shared with the p38 cascade, and can all phosphorylate and activate the kinases at the MAP3K tier (88). Again, most of the MAP3Ks are shared with those of the p38 cascade, although at this stage, the MAP3Ks ASK2, LZK1, MLK1, and MEKK4 have been reported only in the JNK cascade.

Next, the activated MAP3Ks transmit the signals to kinases at the MAPKK level, which are mainly MKK4 and MKK7 (89–93) but may include (to a much lesser extent) also MKK3/6 (65). Interestingly, MKK7 seems to express a particularly high number (~6) of alternatively spliced isoforms (94)) with distinct activities, and those were suggested to extent the specificity of the response stress stimuli in different cell lines or compartments. As the other MAPKKs, the main JNK kinases (MKK4, MKK7) are activated by phosphorylation of the typical Ser-Xaa-Ala-Xaa-Ser/Thr motif (Ser 198, Thr 202 in MKK7) and are then able to transmit the signal further to the JNK level. Three genes (JNK1–3, SAPK1s), encode the JNK proteins, which are translated from total of ten JNK alternatively spliced transcripts (95, 96). Interestingly, all of them are translated into either 46 or ~54 kDa proteins. The smaller p46 JNKs are referred to as JNK1α1, JNK1β1, JNK2α1, JNK2β1, and JNK3α1. The p54 JNK proteins are referred to as JNK1α2, JNK1β2, JNK2α2, JNK2β2, and JNK3α2, but the latter can also appear as a 52 kDa protein. It should be noted, however, that in most systems the 46 kDa proteins are mainly JNK1 isoforms, the 54 kDa proteins are mainly JNK2, the 52 kDa band represents only JNK3, and this is often the simplified nomenclature used for of the three JNK bands upon an SDS-PAGE.

The activation loop of JNKs contains a Pro in the Xaa position of the Thr-Xaa-Tyr motif, and as with the other MAPKs both Thr and Tyr need to be phosphorylated to achieve activation (Thr 183 and Tyr185 in human JNK2). Only one MAPKAPK, MAPKAPK3 (97), is currently reported as a component of the MAPK cascade. Another kinase that has been reported to act downstream of JNK was RSK1 (98), but this kinase does not seem to be a JNK target under most conditions, and therefore it is not considered as a genuine component of the cascade. In addition, only a small number of targets have been identified for JNKs in the cytosol (98–100), while this cascade appears to be a major regulator of nuclear processes, in particular transcription. Thus, shortly after activation, like the other MAPKs, JNKs translocate into the nucleus where they usually physically associate with their targets (e.g. transcription factors such as c-Jun, ATF, Elk1; (101)) and activate them.

Despite the pronounced similarities and shared components of the JNK and p38 cascades, they clearly transmit separated signals, and often regulate distinct cellular processes. The specificity of these cascades and the identity of the components that participate in each signaling event seem to be regulated particularly by scaffold proteins that bring the required components to close proximity to each other (101). These scaffolds may also bring the cascade's components to the vicinity of its activator and/or downstream target to secure proper signal transmission. This is often reflected in the identity of the upstream activators, and it is now assumed that the p38 cascade requires mainly the small GTPases to MAP3K activation, while the activation of the JNK cascade is induced mostly by adaptor protein and MAP4Ks. Other parameters that determine the signaling specificity such as subcellular localization and regulation by phosphatases that determine the duration and strength of the signals (5) seem to be shared by other MAPK cascades and will be covered in a special section below.

5. The ERK5 Cascade

Another MAPK cascade is that of ERK5, which is significantly less studied than the other three MAPK cascades (for recent review see (102–106)). This cascade was initially thought to be activated by stress-related stimuli similar to those activating other SAPKs. Indeed, it was found that ERK5 is activated by oxidative stress and hyperosmolarity (16). However, it is clear today that the ERK5 cascade can be activated equally by mitogens (107), confirming its central role in many stress- and mitogen-induced cellular processes (108). The mechanism of upstream activation of the cascade has not been fully elucidated yet, but can include activation of protein Tyr kinases (109) that either transmit their signals to the adaptor protein Lad1 (110), or to WNK1 that seem to act as a MAP4K in this cascade (111). Those components activate the kinases in the MAP3K level, which are MEKK2/3 (112, 113), and possibly also TPL2 (114) and MLTK (115).

The MAP3K tier kinases then phosphorylate and activate the two alternative spliced MAPKK isoforms MEK5α (50 kDa) and MEK5β (40 kDa; (15, 116)), which are both active kinases. As other MAPKKs, this phosphorylation occurs on the Ser and Thr residues within the Ser-Xaa-Ala-Xaa-Thr activation motif of MEK5s (Ser311 and Thr315 in human MEK5α). The MEK5s then activate the MAPK, ERK5 by phosphorylating it on both Thr and Tyr residues within the sequence Thr-Glu-Tyr (Thr218 and Tyr220 in human ERK5), which is very similar to that of ERK1/2. However, in spite of the similarity in the activation motif, ERK5 cannot be phosphorylated by MEK1/2, and MEK5 cannot phosphorylate ERK1/2. There is one gene that encodes the MAPK component of the cascade, ERK5, which encodes a main protein of 110 kDa, a size that gave this kinase its other name – big MAPK1 (BMK1; (15, 16)). Three alternatively spliced variants have been identified for mouse ERK5, including the full length ERK5a, as well as the shorter ERK5b and ERK5c that do not seem to demonstrate any significant catalytic activity (117). As other MAPKs, ERK5 can be localized in the cytoplasm of resting cells and translocates to the nucleus upon stimulation (118). However, in some cell lines, ERK5 seems to be localized mainly in the nucleus, where its activation is mediated by nuclear MEK5 and translocating MEKK5 (119). The differences between these two modes of activation are not fully elucidated yet and need further clarifications.

Upon stimulation, ERK5 phosphorylates the serum and glucocorticoid-activated kinase (SGK; (120)), which may serve as a MAPKAPK of this cascade, and thereby complete a five-tiered pathway. However, this does not seem to be the main substrate of the cascade, as several transcription factors including c-Myc (121), MEF2 family members (122, 123), c-Fos (124), and possibly also SAP1a (124) have been identified as ERK5 substrates as well. Interestingly, unlike the dogma for the other MAPKs, ERK5 was found to influence transcription also through direct protein–protein interactions that are mediated by its C-terminal noncatalytic half (e.g., MEF2C, (125)). More importantly, this region of the ERK5 seems to exhibit intrinsic transcriptional activity, which was shown to activate Nur77 gene by binding to the MEF2 site of this gene, making the ERK5 a unique dual activity protein that, unlike other MAPKs, catalyzes two independent activities (126).

6. MAPK-Like, Cascade-Independent Protein Kinases

The MAPKs described above share a common mechanism of activation, which involves a multi-tiered cascade-induced phosphorylation of Thr and Tyr residues of a Thr-Xaa-Tyr motif within their activation loop. Interestingly, several other protein kinases with sequence similarity and even with Thr-Xaa-Tyr motif have been identified over the years. However, as of today, these protein kinases do not seem to be activated within a canonical MAPK signaling cascade, and therefore are not considered as genuine members of the family (for review see (6, 17)). One such putative MAPK family-member is the gene product of MAPK15, which seems to significantly differ between humans and rodents, and therefore was named ERK7 (63 kDa (127)) for the rodent protein, and ERK8 (62 kDa; (127, 128)) for the human one. Although it has the signature Thr-Glu-Tyr activation motif of ERK1/2/5, ERK7/8 are not significantly activated by extracellular stimuli that typically activate MAPKs, and no MAPKK has been identified for these kinases (129). Instead, ERK7 has an appreciable activity in serum-starved cells, which may be mediated by autoactivation, indicating that it might operate as a constitutively active protein kinase (130). This activity is likely to be regulated by the ubiquitin–proteosome pathway (131), and can also be influenced by constitutively active RTKs (132) and DNA damage (133). In addition, it was shown that, as all other MAPKs (134), ERK7/8 are a Pro-directed kinase (Pro-Xaa-Ser/Thr-Pro, or just Ser/Thr-Pro (129)). However, only few targets (e.g., estrogen receptor (135) and glucocorticoid receptor (136)) have been identified for these protein kinases, and more studies are required in order to elucidate the full array of their physiological functions.

Another protein kinase that has been implicated in the MAPK tier is ERK3, which exhibits about 50% homology to ERK1/2 (9). However, this protein kinase does not contain the characteristic Thr-Xaa-Tyr motif, and therefore is not considered as a genuine MAPK. Rather, ERK3 seems to be activated by phosphorylation of a Ser residue (Ser189), localized next to Glu and Gly within its activation loop, by a protein kinase that has not been fully identified yet (137). The physiological functions, mode of activation, and targets of ERK3 have not been fully elucidated yet, although it is clear that it is readily regulated by stress conditions and by the proteosome–ubiquitin pathway (138). Interestingly, it was also reported that ERK3 can directly activate the MAPKAPK tier kinase MK5/PRAK (139, 140), indicating that, although ERK3 is not a MAPK by itself, it can interact with the MAPK cascades by affecting their downstream components (141).

As yet additional MAPK-like enzyme with many similarities to ERK3 is the 97 kDa ERK4 (47, 142). It should be noted that this protein kinase is completely distinct from the 46 kDa protein that was initially termed ERK4 by the Cobb group (143), which was later identified as the rodent ERK1b (45). As much as ERK3, ERK4 is activated by phosphorylation of an activation loop Ser residue within a Ser-Glu-Gly sequence, and its upstream kinase is not known. This similarity is extended by the ability of this kinase to phosphorylate and activate the MK5/PRAK, and by its nuclear localization. Thus, the ERK3 and ERK4 constitute a subgroup of kinases that is distinct from the other MAPKs and its physiological function is not clear as yet. Finally, several other kinases with a Thr-Xaa-Tyr motif in their activation loop, where the middle residue in this motif is distinct of Glu, Pro or Gly of the other MAPKs have been identified. These protein kinases were termed cousins of MAPKs (144), but since there is no indication that their Thr and Tyr residues are phosphorylated under any conditions, these protein kinases are not considered as MAPKs, and therefore are not specifically described in the current review.

7. Regulation and Specificity-Determination of MAPKs

As mentioned above, each MAPK signaling cascade is involved in the mediation and regulation of a large number of many distinct and even opposing cellular processes. This diversity raises the question as to how is the specificity of each of the four signaling cascades regulated (5, 7, 145–148). One intuitive answer could have been that the recognition and phosphorylation of distinct substrates is critical for this feature. However, the consensus phosphorylation sites (134, 149), and the protein–protein interaction domains (150) seem to be shared by all MAPKs. In addition, the MAPKs induce phosphorylation of a large number of proteins, as it was reported that ERK1/2 have no less than 160 substrates (48), and the number of substrates of p38s and JNKs is likely to be similar. Moreover, the distinct MAPKs seem to have common MAP4K and MAP3K, yet activation of these common components does not result in the simultaneous activation of all MAPKs. Therefore, other mechanisms are required to induce the specificity of the different MAPKs. As of today, five mechanisms for determination of MAPK specificity have been proposed (5, 148), including: (1) Distinct duration and strength of the signal. (2) Interaction with various scaffold proteins that direct components of the MAPK cascades to distinct upstream components and downstream substrates. (3) Interaction between the MAPK cascades or with other signaling pathways that are activated or inhibited simultaneously with the MAPK cascades. (4) Distinct subcellular localizations that may compartmentalize the MAPK cascade components and their targets in certain organelles or other cellular regions. (5) Presence of multiple components with distinct specificities in each level of the cascade. Here we briefly describe the first four points, and elaborate in a separate chapter on the importance of multiple components.

7.1. Duration and Strength of the Signals

The duration and strength of the signal was the first mechanism suggested to explain the signaling specificity of ERK (151). ERK activity is elevated within minutes (2–30) after stimulation and then declines back to basal levels. Fast decline (within 15–40 min) gives rise to an activation kinetics named "transient activation," while a slower decrease back to basal levels (40–180 min) is termed "sustained activation." These differences are mainly regulated by phosphatases that counterbalance the activatory phosphorylations. Because simultaneous phosphorylation of Tyr and Thr residues is required for the activity of MAPKs their full inactivation can be achieved by the removal of phosphates from either one of the regulatory residues, or from both of them together. Thus, protein Ser/Thr phosphatase, protein Tyr phosphatase, and dual specificity phosphatase (MKPs) all act as MAPK phosphatases to directly determine the strength and duration of the signals (152). The first example for the importance of duration in determining MAPK-dependent processes came from studies on PC12 cells in which EGF stimulation induces transient activation of ERKs in the cytoplasm that is essential for proliferation. On the other hand, NGF treatment induces sustained ERK activation and nuclear accumulation that are both required for PC12 cells differentiation (151, 153). Similar effects of varying duration of ERK and other MAPK activities have been reported in other systems (154), indicating that this effect is one of the major specificity determining mechanisms.

7.2. Scaffolding Interactions

Scaffold proteins play crucial roles in many aspects of regulation of MAPK cascades (155–158). These proteins allow formation of multicomponent complexes that are important steps in the regulation of all MAPKs. Thus, scaffolds, such as KSR1, IQGAP1, and Sef1 play pivotal role in the regulation of ERKs (159, 160), JIP participates in the regulation of JNK and possibly also p38 (158, 161), and ERK5 seems to be regulated by scaffold proteins in the nucleus (119). These scaffold proteins have several functions, including: (1) Facilitation of signaling rate by bringing distinct cascade components to close proximity to each other. (2) Determination of signaling specificity by complexing proper signaling components, especially from the MAP4K and MAP3K tiers, upon distinct stimulation. (3) Directing MAPK components to their proper subcellular localization, upstream activators, downstream targets, and various regulators (162). (4) Stabilizing

MAPK components due to protection from phosphatases and proteinases. (5) Determining signaling threshold. All these effects of scaffold proteins make them key regulators of signaling specificity in various systems and upon distinct extracellular and intracellular stimulations.

7.3. Cross-Talk with Other Cascades

Cross-talk and interplays between the MAPK cascades and with other signaling components is another important regulatory mechanism in the determination of signaling specificity. These mechanisms can be mediated by interactions and modulation of activity of components in distinct MAPK cascades or by combinatorial effects of the signaling pathways on downstream targets such as transcription factors (163, 164). Examples for cross-talks between cascades is the phosphorylation of Ser 298 of MEK1 by PAK1 that acts in a signaling cascade downstream of the small GTPase Rac1 (165), and the inhibitory effect of the PI3K–AKT pathway on the ERK cascade (166, 167). Example for the other type of cross-talk, which simultaneously affects a downstream compound, is the activation of IL_2 gene promoter (168). It was shown that this promoter requires binding of quite a few transcription factors for its activation. Interestingly some of the factors lie downstream of distinct signaling cascades including AP1 downstream of the JNK cascade, CREB downstream of PKA, or ERK cascades and NFkB that is a representative of as yet another signaling pathway. Activation of several of these transcription factors upon unique stimulation is required for full promoter activity, while activity of just one of them causes a limited activation and may induce distinct physiological function. These mechanisms clearly determine the signaling specificity by changing the output of the cascade at various tiers or downstream targets of MAPK signaling.

7.4. Distinct Subcellular Localization of Components of the Cascades

As mentioned above, in resting cells, most components of the four MAPK cascades are localized preferentially in the cytoplasm (26, 84, 147, 169). Upon stimulation, some of the components remain cytosolic (e.g., most MAPK3K and MAPKK components), while many of the MAPK and MAPKAPK components (p38 and ERK5 may be constantly nuclear in certain cells) seem to change their localization and are directed mainly to the nucleus, but also to other compartments where they execute their action upon distinct stimulations. The mechanism of nuclear translocation of ERK1/2 and MEK1/2 has been recently elucidated, and found to involve phosphorylation of a novel nuclear translocation signal that allows interaction with Importin7 that allow their penetration through the nuclear pores (170). It is likely that similar mechanism may play a role in the subcellular localization features of components of the other MAPK cascades. Importantly, a small portion of the different components seems to translocate to various compartments including mitochondria, Golgi surface, ER, and endosomes. These translocations are clearly important for the determination of physiological activity and specificity of the MAPK cascades (171, 172). Therefore, subcellular localization and compartmentalization joins the other mechanism described above as a major factor in the specificity-determination of the ERK and other MAPK cascades.

8. Multiple Components in Each Tier Contribute to MAPK Cascade's Specificity

All tiers of the MAPK cascade contain several, or even a large number of distinct protein kinases, that can be either products of distinct genes or alternatively spliced isoforms. As mentioned above, the total number of genes encoding components of the MAPK cascade seems to reach ~70, and together with the multiple alternatively spliced isoforms the total number of components currently known is close to 200, a number that may be increased upon identification of additional spliced isoforms. Assuming that the components in the various tiers of each cascade are independently regulated and have unique function, the number of output signals that can be generated by the MAPK system is huge. This assumption holds true even if some of the gene products are functionally redundant, as might be the case for some of the very similar components (e.g., ERK1/2 (173)). Therefore, the existence of various components with distinct function or regulation is one of the most important mechanisms for determination and extension of signaling specificity by the MAPK cascades.

The MAP4K and MAP3K tiers of each of the cascades are characterized by a large number of components that can each induce activation of their downstream MAPKs. Moreover, the different MAP3K components may act in many cases as common activators of more than one cascade, and few of them (e.g., TPL2 (114), MLTKs (115)) seem to activate all four of them. However, not all MAPKs are simultaneously stimulated upon activation of one such common MAP3K, and distinct MAP3Ks can induce similar activations of distinct MAPKs, without sharing identical downstream targets. This could be mediated, at least in part, by the ability of the different MAP3Ks to activate several distinct signaling cascades. Among others, this is exemplified by the ability of MEKK1 to activate mainly JNK, but under some conditions also ERK, p38, and even NFkB, while MEKK2 activates similarly JNK and p38 (64). As mentioned above, the identity of the participating components, as well as their specificity, is determined in large by scaffold proteins including JNK interacting protein 1–4 (JIP1–4) and others (158, 174). These proteins exhibit different affinities to distinct components in each tier of the cascade and direct them to their upstream effectors and downstream targets. For example, the JIP proteins seem to have different affinities to various MAP3K proteins, although they all seem to bind MKK7 and JNK1–3 (175, 176). However, JIP2 can preferentially bind to the p38 cascade component MKK3 and p38 and direct them to their upstream components Tiam1 and Ras–GRF1 (177). These distinct affinities as well as varying subcellular localizations and protection from phosphatases clearly emphasize the ability of scaffold proteins to pick up the proper components necessary for directing signals to their right destinations and to regulate distinct functions.

In the ERK cascade, it was initially thought that the various components form a linear cascade, with similar kinetics of activation and targets (178). Indeed, even today it seems that diversity in isoform functioning in this cascade is smaller than that in the JNK and p38 cascade, although increasing number of studies indicate that some differences between the isoforms do exist. Thus, the components at the MAP3K tier seem to cause activation of the ERK cascade upon different stimulation and in different cell types. For example, the Raf kinases seem to transmit mostly mitogenic signals, while MEKK1 and MLTK are involved primarily in the activation of ERKs in response to stress (179). Interestingly, although the catalytic activity of the three Raf kinases is very similar, they were shown to transmit distinct signals due to their differential expression and mode of regulation (180). An example for distinct physiological activities regulated by the Raf kinases is their involvement in mitotic progression, where it was shown that B-Raf is responsible for spindle formation, without involvement of the coexisting Raf1 (181). However, no scaffold proteins have been attributed yet to specific Raf kinases under distinct conditions, and therefore their mode of specificity needs further clarification. In addition, the extensive sequence similarity between the components in the next tier (MEK1/2), and their identical substrate recognition, led to the initial conclusion that the isoforms in this tier are functionally redundant. However, soon after, it became clear that the two MEKs are regulated differently, especially due to multiple phosphorylation of the Pro-reach domain of MEK1 that does not appear in the sequence of MEK2 (182–184). This differential phosphorylation results in distinct kinetics of activations, and thereby in a modified duration of the signals, which in turn is able to determine specificity. Thus, the different MAP3K and MAPKK components of the ERK cascade are clearly able to contribute to the extension and determination of its signaling specificity.

In the MAPK tier, ERK1 and ERK2 are very similar proteins as well, with about 75% similarity between them (9). As for MEK1/2, it was initially assumed that they are functionally redundant, as numerous studies revealed that the two are expressed in essentially all cells, and share similar activation kinetics and set of substrates (185). However, some differences between the two isoforms have been reported under certain restricted conditions. Thus, while ERK1-deficient mice are viable and mostly normal (186), ERK2-deficient mice die early in development (187–189). Although these differences might have been caused by differences in ERK1/2 properties, it is now thought that they are mostly due to differences in expression levels of the two proteins (173). Nonetheless, other studies indicated that the changes between ERK1 and ERK2 do occur in some systems, including differential participation of ERK1 and ERK2 in cell cycle progression (190). Moreover, it was shown that during Ras-dependent signaling in fibroblasts, ERK1 inhibits cell proliferation, while ERK2 promotes it (191). However, these results were strongly objected by another group that did not find any differences between the two ERK isoforms in regulating proliferation (192). Thus, this controversy may suggest that the differences between the ERKs are minor and are not manifested under all conditions.

Although the differences between the main MEK and ERK isoforms do not seem to be very pronounced and are still debatable, their alternatively spliced isoforms do seem to exhibit specific functions. Thus, it has recently been shown that the human alternatively spliced isoform of ERK1, namely ERK1c (46) is involved in the regulation of Golgi fragmentation without the involvement of ERK1/2 (193). Importantly, the activation of ERK1c during that process was solely mediated by the upstream MEK1b (43), giving rise to a unique ERK subroute of the MEK1b–ERK1c isoforms that function distinctly from the main rout of MEK1/2–ERK1/2, in regulating processes such as Golgi fragmentation. The existence of this subroute indicates that the alternatively spliced isoforms participate in expending and determining the substrate specificity of the ERK cascade under various conditions. This effect of alternatively spliced isoforms does not seem to be unique to the ERK cascade, as differences in regulation by alternatively spliced isoforms seem to be pronounced also in the JNK and p38 cascades (94, 96, 194, 195). Moreover, in the ERK5 cascade, the short, alternatively spliced forms of ERK5 were suggested to act as dominant negative isoforms, and thereby to modulate the activity of the cascade under certain conditions (117). Taken together, all these evidence clearly support a major role of the multiple components in each tier in regulating and determining signaling specificity of the MAPK signaling.

In summary, we covered here in detail the structure of all four genuine MAPK cascades, which are those of ERK1/2, JNK, p38, and ERK5. Each cascade is composed of a sequential activation of 3–5 tiers of protein Ser/Thr kinases. The core of the MAPK cascade is composed of three tiers, MAPK, MAPKK, and MAP3K, and two other tiers MAP4K and MAPKAPK can be included in each of the cascades dependent on the cell line and the nature of activation. These MAPK cascades are activated by a large number of extracellular stimuli and thereby govern essentially all stimulated cellular processes. The fact that the same cascades can regulate so many different and even opposing signaling processes raises the question as to how is their specificity determined. As of today, five mechanisms for determination of the specificity of each MAPK cascade have been proposed, including: duration and strength of the signal, multiple components in each tier, scaffolding interactions, interplay with other cascades, and proteins that direct the components of the MAPK cascades to distinct upstream regulators and downstream substrates, and compartmentalization. Mis-direction of the signals via the MAPK cascade can lead to diseases such as cancer and diabetes, and therefore understanding their structure and regulation can enhance the developments of drugs aimed to combat them.

Acknowledgment

This work was supported by grants from the Mario Negri–Weizmann collaborative fund and from the EU Sixth Framework Program under the SIMAP (IST-2004-027265) and GROWTHSTOP (LSHC CT-2006-037731). RS is an Incumbent of the Yale S. Lewine and Ella Miller Lewine professorial chair for cancer research.

References

1. Campbell, J. S., Seger, R., Graves, J. D., Graves, L. M., Jensen, A. M., and Krebs, E. G. (1995) The MAP kinase cascade. *Recent Prog Horm Res* **50,** 131–59.
2. Dhanasekaran, D. N., and Johnson, G. L. (2007) MAPKs: function, regulation, role in cancer and therapeutic targeting. *Oncogene* **26,** 3097–9.
3. Avruch, J. (2007) MAP kinase pathways: the first twenty years. *Biochim Biophys Acta* **1773,** 1150–60.
4. Raman, M., Chen, W., and Cobb, M. H. (2007) Differential regulation and properties of MAPKs. *Oncogene* **26,** 3100–12.
5. Shaul, Y. D., and Seger, R. (2007) The MEK/ERK cascade: from signaling specificity to diverse functions. *Biochim Biophys Acta* **1773,** 1213–26.
6. Pimienta, G., and Pascual, J. (2007) Canonical and alternative MAPK signaling. *Cell Cycle* **6,** 2628–32.
7. Krishna, M., and Narang, H. (2008) The complexity of mitogen-activated protein kinases (MAPKs) made simple. *Cell Mol Life Sci* **65,** 3525–44.
8. Tidyman, W. E., and Rauen, K. A. (2009) The RASopathies: developmental syndromes of Ras/MAPK pathway dysregulation. *Curr Opin Genet Dev* **19,** 230–6.
9. Boulton, T. G., Nye, S. H., Robbins, D. J., Ip, N. Y., Radziejewska, E., Morgenbesser, S. D., et al. (1991) ERK's: a family of protein-serine/threonine kinases that are activated and tyrosine phosphorylated in response to insulin and NGF. *Cell* **65,** 663–75.
10. Derijard, B., Hibi, M., Wu, I. H., Barrett, T., Su, B., Deng, T., et al. (1994) JNK1: a protein kinase stimulated by UV light and Ha-Ras that binds and phosphorylates the c-Jun activation domain. *Cell* **76,** 1025–7.
11. Kyriakis, J. M., Banerjee, P., Nikolakaki, E., Dai, T., Rubie, E. A., Ahmad, M. F., et al. (1994)

The stress-activated protein kinase subfamily of c-Jun kinases. *Nature* **369,** 156–60.

12. Freshney, N. W., Rawlinson, L., Guesdon, F., Jones, E., Cowley, S., Hsuan, J., and Saklatvala, J. (1994) Interleukin-1 activates a novel protein kinase cascade that results in the phosphorylation of Hsp27. *Cell* **78,** 1039–49.
13. Han, J., Lee, J. D., Bibbs, L., and Ulevitch, R. J. (1994) A MAP kinase targeted by endotoxin and hyperosmolarity in mammalian cells. *Science* **265,** 808–11.
14. Rouse, J., Cohen, P., Trigon, S., Morange, M., Alonso-Llamazares, A., Zamanillo, D., et al. (1994) A novel kinase cascade triggered by stress and heat shock that stimulates MAPKAP kinase-2 and phosphorylation of the small heat shock proteins. *Cell* **78,** 1027–37.
15. Zhou, G., Bao, Z. Q., and Dixon, J. E. (1995) Components of a new human protein kinase signal transduction pathway. *J Biol Chem* **270,** 12665–9.
16. Lee, J. D., Ulevitch, R. J., and Han, J. (1995) Primary structure of BMK1: a new mammalian map kinase. *Biochem Biophys Res Commun* **213,** 715–24.
17. Coulombe, P., and Meloche, S. (2007) Atypical mitogen-activated protein kinases: structure, regulation and functions. *Biochim Biophys Acta* **1773,** 1376–87.
18. Bacus, S. S., Gudkov, A. V., Lowe, M., Lyass, L., Yung, Y., Komarov, A. P., et al. (2001) Taxol-induced apoptosis depends on MAP kinase pathways (ERK and p38) and is independent of p53. *Oncogene* **20,** 147–55.
19. Schwabe, R. F., Bradham, C. A., Uehara, T., Hatano, E., Bennett, B. L., Schoonhoven, R., and Brenner, D. A. (2003) c-Jun-N-terminal kinase drives cyclin D1 expression and proliferation during liver regeneration. *Hepatology* **37,** 824–32.
20. Dhillon, A. S., Hagan, S., Rath, O., and Kolch, W. (2007) MAP kinase signalling pathways in cancer. *Oncogene* **26,** 3279–90.
21. Zick, Y. (2005) Ser/Thr phosphorylation of IRS proteins: a molecular basis for insulin resistance. *Sci STKE* **2005,** pe4.
22. Jeffrey, K. L., Camps, M., Rommel, C., and Mackay, C. R. (2007) Targeting dual-specificity phosphatases: manipulating MAP kinase signalling and immune responses. *Nat Rev Drug Discov* **6,** 391–403.
23. Torii, S., Nakayama, K., Yamamoto, T., and Nishida, E. (2004) Regulatory mechanisms and function of ERK MAP kinases. *J Biochem (Tokyo)* **136,** 557–61.
24. McCubrey, J. A., Milella, M., Tafuri, A., Martelli, A. M., Lunghi, P., Bonati, A., et al. (2008) Targeting the Raf/MEK/ERK pathway with small-molecule inhibitors. *Curr Opin Investig Drugs* **9,** 614–30.
25. Aoki, Y., Niihori, T., Narumi, Y., Kure, S., and Matsubara, Y. (2008) The RAS/MAPK syndromes: novel roles of the RAS pathway in human genetic disorders. *Hum Mutat* **29,** 992–1006.
26. Mebratu, Y., and Tesfaigzi, Y. (2009) How ERK1/2 activation controls cell proliferation and cell death is subcellular localization the answer? *Cell Cycle* **8,** 1168–75.
27. Marmor, M. D., Skaria, K. B., and Yarden, Y. (2004) Signal transduction and oncogenesis by ErbB/HER receptors. *Int J Radiat Oncol Biol Phys* **58,** 903–13.
28. Naor, Z., Benard, O., and Seger, R. (2000) Activation of MAPK cascades by G-protein-coupled receptors: the case of gonadotropin-releasing hormone receptor. *Trends Endocrinol Metab* **11,** 91–9.
29. Rane, S. G. (1999) Ion channels as physiological effectors for growth factor receptor and Ras/ERK signaling pathways. *Adv Second Messenger Phosphoprotein Res* **33,** 107–27.
30. Kyriakis, J. M., App, H., Zhang, F. X., Banerjee, P., Brautigan, D. L., Rapp, U. R., and Avruch, J. (1992) Raf-1 activates MAP kinase-kinase. *Nature* **358,** 417–21.
31. Barkoff, A., Ballantyne, S., and Wickens, M. (1998) Meiotic maturation in Xenopus requires polyadenylation of multiple mRNAs. *EMBO J* **17,** 3168–75.
32. Wellbrock, C., Karasarides, M., and Marais, R. (2004) The RAF proteins take centre stage. *Nat Rev Mol Cell Biol* **5,** 875–85.
33. Kolch, W., Heidecker, G., Kochs, G., Hummel, R., Vahidi, H., Mischak, H., et al. (1993) Protein kinase C alpha activates RAF-1 by direct phosphorylation. *Nature* **364,** 249–52.
34. Chadee, D. N., and Kyriakis, J. M. (2004) MLK3 is required for mitogen activation of B-Raf, ERK and cell proliferation. *Nat Cell Biol* **6,** 770–6. Epub 2004 Jul 18.
35. Ahn, N. G., Seger, R., Bratlien, R. L., Diltz, C. D., Tonks, N. K., and Krebs, E. G. (1991) Multiple components in an epidermal growth factor-stimulated protein kinase cascade. *In vitro* activation of myelin basic protein/microtubule-associated protein-2 kinase. *J Biol Chem* **266,** 4220–7.
36. Gomez, N., and Cohen, P. (1991) Dissection of the protein kinase cascade by which nerve

growth factor activates MAP kinases. *Nature* **353,** 170–3.
37. Crews, C. M., Alessandrini, A., and Erikson, R. L. (1992) The primary structure of MEK, a protein kinase that phosphorylates the ERK gene product. *Science* **258,** 478–80.
38. Seger, R., Seger, D., Lozeman, F. J., Ahn, N. G., Graves, L. M., Campbell, J. S., et al. (1992) Human T-cell Map kinase kinases are related to yeast signal transduction kinases. *J Biol Chem* **267,** 25628–31.
39. Zheng, C. F., and Guan, K. L. (1993) Properties of MEKs, the kinases that phosphorylate and activate the extracellular signal-regulated kinases. *J Biol Chem* **268,** 23933–9.
40. Alessi, D. R., Saito, Y., Campbell, D. G., Cohen, P., Sithanandam, G., Rapp, U., et al. (1994) Identification of the sites in MAP kinase kinase-1 phosphorylated by p74raf-1. *EMBO J* **13,** 1610–9.
41. Seger, R., Seger, D., Reszka, A. A., Munar, E. S., Eldar-Finkelman, H., Dobrowolska, G., et al. (1994) Over-expression of Mitogen-Activated Protein Kinase Kinase (MAPKK) and its mutants in NIH-3T3 cells: evidence that MAPKK's involvement in cellular proliferation is regulated by phosphorylation of serine residues in its kinase subdomains VII and VIII. *J Biol Chem* **269,** 25699–709.
42. Seger, R., Ahn, N. G., Posada, J., Munar, E. S., Jensen, A. M., Cooper, J. A., et al. (1992) Purification and characterization of MAP kinase activator(s) from epidermal growth factor stimulated A431 cells. *J Biol Chem* **267,** 14373–81.
43. Shaul, Y. D., Gibor, G., Plotnikov, A., and Seger, R. (2009) Specific phosphorylation and activation of ERK1c by MEK1b: a unique route in the ERK cascade. *Genes Dev* **23,** 1779–90.
44. Ray, L. B., and Sturgill, T. W. (1987) Characterization of insulin-stimulated microtubule-associated protein kinase. Rapid isolation and stabilization of a novel serine/threonine kinase from 3T3-L1 cells. *Proc Natl Acad Sci U S A* **84,** 1502–6.
45. Yung, Y., Yao, Z., Hanoch, T., and Seger, R. (2000) ERK1b, a 46-kDa ERK isoform that is differentially regulated by MEK. *J Biol Chem* **275,** 15799–808.
46. Aebersold, D. M., Shaul, Y. D., Yung, Y., Yarom, N., Yao, Z., Hanoch, T., and Seger, R. (2004) Extracellular signal-regulated kinase 1c (ERK1c), a novel 42-kilodalton ERK, demonstrates unique modes of regulation, localization, and function. *Mol Cell Biol* **24,** 10000–15.
47. Gonzalez, F. A., Raden, D. L., Rigby, M. R., and Davis, R. J. (1992) Heterogeneous expression of four MAP kinase isoforms in human tissues. *FEBS Lett* **304,** 170–8.
48. Yoon, S., and Seger, R. (2006) The extracellular signal-regulated kinase: multiple substrates regulate diverse cellular functions. *Growth Factors* **24,** 21–44.
49. Roux, P. P., and Blenis, J. (2004) ERK and p38 MAPK-Activated Protein Kinases: a Family of Protein Kinases with Diverse Biological Functions. *Microbiol Mol Biol Rev* **68,** 320–44.
50. Sturgill, T. W., Ray, L. B., Erikson, E., and Maller, J.L. (1988) Insulin-stimulated MAP-2 kinase phosphorylates and activates ribosomal protein S6 kinase II. *Nature* **334,** 715–8.
51. Deak, M., Clifton, A. D., Lucocq, L. M., and Alessi, D. R. (1998) Mitogen- and stress-activated protein kinase-1 (MSK1) is directly activated by MAPK and SAPK2/p38, and may mediate activation of CREB. *EMBO J* **17,** 4426–41.
52. Fukunaga, R., and Hunter, T. (1997) MNK1, a new MAP kinase-activated protein kinase, isolated by a novel expression screening method for identifying protein kinase substrates. *EMBO J* **16,** 1921–33.
53. Waskiewicz, A. J., Flynn, A., Proud, C. G., and Cooper, J. A. (1997) Mitogen-activated protein kinases activate the serine/threonine kinases Mnk1 and Mnk2. *EMBO J* **16,** 1909–20.
54. Eldar-Finkelman, H., Seger, R., Vandenheede, J. R., and Krebs, E. G. (1995) Inactivation of glycogen synthase kinase-3 by epidermal growth factor is mediated by mitogen-activated protein kinase/p90 ribosomal protein S6 kinase signaling pathway in NIH/3T3 cells. *J Biol Chem* **270,** 987–90.
55. Sapkota, G. P., Kieloch, A., Lizcano, J. M., Lain, S., Arthur, J. S., Williams, M. R., et al. (2001) Phosphorylation of the protein kinase mutated in Peutz-Jeghers Cancer syndrome, LKB1/STK11, at Ser431 by p90RSK and cAMP-dependent protein kinase, but not its farnesylation at Cys433, is essential for LKB1 to suppress cell growth. *J Biol Chem* **276,** 19469–82.
56. Cuenda, A., and Rousseau, S. (2007) p38 MAP-kinases pathway regulation, function and role in human diseases. *Biochim Biophys Acta* **1773,** 1358–75.
57. Han, J., and Sun, P. (2007) The pathways to tumor suppression via route p38. *Trends Biochem Sci* **32,** 364–71.

58. Loesch, M., and Chen, G. (2008) The p38 MAPK stress pathway as a tumor suppressor or more? *Front Biosci* **13,** 3581–93.
59. Thornton, T. M., and Rincon, M. (2009) Non-classical p38 map kinase functions: cell cycle checkpoints and survival. *Int J Biol Sci* **5,** 44–51.
60. Cohen, P. (2009) Targeting protein kinases for the development of anti-inflammatory drugs. *Curr Opin Cell Biol* **21,** 317–24.
61. Cohen, D. M. (2005) SRC family kinases in cell volume regulation. *Am J Physiol Cell Physiol* **288,** C483–93.
62. Hall, A. (2005) Rho GTPases and the control of cell behaviour. *Biochem Soc Trans* **33,** 891–5.
63. Dan, I., Watanabe, N. M., and Kusumi, A. (2001) The Ste20 group kinases as regulators of MAP kinase cascades. *Trends Cell Biol* **11,** 220–30.
64. Uhlik, M. T., Abell, A. N., Cuevas, B. D., Nakamura, K., and Johnson, G. L. (2004) Wiring diagrams of MAPK regulation by MEKK1, 2, and 3. *Biochem Cell Biol* **82,** 658–63.
65. Derijard, B., Raingeaud, J., Barrett, T., Wu, I. H., Han, J., Ulevitch, R. J., and Davis, R. J. (1995) Independent human MAP-kinase signal transduction pathways defined by MEK and MKK isoforms [published erratum appears in Science 1995 Jul 7;269(5220):17]. *Science* **267,** 682–5.
66. Han, J., Lee, J. D., Jiang, Y., Li, Z., Feng, L., and Ulevitch, R. J. (1996) Characterization of the structure and function of a novel MAP kinase kinase (MKK6). *J Biol Chem* **271,** 2886–91.
67. Cuenda, A., Alonso, G., Morrice, N., Jones, M., Meier, R., Cohen, P., and Nebreda, A. R. (1996) Purification and cDNA cloning of SAPKK3, the major activator of RK/p38 in stress- and cytokine-stimulated monocytes and epithelial cells. *EMBO J* **15,** 4156–64.
68. Dashti, S. R., Efimova, T., and Eckert, R. L. (2001) MEK7-dependent activation of p38 MAP kinase in keratinocytes. *J Biol Chem* **276,** 8059–63.
69. Lee, J. C., Laydon, J. T., McDonnell, P. C., Gallagher, T. F., Kumar, S., Green, D., et al. (1994) A protein kinase involved in the regulation of inflammatory cytokine biosynthesis. *Nature* **372,** 739–46.
70. Jiang, Y., Chen, C., Li, Z., Guo, W., Gegner, J. A., Lin, S., and Han, J. (1996) Characterization of the structure and function of a new mitogen-activated protein kinase (p38beta). *J Biol Chem* **271,** 17920–6.
71. Li, Z., Jiang, Y., Ulevitch, R. J., and Han, J. (1996) The primary structure of p38 gamma: a new member of p38 group of MAP kinases. *Biochem Biophys Res Commun* **228,** 334–40.
72. Mertens, S., Craxton, M., and Goedert, M. (1996) SAP kinase-3, a new member of the family of mammalian stress-activated protein kinases. *FEBS Lett* **383,** 273–6.
73. Goedert, M., Cuenda, A., Craxton, M., Jakes, R., and Cohen, P. (1997) Activation of the novel stress-activated protein kinase SAPK4 by cytokines and cellular stresses is mediated by SKK3 (MKK6); comparison of its substrate specificity with that of other SAP kinases. *EMBO J* **16,** 3563–71.
74. Stokoe, D., Campbell, D. G., Nakielny, S., Hidaka, H., Leevers, S. J., Marshall, C., and Cohen, P. (1992) MAPKAP kinase-2; a novel protein kinase activated by mitogen-activated protein kinase. *EMBO J* **11,** 3985–94.
75. McLaughlin, M. M., Kumar, S., McDonnell, P. C., Van Horn, S., Lee, J. C., Livi, G. P., and Young, P. R. (1996) Identification of mitogen-activated protein (MAP) kinase-activated protein kinase-3, a novel substrate of CSBP p38 MAP kinase. *J Biol Chem* **271,** 8488–92.
76. New, L., Jiang, Y., Zhao, M., Liu, K., Zhu, W., Flood, L. J., et al. (1998) PRAK, a novel protein kinase regulated by the p38 MAP kinase. *EMBO J* **17,** 3372–84.
77. Ni, H., Wang, X. S., Diener, K., and Yao, Z. (1998) MAPKAPK5, a novel mitogen-activated protein kinase (MAPK)-activated protein kinase, is a substrate of the extracellular-regulated kinase (ERK) and p38 kinase. *Biochem Biophys Res Commun* **243,** 492–6.
78. Kramer, R. M., Roberts, E. F., Um, S. L., Borsch-Haubold, A. G., Watson, S. P., Fisher, M. J., and Jakubowski, J. A. (1996) p38 mitogen-activated protein kinase phosphorylates cytosolic phospholipase A2 (cPLA2) in thrombin-stimulated platelets. Evidence that proline-directed phosphorylation is not required for mobilization of arachidonic acid by cPLA2. *J Biol Chem* **271,** 27723–9.
79. Ben-Levy, R., Hooper, S., Wilson, R., Paterson, H. F., and Marshall, C. J. (1998) Nuclear export of the stress-activated protein kinase p38 mediated by its substrate MAPKAP kinase-2. *Curr Biol* **8,** 1049–57.
80. Johnson, G. L., and Nakamura, K. (2007) The c-jun kinase/stress-activated pathway: regulation, function and role in human disease. *Biochim Biophys Acta* **1773,** 1341–8.
81. Wang, X., Destrument, A., and Tournier, C. (2007) Physiological roles of MKK4 and

MKK7: insights from animal models. *Biochim Biophys Acta* **1773,** 1349–57.

82. Weston, C. R., and Davis, R. J. (2007) The JNK signal transduction pathway. *Curr Opin Cell Biol* **19,** 142–9.
83. Bogoyevitch, M. A., and Arthur, P. G. (2008) Inhibitors of c-Jun N-terminal kinases: JuNK no more? *Biochim Biophys Acta* **1784,** 76–93.
84. Dhanasekaran, D.N., and Reddy, E.P. (2008) JNK signaling in apoptosis. *Oncogene* **27,** 6245–51.
85. Baker, S.J., and Reddy, E.P. (1998) Modulation of life and death by the TNF receptor superfamily. *Oncogene* **17,** 3261–70.
86. Coso, O.A., Chiariello, M., Yu, J.C., Teramoto, H., Crespo, P., Xu, N., et al. (1995) The small GTP-binding proteins Rac1 and Cdc42 regulate the activity of the JNK/SAPK signaling pathway. *Cell* **81,** 1137–46.
87. Strange, K., Denton, J., and Nehrke, K. (2006) Ste20-type kinases: evolutionarily conserved regulators of ion transport and cell volume. *Physiology (Bethesda)* **21,** 61–8.
88. Craig, E. A., Stevens, M. V., Vaillancourt, R. R., and Camenisch, T. D. (2008) MAP3Ks as central regulators of cell fate during development. *Dev Dyn* **237,** 3102–14.
89. Sanchez, I., Hughes, R. T., Mayer, B. J., Yee, K., Woodgett, J. R., Avruch, J., et al. (1994) Role of SAPK/ERK kinase-1 in the stress-activated pathway regulating transcription factor c-Jun. *Nature* **372,** 794–8.
90. Yan, M., Dai, T., Deak, J. C., Kyriakis, J. M., Zon, L. I., Woodgett, J. R., and Templeton, D. J. (1994) Activation of stress-activated protein kinase by MEKK1 phosphorylation of its activator SEK1. *Nature* **372,** 798–800.
91. Tournier, C., Whitmarsh, A. J., Cavanagh, J., Barrett, T., and Davis, R. J. (1997) Mitogen-activated protein kinase kinase 7 is an activator of the c-Jun NH2-terminal kinase. *Proc Natl Acad Sci U S A* **94,** 7337–42.
92. Holland, P. M., Suzanne, M., Campbell, J. S., Noselli, S., and Cooper, J. A. (1997) MKK7 is a stress-activated mitogen-activated protein kinase kinase functionally related to hemipterous. *J Biol Chem* **272,** 24994–8.
93. Toyoshima, F., Moriguchi, T., and Nishida, E. (1997) Fas induces cytoplasmic apoptotic responses and activation of the MKK7-JNK/SAPK and MKK6-p38 pathways independent of CPP32-like proteases. *J Cell Biol* **139,** 1005–15.
94. Tournier, C., Whitmarsh, A. J., Cavanagh, J., Barrett, T., and Davis, R. J. (1999) The MKK7 gene encodes a group of c-Jun NH2-terminal kinase kinases. *Mol Cell Biol* **19,** 1569–81.
95. Gupta, S., Barrett, T., Whitmarsh, A. J., Cavanagh, J., Sluss, H. K., Derijard, B., and Davis, R. J. (1996) Selective interaction of JNK protein kinase isoforms with transcription factors. *EMBO J* **15,** 2760–70.
96. Dreskin, S. C., Thomas, G. W., Dale, S. N., and Heasley, L. E. (2001) Isoforms of Jun kinase are differentially expressed and activated in human monocyte/macrophage (THP-1) cells. *J Immunol* **166,** 5646–53.
97. Zakowski, V., Keramas, G., Kilian, K., Rapp, U. R., and Ludwig, S. (2004) Mitogen-activated 3p kinase is active in the nucleus. *Exp Cell Res* **299,** 101–9.
98. Zhang, Y., Zhong, S., Dong, Z., Chen, N., Bode, A. M., and Ma, W. (2001) UVA induces Ser381 phosphorylation of p90RSK/MAPKAP-K1 via ERK and JNK pathways. *J Biol Chem* **276,** 14572–80.
99. Deng, X., Xiao, L., Lang, W., Gao, F., Ruvolo, P., and May, W. S., Jr. (2001) Novel role for JNK as a stress-activated Bcl2 kinase. *J Biol Chem* **276,** 25.
100. Gdalyahu, A., Ghosh, I., Levy, T., Sapir, T., Sapoznik, S., Fishler, Y., et al. (2004) DCX, a new mediator of the JNK pathway. *EMBO J* **23,** 823–32.
101. Davis, R. J. (2000) Signal transduction by the JNK group of MAP kinases. *Cell* **103,** 239–52.
102. Hayashi, M., and Lee, J. D. (2004) Role of the BMK1/ERK5 signaling pathway: lessons from knockout mice. *J Mol Med* **82,** 800–8.
103. Cavanaugh, J. E. (2004) Role of extracellular signal regulated kinase 5 in neuronal survival. *Eur J Biochem* **271,** 2056–9.
104. Wang, X., and Tournier, C. (2006) Regulation of cellular functions by the ERK5 signalling pathway. *Cell Signal* **18,** 753–60.
105. Nishimoto, S., and Nishida, E. (2006) MAPK signalling: ERK5 versus ERK1/2. *EMBO Rep* **7,** 782–6.
106. Sumimoto, H., Kamakura, S., and Ito, T. (2007) Structure and function of the PB1 domain, a protein interaction module conserved in animals, fungi, amoebas, and plants. *Sci STKE* **2007,** re6.
107. Kato, Y., Tapping, R. I., Huang, S., Watson, M. H., Ulevitch, R. J., and Lee, J. D. (1998) Bmk1/Erk5 is required for cell proliferation induced by epidermal growth factor. *Nature* **395,** 713–6.

108. Kato, Y., Chao, T. H., Hayashi, M., Tapping, R. I., and Lee, J. D. (2000) Role of BMK1 in regulation of growth factor-induced cellular responses. *Immunol Res* **21,** 233–7.

109. Abe, J., Takahashi, M., Ishida, M., Lee, J. D., and Berk, B. C. (1997) c-Src is required for oxidative stress-mediated activation of big mitogen-activated protein kinase 1. *J Biol Chem* **272,** 20389–94.

110. Sun, W., Wei, X., Kesavan, K., Garrington, T. P., Fan, R., Mei, J., et al. (2003) MEK kinase 2 and the adaptor protein Lad regulate extracellular signal-regulated kinase 5 activation by epidermal growth factor via Src. *Mol Cell Biol* **23,** 2298–308.

111. Xu, B. E., Stippec, S., Lenertz, L., Lee, B. H., Zhang, W., Lee, Y. K., and Cobb, M. H. (2004) WNK1 activates ERK5 by an MEKK2/3-dependent mechanism. *J Biol Chem* **279,** 7826–31.

112. Chao, T. H., Hayashi, M., Tapping, R. I., Kato, Y., and Lee, J. D. (1999) MEKK3 directly regulates MEK5 activity as part of the big mitogen-activated protein kinase 1 (BMK1) signaling pathway. *J Biol Chem* **274,** 36035–8.

113. Garrington, T. P., Ishizuka, T., Papst, P. J., Chayama, K., Webb, S., Yujiri, T., et al. (2000) MEKK2 gene disruption causes loss of cytokine production in response to IgE and c-Kit ligand stimulation of ES cell-derived mast cells. *EMBO J* **19,** 5387–95.

114. Chiariello, M., Marinissen, M. J., and Gutkind, J. S. (2000) Multiple mitogen-activated protein kinase signaling pathways connect the cot oncoprotein to the c-jun promoter and to cellular transformation. *Mol Cell Biol* **20,** 1747–58.

115. Gotoh, I., Adachi, M., and Nishida, E. (2001) Identification and characterization of a novel MAP kinase kinase kinase, MLTK. *J Biol Chem* **276,** 4276–86. Epub 2000 Oct 19.

116. English, J. M., Vanderbilt, C. A., Xu, S., Marcus, S., and Cobb, M. H. (1995) Isolation of MEK5 and differential expression of alternatively spliced forms. *J Biol Chem* **270,** 28897–902.

117. Yan, C., Luo, H., Lee, J. D., Abe, J., and Berk, B. C. (2001) Molecular cloning of mouse ERK5/BMK1 splice variants and characterization of ERK5 functional domains. *J Biol Chem* **276,** 10870–8.

118. Kondoh, K., Terasawa, K., Morimoto, H., and Nishida, E. (2006) Regulation of nuclear translocation of extracellular signal-regulated kinase 5 by active nuclear import and export mechanisms. *Mol Cell Biol* **26,** 1679–90.

119. Raviv, Z., Kalie, E., and Seger, R. (2004) MEK5 and ERK5 are localized in the nuclei of resting as well as stimulated cells, while MEKK2 translocates from the cytosol to the nucleus upon stimulation. *J Cell Sci* **117,** 1773–84.

120. Hayashi, M., Tapping, R. I., Chao, T. H., Lo, J. F., King, C. C., Yang, Y., and Lee, J. D. (2001) BMK1 mediates growth factor-induced cell proliferation through direct cellular activation of serum and glucocorticoid-inducible kinase. *J Biol Chem* **276,** 8631–34.

121. English, J. M., Pearson, G., Baer, R., and Cobb, M. H. (1998) Identification of substrates and regulators of the mitogen-activated protein kinase ERK5 using chimeric protein kinases. *J Biol Chem* **273,** 3854–60.

122. Yang, C. C., Ornatsky, O. I., McDermott, J. C., Cruz, T. F., and Prody, C. A. (1998) Interaction of myocyte enhancer factor 2 (MEF2) with a mitogen-activated protein kinase, ERK5/BMK1. *Nucleic Acids Res* **26,** 4771–7.

123. Kato, Y., Kravchenko, V. V., Tapping, R. I., Han, J., Ulevitch, R. J., and Lee, J. D. (1997) BMK1/ERK5 regulates serum-induced early gene expression through transcription factor MEF2C. *EMBO J* **16,** 7054–66.

124. Kamakura, S., Moriguchi, T., and Nishida, E. (1999) Activation of the protein kinase ERK5/BMK1 by receptor tyrosine kinases. Identification and characterization of a signaling pathway to the nucleus. *J Biol Chem* **274,** 26563–71.

125. Suzaki, Y., Yoshizumi, M., Kagami, S., Koyama, A. H., Taketani, Y., Houchi, H., et al. (2002) Hydrogen peroxide stimulates c-Src-mediated big mitogen-activated protein kinase 1 (BMK1) and the MEF2C signaling pathway in PC12 cells: potential role in cell survival following oxidative insults. *J Biol Chem* **277,** 9614–21.

126. Kasler, H. G., Victoria, J., Duramad, O., and Winoto, A. (2000) ERK5 is a novel type of mitogen-activated protein kinase containing a transcriptional activation domain. *Mol Cell Biol* **20,** 8382–9.

127. Abe, M. K., Kuo, W. L., Hershenson, M. B., and Rosner, M. R. (1999) Extracellular signal-regulated kinase 7 (ERK7), a novel ERK with a C-terminal domain that regulates its activity, its cellular localization, and cell growth. *Mol Cell Biol* **19,** 1301–12.

128. Abe, M. K., Saelzler, M. P., Espinosa, R., 3rd, Kahle, K. T., Hershenson, M. B., Le Beau, M. M., and Rosner, M. R. (2002) ERK8, a new member of the mitogen-activated

protein kinase family. *J Biol Chem* **277,** 16733–43.

129. Kleveric, I. V., Stafford, M. J., Morrice, N., Peggie, M., Morton, S., and Cohen, P. (2006) Characterization of the reversible phosphorylation and activation of ERK8. *Biochem J* **394,** 365–73.
130. Abe, M. K., Kahle, K. T., Saelzler, M. P., Orth, K., Dixon, J. E., and Rosner, M. R. (2001) ERK7 is an autoactivated member of the MAP kinase family. *J Biol Chem* **276,** 21272–9.
131. Kuo, W. L., Duke, C. J., Abe, M. K., Kaplan, E. L., Gomes, S., and Rosner, M. R. (2004) ERK7 expression and kinase activity is regulated by the ubiquitin-proteosome pathway. *J Biol Chem* **279,** 23073–81.
132. Iavarone, C., Acunzo, M., Carlomagno, F., Catania, A., Melillo, R. M., Carlomagno, S. M., et al. (2006) Activation of the Erk8 mitogen-activated protein (MAP) kinase by RET/PTC3, a constitutively active form of the RET proto-oncogene. *J Biol Chem* **281,** 10567–76.
133. Kleveric, I. V., Martin, N. M., and Cohen, P. (2009) Regulation of the activity and expression of ERK8 by DNA damage. *FEBS Lett* **583,** 680–4.
134. Alvarez, E., Northwood, I. C., Gonzalez, F. A., Latour, D. A., Seth, A., Abate, C., et al. (1991) Pro-Leu-Ser/Thr-Pro is a consensus primary sequence for substrate protein phosphorylation. Characterization of the phosphorylation of c-myc and c-jun proteins by an epidermal growth factor receptor threonine 669 protein kinase. *J Biol Chem* **266,** 15277–85.
135. Henrich, L. M., Smith, J. A., Kitt, D., Errington, T. M., Nguyen, B., Traish, A. M., and Lannigan, D. A. (2003) Extracellular signal-regulated kinase 7, a regulator of hormone-dependent estrogen receptor destruction. *Mol Cell Biol* **23,** 5979–88.
136. Saelzler, M. P., Spackman, C. C., Liu, Y., Martinez, L. C., Harris, J. P., and Abe, M. K. (2006) ERK8 down-regulates transactivation of the glucocorticoid receptor through Hic-5. *J Biol Chem* **281,** 16821–32.
137. Cheng, M., Zhen, E., Robinson, M. J., Ebert, D., Goldsmith, E., and Cobb, M. H. (1996) Characterization of a protein kinase that phosphorylates serine 189 of the mitogen-activated protein kinase homolog ERK3. *J Biol Chem* **271,** 12057–62.
138. Zimmermann, J., Lamerant, N., Grossenbacher, R., and Furst, P. (2001) Proteasome- and p38-dependent regulation of ERK3 expression. *J Biol Chem* **276,** 10759–66.
139. Seternes, O. M., Mikalsen, T., Johansen, B., Michaelsen, E., Armstrong, C. G., Morrice, N. A., et al. (2004) Activation of MK5/PRAK by the atypical MAP kinase ERK3 defines a novel signal transduction pathway. *EMBO J* **23,** 4780–91.
140. Schumacher, S., Laass, K., Kant, S., Shi, Y., Visel, A., Gruber, A. D., et al. (2004) Scaffolding by ERK3 regulates MK5 in development. *EMBO J* **23,** 4770–9.
141. Perander, M., Keyse, S. M., and Seternes, O. M. (2008) Does MK5 reconcile classical and atypical MAP kinases? *Front Biosci* **13,** 4617–24.
142. Zhu, A. X., Zhao, Y., Moller, D. E., and Flier, J. S. (1994) Cloning and characterization of p97MAPK, a novel human homolog of rat ERK-3. *Mol Cell Biol* **14,** 8202–11.
143. Boulton, T. G., and Cobb, M. H. (1991) Identification of multiple extracellular signal-regulated kinases (ERKs) with antipeptide antibodies. *Cell Regul* **2,** 357–71.
144. Miyata, Y., and Nishida, E. (1999) Distantly related cousins of MAP kinase: biochemical properties and possible physiological functions. *Biochem Biophys Res Commun* **266,** 291–5.
145. O'Neill, E., and Kolch, W. (2004) Conferring specificity on the ubiquitous Raf/MEK signalling pathway. *Br J Cancer* **90,** 283–8.
146. Bardwell, L. (2006) Mechanisms of MAPK signalling specificity. *Biochem Soc Trans* **34,** 837–41.
147. Murphy, L. O., and Blenis, J. (2006) MAPK signal specificity: the right place at the right time. *Trends Biochem Sci* **31,** 268–75.
148. Zehorai, E., Yao, Z., Plotnikov, A., and Seger, R. (2010) The subcellular localization of MEK and ERK-A novel nuclear translocation signal (NTS) paves a way to the nucleus. *Mol Cell Endocrinol* **314,** 213–20.
149. Songyang, Z., Lu, K. P., Kwon, Y. T., Tsai, L. H., Filhol, O., Cochet, C., et al. (1996) A structural basis for substrate specificities of protein Ser/Thr kinases: primary sequence preference of casein kinases I and II, NIMA, phosphorylase kinase, calmodulin-dependent kinase II, CDK5, and Erk1. *Mol Cell Biol* **16,** 6486–93.
150. Tanoue, T., Adachi, M., Moriguchi, T., and Nishida, E. (2000) A conserved docking motif in MAP kinases common to substrates, activators and regulators. *Nat Cell Biol* **2,** 110–6.
151. Marshall, C. J. (1995) Specificity of receptor tyrosine kinase signaling: transient versus sustained extracellular signal-regulated kinase activation. *Cell* **80,** 179–85.

152. Yao, Z., and Seger, R. (2004) The molecular mechanism of MAPK/ERK inactivation. *Curr Genomics* **5,** 385–93.
153. Peraldi, P., Scimeca, J., Filloux, C., and Van Obberghen, E. (1993) Regulation of extracellular signal regulated protein kinase-1 (ERK1; pp44/mitogen-activated protein kinase) by epidermal growth factor and nerve growth factor in PC12 cells: implication of ERK1 inhibitory activities. *Endocrinology* **132,** 2578–85.
154. Owens, D. M., and Keyse, S. M. (2007) Differential regulation of MAP kinase signalling by dual-specificity protein phosphatases. *Oncogene* **26,** 3203–13.
155. Turner, C. E. (2000) Paxillin interactions. *J Cell Sci* **113 Pt 23,** 4139–40.
156. Kolch, W. (2005) Coordinating ERK/MAPK signalling through scaffolds and inhibitors. *Nat Rev Mol Cell Biol* **6,** 827–37.
157. Pullikuth, A. K., and Catling, A. D. (2007) Scaffold mediated regulation of MAPK signaling and cytoskeletal dynamics: a perspective. *Cell Signal* **19,** 1621–32.
158. Whitmarsh, A. J. (2006) The JIP family of MAPK scaffold proteins. *Biochem Soc Trans* **34,** 828–32.
159. Morrison, D. K., and Davis, R. J. (2003) Regulation of MAP kinase signaling modules by scaffold proteins in mammals. *Annu Rev Cell Dev Biol* **19,** 91–118.
160. Casar, B., Pinto, A., and Crespo, P. (2009) ERK dimers and scaffold proteins: unexpected partners for a forgotten (cytoplasmic) task. *Cell Cycle* **8,** 1007–13.
161. Takaesu, G., Kang, J. S., Bae, G. U., Yi, M. J., Lee, C. M., Reddy, E. P., and Krauss, R. S. (2006) Activation of p38alpha/beta MAPK in myogenesis via binding of the scaffold protein JLP to the cell surface protein Cdo. *J Cell Biol* **175,** 383–8.
162. Casar, B., Pinto, A., and Crespo, P. (2008) Essential role of ERK dimers in the activation of cytoplasmic but not nuclear substrates by ERK-scaffold complexes. *Mol Cell* **31,** 708–21.
163. Raman, M., and Cobb, M. H. (2003) MAP kinase modules: many roads home. *Curr Biol* **13,** R886–8.
164. Junttila, M. R., Li, S. P., and Westermarck, J. (2008) Phosphatase-mediated crosstalk between MAPK signaling pathways in the regulation of cell survival. *FASEB J* **22,** 954–65.
165. Frost, J. A., Steen, H., Shapiro, P., Lewis, T., Ahn, N., Shaw, P. E., and Cobb, M. H. (1997) Cross-cascade activation of ERKs and ternary complex factors by Rho family proteins. *EMBO J* **16,** 6426–38.
166. Rommel, C., Clarke, B. A., Zimmermann, S., Nunez, L., Rossman, R., Reid, K., et al. (1999) Differentiation stage-specific inhibition of the Raf-MEK-ERK pathway by Akt. *Science* **286,** 1738–41.
167. Zimmermann, S., and Moelling, K. (1999) Phosphorylation and regulation of Raf by Akt (protein kinase B). *Science* **286,** 1741–4.
168. Crispin, J. C., and Tsokos, G. C. (2009) Transcriptional regulation of IL-2 in health and autoimmunity. *Autoimmun Rev* **8,** 190–5.
169. Kondoh, K., Torii, S., and Nishida, E. (2005) Control of MAP kinase signaling to the nucleus. *Chromosoma* **114,** 86–91.
170. Chuderland, D., Konson, A., and Seger, R. (2008) Identification and characterization of a general nuclear translocation signal in signaling proteins. *Mol Cell* **31,** 850–61.
171. Inder, K., Harding, A., Plowman, S. J., Philips, M. R., Parton, R. G., and Hancock, J. F. (2008) Activation of the MAPK module from different spatial locations generates distinct system outputs. *Mol Biol Cell* **19,** 4776–84.
172. Casar, B., Arozarena, I., Sanz-Moreno, V., Pinto, A., Agudo-Ibanez, L., Marais, R., et al. (2009) Ras subcellular localization defines extracellular signal-regulated kinase 1 and 2 substrate specificity through distinct utilization of scaffold proteins. *Mol Cell Biol* **29,** 1338–53.
173. Lefloch, R., Pouyssegur, J., and Lenormand, P. (2009) Total ERK1/2 activity regulates cell proliferation. *Cell Cycle* **8,** 705–11.
174. Handley, M. E., Rasaiyaah, J., Chain, B. M., and Katz, D. R. (2007) Mixed lineage kinases (MLKs): a role in dendritic cells, inflammation and immunity? *Int J Exp Pathol* **88,** 111–26.
175. Lee, C. M., Onesime, D., Reddy, C. D., Dhanasekaran, N., and Reddy, E. P. (2002) JLP: a scaffolding protein that tethers JNK/p38MAPK signaling modules and transcription factors. *Proc Natl Acad Sci U S A* **99,** 14189–94.
176. Song, J. J., and Lee, Y. J. (2007) Differential activation of the JNK signal pathway by UV irradiation and glucose deprivation. *Cell Signal* **19,** 563–72.
177. Buchsbaum, R. J., Connolly, B. A., and Feig, L. A. (2002) Interaction of Rac exchange factors Tiam1 and Ras-GRF1 with a scaffold for the p38 mitogen-activated protein kinase cascade. *Mol Cell Biol* **22,** 4073–85.

178. Seger, R., and Krebs, E. G. (1995) The MAPK signaling cascade. *FASEB J* **9,** 726–35.
179. Rubinfeld, H., and Seger, R. (2005) The ERK cascade: a prototype of MAPK signaling. *Mol Biotechnol* **31,** 151–74.
180. Dhillon, A. S., von Kriegsheim, A., Grindlay, J., and Kolch, W. (2007) Phosphatase and feedback regulation of Raf-1 signaling. *Cell Cycle* **6,** 3–7.
181. Borysova, M. K., Cui, Y., Snyder, M., and Guadagno, T. M. (2008) Knockdown of B-Raf impairs spindle formation and the mitotic checkpoint in human somatic cells. *Cell Cycle* **7,** 2894–901.
182. Rossomando, A. J., Dent, P., Sturgill, T. W., and Marshak, D. R. (1994) Mitogen-activated protein kinase kinase 1 (MKK1) is negatively regulated by threonine phosphorylation. *Mol Cell Biol* **14,** 1594–602.
183. Beeser, A., Jaffer, Z. M., Hofmann, C., and Chernoff, J. (2005) Role of group A p21-activated kinases in activation of extracellular-regulated kinase by growth factors. *J Biol Chem* **280,** 36609–15.
184. Catalanotti, F., Reyes, G., Jesenberger, V., Galabova-Kovacs, G., de Matos Simoes, R., Carugo, O., and Baccarini, M. (2009) A Mek1-Mek2 heterodimer determines the strength and duration of the Erk signal. *Nat Struct Mol Biol* **16,** 294–303.
185. Shaul, Y. D., and Seger, R. (2005) Methods in MAPK signaling. *Curr Protoc Cell Biol* **14.3, Suppl 28,** 1–33.
186. Pages, G., Guerin, S., Grall, D., Bonino, F., Smith, A., Anjuere, F., et al. (1999) Defective thymocyte maturation in p44 MAP kinase (Erk 1) knockout mice. *Science* **286,** 1374–7.
187. Saba-El-Leil, M. K., Vella, F. D., Vernay, B., Voisin, L., Chen, L., Labrecque, N., et al. (2003) An essential function of the mitogen-activated protein kinase Erk2 in mouse trophoblast development. *EMBO Rep* **4,** 964–8.
188. Hatano, N., Mori, Y., Oh-hora, M., Kosugi, A., Fujikawa, T., Nakai, N., et al. (2003) Essential role for ERK2 mitogen-activated protein kinase in placental development. *Genes Cells* **8,** 847–56.
189. Yao, Y., Li, W., Wu, J., Germann, U. A., Su, M. S., Kuida, K., and Boucher, D. M. (2003) Extracellular signal-regulated kinase 2 is necessary for mesoderm differentiation. *Proc Natl Acad Sci U S A* **100,** 12759–64.
190. Liu, X., Yan, S., Zhou, T., Terada, Y., and Erikson, R. L. (2004) The MAP kinase pathway is required for entry into mitosis and cell survival. *Oncogene* **23,** 763–76.
191. Vantaggiato, C., Formentini, I., Bondanza, A., Bonini, C., Naldini, L., and Brambilla, R. (2006) ERK1 and ERK2 mitogen-activated protein kinases affect Ras-dependent cell signaling differentially. *J Biol* **5,** 14.
192. Lefloch, R., Pouyssegur, J., and Lenormand, P. (2008) Single and combined silencing of ERK1 and ERK2 reveals their positive contribution to growth signaling depending on their expression levels. *Mol Cell Biol* **28,** 511–27.
193. Shaul, Y. D., and Seger, R. (2006) ERK1c regulates Golgi fragmentation during mitosis. *J Cell Biol* **172,** 885–97.
194. Zervos, A. S., Faccio, L., Gatto, J. P., Kyriakis, J. M., and Brent, R. (1995) Mxi2, a mitogen-activated protein kinase that recognizes and phosphorylates Max protein. *Proc Natl Acad Sci U S A* **92,** 10531–4.
195. Cameron, S. J., Abe, J., Malik, S., Che, W., and Yang, J. (2004) Differential role of MEK5alpha and MEK5beta in BMK1/ERK5 activation. *J Biol Chem* **279,** 1506–12.
196. Pombo, C. M., Kehrl, J. H., Sanchez, I., Katz, P., Avruch, J., Zon, L. I., et al. (1995) Activation of the SAPK pathway by the human STE20 homologue germinal centre kinase. *Nature* **377,** 750–4.
197. Shi, C. S., and Kehrl, J. H. (1997) Activation of stress-activated protein kinase/c-Jun N-terminal kinase, but not NF-kappaB, by the tumor necrosis factor (TNF) receptor 1 through a TNF receptor-associated factor 2- and germinal center kinase related-dependent pathway. *J Biol Chem* **272,** 32102–7.
198. Diener, K., Wang, X. S., Chen, C., Meyer, C. F., Keesler, G., Zukowski, M., et al. (1997) Activation of the c-Jun N-terminal kinase pathway by a novel protein kinase related to human germinal center kinase. *Proc Natl Acad Sci U S A* **94,** 9687–92.
199. Kiefer, F., Tibbles, L. A., Anafi, M., Janssen, A., Zanke, B. W., Lassam, N., et al. (1996) HPK1, a hematopoietic protein kinase activating the SAPK/JNK pathway. *EMBO J* **15,** 7013–25.
200. Dan, I., Watanabe, N. M., Kobayashi, T., Yamashita-Suzuki, K., Fukagaya, Y., Kajikawa, E., et al. (2000) Molecular cloning of MINK, a novel member of mammalian GCK family kinases, which is up-regulated during postnatal mouse cerebral development. *FEBS Lett* **469,** 19–23.
201. Graves, J. D., Gotoh, Y., Draves, K. E., Ambrose, D., Han, D. K., Wright, M., et al.

(1998) Caspase-mediated activation and induction of apoptosis by the mammalian Ste20-like kinase Mst1. *EMBO J* **17,** 2224–34.

202. Lin, J. L., Chen, H. C., Fang, H. I., Robinson, D., Kung, H. J., and Shih, H. M. (2001) MST4, a new Ste20-related kinase that mediates cell growth and transformation via modulating ERK pathway. *Oncogene* **20,** 6559–69.

203. Nakano, K., Yamauchi, J., Nakagawa, K., Itoh, H., and Kitamura, N. (2000) NESK, a member of the germinal center kinase family that activates the c-Jun N-terminal kinase pathway and is expressed during the late stages of embryogenesis. *J Biol Chem* **275,** 20533–9.

204. Su, Y. C., Han, J., Xu, S., Cobb, M., and Skolnik, E. Y. (1997) NIK is a new Ste20-related kinase that binds NCK and MEKK1 and activates the SAPK/JNK cascade via a conserved regulatory domain. *EMBO J* **16,** 1279–90.

205. Chen, W., Yazicioglu, M., and Cobb, M. H. (2004) Characterization of OSR1, a member of the mammalian Ste20p/GCK subfamily. *J Biol Chem* **279,** 11129–36.

206. Zhang, S., Han, J., Sells, M. A., Chernoff, J., Knaus, U. G., Ulevitch, R. J., and Bokoch, G. M. (1995) Rho family GTPases regulate p38 mitogen-activated protein kinase through the downstream mediator Pak1. *J Biol Chem* **270,** 23934–6.

207. Dan, C., Nath, N., Liberto, M., and Minden, A. (2002) PAK5, a new brain-specific kinase, promotes neurite outgrowth in N1E-115 cells. *Mol Cell Biol* **22,** 567–77.

208. Liu, Z. G., Hsu, H., Goeddel, D. V., and Karin, M. (1996) Dissection of TNF receptor 1 effector functions: JNK activation is not linked to apoptosis while NF-kappaB activation prevents cell death. *Cell* **87,** 565–76.

209. Johnston, A. M., Naselli, G., Gonez, L. J., Martin, R. M., Harrison, L. C., and DeAizpurua, H. J. (2000) SPAK, a STE20/SPS1-related kinase that activates the p38 pathway. *Oncogene* **19,** 4290–7.

210. Sabourin, L. A., and Rudnicki, M. A. (1999) Induction of apoptosis by SLK, a Ste20-related kinase. *Oncogene* **18,** 7566–75.

211. Fu, C. A., Shen, M., Huang, B. C., Lasaga, J., Payan, D. G., and Luo, Y. (1999) TNIK, a novel member of the germinal center kinase family that activates the c-Jun N-terminal kinase pathway and regulates the cytoskeleton. *J Biol Chem* **274,** 30729–37.

212. Ichijo, H., Nishida, E., Irie, K., ten Dijke, P., Saitoh, M., Moriguchi, T., et al. (1997) Induction of apoptosis by ASK1, a mammalian MAPKKK that activates SAPK/JNK and p38 signaling pathways. *Science* **275,** 90–4.

213. Wang, X. S., Diener, K., Tan, T. H., and Yao, Z. (1998) MAPKKK6, a novel mitogen-activated protein kinase kinase kinase, that associates with MAPKKK5. *Biochem Biophys Res Commun* **253,** 33–7.

214. Fan, G., Merritt, S. E., Kortenjann, M., Shaw, P. E., and Holzman, L. B. (1996) Dual leucine zipper-bearing kinase (DLK) activates p46SAPK and p38mapk but not ERK2. *J Biol Chem* **271,** 24788–93.

215. Sakuma, H., Ikeda, A., Oka, S., Kozutsumi, Y., Zanetta, J. P., and Kawasaki, T. (1997) Molecular cloning and functional expression of a cDNA encoding a new member of mixed lineage protein kinase from human brain. *J Biol Chem* **272,** 28622–9.

216. Deacon, K., and Blank, J. L. (1997) Characterization of the mitogen-activated protein kinase kinase 4 (MKK4)/c-Jun NH2-terminal kinase 1 and MKK3/p38 pathways regulated by MEK kinases 2 and 3. MEK kinase 3 activates MKK3 but does not cause activation of p38 kinase in vivo. *J Biol Chem* **272,** 14489–96.

217. Blank, J. L., Gerwins, P., Elliott, E. M., Sather, S., and Johnson, G. L. (1996) Molecular cloning of mitogen-activated protein/ERK kinase kinases (MEKK) 2 and 3. Regulation of sequential phosphorylation pathways involving mitogen-activated protein kinase and c-Jun kinase. *J Biol Chem* **271,** 5361–8.

218. Gerwins, P., Blank, J. L., and Johnson, G. L. (1997) Cloning of a novel mitogen-activated protein kinase kinase kinase, MEKK4, that selectively regulates the c-Jun amino terminal kinase pathway. *J Biol Chem* **272,** 8288–95.

219. Xu, Z., Maroney, A. C., Dobrzanski, P., Kukekov, N. V., and Greene, L. A. (2001) The MLK family mediates c-Jun N-terminal kinase activation in neuronal apoptosis. *Mol Cell Biol* **21,** 4713–24.

220. Hirai, S., Katoh, M., Terada, M., Kyriakis, J. M., Zon, L. I., Rana, A., et al. (1997) MST/MLK2, a member of the mixed lineage kinase family, directly phosphorylates and activates SEK1, an activator of c-Jun N-terminal kinase/stress-activated protein kinase. *J Biol Chem* **272,** 15167–73.

221. Rana, A., Gallo, K., Godowski, P., Hirai, S., Ohno, S., Zon, L., et al. (1996) The mixed lineage kinase SPRK phosphorylates and

activates the stress-activated protein kinase activator, SEK-1. *J Biol Chem* **271,** 19025–8.

222. Gallo, K. A., and Johnson, G. L. (2002) Mixed-lineage kinase control of JNK and p38 MAPK pathways. *Nat Rev Mol Cell Biol* **3,** 663–72.

223. Posada, J., Yew, N., Ahn, N. G., Vande-Woude, G. F., and Cooper, J. A. (1993) Mos stimulates MAP kinase in *Xenopus* oocytes and activates a MAP kinase kinase *in vitro*. *Mol Cell Biol* **13,** 2546–52.

224. Hagemann, C., and Rapp, U. R. (1999) Isotype-specific functions of Raf kinases. *Exp Cell Res* **253,** 34–46.

225. Peraldi, P., Frodin, M., Barnier, J. V., Calleja, V., Scimeca, J. C., Filloux, C., et al. (1995) Regulation of the MAP kinase cascade in PC12 cells: B-Raf activates MEK-1 (MAP kinase or ERK kinase) and is inhibited by cAMP. *FEBS Lett* **357,** 290–6.

226. Moriguchi, T., Kuroyanagi, N., Yamaguchi, K., Gotoh, Y., Irie, K., Kano, T., et al. (1996) A novel kinase cascade mediated by mitogen-activated protein kinase kinase 6 and MKK3. *J Biol Chem* **271,** 13675–9.

227. Hutchison, M., Berman, K. S., and Cobb, M. H. (1998) Isolation of TAO1, a protein kinase that activates MEKs in stress- activated protein kinase cascades [In Process Citation]. *J Biol Chem* **273,** 28625–32.

228. Chen, Z., Hutchison, M., and Cobb, M. H. (1999) Isolation of the protein kinase TAO2 and identification of its mitogen-activated protein kinase/extracellular signal-regulated kinase kinase binding domain. *J Biol Chem* **274,** 28803–7.

229. Manning, G., Whyte, D. B., Martinez, R., Hunter, T., and Sudarsanam, S. (2002) The protein kinase complement of the human genome. *Science* **298,** 1912–34.

230. Salmeron, A., Ahmad, T. B., Carlile, G. W., Pappin, D., Narsimhan, R. P., and Ley, S. C. (1996) Activation of MEK-1 and SEK-1 by Tpl-2 proto-oncoprotein, a novel MAP kinase kinase kinase. *EMBO J* **15,** 817–26.

231. Boulton, T. G., Yancopoulos, G. D., Gregory, J. S., Slaughter, C., Moomaw, C., Hsu, J., and Cobb, M. H. (1990) An insulin-stimulated protein kinase similar to yeast kinases involved in cell cycle control. *Science* **249,** 64–7.

232. Qian, Z., Okuhara, D., Abe, M. K., and Rosner, M. R. (1999) Molecular cloning and characterization of a mitogen-activated protein kinase-associated intracellular chloride channel. *J Biol Chem* **274,** 1621–7.

233. Hibi, M., Lin, A., Smeal, T., Minden, A., and Karin, M. (1993) Identification of an oncoprotein- and UV-responsive protein kinase that binds and potentiates the c-Jun activation domain. *Genes Dev* **7,** 2135–48.

234. Kallunki, T., Su, B., Tsiegelny, I., Sluss, H. K., Derijard, B., Moore, G., et al. (1994) JNK2 contains a specificity-determining region responsible for efficient c-Jun binding and phosphorylation. *Genes Dev* **8,** 2996–3007.

235. Yntema, H. G., van den Helm, B., Kissing, J., van Duijnhoven, G., Poppelaars, F., Chelly, J., et al. (1999) A novel ribosomal S6-kinase (RSK4; RPS6KA6) is commonly deleted in patients with complex X-linked mental retardation. *Genomics* **62,** 332–43.

236. Caivano, M., and Cohen, P. (2000) Role of mitogen-activated protein kinase cascades in mediating lipopolysaccharide-stimulated induction of cyclooxygenase-2 and IL-1 beta in RAW264 macrophages. *J Immunol* **164,** 3018–25.

237. Jones, S. W., Erikson, E., Blenis, J., Maller, J. L., and Erikson, R. L. (1988) A Xenopus ribosomal protein S6 kinase has two apparent kinase domains that are each similar to distinct protein kinases. *Proc Natl Acad Sci U S A* **85,** 3377–81.

238. Alcorta, D. A., Crews, C. M., Sweet, L. J., Bankston, L., Jones, S. W., and Erikson, R. L. (1989) Sequence and expression of chicken and mouse rsk: homologs of Xenopus laevis ribosomal S6 kinase. *Mol Cell Biol* **9,** 3850–9.

239. Lavoinne, A., Erikson, E., Maller, J. L., Price, D. J., Avruch, J., and Cohen, P. (1991) Purification and characterisation of the insulin-stimulated protein kinase from rabbit skeletal muscle; close similarity to S6 kinase II. *Eur J Biochem* **199,** 723–8.

240. Zhao, Y., Bjorbaek, C., Weremowicz, S., Morton, C. C., and Moller, D. E. (1995) RSK3 encodes a novel pp90rsk isoform with a unique N-terminal sequence: growth factor-stimulated kinase function and nuclear translocation. *Mol Cell Biol* **15,** 4353–63.

241. Webster, M. K., Goya, L., Ge, Y., Maiyar, A. C., and Firestone, G. L. (1993) Characterization of sgk, a novel member of the serine/threonine protein kinase gene family which is transcriptionally induced by glucocorticoids and serum. *Mol Cell Biol* **13,** 2031–40.

Chapter 2

Determination of ERK Activity: Anti-phospho-ERK Antibodies and In Vitro Phosphorylation

Shiri Procaccia, Sarah Kraus, and Rony Seger

Abstract

The ERK signaling cascade is composed of several protein kinases that sequentially activate each other by phosphorylation. This pathway is a central component of a complex signaling network that regulates important cellular processes including proliferation, differentiation, and survival. In most of these cases, the ERK cascade is activated downstream of the small GTPase Ras that, upon activation, recruits and activates the first tier in the cascade, which contains the Raf kinases. Afterward the signal is further transmitted by MEKs, ERKs, and often RSKs in the MAPKK, MAPK, and MAPKAPKs tiers of the cascade, respectively. ERKs and RSKs can further disseminate the signal by phosphorylating and modulating the activity of a large number of regulatory proteins including transcription factors and chromatin modifying enzymes. Understanding the mechanisms of activation and the regulation of the various components of this cascade will enhance our insight into the regulation of the ERK-dependent cellular processes in normal cells or of their malfunctioning in various diseases, including cancer. In this chapter, we describe methods used to determine the activity of ERKs, which upon slight modifications can also be used for the study of other signaling kinases, either within the cascade or in other pathways. These methods have been successfully applied to study the ERK signaling cascades in a variety of tissue-cultured cell lines, homogenized animal organs, and lower organisms. As such, the use of these methods should expand our knowledge on the regulation of many distinct systems and upon induction of various stimulations.

Key words: Raf, MEK, ERK phosphorylation

1. Introduction

The mitogen-activated protein kinases (MAPKs) are a family of protein serine/threonine kinases that operate within specific signaling pathways termed MAPK cascades (for reviews see first chapter of this book and references therein). Each MAPK cascade is composed of three core tiers (MAP3K, MAPKK, MAPK), which are occasionally accompanied by additional upstream

Rony Seger (ed.), *MAP Kinase Signaling Protocols: Second Edition*, Methods in Molecular Biology, vol. 661,
DOI 10.1007/978-1-60761-795-2_2,

(MAP4K) and downstream (MAPKAPK2) tiers. Each of the tiers is composed of several dedicated protein kinases, which are either products of distinct genes (e.g., ERK1 and ERK2) or alternative spliced isoforms (e.g., ERK1a, b, c). Upon stimulation, the kinases in each level phosphorylate and activate some or all of the protein kinases in the next level, thus participating in the facilitation, amplification, and specificity-determination of the transmitted signals. In many cases, enzymes at a given tier of the cascade share a common phosphorylation site, such as the Thr-Xaa-Tyr motif for MAPKs. Eventually, the transmitted signals are passed on from the MAPK and MAPKAPK tier components to several regulatory proteins that essentially govern all stimulated cellular processes including proliferation, differentiation, response to stress, and others.

Five distinct MAPK signaling cascades have been identified so far and these are termed according to the components in the MAPK tier of the cascades. These cascades are: (1) Extracellular signal-regulated kinase 1/2 (ERK1/2, ERKs; (1)); (2) Jun N-terminal kinase (JNK; also termed SAPK1 (2, 3)); (3) p38MAPK (p38; also termed SAPK2–4; (4–6)); and (4) ERK5 (also termed BMK1 (7, 8)). Additional kinases termed ERK7 and ERK8, which also contain the Thr-Xaa-Tyr motif and share a sequence homology with other MAPKs, represent an additional group of MAPKs (9, 10). However, the mechanism of activation of these kinases is not fully understood, but it does not seem to include phosphorylation by upstream protein kinases. Therefore, at this stage the ERK7/8 are not considered a genuine MAPK cascade. Finally, ERK3 and ERK4 that present up to 50% identity to ERK1/2 do not contain a Thr-Xaa-Tyr domain in their activation loop (11), and therefore are not considered to act as genuine MAPKs as well. The different groups of MAPKs seem to differ in their physiological activities; the ERKs usually play a role in proliferation, whereas the other cascades seem to respond mainly to stress and are involved in apoptosis.

The amount of transmitted signals via each MAPK cascade is important for studying intracellular signaling. Usually, the activity of one component of the MAPK level of each cascade (ERK1/2, JNK, p38MAPK, etc.) is a sufficient indicator of the transmitted signal. However, for certain studies the activity of additional components within the cascades must be determined in order to understand the actual fate of the signal. For example, JNKs can be activated by several components in the MAPKK (MKK 4 and 7 (12, 13)) and MAP3K (e.g., MEKK1–4 (14)) tires, which seem to be held together by specific scaffold proteins (15). Similar diverse scaffolding interactions seem to determine the spatial and temporal parameters of the other MAPK cascades as well (16). Since such signaling complexes appear to operate simultaneously in response to certain stimuli, the study of several levels within the cascade is necessary so as to evaluate the amount of signal in the different branches of the JNK cascade that are formed by the different complexes.

In this chapter, we mainly describe methods that are used to follow the activity of the ERK cascade, which is usually referred to as a prototype for the MAPK system. This cascade is principally activated by mitogens, hormones, and differentiation factors, but also by various stress and other stimuli. These cues activate the ERK cascade by various mechanisms, which mostly culminate in activation of small GTPases (e.g., Ras, (17)). These important signaling components can by themselves or in coordination with other regulatory proteins (e.g., PKC, (18)) further transmit the signal to the MAP3K tier of the cascade that includes mostly Raf kinases (19). Several other MAP3Ks of the ERK cascade (e.g., Tpl2, MOS and MEKK1) function under more specific conditions (20). Thereafter, the signal is transmitted down the cascade through the MAPKKs MEK1, MEK1b, and MEK2, which are dual specificity protein kinases that demonstrate a high selectivity toward ERK1/2 and their alternative spliced isoforms in the MAPK tier (21). Upon their phosphorylation on the Tyr-Xaa-Thr motif, the ERKs are activated, and, in turn, phosphorylate hundreds of regulatory proteins, either in the cytoplasm or upon translocation to the nucleus (22, 23), they activate transcription factors and other enzymes there. The MAPKAPK tier of the ERK cascade includes RSK1–4, which are specific to ERK1/2, and also MSK1, MNK1/2, and possibly MK3/5, which are activated by p38 as well (24).

Since most components of the MAPK cascades belong to the large family of protein kinases, singling out the activity of the studied protein kinase is essential. Several methods have been developed over the years to detect the activity of components of the MAPK cascades. One of the first methods used for the detection of protein kinases in growth factor signaling employed fractionation by a MonoQ fast protein liquid chromatography (FPLC; (21, 25)). This method involved examination of the resulting fractions of the MonoQ column for protein kinase activity. Since fractionation with the MonoQ column is extremely reproducible, kinases that are activated upon stimulation can be detected by comparing the elution profiles of kinases from both activated and nonactivated cells. The fact that the protein kinases are eluted from the column is advantageous, as it allows determination of the actual kinase activity in solution rather than on any solid support. However, since separation of the various protein kinases is not always complete, and since this method is laborious in nature, it is not widely used and hence will not be described here.

Another method that is used for the detection of novel protein kinases is the in-gel kinase assay (26). This technique involves co-polymerization of a given substrate in a sodium dodecyl sulfate (SDS)-polyacrylamide gel used for electrophoresis (SDS-PAGE) of the samples of interest on the co-polymerized gel, and in-gel phosphorylation of the embedded substrate in the presence of (γ^{32}P)-ATP. The advantage of this method is that it reveals the molecular weight of the detected kinases, assisting in their identification. The disadvantages of this procedure lie in the inability of certain protein kinases to renature, the length of the procedure and the narrow linear range of activities for the embedded kinases. Also this method is not widely used, and therefore will not be covered here.

Since both MonoQ fractionation and in-gel kinase assay methods are lengthy and not always accurate, more specific and convenient methods are recommended for the characterization of a given protein kinase. These methods often require the use of specific reagents, such as antibodies and affinity reagents, for the isolation of the protein kinase of interest. Two additional important methods described hereinafter detect kinases' activity by either anti-phosphorylated ERK antibodies (27) or by immunoprecipitation with specific antibodies followed by an in vitro kinase reaction (28). The detection of kinase activity based on slower mobility of activated kinases upon SDS-PAGE ("gel-shift," "upshift") is not recommended because it does not always correlate with the actual enzymatic activity, as was shown for ERKs (29) and for Raf-1 (30). It should also be noted that although affinity techniques (including immunoprecipitation) are often used, the attachment to a solid support that occurs in these methods might interfere with the full catalytic activity of the studied protein kinases. Thus, although these methods can give a good estimation regarding the relative activity of the kinase in question, it cannot be used when accurate kinetic data is required.

Several points have to be considered before attempting to determine the activity of any component in the MAPK cascades. One of the most influential parameters is the *method of protein extraction*. The methods of choice should extract the protein kinases from the proper cellular compartment, and preserve their active form if necessary, while decreasing the amount of nonrelevant kinases. For example, activated Raf-1 can be present in mitochondria membranes, which may not be disrupted by some extraction procedures, but are disrupted if RIPA Buffer is used. Several methods have been developed for the proper extraction of MAPK components. Sonication, which disrupts the plasma membrane but does not solubilize it, is used to produce extracts that contain both cytoplasmic and some nuclear fraction. Solubilization by detergents (e.g., Triton X-100; NP-40) usually extracts proteins from the membrane and cytoplasm, although including SDS and deoxycholate among the detergents can extract proteins also from the nuclear compartment. On the other hand, cellular extraction by addition of hot SDS-PAGE sample buffer is not recommended, because it frees chromatin, which is physically hard to handle. Extraction by freeze–thawing is also not recommended, because of protein phosphatases that may act at low temperatures.

Another important consideration is the *inhibition of proteinases and/or protein phosphatases*, which are released from cellular organelles upon solubilization. Addition of specific inhibitors of phosphatases and proteinases to the extraction buffers, and extraction at low temperatures minimize the effect of these enzymes. However, since phosphatases are usually efficient enzymes, extractions should be performed as fast as possible even if these precautions are taken. Furthermore, the *quality of the antibodies* employed is of great importance for the success of the various procedures described below. These antibodies should recognize only the desired protein kinase, and not isoforms or nonrelevant enzymes. When in vitro kinase activity is determined, those antibodies should also not interfere with the catalytic activity of the tested enzymes. Other parameters that should be considered for accurate comparison of protein kinase activity are as follows: (1) *amount of total proteins* for each assay, (2) the dilution and *amount of the antibodies*, (3) *starvation* of the cells before activation, (4) optimal *length of stimulation*, and (5) the *linear dynamic range* of the phosphorylation reaction. Recommended amounts and concentrations are mentioned below; however, these should always be optimized for the particular cell line, stimuli, and MAPK component.

In this chapter, the main method described for detecting MAPK signaling is the determination of a MAPK activity by anti-phospho antibodies. This method takes advantage of the fact that most MAPK components are activated by phosphorylation, as mentioned above. Western blot analysis with both anti-phospho-MAPK antibody and with the general antibody provides information on the specific and total activity of most MAPKs in a given fraction. This is a very comfortable method although, since it actually detects the phosphorylation by upstream components and dephosphorylation by phosphatases, it does not always reflect the actual activity of the tested kinase. The second assay described hereafter involves immunoprecipitation and in vitro kinase assay. This method is also quite convenient; its disadvantage is that the kinase activity might be influenced by the solid support. An alternative method, using affinity reagents for the isolation of MAPK components, such as JNK, is described in other chapters of this book.

2. Materials

2.1. Cell Culture and Protein Extraction

All solutions should be prepared in distilled/deionized water.

1. Dulbecco modified Eagle's medium (DMEM) (Gibco-BRL).
2. Fetal calf serum (FCS) (Gibco-BRL), glutamine solution (Biological industries, Beit Haemek, Israel), and antibiotics (Biolab, Jerusalem, Israel) stored in aliquots at -20°C.
3. Trypsin–EDTA (Sigma).
4. Stimulant: 50 μg/ml epidermal growth factor (EGF) (Sigma) in EGF buffer (PBS containing 0.5 mg/ml bovine serum albumin (BSA) (Sigma)).
5. 10× phosphate-buffered saline (PBS), calcium- and magnesium free (Gibco-BRL). Prepare 1× ice-cold PBS.
6. Homogenization buffer (Buffer H) with protease inhibitors: 50 mM β-glycerophosphate (Sigma), pH 7.3, 1.5 mM EGTA, 1.0 mM EDTA, 1.0 mM dithiothreitol (DTT) (Sigma), 0.1 mM sodium orthovanadate, 1.0 mM benzamidine (Sigma), 10 μg/ml aprotinin, 10 μg/ml leupeptin, 2.0 μg/ml pepstatin-A.
7. Buffer A: 50 mM β-glycerophosphate, pH 7.3, 1.5 mM EGTA, 1.0 mM EDTA, 1.0 mM DTT, 0.1 mM sodium orthovanadate. Prepare 10× stock solution (without DTT) and store at -20°C. Prior to use add freshly prepared DTT.
8. Bradford reagent (Coomassie protein assay reagent, Pierce).

2.2. SDS-Polyacrylamide Gel Electrophoresis

1. Gel electrophoresis apparatus and power supply.
2. X4 Laemmli reducing sample buffer: 0.2 M Tris–HCl, pH 6.8, 40% (v/v) glycerol, 8% (w/v) SDS, 8% (v/v) β-mercaptoethanol, and 0.2% (w/v) bromophenol blue. Store aliquoted at -20°C.
3. Prestained molecular-weight protein markers.
4. Acrylamide (30%):bisacrylamide (0.8%) solution.
5. Lower (separating) buffer: 1.5 M Tris–HCl, pH 8.8.
6. Upper (stacking) buffer: 0.5 M Tris–HCl, pH 6.8.
7. Tetramethylethylenediamine (TEMED).
8. 10% ammonium persulfate (APS).
9. Running buffer: 25 mM Tris, 192 mM glycine, 0.1% SDS, pH 8.3.
10. Staining solution: 40% methanol, 7% acetic acid, 0.005% bromophenol blue.
11. Destaining solution: 15% isopropanol, 7% acetic acid.

2.3. Western Blot Analysis

1. Transfer apparatus.
2. Transfer buffer: 15 mM Tris, 120 mM glycine, approximate pH 8.8.
3. Nitrocellulose membrane (Protran BA 85, Schleicher & Schuell).
4. Whatman paper 3 mm.
5. Washing buffer (TBS-T): 20 mM Tris–HCl, pH 7.5, 150 mM NaCl, 0.05% Tween-20.
6. Blocking solution: 2% (w/v) BSA in washing buffer.
7. Primary antibody appropriate for signaling MAPK of interest (e.g., monoclonal anti-diphospho-ERKs and polyclonal anti-general ERKs (from Sigma Israel)) and secondary antibody (alkaline phosphatase (AP) or horseradish peroxidase (HRP)-conjugated anti-mouse or anti-rabbit Fab antibodies (Jackson Laboratories)) diluted in washing buffer to appropriate dilutions.
8. Enhanced chemiluminescence (ECL): Commercial kits are available (Amersham; Pierce; Bio-Rad). Otherwise, ECL solutions can be made by mixing equal volumes of solution A (2.5 mM Luminol, 400 μM *p*-coumaric acid in 100 mM Tris, pH 8.5) and solution B (5.4 mM H_2O_2 in 100 mM Tris, pH 8.5).
9. Alkaline phosphatase (AP)-based detection assay: NBT/BCIP visualization solution is comprised of 10 ml AP substrate buffer (100 mM Tris–HCl, pH 9.5, 100 mM NaCl, and 5.0 mM $MgCl_2$) containing 66 μl NBT (50 mg/ml, Promega) and 33 μl BCIP (50 mg/ml, Promega).

2.4. Immuno-precipitation

1. Antibodies for immunoprecipitation (e.g., anti-ERK C-terminus; C-16 Santa Cruz CA).
2. Protein A-Sepharose.
3. 0.5 M LiCl in 0.1 M Tris, pH 8.0.
4. Radioimmune precipitation (RIPA) buffer: 137 mM NaCl, 20 mM Tris, pH 7.4, 10% (v/v) glycerol, 1% (v/v) Triton X-100, 0.5% (w/v) deoxycholate, 0.1% (w/v) sodium dodecyl sulfate (SDS), 2.0 mM EDTA, 1.0 mM phenylmethylsulphonyl fluoride (PMSF), and 20 μM leupeptin.
5. Buffer A (see Subheading 2.2, item 8).
6. 3× Reaction Mixture (RM) (with ($\gamma^{32}P$)-ATP): 75 mM β-glycerophosphate, pH 7.3, 100 μM ($\gamma^{32}P$)-ATP (~4,000 cpm/pmol) (Amersham or NEN), 0.3 mM unlabeled ATP, 30 mM $MgCl_2$, 2.5 mg/ml BSA, 1.5 mM DTT, 3.75 mM EGTA, 0.15 mM sodium orthovanadate, 30 μM calmidazolium (Calbiochem), 6 μM PKI peptide (Calbiochem).

7. Substrate: 2 mg/ml myelin basic protein (MBP, bovine brain, Sigma).
8. Perspex shielding for radioactive work.

3. Methods

3.1. Cell Culture

Cultured cells (Rat1 or any other cell types, see Note 1) are maintained in growth medium (e.g., DMEM) supplemented with 10% heat-inactivated FCS, 1% glutamine and an antibiotic mixture added to a final concentration of 100 units/ml penicillin and 100 mg/ml streptomycin. Heat inactivation of FCS is performed by heating it for 45 min at 56°C. Cells are periodically harvested with trypsin–EDTA from confluent cultures. Prior to stimulation the cells are serum starved in starvation medium (DMEM containing 0.1% FCS) for 14–20 h. The cells should not be removed from the incubator or handled in any other way at least 4 h before stimulation to avoid activation of the MAPKs due to varying physical conditions (e.g., low temperature).

3.2. Preparation of Cell Extracts

One of the most important parameters for the successful determination of ERK-activation is the proper extraction of the protein from the examined cell lines or tissues. We describe here an extraction by sonication, which is useful for cytoplasmic and nuclear proteins. However, other methods of extractions (e.g., by detergent) can be used as well (see Note 2), provided that inhibitors of phosphatases and proteinases are included in the extraction buffer at 4°C. The example used here for EGF stimulation of Rat1 cells can be employed with minor modifications for most cell types and stimuli.

1. Grow cells (6 cm tissue culture plates) in DMEM containing 10% FCS to subconfluency (~0.5×10^6 cells/plate) in a tissue culture incubator (37°C, 5% CO_2).
2. Starve cells (14–20 h) in starvation medium (2 ml/plate).
3. Stimulate the cells by incubating them with 2 μl EGF (final concentration can vary between 5 and 100 μM) for various time points. Control plates should be treated with EGF buffer alone for the same times as for the EGF-treatment.
4. At the appropriate time interval, remove the medium from the plates. Then, rinse the plates twice with ice-cold PBS and once with ice-cold Buffer A (5 ml each). Since the arrest and slowing down of biological processes is desired at this stage, it is recommended to place the plates on ice.
5. Add 300 μl of ice-cold Buffer H to each plate, tilt the plate gently and scrape the cells using a plastic scraper. Transfer the cells to labeled, pre-cooled 1.5 ml plastic eppendorf tubes.

6. Disrupt the cells by sonication (two 7 s 50 W pulses) on ice.
7. Centrifuge the cellular extracts at 14,000 × *g* for 15 min at 4°C. The supernatant contains the cytoplasmic extracts to be examined for phosphorylation (see Note 5), transfer to new pre-cooled, test tubes.
8. Take aliquots (5–10 μl) from the resulting supernatants for protein determination. Store the remainder of each cytoplasmic extract on ice until needed.
9. Dilute the samples (usually 1:20) to make sure that the protein concentration is within the dynamic range of the detection (within the concentration of the used standards) and proceed as follows:
   (a) Put 10 μl of each of the protein standards (5, 10, 20, 50, 100 and 200 μg/ml BSA in Buffer H) into at least two wells of a flat-bottom 96-well microplate.
   (b) Put 10 μl of each of the diluted samples in duplicates. Add 190 μl of Bradford reagent to all wells.
   (c) Place the microplate in a microplate reader and determine the optical density of the samples at 595 nM. From the optical densities, calculate the protein concentrations of the samples.
10. Equal amounts of cell extract from each of the above treatments (Subheading 3.2, step 3) are used for Western blotting (usually 20 μg protein/sample), immunoprecipitation (usually 300 μg).
11. For Western blot analysis, add to each of the samples 1/3 volume of 4× sample buffer, mix the contents, boil for 3 min, and spin for 1 min at 14,000 × *g*. For immunoprecipitation the cytoplasmic extracts are incubated with the antibodies as described below.

3.3. Western Blot Analysis and Antibodies

1. For Western blot analysis, proteins are first separated by 10 or 12% SDS-PAGE. To prepare the gel, first assemble glass plates and spacers in a minigel apparatus (Bio-Rad). Prepare 10% polyacrylamide separating gel (10 ml) by mixing 3.3 ml acrylamide stock solution, 2.5 ml of Lower Buffer, 4.2 ml of water, 100 μl APS and 10 μl TEMED. Insert ~7.5 ml into the glass plates. Overlay separating gel with water and allow gel to polymerize.
2. Prepare 5 ml of 3% polyacrylamide stacking gel by mixing 750 μl acrylamide stock solution, 1.25 ml of Upper Buffer, 3.0 ml of water, 100 μl APS, and 10 μl TEMED. Cast the gel, insert comb, and allow polymerization. Assemble gel in apparatus and add running buffer.

3. Load the samples prepared above and a prestained protein marker on the gel and run the gel at 150 V. Once the dye front of the SDS-PAGE has reached the end of the gel, remove the gel from the apparatus, and proceed with the transfer step.
4. Prewet (soak) the nitrocellulose membrane in transfer buffer.
5. Fill the transfer apparatus with transfer buffer. Make a sandwich of the SDS-gel, nitrocellulose membrane and the transfer pads by putting a wet (transfer buffer) 3 mm piece of Whatman paper on a wet pad, the gel on top of the Whatman paper, the wet nitrocellulose membrane on top of the gel, and the other wet 3 mm Whatman paper on top of the nitrocellulose membrane.
6. Remove any air bubbles from between the different layers of the transfer sandwich by gently rolling a 10 ml pipette over the sandwich. Place the other wet pad on top of the transfer sandwich. Make sure air bubbles are not trapped between the gel and the other components.
7. Place the sandwich containing the SDS-gel and nitrocellulose membrane into the buffer-filled transfer apparatus. The nitrocellulose membrane should face the side with the cathode and the SDS-gel should face the side with the anode. Connect the apparatus to a power supply and start the current (200 mA constant current, 90 min, preferably with a cooling device). Methanol or 0.05% SDS are sometimes included in the transfer buffer; their inclusion will require different transfer conditions.
8. At the end of the transfer period, turn off the power supply and remove the nitrocellulose membrane from the transfer sandwich. Rinse the nitrocellulose membrane with transfer buffer to remove any adhering pieces of gel and place the membrane in a flat container.
9. Incubate the nitrocellulose membrane in blocking solution for 60 min at room temperature.
10. Incubate the blot with the first antibody (monoclonal anti-active ERKs antibody, diluted according to the manufacturer recommendations). This incubation can be done either overnight at 4°C, 30 min at 37°C, or 1–2 h at room temperature.
11. Wash the blot in the flat container at least three times for 15 min each with TBS-T at 23°C.
12. Incubate the blot with second antibody (AP/ECL-conjugated goat anti-mouse IgG diluted according to the manufacturer instructions in TBS-T) for 45 min at room temperature.

13. Wash the blot at least three times for 10 min each with TBS-T.
14. Use an AP/ECL detection protocol to detect phosphorylated ERKs.
15. After detecting the phosphorylated ERKs it is recommended to determine whether there is an equal amount of ERKs using an anti-general ERK antibody. It should be noted that the different antibodies may interfere with the detection of each other and therefore, either additional identical blot or a stripping step are required. For the second staining of the same nitrocellulose, incubate it in blocking solution for 30 min at room temperature.
16. Incubate the blot with the "new" first antibody (polyclonal anti-general ERK antibody). Develop as above with HRP/AP system that had not been used for the first step and appropriate ECL/AP system.
17. Two or three bands are usually stained by the antibodies. When two bands appear, these are the p42 ERK2 and p44 ERK1. In some cell lines and tissues a third band at 46 kDa is detected (ERK1b). The intensity of staining of the bands is elevated and this reflects their time course of regulatory phosphorylation upon stimulation (Fig. 1), while the amount of the ERKs as detected by the anti-general ERK antibody is not changed for up to 2 h of stimulation (Fig. 1).

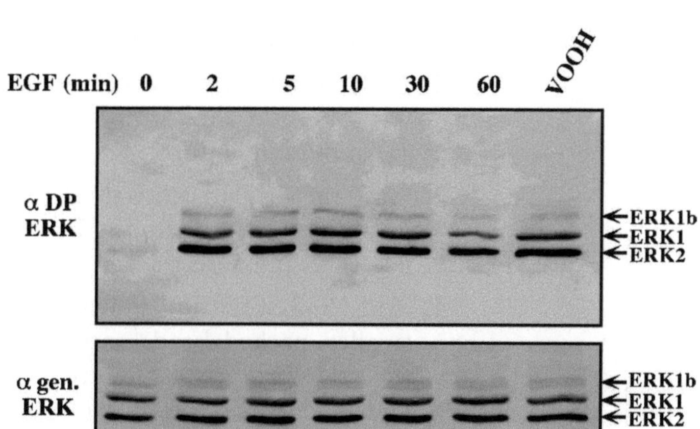

Fig. 1. Detection of ERK activity by Western Blotting with anti-diphospho ERK antibody. Subconfluent Rat1 cells were serum starved (DMEM + 0.1% FCS, 18 h) and then treated with either EGF (50 ng/ml) for the indicated times, VOOH (100 μM sodium orthovanadate and 200 μM H_2O_2) for 15 min or left untreated (Basal control). Cytoplasmic extracts were prepared as described. Samples (20 μg) were prepared, separated by a 12% SDS-PAGE and blotted with either the anti-diphospho ERK antibody (*upper panel*) or with anti-general ERK (phosphorylated + nonphosphorylated) antibody (*lower panel*). This was followed by development with the AP system. The site of ERK2, ERK1, and ERK1b is indicated.

3.4. Determination of ERKs Activity by Immunoprecipitation

This method to determine ERKs activity involves the isolation of the enzyme using immunoprecipitation with specific antibodies and then performing a phosphorylation reaction in vitro. Although ERKs are used here as an example, if appropriate reagents are used, this protocol can be performed with most MAPK isoforms and other components of the MAPK cascade. This protocol facilitates a fast and efficient isolation of the kinase of interest and its reliable quantification by a phosphorylation reaction. For immunoprecipitation, specific antibodies directed to the C-terminal domain of the ERKs are used. The quality and specificity of the antibodies used for the immunoprecipitation protocol is particularly important. Usually, anti-C-terminal ERKs antibodies are used and those do not interfere with the enzymatic activity of the kinase tested.

In this assay, the amount of proteins in the different samples and the dilution of antibodies should be optimized to avoid nonspecific recognition of excess proteins. The stringent washings of the immunoprecipitates are necessary to avoid nonspecific precipitation of contaminant kinases. In addition, this assay is performed while the enzyme is still on the beads, and therefore the results obtained do not accurately reflect the specific activity of ERKs (qualitative and not quantitative). For accurate kinetic data, it is possible to elute the protein kinase from the immunoprecipitating beads (or isolate them by other means) and then determined their activity in solution. The protocol is as follows:

1. As above, the assay is described for six samples. The protein A-Sepharose beads described are supplied as a dry powder; in case the beads are preswollen, proceed from step 4.
2. Place Protein A-Sepharose beads (~150 μl) in a 1.5 ml plastic test tube, add 1 ml of PBS, and let the beads swell for 10 min at room temperature.
3. Wash the swollen beads three times with 1 ml PBS (resuspend in buffer and centrifuge (1 min, 14,000 × *g*), at room temperature). Discard the supernatant.
4. Add 15 μl of the antibodies to be conjugated to 120 μl of the swollen packed beads and 365 μl of PBS (final volume of 0.5 ml). Rotate the mixture (1 h, room temperature) on an end-by-end rotator to allow the antibodies to bind to the Protein A (this can be done at 4°C, 16 h). The volumes listed here should be sufficient for eight reactions, but because of the density of the beads, will probably only be sufficient for six or seven reactions.
5. Wash the beads once with 1 ml ice-cold PBS and then three times with 1 ml ice-cold Buffer H (all at 4°C). Resuspend the washed beads in an equal volume of ice-cold buffer H (~250 μl for ~250 μl of beads). Either use the antibody-conjugated beads immediately, or store at 4°C until used. It is best to use the conjugated beads within 3 days of preparation.

6. Add 30 μl of the antibody-conjugated bead suspension (15 μl net) to 300 μl sample of cytoplasmic extract containing 50–500 μg total protein (in Buffer H) in pre-cooled 1.5 ml plastic test tubes. Rotate end to end for 2 h at 4°C. Although this is not always necessary, we recommend using equal amounts of protein in each of the samples to be immunoprecipitated to avoid inaccuracy.
7. Centrifuge the incubation mixture (1 min, 14,000 × *g* at 4°C). Remove and discard the incubation supernatant from the antibody-conjugated beads. Wash the beads once with 1 ml ice-cold RIPA buffer, twice with ice-cold 0.5 M LiCl, and twice with 1 ml ice-cold Buffer A. As been previously mentioned, these stringent washes are important, because they remove "sticky" protein kinases that might interact nonspecifically with the Protein A beads.
8. After the last washing step, remove Buffer A completely from the conjugated beads and resuspend the pellets of the beads in 15 μl of double distilled water.
9. At this stage, prepare your working bench for working with a small amount of radioactivity and add 10 μl of 3× RM to each tube.
10. Start the phosphorylation reaction by adding 5 μl of the phosphorylation substrate (MBP, 2 mg/ml) or other substrate to the tube and placing the mixture in a thermomixer at 30°C.
11. Incubate 10–20 min at 30°C with either constant or frequent shaking. If a thermomixer is not available, a water bath or other heating device can be used.
12. End the phosphorylation reactions by adding 10 μl X4 sample buffer to each tube. Boil, centrifuge (1 min at 14,000 × *g*), and load the supernatants on a 15% SDS-PAGE gel.
13. When the front dye of the gel reaches about 0.5 cm from the bottom of the gel, stop the current. To remove the excess of free radiolabeled ATP, which migrates just in front of the bromophenol blue, cut out the part of the gel below the dye. This will considerably reduce the amount of radioactivity in the gel.
14. Transfer the separated proteins onto a nitrocellulose paper using a blotting apparatus as described above (Subheading 3.2). Wash briefly with distilled water and let dry. An alternative way would be to stain, destain, and dry the gel on a Whatman 3 mm paper, but this procedure does not allow further detection of proteins in the gel as described for the immunoprecipitated ERKs in step 17.
15. Expose the gel in a phosphoimager or on X-ray film (at −80°C) Band should appear at 16–21 kDa, which is the molecular weight of the four MBP isoforms.

16. To make sure that equal amount of ERKs was immunoprecipitated in each treatment, the nitrocellulose can be then blocked with BSA, overlayed with anti-general ERK antibodies and developed (see Subheading 3.2). Special precaution should be taken because of the radioactivity.
17. Upon exposure on an X-ray film, phosphorylation is detected on a group of bands at 12–21 kDa, which are the different isoforms of MBP (Fig. 2). The intensity of phosphorylation in each time is changed and reflects the time course of activation of the ERKs. When the amount of ERKs is detected by the anti-general ERK antibody, primarily ERK2 can be detected at 44 kDa (Fig. 2).

3.5. Summary

Over the past years, increasing knowledge on the biochemical properties, as well as the development of specific pharmacological tools have led to a marked improvement in the methods employed for the analysis and assay of MAP kinases.

The ERK signaling cascade is part of a complex network of signaling pathways that enable cells to respond to varying conditions and extracellular signals in a regulated and coordinated manner. Thus, understanding the mechanisms by which the components of this cascade are regulated, targeted intracellularly and linked to other signaling pathways will enhance our insight into the regulatory networks that control the cellular response to a particular agonist.

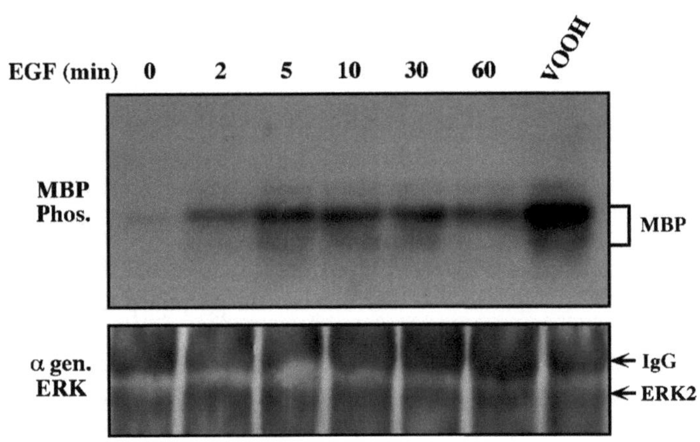

Fig. 2. Detection of ERK activity using in vitro *kinase assay*. Subconfluent Rat1 cells were serum starved (DMEM + 0.1% FCS, 18 h) and then treated with either EGF (50 ng/ml) for the indicated times, VOOH (100 μM sodium orthovanadate and 200 μM H_2O_2) for 15 min or left untreated (Basal control). Cytoplasmic extracts were prepared as described. For immunoprecipitation and in vitro kinase assay proteins (300 μg) were incubated either with 30 μl of anti-ERK C-terminus antibody-conjugated protein A beads. Phosphorylation reaction on MBP was performed as described and terminated with boiling in sample buffer. The proteins were then separated by 15% SDS-PAGE, blotted onto nitrocellulose and subjected to autoradiography (*upper panel*), and to anti-general ERK antibody (*lower panel*). The sites of MBP and of ERK2 are indicated.

The methods given in this chapter have been successfully applied to study the ERK signaling cascades in a variety of tissue-cultured cell lines, homogenized animal organs and lower organisms.

4. Notes

The protocols given in this chapter are those used in our laboratory. The following notes may be of interest:

1. The methodology is suitable for a wide range of cell types and agonists. The methods described here were originally developed for cells grown in monolayer cultures. It may be necessary to adapt them for other (nonadherent) cell types. Although this protocol describes EGF stimulation of cells, this procedure, with minor changes, can be used for most extracellularly stimulated cells.
2. One of the parameters that should be considered before activation of cells by any stimulus is the serum starvation (serum-deprivation), which is usually done in 0.1% serum or sometimes even without serum at all. The aim of this starvation, which makes the cells quiescent, is to significantly reduce the amount of inducible MAPK phosphatases and to obtain a lower basal activity and thus extend the possible fold activation. For most cells, this can be achieved within 14 h. Starvation for too long, or any change in temperature or pH, may be stressful to the cells, and thereby induce activation of one or more signaling pathways.
3. The optimal length of stimulation (subheading 3.2, item 3) may vary between stimuli, cells, and other conditions. Thus appropriate time courses for each kinase should be determined to obtain an accurate stimulation by various stimuli.
4. Positive and negative controls are very important for the success of the experiments described under subheading 3.2. Negative control is a plate that was not exposed to any stimulant of exposed to the vehicle used to dissolve the stimulant. Because of its important as a base-line for the whole experiment, we use as negative controls either two plates or one plate for each time point and concentration. If the influence of the stimulating agent on the cells is not yet known it is recommended to include a positive control in each experiment, such as peroxovanadate (VOOH), which nonspecifically activates many signaling events (31).
5. As mentioned above, the method of protein extraction is an important parameter in the determination of activity of any cellular enzyme. Since MAPKs are localized within cells, the cellular membranes must be disrupted to access the

desired targets. The protein kinases of interest must then be obtained and preserved in their active form, while decreasing the amount of nonrelevant kinases. For more details see introduction.

6. Special consideration should be given to the composition of Buffer H (subheading 3.2 item 5 and ref 32). It is recommended to use β-glycerophosphate, which serves both as a buffer and as a general phosphatase inhibitor, rather than TRIS or HEPES. Sodium orthovanadate is used to inhibit tyrosine phosphatases and the mixture of pepstatin-A, aprotinin, leupeptin, and benzamidine are used to inhibit proteases. This buffer, when cold, blocks most of the phosphatase and proteinase activities in cell extracts. Addition of specific inhibitors of phosphatases and proteases and extraction at low temperatures, minimize the effect of these enzymes. However, since phosphatases are usually efficient enzymes, even if these precautions are taken, extractions should be performed as fast as possible.
7. In the Western blot step (subheading 3.3), the efficiency of protein transfer is usually monitored visually by the transfer of prestained protein markers from the gel to the nitrocellulose membrane. The total amount of protein transferred can also be detected by staining the nitrocellulose membrane with Ponceau red. However, since the total amount of nonphosphorylated protein is determined by general antibodies as described, staining with Ponceau red is probably not essential for this particular protocol.
8. For blocking of the nitrocellulose membrane (subheading 3.3, item 9), we are usually using BSA. Although BSA is considered relatively expensive, it is often used as a blocking solution in the Western blot procedure. The use of nonfat dry milk is not always recommended, because it can cause high background due to phosphotyrosine-containing proteins in the milk or it may contain phosphatases.
9. The successful use of sequence-specific anti-phosphoprotein antibodies (described in subheading 3.3, item 10) relies on their specificity for the phosphorylated form of the examined protein (see anti-MAPK antibodies below). Monoclonal antibodies, which usually confer better specificity than polyclonal ones, are considered as a reliable tool for distinguishing phosphorylated from nonphosphorylated forms of proteins, although affinity-purified polyclonal antibodies can be used as well.
10. For accurate comparison of the amounts of phosphoproteins (subheading 3.3, item 14), detection should be performed in the linear range of the detection system. Thus, the amount of protein loaded on the gel, the concentration of primary and

secondary antibodies, and the time of ECL exposure should be optimized in order to reach linearity. Alternatively, a standard curve with the proteins of interest can be made and serial dilutions of the cellular extracts of each treatment can be loaded to the SDS-PAGE. The blotting detection systems, such as, ECL- ^{125}I-, AP-, or biotin-conjugated antibodies should be chosen carefully. Usually, ECL has the narrowest linear range of these systems whereas ^{125}I-antibodies have a relatively broad range. The AP detection system, which has moderate linear range, is usually used for the types of experiments described here, because it is a convenient method.

11. Immunoprecipitation methods (subheading 3.4) may vary in the order in which the antibodies and protein A are added to the cell extracts. In the protocol described here, the antibodies are conjugated to protein A beads, and only then added to the cytoplasmic extracts. This procedure minimizes the time of incubation of the samples with the antibodies, and thereby, minimizes exposure of the desired kinases to phosphatases and proteinases in the extracts. Furthermore, this procedure ensures that only antibodies recognized by protein A will be used for the immunoprecipitation. In this case, antibodies that are not recognized by protein A are able to bind the desired antigen, but then cannot be precipitated when protein A beads are added, and therefore the efficiency of immunoprecipitation is reduced.
12. The antibodies used for immunoprecipitation (subheading 3.4, item 4) should not mask the kinase activity of the MAPKs. A specific antibody directed to the C-terminus of ERKs is usually used for this purpose. If the nature of the antibody is not known, it is recommended to use nonrelated antibody in parallel to the examined antibody a control for the efficiency of the immunoprecipitation.
13. In the in vitro phosphorylation step (subheading 3.4, item 10), the composition of the reaction mixture (3× RM) is important for optimal ERKs activity. The most important components of the reaction mixture are the Mg^{2+} and (γ^{32}P)-ATP, which are essential for the phosphorylation reaction. We recommend the use of 100 μM ATP with ~4,000 cpm/pmol of the labeled ATP, which provides an extended linear range and reproducible results. When the enzymatic activity of the kinases is very low, which makes detection of phosphorylation difficult, the concentration of cold ATP should be reduced to 10–20 μM and the amount of radioactive material elevated. Addition of labeled ATP alone is not recommended because this will result in a nanomolar concentration of ATP, which is much below the Km for ATP and may lead to nonspecific phosphorylation. As previously mentioned, the β-glycerophosphate in the reaction mixture serves as a buffer, but can also inhibit

residual phosphatases that may have nonspecifically bound to the beads. The BSA serves as a carrier protein but when purity is required, it can be eliminated. The EGTA chelates Ca^{2+}, which may interfere with some kinase activities, DTT keeps the proteins reduced and sodium orthovanadate inhibits tyrosine phosphatases. Additional protein kinase inhibitors may be included in the mixture, such as calmidazolium (calmodulin antagonist) and PKI peptide (PKA inhibitor).

14. Substrates used in the phosphorylation reaction (subheading 3.4, item 10) should be well phosphorylated by the desired kinases to allow accurate detection of the phosphorylation reaction. MBP can serve as a good, nonspecific substrate for many protein kinases including ERKs, although it is probably not a physiological substrate for any MAPK. However, more specific substrates are often used and those include the purified, recombinant RSK, MNK or Elk1 and peptides made according to the phosphorylation sites on this protein.
15. As mentioned above, the determination of enzymatic activity (subheading 3.4, items 13 and 14) when enzymes (in this case MAPKs) are bound to beads is not always accurate. One solution for this problem is to release the kinase(s) of interest from the beads, by adding excess immunizing peptide. The phosphorylation reaction can then be performed without the interference of the beads, and the activity can be measured by a "paper assay" (22).
16. As mentioned for the Western blot technique, for accurate comparison of the activities of protein kinases (subheading 3.4, items 13 and 14), detection should be performed in the linear range of the phosphorylation reaction. Thus, the amount of protein used for immunoprecipitation, the concentration of antibodies, the length of the phosphorylation reaction and the exposure to X-ray film or to the phosphoimager should be optimized in order to reach linearity. If necessary, a standard curve with the protein kinases of interest can be made, and serial dilutions of the cytoplasmic extracts or a time course of the phosphorylation can be used to ensure one is working in a linear range.

Acknowledgments

This work was supported by grants from the Mario Negri–Weizmann collaborative fund and from the EU Sixth Framework Program under the SIMAP (IST-2004-027265) and GROWTHSTOP (LSHC CT-2006-037731). RS is an Incumbent of the Yale S. Lewine and Ella Miller Lewine professorial chair for cancer research.

References

1. Boulton, T. G., Yancopoulos, G. D., Gregory, J. S., Slaughter, C., Moomaw, C., Hsu, J., and Cobb, M. H. (1990) An insulin-stimulated protein kinase similar to yeast kinases involved in cell cycle control. *Science* **249,** 64–7.
2. Hibi, M., Lin, A., Smeal, T., Minden, A., and Karin, M. (1993) Identification of an oncoprotein- and UV-responsive protein kinase that binds and potentiates the c-Jun activation domain. *Genes Dev* **7,** 2135–48.
3. Kyriakis, J. M., Banerjee, P., Nikolakaki, E., Dai, T., Rubie, E. A., Ahmad, M. F., et al. (1994) The stress-activated protein kinase subfamily of c-Jun kinases. *Nature* **369,** 156–60.
4. Freshney, N. W., Rawlinson, L., Guesdon, F., Jones, E., Cowley, S., Hsuan, J., and Saklatvala, J. (1994) Interleukin-1 activates a novel protein kinase cascade that results in the phosphorylation of Hsp27. *Cell* **78,** 1039–49.
5. Han, J., Lee, J. D., Bibbs, L., and Ulevitch, R. J. (1994) A MAP kinase targeted by endotoxin and hyperosmolarity in mammalian cells. *Science* **265,** 808–11.
6. Rouse, J., Cohen, P., Trigon, S., Morange, M., Alonso-Llamazares, A., Zamanillo, D., et al. (1994) A novel kinase cascade triggered by stress and heat shock that stimulates MAPKAP kinase-2 and phosphorylation of the small heat shock proteins. *Cell* **78,** 1027–37.
7. Lee, J. D., Ulevitch, R. J., and Han, J. (1995) Primary structure of BMK1: a new mammalian map kinase. *Biochem Biophys Res Commun* **213,** 715–24.
8. Zhou, G., Bao, Z. Q., and Dixon, J. E. (1995) Components of a new human protein kinase signal transduction pathway. *J Biol Chem* **270,** 12665–69.
9. Abe, M. K., Kahle, K. T., Saelzler, M. P., Orth, K., Dixon, J. E., and Rosner, M. R. (2001) ERK7 is an autoactivated member of the MAP kinase family. *J Biol Chem* **276,** 21272–79.
10. Abe, M. K., Saelzler, M. P., Espinosa, R., 3rd, Kahle, K. T., Hershenson, M. B., Le Beau, M. M., and Rosner, M. R. (2002) ERK8, a new member of the mitogen-activated protein kinase family. *J Biol Chem* **277,** 16733–43.
11. Boulton, T. G., Nye, S. H., Robbins, D. J., Ip, N. Y., Radziejewska, E., Morgenbesser, S. D., et al. (1991) ERK's: a family of protein-serine/threonine kinases that are activated and tyrosine phosphorylated in response to insulin and NGF. *Cell* **65,** 663–75.
12. Derijard, B., Raingeaud, J., Barrett, T., Wu, I. H., Han, J., Ulevitch, R. J., and Davis, R. J. (1995) Independent human MAP-kinase signal transduction pathways defined by MEK and MKK isoforms (published erratum appears in Science 1995 Jul 7;269(5220):17). *Science* **267,** 682–85.
13. Tournier, C., Whitmarsh, A. J., Cavanagh, J., Barrett, T., and Davis, R. J. (1997) Mitogen-activated protein kinase kinase 7 is an activator of the c-Jun NH2-terminal kinase. *Proc Natl Acad Sci U S A* **94,** 7337–42.
14. Yan, M., Dai, T., Deak, J. C., Kyriakis, J. M., Zon, L. I., Woodgett, J. R., and Templeton, D. J. (1994) Activation of stress-activated protein kinase by MEKK1 phosphorylation of its activator SEK1. *Nature* **372,** 798–800.
15. Pawson, T., and Scott, J. D. (1997) Signaling through scaffold, anchoring, and adaptor proteins. *Science* **278,** 2075–80.
16. Kolch, W. (2005) Coordinating ERK/MAPK signalling through scaffolds and inhibitors. *Nat Rev Mol Cell Biol* **6,** 827–37.
17. Zhang, X. F., Settleman, J., Kyriakis, J. M., Takeuchi, S. E., Elledge, S. J., Marshall, M. S., et al. (1993) Normal and oncogenic p21ras proteins bind to the amino-terminal regulatory domain of c-Raf-1. *Nature* **364,** 308–13.
18. Kolch, W., Heidecker, G., Kochs, G., Hummel, R., Vahidi, H., Mischak, H., et al. (1993) Protein kinase C alpha activates RAF-1 by direct phosphorylation. *Nature* **364,** 249–52.
19. Kyriakis, J. M., App, H., Zhang, F. X., Banerjee, P., Brautigan, D. L., Rapp, U. R., and Avruch, J. (1992) Raf-1 activates MAP kinase-kinase. *Nature* **358,** 417–21.
20. Shaul, Y. D., and Seger, R. (2007) The MEK/ERK cascade: From signaling specificity to diverse functions. *Biochim Biophys Acta* **1773,** 1213–26.
21. Ahn, N. G., Seger, R., Bratlien, R. L., Diltz, C. D., Tonks, N. K., and Krebs, E. G. (1991) Multiple components in an epidermal growth factor-stimulated protein kinase cascade. *In vitro* activation of myelin basic protein/microtubule-associated protein-2 kinase. *J Biol Chem* **266,** 4220–27.
22. Chen, R. H., Sarnecki, C., and Blenis, J. (1992) Nuclear localization and regulation of erk- and rsk-encoded protein kinases. *Mol Cell Biol* **12,** 915–27.
23. Chuderland, D., Konson, A., and Seger, R. (2008) Identification and characterization of

a general nuclear translocation signal in signaling proteins. *Mol Cell* **31,** 850–61.

24. Roux, P. P., and Blenis, J. (2004) ERK and p38 MAPK-activated protein kinases: a family of protein kinases with diverse biological functions. *Microbiol Mol Biol Rev* **68,** 320–44.
25. Ahn, N. G., and Krebs, E. G. (1990) Evidence for an epidermal growth factor- stimulated protein kinase cascade in Swiss 3T3 cells. Activation of serine peptide kinase activity by myelin basic protein kinases in vitro. *J Biol Chem* **265,** 11495–501.
26. Alroy, I., Soussan, L., Seger, R., and Yarden, Y. (1999) Neu differentiation factor stimulates phosphorylation and activation of the Sp1 transcription factor. *Mol Cell Biol* **19,** 1961–72.
27. Yung, Y., Dolginov, Y., Yao, Z., Rubinfeld, H., Michael, D., Hanoch, T., et al. (1997) Detection of ERK activation by a novel monoclonal antibody. *FEBS Lett* **408,** 292–96.
28. Jaaro, H., Rubinfeld, H., Hanoch, T., and Seger, R. (1997) Nuclear translocation of mitogen-activated protein kinase kinase (MEK1) in response to mitogenic stimulation. *Proc Natl Acad Sci U S A* **94,** 3742–47.
29. Yao, Z., Dolginov, Y., Hanoch, T., Yung, Y., Ridner, G., Lando, Z., et al. (2000) Detection of partially phosphorylated forms of ERK by monoclonal antibodies reveals spatial regulation of ERK activity by phosphatases. *FEBS Lett* **468,** 37–42.
30. Force, T., Bonventre, J. V., Heidecker, G., Rapp, U., Avruch, J., and Kyriakis, J. M. (1994) Enzymatic characteristics of the c-Raf-1 protein kinase. *Proc Natl Acad Sci U S A* **91,** 1270–74.
31. Zhao, Z., Tan, Z., Diltz, C. D., You, M., and Fischer, E. H. (1996) Activation of mitogen-activated protein (MAP) kinase pathway by pervanadate, a potent inhibitor of tyrosine phosphatases. *J Biol Chem* **271,** 22251–55.
32. Ahn, N. G., Weiel, J. E., Chan, C. P., and Krebs, E. G. (1990) Identification of multiple epidermal growth factor-stimulated protein serine/threonine kinases from Swiss 3T3 cells. *J Biol Chem* **265,** 11487–94.

Chapter 3

Activation of SAPK/JNKs In Vitro

Deborah N. Chadee and John M. Kyriakis

Abstract

The stress-activated protein kinase/c-jun N-terminal kinases (SAPK/JNKs) are mitogen-activated protein kinases (MAPKs) that are activated by stressful and inflammatory stimuli and regulate cellular responses such as proliferation, differentiation, and apoptosis. The SAPK/JNKs are phosphorylated and activated by the MAP kinase kinases (MAP2Ks), SEK1/MKK4 and MKK7. These MAP2Ks are phosphorylated and activated by upstream stress-activated MAPK kinase kinases (MAP3Ks). Upon activation, SAPK/JNKs translocate to the nucleus and phosphorylate transcription factors, ultimately resulting in the modulation of gene expression. We have analyzed the activation of SAPK/JNK and stress-activated MAP3Ks using in vitro kinase assays. In addition, we have studied the role of different MAP3Ks in SAPK/JNK signaling by silencing specific MAP3K expression with RNAi and then analyzing the effect on activation of SAPK/JNKs and other MAPKs.

Key words: SAPK, JNK, MAP kinase, Signal transduction, Kinase assay, RNAi

1. Introduction

The stress-activated protein kinase/c-jun N-terminal kinases (SAPK/JNKs) are a subfamily of the mitogen-activated protein kinases (MAPKs) that are activated preferentially by stressful and inflammatory stimuli including tumor necrosis factor α (TNFα), interleukin 1 (IL-1) and related cytokines, pathogen-associated molecular patterns (PAMPs), heat shock, osmotic shock, UV-C radiation, endothelin, and anisomycin (Table 1) (1). In response to these stimuli, the SAPK/JNK signaling pathway regulates a range of cellular events that include proliferation, differentiation, apoptosis, and inflammation. SAPK/JNKs are activated by Thr/Tyr phosphorylation by upstream MAPK kinases (MAP2Ks), MKK4, and MKK7. The MAP2Ks are activated by Ser/Thr

Rony Seger (ed.), *MAP Kinase Signaling Protocols: Second Edition*, Methods in Molecular Biology, vol. 661,
DOI 10.1007/978-1-60761-795-2_3,

Table 1
Stimuli that activate SAPK/JNK in vivo[a]

Stimulus (references)	Amount and duration of treatment	Responsive cell lines	Notes
UV-C (9–11)	40 J/m^2, 20 per s	All adherent cells	Wash cells with 1× PBS before exposure
TNF (12)	2–100 ng/ml, 1–60 min	CCD-18 Co, HEK-293, COS, U937, HL-60, Jurkat, L929, HeLa, SKOV3, primary thymocytes or hepatocytes	
IL-1 (12)	2–100 ng/ml, 1–60 min	EL-4, HepG2, Jurkat, HL-60, KB, SKOV3	
Osmotic stress (Sorbitol) (13)	500 mM	All cells	
Anisomycin (11)	1–10 μg/ml, 1–60 min	All cells	
Endothelin (14, 15)	100–200 nM, 2–60 min	Endothelial, vascular smooth muscle, cardiomyocyte	
Methyl methane sulfonate (MMS) (16)	1 mM, 3 h	All cells	
Oxidant stress (H_2O_2) (17)	2–20 mM, 2–60 min	All cells	
Angiotensin II (14)	20–500 nM, 2–60 min	Endothelial, mesangial, pulmonary fibroblast, vascular smooth muscle, cardiomyocyte	Choose a cell type that expresses the Angiotensin II receptor and is known to induce a biologic effect
Ischemia/reperfusion (18)	40 min ischemia, 2–120 min reperfusion	Kidney, heart, brain	SAPK/JNK activated only in reperfusion
cis-platinum or (ara-C) (16)	50–300 μM, 2–6 h	All cells	
Heat shock (11, 19)	42°C, 30 min	All cells	
TGF-β (20)	2–5 ng/ml, 16 h	HepG2	
Bacterial lipopolysaccharide (21, 22)	100 ng/ml, 5–30 min	Primary macrophage, dendritic, RAW264.7 macrophage-like cells, HL-60 (differentiated to macrophages), Jurkat	

(continued)

Table 1 (continued)

Stimulus (references)	Amount and duration of treatment	Responsive cell lines	Notes
Bacterial peptidoglycan (21, 22)	100 ng/ml, 5–30 min	Primary macrophage, dendritic, RAW264.7 macrophage-like cells, HL-60 (differentiated to macrophages), Jurkat	
CpG-DNA (22)	100 nm, 5–30 min	Primary macrophage, dendritic, RAW264.7 macrophage-like cells, HL-60 (differentiated to macrophages), Jurkat	

[a]This list of stimuli is not inclusive but serves as a guide for stimuli that activate SAPK/JNK

phosphorylation by a group of over 20 upstream MAP3Ks (MAPK kinase kinases) (1). Some of the MAP3Ks that activate the SAPK/JNK pathway include the mitogen- and extracellular signal-regulated kinase kinase kinase (MEKK), mixed lineage kinase (MLK), apoptosis signal-regulating kinase (ASK), and TGFβ-activated kinase (TAK) (1).

Upon activation, SAPK/JNKs regulate gene expression by directly phosphorylating and activating transcription factors. Specifically, activated SAPK/JNKs undergo dimerization and translocate to the nucleus, where they phosphorylate the transcription factors, c-Jun and ATF-2 (1). c-Jun heterodimerizes with c-Fos to form the AP-1 transcription complex. AP-1 is involved in activation of transcription of specific genes that regulate stress, proliferation, and inflammation (1). SAPK/JNK can also phosphorylate other transcription factors, such as Elk-1 and NFATc1 (also called NFAT2), and cytoplasmic proteins, such as 14-3-3 and the Bcl2 family members, Bim and Bmf (1, 2).

Similar to ERK and p38 MAPKs, SAPK/JNKs are proline-directed kinases and phosphorylate substrates on Ser or Thr residues that are N-terminal to a Pro residue (S/T-P) (1). Substrate specificity is also regulated by the presence of docking sites on the substrate that interact with specific regions of the SAPK/JNK enzyme and allow for high affinity enzyme–substrate interactions (3).

SAPK/JNKs are encoded by three different genes (SAPKα/JNK2, SAPKβ/JNK3, and SAPKγ/JNK1) that are alternatively spliced to give ten SAPK/JNK isoforms (Table 2) (1). SAPKγ/

Table 2
SAPK/JNK nomenclature[a]

Name	Alternate names	Mammalian genome nomenclature	Genbank accession numbers
SAPKα	JNK2, SAPK1a	MAPK9	(m) BC028341 (h) BC032539
SAPK-p54α1	JNK2β2	MAPK9	(m) NM_016961 (h) NM_139070
SAPK-p54α2	JNK2α2	MAPK9	(m) NM_207692 (h) NM_002752
SAPK-p46α1	JNK2β1	MAPK9	(h) NM_139069
SAPK-p46α2	JNK2α1	MAPK9	(h) NM_139068
SAPK-β	JNK3, SAPK1β	MAPK10	(h) BC065516 (m) BC046625
SAPK-p54β1	JNK3β2	MAPK10	(m) NM_001081567
SAPK-p54β2	JNK3α2	MAPK10	(h) NM_138982 (m) AB096077
SAPK-p46β1	JNK3β1	MAPK10	(m) NM_009158
SAPK-p46β2	JNK3α1	MAPK10	(h) NM_002753 (m) AB096076
SAPKγ	JNK1, SAPK1c	MAPK8	(m) NM_016700 (h) BC130572
SAPK-p54γ1	JNK1β2	MAPK8	(h) NM_139047
SAPK-p54γ2	JNK1α2	MAPK8	(h) NM_139049
SAPK-p46γ1	JNK1β1	MAPK8	(h) NM_139046
SAPK-p46γ2	JNK1α1	MAPK8	(h) NM_002750

[a](h)-human, (m)-mouse

JNK1 and SAPKβ/JNK2 are widely expressed, while SAPKβ/JNK3 expression is limited to the brain, heart, and testis (1).

SAPK/JNK signaling specificity is also controlled by scaffold proteins that interact with specific MAP3Ks, MAP2Ks, and MAPKs and organize JNK signaling modules. In addition, they serve to insulate JNK signaling modules from activation by inappropriate stimuli. Some scaffold proteins that specifically regulate JNK signaling include the JIP (JNK Interaction Protein) 1–4, plenty of SH3s (POSH), and IκB kinase complex-associated protein (IKAP) (1, 4, 5).

2. Materials

2.1. SAPK/JNK Assay

2.1.1. Plasmids and DNA Transfection

1. pEBG-SAPKβ/JNK3 is a plasmid for expression of glutathione-S-transferase (GST)–SAPK/JNK in mammalian cells. The pEBG vector has a human EF-1α promoter that drives the expression of an N-terminal glutathione S-transferase (GST) tag. Additional constructs include, but are not limited to pCMV-FLAG-SAPKβ/JNK3 or pMT3-HA-SAPKβ/JNK3. Other SAPK/JNK isoforms can be expressed in similar vectors. GST–SAPK/JNK is ideal for use in multistep kinase assays (i.e., for MAP2Ks, MAP3Ks), whereas the others are more suited for analysis of SAPK/JNK activation. Endogenous JNK can also be immunoprecipitated (Santa Cruz Biotechnology or Cell Signaling Technology antibodies) and assayed as described below. SAPK/JNK expression constructs can be obtained from Addgene, Inc. for a small fee.
2. For transfections of plasmid DNA, we use Lipofectamine Reagent (Invitrogen), but calcium phosphate transfection or other transfection methods can also be used.

2.1.2. Cell Culture

1. Cells of interest.
2. Dulbecco's modification of Eagle's medium (DMEM) (Mediatech), RPMI 1640 (Mediatech), or appropriate media for cells of interest, supplemented with 10% fetal bovine serum (FBS) (Hyclone).
3. Specific stimuli for activation of endogenous or overexpressed SAPK/JNK (Table 1).
4. 10-cm tissue culture dishes.

2.1.3. Cell Lysis

1. Lysis buffer: 20 mM Hepes, pH 7.4, 2 mM EGTA, 1 mM DTT, 50 mM β-glycerophosphate, 1 mM Na_3VO_4 (Tyr phosphatase inhibitor that is prepared from a 20 mM stock boiled for 2 min and cooled to room temperature), 1% Triton X-100, 10% glycerol, 2 μM leupeptin, 2 μg/ml aprotinin, and 400 μM phenylmethylsulfonyl fluoride (PMSF). Add PMSF to the buffer immediately before the cell lysis because the half-life of PMSF in aqueous solutions is approximately 30 min.

2.1.4. Immunoprecipitation and GST-Pulldown

1. Use glutathione-agarose beads (Pierce) for GST-pulldown of overexpressed GST-SAPK/JNK. Anti-FLAG or HA can be used to isolate the cognate tagged SAPK/JNK construct.
2. Use an antibody that recognizes endogenous SAPK/JNK such as JNK (C-17) antibody (Santa Cruz Biotechnology) and protein-G sepharose (Pierce), if assaying endogenous

SAPK/JNK activity. Similar antibodies are available from Cell Signaling Technology.

3. High salt wash buffer: 500 mM LiCl, 100 mM Tris–HCl, pH 7.6, 0.1% Triton X-100, and 1 mM DTT.
4. Wash buffer: 100 mM Tris–HCl, pH 7.6, 0.1% Triton X-100, and 1 mM DTT.

2.1.5. Kinase Assay

1. Assay buffer: 20 mM MOPS, pH 7.2, 2 mM EGTA, 10 mM $MgCl_2$, 1 mM DTT, and 0.1% Triton X-100.
2. Purified GST–c-Jun(1–135) protein can be used as a substrate for assaying SAPK/JNK activity. GST–c-Jun(1–135) consists of amino acids 1–135 of c-Jun and is expressed from the pGEX-KG bacterial expression vector. The GST–c-Jun(1–135) fusion protein is approximately 45 kDa and can be expressed in bacterial cells and purified using standard methods for GST fusion protein purification. The protein can be eluted off the glutathione beads with 0.5 M glutathione, and the samples can be dialyzed overnight at 4°C with dialysis buffer (50 mM Tris–HCl, pH 7.5, 2 mM EGTA, 1 mM DTT, 0.1% Triton X-100, and 50% glycerol). After dialysis, the purified GST–c-Jun protein can be stored in this buffer at –20°C, until needed.
3. 5× $MgCl_2$/ATP stock: 50 mM $MgCl_2$, 0.5 mM ATP, and 1,000 μCi (γ-32P)-ATP (2,000 c.p.m./pmol) in assay buffer.
4. 4× SDS loading buffer: 0.2 M Tris–HCl, pH 6.8, 40% glycerol, 8% SDS, 0.1 M DTT, and 0.2% bromophenol blue.
5. Water bath or Eppendorf Thermomixer.

2.2. Analysis of Phosphorylated SAPK/JNK Using Phospho-Specific Antibodies

2.2.1. Cell Culture and Treatments

1. Cells of interest.
2. DMEM, RPMI, or appropriate media for the cells of interest, supplemented with 10% FBS.
3. Specific stimuli for activation of endogenous or overexpressed SAPK/JNK (Table 1).
4. 6-cm tissue culture dishes.

2.2.2. Cell Lysis

1. Lysis buffer: As in the Subheading 2.1.3, item 4.
2. 4× SDS loading buffer (see Subheading 2.1.5, item 4).

2.2.3. Immunoblotting

1. PVDF membrane.
2. Primary phospho-specific JNK antibody (Cell Signaling Technology) and total JNK antibody (Santa Cruz Biotechnology, Cell Signaling Technology).
3. Secondary antibody conjugated with horseradish peroxidase (Bio-Rad).

4. Blocking solution: 1× PBS with 5% (w/v) nonfat dry milk.
5. Antibody buffer: 1× PBS with 0.05% Tween-20 and 5% BSA for phospho-specific antibodies, and 1× PBS with 0.05% Tween-20 and 5% nonfat dry milk for all other antibodies.
6. Wash buffer: 1× PBS with 0.05% Tween-20.
7. Kits for enhanced chemiluminescence are commercially available from Pierce, Amersham, Bio-Rad, and Millipore.

2.3. MAP3K Assay

2.3.1. Plasmids and DNA Transfection

1. MAP3K cDNA of interest in mammalian expression vector.
2. Lipofectamine Reagent (Invitrogen) or other transfection reagents for desired method of transfection.

2.3.2. Cell Culture

1. Cells of interest.
2. DMEM, RPMI, or appropriate media for cell line of interest.
3. 10 cm tissue culture dishes.

2.3.3. Cell Lysis

1. Lysis buffer: As in Subheading 2.1.3, item 1.

2.3.4. Immunoprecipitation

1. Use the antibodies that recognize specific epitope tags on expressed protein, or if assaying endogenous MAP3Ks, use appropriate MAP3K antibodies. MAP3K antibodies and epitope tag antibodies can be obtained from Santa Cruz Biotechnology Inc. and Cell Signaling Technology.
2. Protein-G sepharose (1:1 solution in lysis buffer).
3. High salt wash buffer: 500 mM LiCl, 100 mM Tris–HCl, pH 7.6, 0.1% Triton X-100, and 1 mM DTT.
4. Wash buffer: 100 mM Tris–HCl, pH 7.6, 0.1% Triton X-100, and 1 mM DTT.

2.3.5. Kinase Assay

1. Assay buffer: 20 mM MOPS, pH 7.2, 2 mM EGTA, 10 mM MgCl2, 1 mM DTT, and 0.1% Triton X-100.
2. Purified GST–SEKl/MKK4(KR) protein can be used as a substrate for MAP3K assays. GST–SEKl/MKK4(KR) fusion protein is expressed in the pEBG vector and has a Lys to Arg mutation at amino acid 129 (pEBG-SEKl/MKK4-K129R). This plasmid can be transfected into HEK293 cells and, the GST–SEKl/MKK4(KR) fusion protein can be purified by GST pull-down. The protein can be eluted off the glutathione-agarose beads with 0.5 M glutathione and the samples can be dialyzed with dialysis buffer as described in the Subheading 2.1.5, item 2. The purified GST-SEKl/MKK4(KR) protein can be stored in this buffer at −20°C, until needed.
3. 5× $MgCl_2$/ATP stock solution (see Subheading 2.1.5).

4. 4× SDS loading buffer (see Subheading 2.1.5).
5. Water bath or Eppendorf Thermomixer.

2.4. siRNA Knockdown of MAP3Ks

2.4.1. Cell Culture

1. DMEM or other appropriate medium for cells of interest. Prepare media containing 10% FBS, and media without FBS.
2. 6-cm tissue culture dishes.

2.4.2. siRNA Oligos and Oligofection: Key MAP3Ks

1. Human siRNA ds oligos sense strand sequences (6, 7):

   MEKK1 – 5′-agaguuuccccagugccuu-3′ (nucleotides 760–779).

   MLK3 – 5′-gcgcgagauccagggucuc-3′ (nucleotides 1,200–1,219).

   MLK2 – 5′-gcuggagauucagcacaug-3′ (nucleotides 1,143–1,162).

   C-RAF – 5′-caucagacaacucuuauug-3′ (nucleotides 567–586).

   MEKK3 – 5′-aagccuuaggauauugcug-3′ (nucleotides 340–359).

   TAK1 – 5′-gaggagccuuuggaguugu-3′ (nucleotides 129–148).

   Note that these sequences come from the coding region. We have used all of these successfully. However, for some studies, you may wish to silence the endogenous enzyme and express a mutant version of the cognate enzyme (for classical "rescue" studies or in situ structure/function analysis). For these purposes, it is useful to select RNAi sequences that come from the 3′-untranslated region. cDNA for the relevant protein can then be expressed in a rescue or structure/function study. The cDNA for the overexpressed protein lacks a 3′-UTR and will not be silenced by the RNAi, while the endogenous protein that has a 3′-UTR will be silenced by the RNAi.
2. Lipofectamine 2000 or Oligofectamine reagent (Invitrogen).

2.4.3. Cell Lysis

1. 2× SDS loading buffer (see Subheading 2.1.5).

3. Methods

3.1. SAPK/JNK Assay

This assay can be used to measure levels of endogenous or overexpressed SAPK/JNK activated by specific stimuli or coexpressed upstream proteins. Cells are treated with specific stimuli or transfected with DNA constructs that express upstream activator proteins. SAPK/JNK is immunoprecipitated and its phosphorylation of GST–c-Jun substrate is analyzed.

1. Seed cells in 10-cm tissue culture dishes.
2. When cells reach approximately 70–80% confluence, treat with appropriate stimuli (Table 1) to activate endogenous SAPK/JNK or overexpressed GST–SAPK/JNK.

3. Remove media and wash cells twice with 1× PBS.
4. Tilt the dish to drain residual PBS and remove as much PBS from the dish as possible.
5. Add 1 ml lysis buffer per dish for cell lysis.
6. Allow cells to solubilize for 10 min on ice.
7. Collect cell lysates by scraping cells, with a cell scraper, into fresh tubes.
8. Centrifuge lysates at 6,000 × *g* for 10 min to remove unbroken cells and nuclei.
9. Discard pellets and keep the supernatants.
10. Perform a protein assay of the supernatants to determine the protein concentration in each sample.
11. For immunoprecipitation of endogenous SAPK/JNK, use 1–2 mg of cell lysate and add 1 μg anti-SAPK/JNK antibody and 20 μl protein-G-sepharose (1:1 solution of beads in lysis buffer) to the lysate. Add 20 μl glutathione-agarose beads to the lysate if performing GST pull-down of overexpressed GST-SAPK/JNK.
12. Incubate samples at 4°C for 3 h with rotation.
13. Collect protein-G-sepharose beads or glutathione-agarose beads by centrifugation at 1,500 × *g* for 5 min.
14. Wash beads three times with lysis buffer, three times with high salt wash buffer, and three timest with assay buffer.
15. After the final wash, leave the beads in 40 μl of assay buffer.
16. Prewarm water bath or Thermomixer to 30°C (we use the Eppendorf Thermomixer).
17. To 40 μl of beads, add 20 μl of GST–c-Jun substrate to 0.3 mg/ml final concentration in assay buffer.
18. Start the reactions by adding 15 μl of 5× $MgCl_2$/ATP stock mixture to give final concentrations of 100 μM ATP and 10 mM $MgCl_2$ in the assay.
19. Place samples in the water bath or Thermomixer (with agitation at 200 rpm if using Eppendorf Thermomixer) for 30 min at 30°C.
20. Stop reactions by adding 25 μl of 4× SDS loading buffer to each sample and boil the samples for 5 min.
21. Separate proteins on 15% SDS polyacrylamide gels.
22. Dry gels or transfer proteins to PVDF membrane.
23. Expose gels or membranes to X-ray film to detect phosphorylated GST–c-Jun(1-135) protein, which migrates at approximately 45 kDa.

3.2. Analysis of Levels of Active SAPK/JNK Using Phospho-Specific Antibodies

This method is useful for analyzing the levels of active endogenous SAPK/JNK that are phosphorylated on residues Thr183/Tyr185. The level of active SAPK/JNK in cells or tissues can be analyzed using commercially available phospho-specific antibodies. We use phospho-specific antibodies from Cell Signaling Technology for these assays. To ensure equal levels of total SAPK/JNK protein in each lane, the PVDF membrane should be stripped and reprobed with an antibody that recognizes total SAPK/JNK protein.

1. Seed cells in 6-cm tissue culture dishes.
2. When cells reach 70–80% confluence, treat them with specific stimuli to activate SAPK/JNK (see Table 1).
3. After treatments, wash cells twice with 1× PBS and then lyse cells with 0.5 ml of cell lysis buffer.
4. Centrifuge the lysates at 6,000 × *g* for 10 min.
5. Discard the pellets and perform a protein assay to quantify the protein concentration of the supernatants.
6. Add SDS sample buffer and boil samples for 5 min.
7. Load equal amounts of protein (between 20 and 50 μg) on an SDS polyacrylamide gel.
8. Pretreat PVDF membrane in methanol for 2 min.
9. Transfer proteins to PVDF membrane at 60 V for 1.5 h.
10. Incubate membranes with 50 ml of blocking buffer for 1 h.
11. Incubate membranes in primary antibody buffer (1:500 dilution of anti-phospho-SAPK/JNK antibody from Cell Signaling Technology) with rotation overnight at 4°C.
12. Wash membranes with wash buffer three times for 10 min each.
13. Incubate membranes in secondary antibody buffer (1:2,000 to 1:5,000 dilution of secondary antibody in buffer) for 1–2 h at room temperature with rotation.
14. Wash membranes with wash buffer three times for 10 min each.
15. Develop membranes using kit for enhanced chemiluminescence.
16. Expose membranes to film or imager to visualize phosphorylated SAPK/JNK proteins.
17. A predominant phospho-SAPK/JNK band at approximately 46 kDa should be observed in the cells treated with stimuli. Another phospho-SAPK/JNK band may also be observed at approximately 54 kDa depending on the stimulus and the cell type.

18. The membranes should also be stripped and reprobed with total SAPK/JNK antibody to determine the level of total SAPK/JNK in the lysates and to compare the levels of phospho vs. total SAPK/JNK in each sample.

3.3. MAP3K Assay

This assay is useful for analysis of stress-activated MAP3K phosphorylation of the substrate GST–SEK1/MKK4. The MAP3K of interest can be expressed in HEK293 cells. The expressed MAP3K protein can be immunoprecipitated and its phosphorylation of GST–SEK1/MKK4(KR) can be analyzed in a kinase assay. The endogenous MAP3K protein can also be immunoprecipitated from cells and assayed in the same manner using GST–SEK1/MKK4-KR as substrate.

1. Seed HEK293 cells in 10-cm tissue culture dishes. When the cells reach approximately 70% confluency, transfect 2–5 μg of MAP3K plasmid of interest. We use Lipofectamine for DNA transfections, but calcium phosphate or other methods are also suitable.
2. 24 h after transfection, wash cells with 1× PBS twice. After washes, tilt over the dish to allow removal of as much residual PBS as possible.
3. Add 1 ml of cell lysis buffer with protease inhibitors to each dish and allow cells to solubilize on ice for 10 min.
4. Scrape cells off dish with cell scraper and collect cell lysates in fresh tubes.
5. Centrifuge the cell lysates at 6,000 × *g* for 10 min.
6. Keep supernatants and transfer to fresh microfuge tubes. Discard pellets containing unbroken cells and nuclei.
7. Perform a protein assay to quantify the amount of total protein in each sample.
8. To the cell lysate (1–2 mg of protein), add 1 μg of MAP3K antibody (if analyzing endogenous MAP3K activity or antibody for specific epitope tag of overexpressed MAP3K) and 40 μl of protein-G-sepharose beads.
9. Incubate samples with rotation at 4°C for 3 h.
10. Centrifuge samples to pellet protein-G-sepharose beads at 1,500 × *g* for 5 min.
11. Discard supernatants and wash beads twice with lysis buffer, twice with high salt wash buffer containing 1 M LiCl, twice with wash buffer, and twice with assay buffer.
12. Leave beads in 40 μl of assay buffer and add 20 μl of assay buffer containing 0.5 μg purified GST–SEK1/MKK4(KR) to each sample.

13. Prewarm water bath or Thermomixer to 30°C. To start the reactions, add 15 μl of 5× $MgCl_2$/ATP stock to give final concentrations of 100 μM ATP and 10 mM $MgCl_2$ in the reaction mixture and allow the reactions to proceed for 30 min at 30°C.
14. Stop the reactions by adding 25 μl of 4× SDS loading buffer to each sample, and boil samples for 5 min.
15. Resolve the proteins on 15% SDS polyacrylamide gels, dry gels, or transfer proteins to PVDF membranes and expose membranes or gels to X-ray film. After the exposure to film, PVDF membranes can be probed with anti-GST antibody to visualize the total GST–SEK(KR) protein in each lane and MAP3K antibody to detect immunoprecipitated MAP3K.
16. Antibodies for detection of specific epitope tags and endogenous MAP3Ks are available commercially. We generally use antibodies from Santa Cruz Biotechnology for immunoprecipitation and detection of endogenous MAP3Ks. For detecting epitope tags, we use FLAG antibody from Stratagene, and we generate HA and Myc monoclonal antibodies from hybridoma clones.

3.4. siRNA Knockdown of MAP3Ks

This method is useful for the silencing of specific MAP3K gene expression in cultured cells. The particular MAP3K of interest can be silenced in cells, and the effect of activation of SAPK/JNK by specific stimuli can be analyzed.

1. Grow HEK293 cells to 70% confluence in 6-cm tissue culture dishes.
2. Add 800 μl of serum-free DMEM medium and 8 μl Lipofectamine 2000 or Oligofectamine (Invitrogen) to a fresh 15 ml sterile tube.
3. Combine 200 μl serum-free DMEM and 50 μM double-stranded, deprotected siRNA duplexes (Dharmacon) for the MAP3K of interest (see Subheading 2.3 for siRNA oligo sequences) in another tube. Include a sample with control siRNA oligo duplexes.
4. Mix Lipofectamine 2000 or Oligofectamine solution and double-stranded siRNA duplex solution together and let stand at room temperature for 20 min.
5. Add 1.7 ml of serum-free medium to each tube (for a total of 2.5 ml).
6. Remove medium from cells and wash cells once with serum-free medium.
7. Remove serum-free medium, add siRNA complexes to the dishes, and return the cells to the incubator.

8. 5 h later, replace the media with 5 ml fresh medium containing 10% FBS. Alternatively, instead of replacing the medium, add 2.5 ml DMEM containing 20% FBS to medium and return cells to incubator.

   If you are performing a "rescue" study, you can transfect the cells with a relevant cDNA construct. Here, we recommend cloning your cDNA into a lenti- or retroviral expression vector to ensure uptake of cDNA by most of the cells.

9. 24–48 h later, treat cells with specific stimuli (Table 1).
10. Wash cells twice with 1× PBS and tilt the dish to allow removal of residual PBS.
11. To measure endogenous SAPK/JNK activation by kinase assay, refer to Subheading 3.1 on analysis of SAPK/JNK activation.
12. To analyze SAPK/JNK activity by Western blotting with phospho-JNK antibodies, refer to Subheading 3.2 for Western blotting procedure.
13. During SAPK/JNK activity assay or Western blotting with phospho-antibodies, it is important to save a small portion of cell extract to analyze total MAP3K protein levels (by Western blotting with appropriate MAP3K antibodies) to determine the efficiency of silencing. In addition, perform immunoblotting of extracts with Actin antibody or any other suitable antibody for a loading control.

4. Notes

1. Select a cell type that responds well to the specific stimulus chosen (refer to Table 1) for analyzing endogenous SAPK/JNK activation.
2. Overexpression of MAP3Ks in the cells can cause a significant toxicity by 48 h after transfection. Therefore, it is better to harvest the cells at 24 h after transfection and transfect low quantities of plasmid DNA (0.2–1.0 μg per 10 cm dish).
3. When analyzing activation of MAP3Ks by other proteins or stimuli, it is important to remember that many of the MAP3Ks oligomerize and become activated when overexpressed in cells. Therefore, transfect the cells with low quantities (less than 1 μg per 10-cm dish) of plasmid DNA. If the basal level of MAP3K activity is still too high to see induction of activity with stimuli, the final concentration of ATP in the assay can be reduced to suboptimal levels of ATP such as 10 μM.

In addition, the duration of the kinase assay can be shortened from 30 to 5 min (8).

4. The siRNA oligos can be purchased from Dharmacon as double-stranded siRNA that has already been duplexed and deprotected and is ready for transfection. Deprotected oligos are less stable, so siRNA oligos can also be ordered as 2′ ACE protected oligos. This allows the investigator to deprotect small quantities of siRNA as needed for experiments. The remaining protected siRNA oligos can be stored at –20°C, until needed.
5. For siRNA oligofection, we have used Oligofectamine or Lipofectamine 2000, and both reagents generally give the good results. It is recommended that Lipofectamine 2000 be used when plasmid DNA is transfected together with the siRNA oligos, for example in a rescue experiment (6).
6. Do not use siRNA oligo concentration greater than 100 nm for the oligofection because this may lead to nonspecific off-target silencing of other genes.
7. Nonspecific oligo control siRNAs are available from Dharmacon. They have been tested and no similarity to the mammalian genome has been found.
8. We usually observe silencing of MAP3K gene expression by 24 h after oligofection. However, if silencing is not complete or is less than 50%, it may be necessary to harvest cells after 48 h instead of 24 h.
9. Silencing of gene expression can be observed up to 5 days after oligofection of the siRNA oligos. Specifically, for MLK3, we observed silencing over a 5-day period in CCD18-Co cells and WI-38 cells (6). If silencing must be sustained over a longer period of time, then a stable cell line expressing the appropriate short hairpin RNA (shRNA) can be generated. This can be done by preparing shRNA oligos that contain the 21-nucleotide siRNA sequence plus additional sequence that allows formation of the hairpin. These shRNA oligos can be cloned into specific shRNA vectors available from Invitrogen.

References

1. Kyriakis, J. M., and Avruch, J. (2001) Mammalian mitogen-activated protein kinase signal transduction pathways activated by stress and inflammation. *Physiol Rev* **81**, 807–69.
2. Lei, K., and Davis, R. J. (2003) JNK phosphorylation of Bim-related members of the Bcl2 family induces Bax-dependent apoptosis. *Proc Natl Acad Sci U S A* **100**, 2432–7.
3. Sharrocks, A. D., Yang, S. H., and Galanis, A. (2000) Docking domains and substrate specificity determination for MAP kinases. *Trends Biochem Sci* **25**, 448–53.
4. Yoshioka, K. (2004) Scaffold proteins in mammalian MAP kinase cascades. *J Biochem* **135**, 657–61.
5. Morrison, D. K., and Davis, R. J. (2003) Regulation of MAP kinase signaling modules

by scaffold proteins in mammals. *Annu Rev Cell Dev Biol* **19**, 91–118.

6. Chadee, D. N., and Kyriakis, J. M. (2004) MLK3 is required for mitogen activation of B-Raf, ERK and cell. *Nat Cell Biol* **6**, 770–6.
7. Zhong, J., and Kyriakis, J. M. (2007) Dissection of a signaling pathway by which pathogen-associated molecular patterns recruit the JNK and p38 MAPKs and trigger cytokine release. *J Biol Chem* **282**, 24246–54.
8. Chadee, D. N., Yuasa, T., and Kyriakis, J. M. (2002) Direct activation of mitogen-activated protein kinase kinase kinase MEKK1 by the Ste20p homologue GCK and the adapter protein TRAF2. *Mol Cell Biol* **22**, 737–49.
9. Hibi, M., Lin, A., Smeal, T., Minden, A., and Karin, M. (1993) Identification of an oncoprotein- and UV-responsive protein kinase that binds and potentiates the c-Jun activation domain. *Genes Dev* **7**, 2135–48.
10. Derijard, B., Hibi, M., Wu, I. H., Barrett, T., Su, B., Deng, T., Karin, M., and Davis, R. J. (1994) JNK1: a protein kinase stimulated by UV light and Ha-Ras that binds and phosphorylates the c-Jun activation domain. *Cell* **76**, 1025–37.
11. Kyriakis, J. M., Banerjee, P., Nikolakaki, E., Dai, T., Rubie, E. A., Ahmad, M. F., Avruch, J., and Woodgett, J. R. (1994) The stress-activated protein kinase subfamily of c-Jun kinases. *Nature* **369**, 156–60.
12. Kyriakis, J. M., Woodgett, J. R., and Avruch, J. (1995) The stress-activated protein kinases. A novel ERK subfamily responsive to cellular stress and inflammatory cytokines. *Ann N Y Acad Sci* **766**, 303–19.
13. Galcheva-Gargova, Z., Derijard, B., Wu, I. H., and Davis, R. J. (1994) An osmosensing signal transduction pathway in mammalian cells. *Science* **265**, 806–8.
14. Makkinje, A., Quinn, D. A., Chen, A., Cadilla, C. L., Force, T., Bonventre, J. V., and Kyriakis, J. M. (2000) Gene 33/Mig-6, a transcriptionally inducible adapter protein that binds GTP-Cdc42 and activates SAPK/JNK. A potential marker transcript for chronic pathologic conditions, such as diabetic nephropathy. Possible role in the response to persistent stress. *J Biol Chem* **275**, 17838–47.
15. Shapiro, P. S., Evans, J. N., Davis, R. J., and Posada, J. A. (1996) The seven transmembrane-spanning receptors for endothelin and thrombin cause proliferation of airway smooth muscle cells and activation of the extracellular regulated kinase and c-Jun NH2-terminal kinase groups of mitogen-activated protein kinases. *J Biol Chem* **271**, 5750–4.
16. Pandey, P., Raingeaud, J., Kaneki, M., Weichselbaum, R., Davis, R. J., Kufe, D., and Kharbanda, S. (1996) Activation of p38 mitogen-activated protein kinase by c-Abl dependent and -independent mechanisms. *J Biol Chem* **271**, 23775–9.
17. Liu, H., Nishitoh, H., Ichijo, H., and Kyriakis, J. M. (2000) Activation of apoptosis signal-regulating kinase 1 (ASK1) by tumor necrosis factor receptor-associated factor 2 requires prior dissociation of the ASK1 inhibitor thioredoxin. *Mol Cell Biol* **20**, 2198–208.
18. Pombo, C. M., Bonventre, J. V., Avruch, J., Woodgett, J. R., Kyriakis, J. M., and Force, T. (1994) The stress-activated protein kinases are major c-Jun amino-terminal kinases activated by ischemia and reperfusion. *J Biol Chem* **269**, 26546–51.
19. Kyriakis, J. M., and Avruch, J. (1996) Sounding the alarm: protein kinase cascades activated by stress and inflammation. *J Biol Chem* **271**, 24313–6.
20. Atfi, A., Djelloul, S., Chastre, E., Davis, R., and Gespach, C. (1997) Evidence for a role of Rho-like GTPases and stress-activated protein kinase/c-Jun N-terminal kinase (SAPK/JNK) in transforming growth factor beta-mediated signaling. *J Biol Chem* **272**, 1429–32.
21. Zhong, J., and Kyriakis, J. M. (2004) Germinal center kinase is required for optimal Jun N-terminal kinase activation by Toll-like receptor agonists and is regulated by the ubiquitin proteasome system and agonist-induced, TRAF6-dependent stabilization. *Mol Cell Biol* **20**, 9165–75.
22. Zhong, J., Gavrilescu, L. C., Molnár, A., Murray, L., Garafalo, S., Kehrl, J. H., Simon, A. R., Van Etten, R. A., and Kyriakis, J. M. (2009) GCK is essential to systemic inflammation and pattern recognition receptor signaling to JNK and p38. *Proc Natl Acad Sci U S A* **106**, 4372–7.

Chapter 4

Activation of p38 and Determination of Its Activity

Huamin Zhou, Jianming Chen, and Jiahuai Han

Abstract

The p38 mitogen-activated protein kinase (MAPK) pathway plays an important role in cellular responses to inflammatory stimuli and environmental stresses. Extracellular stimuli activate kinases upstream of p38, such as MKK3 and MKK6, which subsequently phosphorylate p38. p38 then participates in numerous biological processes by phosphorylating its downstream substrates. Here, our methodology mainly highlights how endogenous or exogenous p38 can be activated and its upstream kinases and downstream substrates identified.

Key words: p38, LPS, MKK3, MKK6, In vitro kinase assay

1. Introduction

Mitogen-activated protein kinases (MAPKs) are ubiquitous serine/threonine protein kinases that mediate key intracellular signaling pathways, translating extracellular signals and leading to nuclear responses. Four distinct subgroups within MAPK signaling cascades have been identified so far: (1) Extracellular signal-regulated kinases (ERKs), (2) c-jun N-terminal- or stress-activated protein kinases (JNKs/SAPKs), (3) ERK5/big MAP kinase 1(BMK1), and (4) the p38 MAP kinases (1). The different subgroups of MAPKs seem to execute different cellular physiological processes.

Like other MAP kinases, p38 family kinases, including four isoforms, p38α, p38β (2), p38γ (3, 4), and p38δ (5, 6), respond to diverse extracellular stimuli, such as growth factors, cytokines, or physical/chemical stress (7). They are activated by dual tyrosine/threonine phosphorylation mediated by MKK3 and MKK6 (8–10), while selective activation of different isoforms by distinct MKKs is

Rony Seger (ed.), *MAP Kinase Signaling Protocols: Second Edition*, Methods in Molecular Biology, vol. 661,
DOI 10.1007/978-1-60761-795-2_4, © Springer Science+Business Media, LLC 2010

observed. p38 kinases have been found to execute their biological functions by phosphorylation of specific target molecules, such as downstream protein kinases, transcription factors, and adaptor proteins. They play essential roles in different physiological and pathological processes, such as inflammation, cell growth, cell differentiation, the cell cycle, and cell death.

In this chapter, we focus on the activation of endogenous or exogenous p38 kinases and the determination of their activity. (1) Endogenous p38 kinase in mammalian cells is activated by lipopolysaccharide (LPS) (11). The cell lysate is then separated by SDS-PAGE gel, and phosphorylation of p38 is detected by Western blotting with anti-phosphotyrosine monoclonal antibodies. (2) The exogenous p38 kinases are activated by cotransfection of p38 and constitutively active MKK3b(E) (351 amino acid forms) or MKK6b(E) (334 amino acid forms) (2, 12) into mammalian cells. p38 kinases are isolated by immunoprecipitation with specific antibodies. An in vitro kinase assay system allows the determination of the activity of p38 kinases using GST-ATF2 as a substrate (13, 14). (3) The recombinant His-MKK3(E), p38 isoforms and appropriate substrate protein MEF2A are expressed in *Escherichia coli* BL21(DE3) and purified, respectively. In the coupling kinase assay, MKK3(E) activates p38, which subsequently phosphorylates MEF2A, and phosphorylated MEF2A is separated on a 10% SDS-PAGE gel and visualized by radioautography (15, 16).

2. Materials

2.1. Cell Culture Suspension and Lysis

1. RPMI1640 medium (Gibco/BRL, Bethesda, MD) supplemented with 10% heat-inactivated fetal bovine serum (FBS, HyClone, Ogden, UT) (56°C, 30 min), 2 mM lc, 1 mM sodium pyruvate, 100 U/mL penicillin, and 100 μg/mL streptomycin.
2. Lipopolysaccharide (LPS): *Salmonella minnesota* Re595 (Calbiochem, La Jolla, CA), is dissolved in cell culture medium at 1 mg/mL, stored in aliquots at −80°C.
3. MY4, an anti-CD14 monoclonal antibody to hCD14 (Coulter Diagnostics, Hialeah, FL).
4. Washing buffer: 10 mM Tris–HCl, 150 mM NaCl, 1 mM Na_3VO_4, pH 7.5 (see Note 1).
5. Lysis buffer A: 20 mM Tris–HCl, 120 mM NaCl, 10% glycerol, 1 mM Na_3VO_4, 2 m EDTA, 1% Triton X-100, 1 m phenylmethylsulfonyl fluoride (PMSF), pH 7.5.
6. Bradford reagent: Coomassie Plus Reagent (Pierce, Rockford, IL).

2.2. SDS-Polyacrylamide Gel Electrophoresis

1. Gel electrophoresis apparatus and power supply.
2. 2× SDS sample buffer, 100 mM Tris–HCl pH 6.8, 4% SDS, 20% glycerol, 0.2% bromophenol blue, 2% δ-mercaptoethanol. Store at room temperature (RT).
3. Lower buffer, 1.5 M Tris-HCl pH 8.8. Store at RT.
4. Stacking buffer, 1.0 M Tris-HCl pH 6.8. Store at RT.
5. 30% acrylamide/bis solution: Dissolve 29 g of acrylamide and 1 g of *N,N′*-methylenebisacrylamide in a total volume 60 mL of H_2O. Heat the solution to 37°C to dissolve the chemicals. Adjust the volume to 100 mL with H_2O and sterile filter. Store the solution in a dark bottle at 4°C (this is a neurotoxin when unpolymerized, so care should be taken to avoid exposure).
6. Ammonium persulfate (APS): prepare 10% solution in water and immediately freeze in single use (200 μL) aliquots at −20°C.
7. Separating gel (12% SDS-PAGE, 15 mL) 4.9 mL H_2O, 6 mL 30% acrylamide/bis solution, 3.8 mL lower buffer (pH 8.8), 150 μL 10% SDS, 150 μL 10% APS and 15 μL TEMED (*N,N,N,N′*-Tetramethylethylenediamine) (Bio-Rad, Hercules, CA).
8. Stacking gel (5% SDS-PAGE, 5 mL), 3.4 mL H_2O, 0.83 mL 30% acrylamide/bis solution, 0.63 mL stacking buffer (pH 6.8), 50 μL 10% SDS, 50 μL 10% APS, and 5 μL TEMED.
9. Running buffer (5×): 125 mM Tris, 960 mM glycine, 0.5% (w/v) SDS, pH 8.3. Store at RT.
10. Prestained molecular weight markers: Kaleidoscope markers (Bio-Rad, Hercules, CA).

2.3. Western Blot Analysis

1. Transfer apparatus (Bio-Rad, Hercules, CA).
2. Nitrocellulose membrane (Millipore, Bedford, MA).
3. Whatman paper (3 mm) (Whatman, Maidstone, UK).
4. Transfer buffer: 48 mM Tris, 39 mM glycine, 0.037% SDS, 20% methanol.
5. 10× TBS: 250 mM Tris-HCl, 1.4 M NaCl, 30 mM KCl, pH 7.4. Store at RT.
6. TBST: 1× TBS, 0.1% Tween-20. Store at RT.
7. Blocking buffer: 5% (w/v) fraction V bovine serum albumin (BSA) in TBST.
8. Primary antibody dilution buffer: TBST supplemented with 2% (w/v) BSA.
9. Secondary antibody: anti-mouse IgG conjugated to horseradish peroxidase (Santa Cruz, CA).

10. Enhanced chemiluminescent (ECL) reagents from Kirkegaard and Perry (Gaithersburg, MD) and Bio-Max ML film (Kodak, Rochester, NY).
11. FB2, an anti-phosphotyrosine monoclonal antibody (mAb), is purified from FB2 hybridoma (ATCC CRL1891) culture supernatant by chromatography on a protein G column.

2.4. Cell Culture and Transfection

1. Dulbecco's Modified Eagle's Medium (DMEM) (Gibco/BRL, Bethesda, MD) supplemented with 10% FBS (HyClone, Ogden, UT).
2. Solution of trypsin (0.25%) and ethylenediamine tetraacetic acid (EDTA) (1 mM) (Gibco/BRL, Bethesda, MD).
3. Phosphate buffered saline (PBS), 20 mM NaCl, 2.68 mM KCl, 10 mM Na_2HPO_4, and 1.76 mM KH_2PO_4 (pH 7.4).
4. 2× HBS (500 mL): 8.0 g NaCl, 0.37 g KCl, 201 mg Na_2HPO_4 $7H_2O$, 1.0 g glucose, 5.0 g HEPES, adjust pH to 7.05 with NaOH and filter sterilize. Store at 4°C.
5. 2 M $CaCl_2$: dissolve 54 g of $CaCl_2$ 6 H_2O in H_2O and filter sterilize. Store at 4°C.

2.5. Immuno-precipitation and In Vitro Kinase Assay

1. Lysis buffer B: PBS (pH 7.4), 1% NP-40, 0.5% sodium deoxycholate, 1 mM Na_3VO_4, 0.1% SDS, 1 mM EDTA, 1 mM EGTA, 20 mM NaF, 1 mM PMSF, and 1 μg/mL (each) of aprotinin, leupeptin, and pepstatin.
2. M2 beads: anti-Flag M2 monoclonal antibody (Sigma, Saint Louis, MO) and coupled to protein G-Sepharose (Pharmacia, Peapack, NJ).
3. Kinase buffer: 10 mM Tris-HCl pH7.4, 150 mM NaCl, 10 mM $MgCl_2$, and 0.5 mM DTT.
4. 5× concentrated electrophoresis sample buffer: 625 mM Tris-HCl, pH 6.8, 10% SDS, 25% glycerol, 0.5% bromophenol blue, and 5% β-mercaptoethanol.
5. Staining solution: 0.25% Coomassie blue in 45% methanol, 10% acetic acid.
6. Destaining solution: 40% methanol and 10% acetic acid.
7. Teflon cell scrapers (Fisher, Pittsburgh, PA).

2.6. Preparation of Recombinant Proteins and Coupling Kinase Assay

1. Isopropyl-β-D-thiogalactopyranoside (IPTG (Calbiochem), San Diego, CA): dissolve 2 g of IPTG in 10 mL of water, sterilize by filtration through a 0.22 μ disposable filter, store at −20°C.
2. B-PER Bacterial Protein Extraction Reagent (Thermo, Rockford, IL), 165 mL, contains 20 mM Tris-HCl buffer, pH 7.5, and a proprietary additive. Store at RT.

3. Nickel Chelated Columns, 5, 1 mL prepacked columns, contains nickel (Ni^{2+}) chelated iminodiacetic acid (IDA) that is covalently immobilized to 4% beaded agarose, 0.02% sodium azide. Store at 4°C.
4. Wash Buffer 1, 45 mL, contains: 35 mM Tris-HCl, 150 mM NaCl, 10 mM imidazole, 5% glycerol, pH 7.2, and 0.5× concentration of proprietary B-PER Reagent additive. Store at 4°C.
5. Wash Buffer 2, 60 mL, contains: 50 mM Tris-HCl, 300 mM NaCl, 25 mM imidazole, and 10% glycerol, pH 6.8. Store at 4°C.
6. Elution Buffer, 45 mL, contains: 50 mM Tris-HCl, 300 mM NaCl, 200 mM imidazole, and 10% glycerol, pH 6.8.

3. Methods

p38α was first characterized as an intracellular protein, rapidly phosphorylated at tyrosine-182 in response to stimulation by LPS, a complex glycolipid residing on the outer membrane of Gram-negative bacteria. CD14, a glycosylphosphatidylinositol-anchored cell surface glycoprotein, functions as a membrane receptor for LPS and binds to LPS to initiate LPS-induced transmembrane signaling. In 70Z/3-hCD14 cells, a murine pre-B cell line transfected with DNA encoding human CD14 (hCD14), LPS specifically activates endogenous p38α (17). If the cells are pretreated with anti-hCD14 monoclonal antibodies, LPS cannot activate endogenous p38α. Activated p38α is detected with anti-phosphotyrosine antibody (see Note 2).

3.1. Cell Culture Suspension and Lysis

1. 70Z/3-hCD14 cells, stably expressing human CD14, are maintained in RPMI1640 medium supplemented with 10% heat-inactivated FBS (56°C, 30 min), 2 mMl-glutamine, 1 mM sodium pyruvate, 100 U/mL penicillin, and 100 μg/mL streptomycin at 37°C in a 5% CO_2/air mixture. The cells are passed at a concentration of 1 to 2×10^5 viable cells/mL. For experimental cells, the culture is at a density of 6×10^6 viable cells/mL in the 35 mm dishes.
2. For some samples, cells are pretreated with the anti-hCD14 monoclonal antibody MY4 for 30 min prior to LPS stimulation.
3. Stimulate the cells with or without pretreatment by incubating them with 1 ng/mL Re595 LPS for the indicated times or with indicated concentrations of Re595 LPS for 15 min (see Note 3).
4. After stimulation, the cells are rapidly chilled on ice, transferred to labeled, precooled 1.5-mL plastic Eppendorf tubes.

Spin at 150 × *g* for 2 min at 4°C and then remove the supernatant. Resuspend the cells with ice-cold washing buffer, wash twice, and then lyse in 0.25 mL of lysis buffer A for 10 min on ice.

5. Disrupt the cells by sonication (10 × 1 s at 35-W pulses) on ice.
6. Centrifuge the cellular extracts at 16,000 × *g* for 30 min at 4°C. The supernatants are transferred to new, precooled test tubes to be examined.
7. Take 5–10 μL from the resulting supernatants for protein concentration determination. Store the remainder of each cell-free lysate on ice until needed.
8. Dilute the samples (usually 1:20) to make sure that the protein concentration is within the linear working range standards and proceed as follows:
   (a) Put 50 μL of each of the protein standards (25, 125, 250, 500, 750, 1,000, 1,500, and 2,000 μg/mL of BSA in water) or sample into at least two test tubes.
   (b) Add 1.5 mL of the Coomassie Plus Reagent to each tube and mix well. Incubate samples for 10 min at RT.
   (c) With the spectrophotometer set to 595 nm, zero the instrument on a cuvette filled only with water. Subsequently, measure the absorbance of Blank (50 μL water plus 1.5 mL of the Coomassie Plus Reagent) and all the samples.
   (d) Subtract the average 595 nm measurement for the Blank replicates from the 595 nm measurement of all other individual standard and sample replicates.
   (e) Prepare a standard curve by plotting the average Blank-corrected 595 nm measurement for each protein standard vs. its concentration in μg/mL. Use the standard curve to determine the protein concentration of each sample.
9. Equal amounts of cell extract from each of the treatments (see step 3) are used for Western blotting (usually 20 μg of protein/sample). Add to each of the samples 1/2 volume of 2× sample buffer, mix the contents, boil for 10 min, and spin for 1 min at 16,000 × *g*.

3.2. SDS-Polyacrylamide Gel Electrophoresis (SDS-PAGE)

1. Proteins are first separated by 12% SDS-PAGE. Prepare a 1.5 mm thick, 12% SDS-PAGE separating gel freshly. Load ~7.5 mL into the assembled glass plates in a minigel apparatus (Bio-Rad), leaving sufficient space for the stacking gel. Overlay the separating gel with water-saturated isopropanol carefully and allow the gel to polymerize in about 30 min. Pour off the overlay, rinse the top of the gel twice with water, and remove the remaining water with the edge of a paper towel.

2. Prepare 5% SDS-PAGE stacking gel freshly. Cast the gel solution directly onto the polymerized separating gel, insert a comb carefully to avoid trapping air bubbles, and allow it to polymerize.
3. After polymerization is complete, assemble the gel in the apparatus, remove the comb, and add running buffer to the upper and lower chambers of the gel unit.
4. Load each sample and a prestained protein marker in a well and run the gel at 100 V for 30 min, then 130 V for the remaining time. Once the dye front of the SDS-PAGE has reached the end of the gel, disconnect the gel unit from the power supply, remove the gel from the apparatus, discard the stacking gel, and proceed with the following transfer step.

3.3. Western Blot Analysis and Antibodies

1. Soak the nitrocellulose membrane in 100% methanol for 10 min.
2. Prepare a tray of transfer buffer that is large enough to lay out a transfer cassette. Soak pads and 3MM Whatman papers in transfer buffer. Make a sandwich with the SDS gel, nitrocellulose membrane, and transfer pads by placing a sheet of 3MM Whatman paper on a pad, the gel on top of the paper, the nitrocellulose membrane on top of the gel, the other 3MM Whatman paper on top of the nitrocellulose membrane, and the other pad on top of the transfer sandwich. Make sure air bubbles are not trapped between the gel and the other components.
3. Place the sandwich containing the SDS gel and nitrocellulose membrane into the buffer-filled transfer apparatus. The nitrocellulose membrane should be between the gel and the anode. Transfer can be accomplished at 350-mA constant current for 4 h (preferably with a cooling device).
4. Once the transfer is finished, remove the nitrocellulose membrane, and then rinse with transfer buffer to remove any adhering pieces of gel.
5. Incubate the membrane in a flat container with blocking buffer for 4 h at RT on a rocking platform (see Note 4).
6. Rinse the membrane with TBST quickly. Incubate the blot with the first antibody (1 pg/mL murine monoclonal anti-phosphotyrosine antibodies, diluted with the primary antibody dilution buffer) for 3 h at 37°C with continuous shaking.
7. Rinse the membrane three times with TBST, once for 15 min and twice for 5 min.
8. A freshly prepared secondary antibody, goat anti-mouse IgG coupled to horseradish peroxidase diluted with TBST, is applied to the membrane for 60 min at RT.

9. Wash four times with TBST and once with TBS for 5 min.
10. Warm 4 mL aliquots of each part of the ECL reagents separately to room temperature. The ECL reagents are mixed together and then immediately added to the membrane once the final wash is completed. Rotate the membrane by hand for 1 min to ensure even coverage.
11. The membrane is removed from the ECL reagents and then placed between the leaves of a cellulose acetate sheet protector that has been cut to the size of an X-ray film cassette.
12. Expose the membrane to X-ray film in a cassette for a suitable exposure time, typically a few minutes. An example of the results produced is shown in Fig. 1 (see Note 5).

3.4. Cell Culture and Transfection

There are two main MAP kinase kinases (MKKs) that are upstream kinases responsible for p38 activation, MKK3 and MKK6. MKK3 and MKK6 exist as two variants with different lengths. The long forms of MKK3 and MKK6, called MKK3b and MKK6b, contain 29 and 56 extra amino acids at the N-termini, respectively, and are the predominant forms in the cells. Constitutively active forms of MKK3b and MKK6b, MKK3b(E) and MKK6b(E), prepared by replacing the double phosphorylation sites Ser^{218} and Thr^{222} or Ser^{207} and Thr^{211} with Glu, can effectively activate p38 without additional extracellular stimuli. Both MKK3b(E) (MKK6b(E)) and wild-type flag-p38 isoform are cotransfected into NIH3T3 fibroblast cells, and then p38 isoform is immunoprecipitated from the cell lysate with M2 beads.

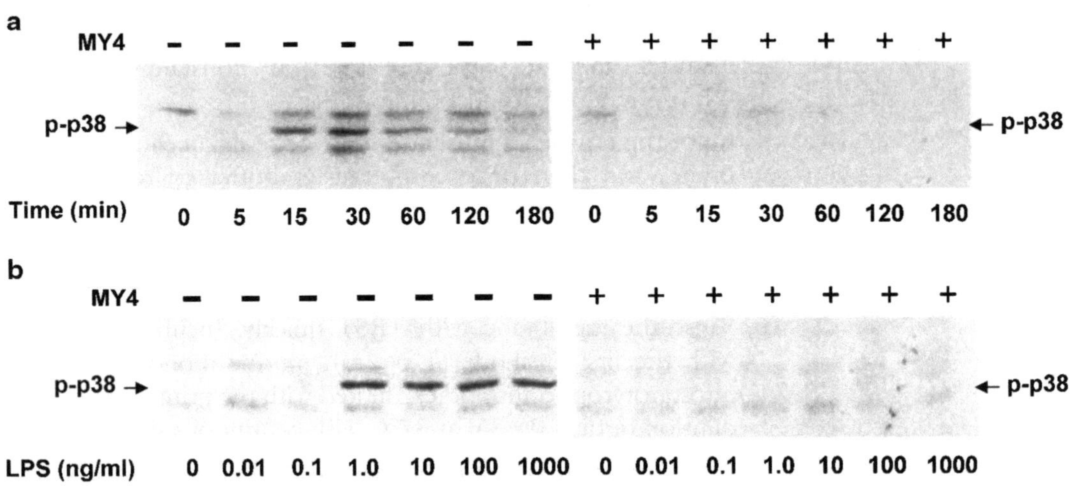

Fig. 1. LPS-induced p38α tyrosine phosphorylation. 70 W3-hCD14 cells were treated with 1 ng/mL Re595 LPS for the indicated times (**a**) or with the indicated concentrations of Re595 LPS for 15 min (**b**). The cells were pretreated with anti-hCD14 monoclonal antibodies MY4 for 30 min before LPS stimulation was noted (+). The cells were lysed and subjected to immunoblotting with anti-phospho-tyrosine antibody. Equal loading of different samples were confirmed by staining the blotted membrane with Ponceau S.

Using GST-ATF2-(1–109) as substrate, the activity of the p38 kinase is determined by in vitro kinase assay.

1. NIH 3 T3 fibroblast cells (see Note 6) are maintained in Dulbecco's modified Eagle's medium (DMEM) supplemented with 10% FBS. Plate cells to 60–70% confluence. One 100-mm culture dish is required for each experimental culture. The transfection efficiency will decrease at a higher confluence. Less than 50% confluence may be fine, but the amount of protein expressed will be low because of the small numbers of cells.
2. One hour prior to transfection, change to medium containing 25 μM chloroquine (from ×1,000 stock in PBS, stored at −20°C). The volume should be 4 mL per dish. (Chloroquine can be omitted, but transfection efficiency will be reduced about twofold).
3. Mix 5 μg of plasmid DNA expressing MKK3b(E) (MKK6b(E)) or the corresponding empty expression vector as control with 5 μg of plasmid DNA expressing p38α (p38β, p38γ, or p38δ) (see Note 7).
4. Add the 10 μg DNA to H_2O (1,095 μL in total) in a 15-mL sterile tube, and then add 155 μL of 2 M $CaCl_2$, followed by 1,250 μL of 2 × HBS (pH 7.15) dropwise. Mix gently and incubate for 15 min at RT.
5. Add this mixture directly to the cells dropwise and evenly sprinkle the droplets over the entire area. Mix gently by rocking the dish back and forth. Notice that the medium turns orange.
6. Incubate for 7–11 h and very fine, dust-like precipitate will be visible. After incubation, rinse the cells once and change to 1% FCS-DMEM medium without chloroquine and continue to culture to confluence.

3.5. Immunoprecipitation and In Vitro Kinase assay

1. Remove the medium from the dish and rapidly rinse the cells once with 5 mL of PBS.
2. Lyse the cells with 1,000 μL cold lysis buffer B in the culture dish, and keep the cells at 4°C for 30 min.
3. Tilt the dish gently and scrape the cells from the dish using a plastic cell scraper. Then transfer the cells to a cold 1.5 mL Eppendorf tube.
4. Sonicate samples for 10 s (10 × 1 s at 35-W pulses) on ice.
5. Spin at top speed (16,000 × *g*) at 4°C for 30 min and collect the supernatant.
6. Wash anti-flag M2 beads with 500 μL of 1× lysis buffer B three times.
7. Dilute 10 μL of beads in 50 μL lysis buffer B.

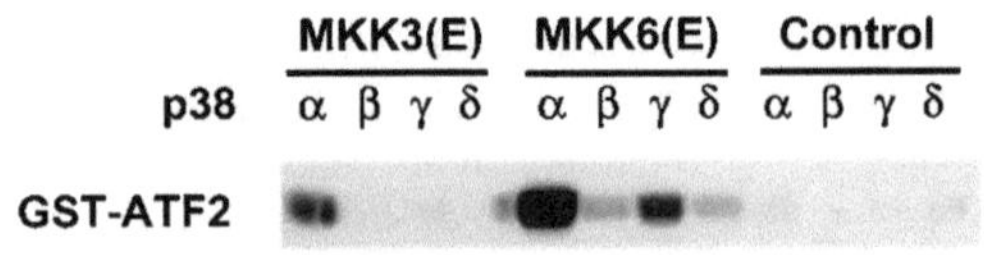

Fig. 2. Activation of p38 isoforms by upstream kinase. NIH 3 T3 fibroblasts were transfected with 5 μg of expression vectors for constitutively active forms of MKK3b and MKK6b, MKK3b(E) and MKK6b(E), together with expression vectors for the p38 isoforms (p38α, p38β, p38γ, and p38δ) (5 μg each). Control cultures were transfected with the corresponding empty vector. Cultures were maintained for 36 h in 10% FCS–DMEM. The activity of the p38 isoforms immunoprecipitated from the cell lysates with anti-flag antibodies was determined by in vitro kinase assay using GST-ATF2-(1-109) as a substrate.

8. To an Eppendorf tube, add 800 μL of lysate and 10 μL of M2 beads.
9. Incubate with end-over-end rotation overnight at 4°C.
10. Spin at low speed (150×*g*) for 3 min and remove the supernatant.
11. Wash with kinase buffer three times.
12. Remove the supernatant by aspiration. Incubate the pellet for 15 min at 37°C with 40 μL of the kinase buffer that contains 25 μM ATP, 2.5 μ Ci[γ-^{32}P]-ATP, and protein substrate GST-ATF2 at 1.0 mg/mL.
13. Add 10 μL of 5× concentrated electrophoresis sample buffer to terminate the reaction. Boil for an additional 5 min.
14. Centrifuge the samples and run the soluble fraction on a 12% SDS-PAGE gel, following the same protocol as described under Subheading 3.2, steps 1–4.
15. Fix and stain the gel, destain and dry using an appropriate gel dryer, and then expose to X-ray film in an X-ray film cassette at −80°C for a suitable time (up to 4 h). Kinase activity will be indicated by a band of phosphorylated protein substrate (Fig. 2).

3.6. Preparation of Recombination Proteins and Its Activity Assay

It is well known that constitutively active long forms of MKK3 and MKK6, MKK3b(E) and MKK6b(E), can also effectively activate p38. In the prokaryote expression system of *E. coli* BL21 (DE3), the recombinant His-MKK3b(E), His-p38 isoforms, and appropriate substrate protein His-MEF2A are expressed and purified separately. In the coupling kinase assay, three recombinant proteins are mixed and incubated for a while. In the reaction mixture, MKK3b(E) activates p38, which subsequently phosphorylates the substrate protein MEF2A or positive control GST-ATF2-(1–109). Phosphorylated substrate proteins are separated on a 12% SDS-PAGE gel and visualized by radioautography (see Note 8).

1. Thaw BL21 (DE3) competent cells (100 μL) on ice. Gently tap tubes to mix the cells. Do not mix cells by pipetting.
2. Add 5–10 μL of the plasmid DNA (0.2–50 ng DNA) to the competent cells. Gently tap tubes to mix. Do not mix cells by pipetting.
3. Incubate the cells on ice for 30 min.
4. Heat shock cells for 90 s in a 42°C water bath; do not shake.
5. Place on ice for 2 min.
6. Spread the cells on LB agar plates containing 100 μg/mL ampicillin.
7. Incubate overnight at 37°C.
8. Grow small cultures from single clones in 2 mL LB medium containing 100 μg/mL ampicillin.
9. Incubate overnight at 37°C.
10. Transfer the small culture to 250 mL of LB medium containing 100 μg/mL ampicillin.
11. Incubate at 37°C until the A600 reaches 0.5.
12. Add IPTG to a final concentration of 1 mM for 5 h.
13. Cells are collected by centrifugation at 3,000 × *g* for 10 min, and the bacterial pellet is resuspended in 10 mL of B-PER Reagent (see Note 9).
14. Sonicate the cell suspension on ice using a sonicator equipped with a microtip. Use six 10 s bursts at 200–300 W with a 10 s cooling period between each burst. Cellular debris is removed by centrifugation at 27,000 × *g* (e.g., 14,000 rpm with a Beckman JA17 rotor) for 30 min (see Note 10).
15. Remove the supernatant (the protein extract) to a new tube.
16. Equilibrate column(s) and buffers to room temperature.
17. Uncap the Nickel-Chelated Column and allow the sodium azide storage solution to drain. The column will stop flowing when the liquid level reaches the top disc.
18. Prepare the resin by adding 10 mL (2 × 5 mL) of B-PER Reagent and allowing it to flow through the column.
19. Apply up to 10 mL of sample (2 × 5 mL) to the column and allow it to flow through the column.
20. Wash the column by adding at least 6 mL (2 × 3 mL) of wash buffer 1 and allowing it to flow through the column.
21. Wash the column by adding at least 9 mL (3 × 3 mL) of wash buffer 2 and allowing it to flow through the column.
22. Elute the 6 His-tagged protein by adding 6 mL (2 × 3 mL) of Elution Buffer and collecting the fractions that emerge. Dialysis (Slide-A-Lyzer dialysis cassettes) can be used for buffer

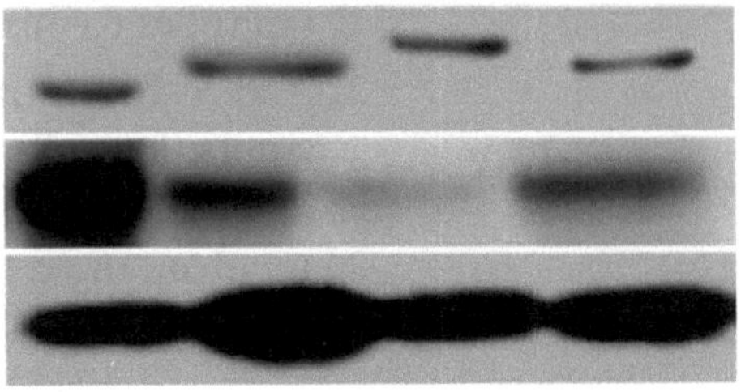

Fig. 3. In vitro phosphorylation of MEF2A by different p38 isoforms. The recombinant His-MKK3b(E), the His-p38 isoforms and His-MEF2A were expressed in *E. coli* BL21 (DE3) and purified, respectively. MKK3b(E), one of the p38 isoforms and His-MEF2A were mixed and incubated for 30 min in a kinase buffer containing γ-^{32}P-ATP. Phosphoraylated substrate proteins were separated on a 12% SDS-PAGE gel and visualized by radioautography. GST-ATF2(1–109) protein, which can be phosphorylated by all p38 isoforms, was used as a positive control.

exchange (replacing elution buffer with kinase buffer). Determine the concentration of protein, following the same protocol described in Subheading 3.1, steps 10a–d.

23. Coupling kinase assays are carried out at 37°C for 30 min, using 1 μg recombinant MKK3b(E), 0.2 μg of recombinant p38 , 5 μg of MEF2A, 250 μM ATP, and 10 μ Ci of [γ-^{32}P] ATP in 20 μL of kinase reaction buffer. Reactions are terminated by the addition of 5 μL of 5× concentrated electrophoresis sample buffer. Reaction products are resolved on a 12% SDS-PAGE gel. Phosphorylated proteins are visualized by radioautography (Fig. 3) (see Note 11).

4. Notes

1. All solutions should be prepared in water with a resistivity of 18.2 MΩ-cm and a total organic content of less than five parts per billion, unless stated otherwise.
2. Anti-phosphotyrosine FB2 monoclonal antibody staining reveals a complex pattern of constitutively tyrosine phosphorylated protein in 70Z/3 cells, but the addition of 1 ng/mL-LPS to the cells results in tyrosine phosphorylation of a protein with an appearent molecular mass of 38 kDa without additional changes in the tyrosine phosphorylation of other proteins. Therefore, only the areas of the Western blot containing p38 is shown. Now that specific antiphospho-p38 antibody (p–p38 (Thr 180/Tyr 182)-rabbit polyclonal IgG, Santa Cruz, CA) is available, it is easier to detect the phosphorylated p38 kinase in a wide range of cell types with different agonists.

3. The time, concentration, or strength of stimulation may vary among cells; therefore, these should always be optimized for particular cell lines or stimuli (16).
4. In the Western blot procedure, 5% BSA in TBST solution is usually used to block the nitrocellulose membrane, although BSA is relatively expensive. Since nonfat dry milk may contain tyrosine phosphorylated protein or phosphatase, and thus makes the background high, it is not often recommended for blocking.
5. Generally, p38α is more abundant than other p38 isoforms in the cells. Extracellular stimuli mainly activate p38α kinase, and anti-phospho-p38 antibodies cannot distinguish the phosphorylated p38α protein from the other p38 isoforms. Thus, if we want to know whether some extracellular signal can activate p38 isoforms other than p38α, the treated cells need to be transfected with a tag-p38 isoform; then the tag-p38 isoform must be immunoprecipitated with anti-tag antibodies, so that the activity of the kinase can be analysed by in vitro kinase assay.
6. It is necessary to select an appropriate cell line, transfection reagent and to optimize the transfection conditions to increase the transfection efficiency. High transfection efficiency and expression level is very important for MKK3b(E) to activate p38.
7. MKK3b is a contiguous upstream kinase. MKK3b(E), the constitutively active form of MKK3b, can activate the p38α kinase without additional extracellular stimuli. Using the experimental procedure mentioned here, we can determine whether some protein is an upstream component of the p38 signal pathway if the dominant positive mutant of this protein is not available, but it is necessary to select an appropriate agonist to stimulate the cells to activate the protein after both genes of this protein and p38 isoform have been cotransfected into the cells and expressed sufficiently (usually 24 h after transfection). The activity of p38 isoform can then be detected.
8. In the in vitro kinase assay system, MKK3b(E) phosphorylates p38 isoforms, which subsequently phosphorylates its substrate. Thus, there are two phosphorylated proteins in the reaction mixture, phosphorylated p38 isoforms and the substrate protein, both of which are labeled with radioactive γ-^{32}P. The molecular weight of the substrate protein should differ from that of p38 isoforms.
9. If desired, add a protease inhibitor cocktail. Do not use protease inhibitors that contain metal chelators, such as EDTA.
10. If the lysate is very viscous, add RNase A (10 μg/mL) and DNase I (5 μg/mL) and incubate on ice for 10–15 min.

11. p38 kinase isoforms can be activated differentially by upstream kinases. For example, both MKK3 and MKK6 can activate p38α, but only MKK6 can activate p38β. MKK4, an upstream kinase, can activate p38α and p38δ in specific cell types. Therefore, a potent upstream activator should be used for a particular p38 isoform, although it may not be the physiologic upstream signal component for the substrate in the particular cell line. It is feasible because we mainly pay attention to whether activated p38 can phosphorylate the substrate or not.

Acknowledgments

This work was supported in part by Grants NSFC30670442 and NSFC30971490 from Chinese National Science Foundation.

References

1. Zarubin,T., and Han, J. (2005) Activation and signaling of the p38 MAP kinase pathway. *Cell Res.* 15, 11–18.
2. Jiang, Y., Chen, C., Li, Z., Guo, W., Gegner, J. A., Lin, S., and Han, J. (1996) Characterization of the structure and function of a new mitogen-activated protein kinase (p38). *J. Biol. Chem.* 271, 17920–17926.
3. Lechner, C., Zahalka, M. A., Giot, J. F., Møller, N. P., and Ullrich, A. (1996) ERK6, a mitogen-activated protein kinase involved in C2C12 myoblast differentiation. *Proc. Natl. Acad. Sci. U.S.A.* 93, 4355–4359.
4. Li, Z., Jiang, Y., Ulevitch, R. J., and Han, J. (1996) The primary structure of p38γ: a new member of p38 group of map kinases. *Biochem. Biophys. Res. Commun.* 228, 334–340.
5. Kumar, S., McDonnell, P. C., Gum, R. J., Hand, A. T., Lee, J. C., and Young, P. R. (1997) Novel homologues of CSBP/p38 MAP kinase: activation, substrate specificity and sensitivity to inhibition by pyridinyl imidazoles. *Biochem. Biophys. Res. Commun.* 235, 533–538.
6. Jiang, Y., Gram, H., Zhao, M., New, L., Gu, J., Feng, L., Padova, F. D., Ulevitch, R. J., and Han, J. (1997)Characterization of the structure and function of the fourth member of p38 group mitogen-activated protein kinases, p38. *J. Biol. Chem.* 272, 30122–30128.
7. Ono, K., and Han, J. (2000) The p38 signal transduction pathway activation and function. *Cell. Signal.* 12, 1–13.
8. Han, J., Lee, J. D., Jiang,Y., Li, Z., Feng, L., and Ulevitch, R. J., (1996) Characterization of the structure and function of a novel MAP kinase kinase (MKK6). *J. Biol. Chem.* 271, 2886–2891.
9. Han, J., Wang, X., Jiang, Y., Ulevitch, R. J., Lin, S. (1997) Identification and characterization of a predominant isoform of human MKK3. *FEBS Lett.* 401, 19–22.
10. Moriguchi, T., Toyoshima, F., Gotoh, Y., Iwamatsu, A., Irie, K., Mori, E., Kuroyanagi, N., Hagiwara, M., Matsumoto, K., and Nishida, E. (1996) Purification and identification of a major activator for p38 from osmotically shocked cells. activation of mitogen-activated protein kinase kinase 6 by osmotic shock, tumor necrosis factor-α, and H_2O_2. *J. Biol. Chem.* 271, 26981–26988.
11. Han, J., Lee, J. D., Tobias, P. S., and Ulevitch R. J. (1993) Endotoxin induces rapid protein tyrosine phosphorylation 70Z/3cells expressing CD14*. *J. Biol. Chem.* 268, 25009–25014.
12. Westermarck, J., Li, S., Kallunki, T., Han, J., and KäHäRi, V. (2001) p38 mitogen-activated protein kinase-dependent activation of protein phosphatases 1 and 2a inhibits MEK1 and MEK2 activity and collagenase 1 (MMP-1) gene expression. *Mol. Cell. Biol.* 21, 2373–2383.
13. Zhao, M., New, L., Kravchenko, V. V., Kato, Y., Gram, H., Padova, F. D., Olson, E. N., Ulevitch, R. J., and Han, J. (1999) Regulation of the MEF2 family of transcription factors by p38. *Mol. Cell. Biol.* 19, 21–30.

14. Han, J., Jiang, Y., Li, Z., Kravchenko, V. V., and Ulevitch, R. J. (1997) Activation of the transcription factor MEF2C by the MAP kinase p38 in inflammation. *Nature.* 386, 296–299.

15. Ouwens, D. M., Ruiter, N. D., van der Zon,G. C. M., Carter, A. P., Schouten, J., van der Burgt, C., Kooistra, K., Bos, J. L., Maassen, J. A., and van Dam, H. (2002) Growth factors can activate ATF2 via a two-step mechanism: phosphorylation of Thr71 through the Ras–MEK–ERK pathway and of Thr69 through Ral GDS–Src–p38. *EMBO J.* 21, 3782–3793.

16. Raingeaud, J., Gupta, S., Dickens, M., and Han, J. (1995) Pro-inflammatory cytokines and environmental stress cause p38 mitogen-activated protein kinase activation by dual phosphorylation on tyrosine and threonine. *J. Biol. Chem.* 270, 7420–7426.

17. Lee, J., Kravchenko, V., Kirklandt, T. N., Han, J., Mackman, N., Moriartyt, A., Leturcqt, D., Tobias, P. S., and Ulevitch, R. J. (1993) Glycosyl-phosphatidylinositol-anchored or integral membrane forms of CD14 mediate identical cellular responses to endotoxin. *Proc. Natl. Acad. Sci. U.S.A.* 90, 9930–9934.

Chapter 5

Activity Assays for Extracellular Signal-Regulated Kinase 5

Kazuhiro Nakamura and Gary L. Johnson

Abstract

Extracellular signal-regulated kinase 5 (ERK5) is also known as big MAPK (BMK1) or MAPK7. ERK5 is 115 kDa in mass and therefore larger than the other MAPKs such as ERK1/2, JNK, and p38. Like other MAPKs, ERK5 is ubiquitously expressed in mammalian cells and is part of a three kinase cascade involving a MAPK kinase (MEK5) and MAPK kinase kinase (primarily MEKK2 and MEKK3). ERK5 is important for proliferative responses to growth factors like epidermal growth factor and stress responses such as hyperosmolarity. Upon stimulation, ERK5 rapidly translocates to the nucleus for the control of transcription. ERK5 is also critical for maintenance of vascular integrity and endothelial cell survival. In this chapter, we define methods used to measure the activation of ERK5 using different biochemical and cell-based assays.

Key words: ERK5, MEK5, MEKK2, MEKK3, Phospho-ERK5 antibody, Nuclear translocation, Endothelial tube formation

1. Introduction

The N-terminus of ERK5 encodes the kinase domain that harbors the activation loop TEY motif that is dually phosphorylated for activation. The C-terminal moiety of ERK5 is unique among MAPKs and plays a transcriptional activation function required for regulation of MEF2C, peroxisome proliferator activated receptor (PPARγ1), c-Fos, and Fra1 (1, 2). ERK5 is strongly activated by stress stimuli, including oxidative stress and hyperosmolarity (3). ERK5 is also activated by EGF and is involved in the control of the G1/S transition in the cell cycle (4). Serum and other growth factors such as NGF in specific cell types also activate ERK5. ERK5 also plays a critical role in the survival of endothelial cells and the maintenance of vascular integrity in adult mice (5). In mouse embryos, ERK5 is required for vascular development (6–8).

Rony Seger (ed.), *MAP Kinase Signaling Protocols: Second Edition*, Methods in Molecular Biology, vol. 661,
DOI 10.1007/978-1-60761-795-2_5, © Springer Science+Business Media, LLC 2010

The MAPK kinase, MEK5, which phosphorylates the activation loop TEY motif, encodes a Phox/Beml (PB1) domain (3). PB1 domains are used for heterodimerization of proteins, and the MAPK kinase kinases, MEKK2 and MEKK3 also encode PB1 domains (9). The MEK5 PB1 domain heterodimerizes with the PB1 domain of MEKK2 or MEKK3 (3, 10), and MEKK2 or MEKK3 and phosphorylate and activate MEK5 in response to different stimuli (9). The PB1 domain-dependent activation of the ERK5 cascade is unique among the MAPK family of kinases (9, 10). Importantly, MEK5 is inhibited by several MEK1/2 inhibitors, including U0126 and PB184352 (11, 12), so caution in interpretation of ERK1/2 versus ERK5 function with these inhibitors is warranted. Herein, we define methods used to measure the activation of ERK5 using different biochemical and cell-based assays.

2. Materials

2.1. Cell Culture

1. Mouse embryonic fibroblasts (MEFs) were grown in IMDM (Life Technologies, Inc) with 10% heat-inactivated fetal calf serum (FCS), 100 U/ml penicillin, and 100 μg/ml streptomycin at 37°C with 7% CO_2. HEK293, COS7, and mouse embryonic endothelial cells (MEECs) were maintained in DMEM supplemented with 10% FCS. Trypsin (0.25%) and EDTA (1 mM) in PBS (GIBCO) is used for the passage of cells.
2. The monoclonal antibody (12CA5) against the hemagglutinin (HA) epitope (Roche Molecular Biochemicals); Anti-Ras antibody (#Y13-259, Abcam); Anti-ERK5 antibody (#E1523, Sigma); the mouse monoclonal antibody for MEK5 (#610957, BD PharMingen); anti-phospho-ERK5 antibody (#3771, Cell Signaling); HRP-donkey anti-rabbit IgG antibody (#711-035-152, Jackson Laboratories); and sheep-anti-mouse IgG antibody (#NA931V, GE Healthcare).
3. Lipofectamine and Plus Reagent (Invitrogen) are used for transfection as per the manufacturer's instruction.
4. MEF2C luciferase reporter gene (3×-MEF2-luc) was constructed with a thymidine kinase promoter with three copies of a high-affinity MEF2 binding site from the desmin gene (13). The pRL-TK Renilla luciferase reporter gene and dual luciferase reporter assay system kit (Promega) were used for the readout. Luminescence for the luciferase activity was measured using a POLARstar Omega plate reader (BMG LabTech).
5. For ERK5 live cell imaging, the ERK5 cDNA was inserted in frame with eYFP using the pEYFP-C1 expression vector

(BD Clontech) for expression of an eYFP–ERK5 fusion protein. Imaging was conducted with a Zeiss Axiovert 200 M inverted microscope with an objective (63× Oil 1.25-numerical aperture, Plan-Neofluar, Zeiss), a 125-W xenon arc lamp (Sutter Instrument Company), digital CCD camera (CoolSNAP HQ, Roper Scientific), and Slidebook 5.0 software (Intelligent Imaging Innovations). The YFP filter set employs a bandpass excitation filter of 500/20 nm, a 515DCLP band beam splitter, and a bandpass emission filter of 535/30 nm (Chroma).

6. siRNA (Dharmacon) was dissolved in RNase-free water. An Amaxa nucleofector II electroporation apparatus (Lonza) was used to introduce siRNA into cells as per the manufacturer's instructions for each cell type used.
7. BD Matrigel Matrix growth factor reduced (BD Biosciences).

2.2. SDS-PAGE and Immunoblotting

1. Solubilizing buffer: 1% NP-40, 10 mM Tris-HCl pH 7.5, 150 mM NaCl, 0.4 mM EDTA, 10 mM NaF, 2 mM Na_3VO_4, 1 μg/ml leupeptin, 1 μg/ml aprotinin, 1 μg/ml α_1-antitrypsin, and 1 mM PMSF.
2. In vitro kinase assay buffer: 0.05% NP-40, 20 mM HEPES pH 7.5, 50 mM NaCl, 2.5 mM $MgCl_2$, 20 mM β glycerophosphate, 0.1 mM Na_3VO_4, and 2 mM DTT.
3. Separating gel stock (4×): 1.5 M Tris–HCl, pH 8.8, 0.4% SDS.
4. Stacking gel stock (4×): 0.5 M Tris–HCl, pH 6.8, 0.4% SDS.
5. 30% Acrylamide/bis solution (29:1 with 3.3% C) (Bio-Rad) and *N,N,N′,N′*-Tetramethylethylenediamine (TEMED) (Sigma). 10% Ammonium persulfate (Bio-Rad) is prepared in water.
6. Running buffer (10×): 250 mM Tris, 1.92 M glycine, and 1% SDS.
7. Nitrocellulose blotting membrane (Whatman).
8. Transfer buffer (10×): 250 mM Tris and 1.92 M glycine. To prepare 1× transfer buffer, mix transfer buffer (10×), methanol, and water at a ratio of 1:2:7.
9. Tris buffered saline (TBS) (20×): 200 mM Tris and 3 M NaCl, pH 8.0. To prepare TBS-T, add Tween-20 at the final concentration of 0.05%.
10. Blocking buffer: 5% BSA (Sigma, essentially γ-globulin-free) and 0.02% NaN_3 in TBS. Primary antibody is added into blocking buffer. The antibody in blocking buffer is reusable for three to five times.

11. Secondary antibody is in 5% non fat dry milk in TBS.
12. Detection is carried out by enhanced chemiluminescent reagents, the Supersignal West Pico detection system (#34080, Pierce), and film (X-OMAT, Kodak).

3. Methods

The most common and straightforward measurement of ERK5 activation involves immunoblotting with either a phospho-specific antibody that measures the phosphorylation of the ERK5 activation loop or measuring the gel mobility shift of ERK5 that correlates with its phosphorylation and activation using anti-ERK5 antibody. Additional assays include in vitro kinase assay of immunoprecipitated ERK5 or a luciferase reporter gene assay using a MEF2C responsive promoter that relies on the fact that ERK5 phosphorylates and activates the transcription factor MEF2C.

3.1. Detection of ERK5 Activation by Using Anti-phospho-ERK5 Antibodies

The most direct measurement of ERK5 activation involves resolving cell lysates on SDS gels, transfer of the proteins to nitrocellulose blotting membrane, and immunoblotting for phosphorylated ERK5 with an anti-phospho-ERK5 antibody. Activation of ERK5 involves the phosphorylation of ^{218}T and ^{220}Y in the activation loop in the ERK5 kinase domain. Phosphorylation of ^{218}T and ^{220}Y is required for ERK5 activation (4), and the phosphorylation of ^{218}T and ^{220}Y is detected using a commercially available antibody that selectively recognizes the dually phosphorylated ERK5 protein (phopsho-ERK5 antibodies, see http://mpr.nci.nih.gov/mpr_proteins/20986497). Four different anti-phospho-ERK5 antibodies are available from different vendors. Our laboratory has used a phospho-ERK5 antibody from Cell Signaling (#3371). Most anti-phospho-ERK5 antibodies have cross-reactivity with phospho-ERK1/2 proteins because of conservation of their activation loop phosphorylation sites. However, ERK5 is a 115 kDa protein and is easily resolved by SDS-PAGE from the 44 and 42 kDa ERK1/2 proteins, allowing easy assay of ERK5 phosphorylation status as a measure of activation. The cross-reactivity, however, prevents these antibodies from being used for immunostaining of phospho-ERK5 in fixed cells. The anti-phospho-ERK5 antibodies commercially available are not particularly strong high-affinity antibodies. The expression of ERK5 is also variable in different cell types, and the anti-phospho-ERK5 antibodies do not always detect phosphorylated ERK5 in cells expressing low levels of the protein. Our experience is that phosphorylation of endogenous ERK5 in total cell lysates resolved by SDS-PAGE is detectable in many cells with stimuli that activate ERK5 (e.g., osmostress in response to sorbitol in mouse embryonic fibroblasts (MEFs)) (see Note 1). In response to other stimuli such as growth factors like EGF, measurement of endogenous ERK5 activation using the anti-phospho-ERK5 antibodies can be difficult. Transfection using an ERK5 expression plasmid for overexpression is often used to study ERK5 regulation. Generally, the overexpressed ERK5 is regulated in the same manner as endogenous ERK5 and can markedly enhance the signal for measurement of ERK5 activation. This is especially useful for cells with modest or low level expression of ERK5. It is often also beneficial to have a lysate from cells transfected with an ERK5 expression plasmid to run in a lane of the SDS-PAGE as a control to identify the endogenous phospho-ERK5 band from cell lysates.

Using a 10% or higher acrylamide gel for SDS-PAGE causes ERK5 to run as a single band (see Note 2). Using a lower percent acrylamide gel (e.g., 8% acrylamide gel, see Note 3) allows the resolution of the phosphorylated and nonphosphorylated ERK5 proteins. This is because the phosphorylated ERK5 protein migrates slower in 8% acrylamide SDS gels than the nonphosphorylated ERK5 protein. This change in migration in SDS-PAGE is seen with many phosphorylated proteins including other MAPKs. This change in migration of phospho-ERK5 can be used to advantage assaying ERK5 activation (Fig. 1, right panels). An anti-ERK5 antibody (note that this is not the anti-phospho-ERK5

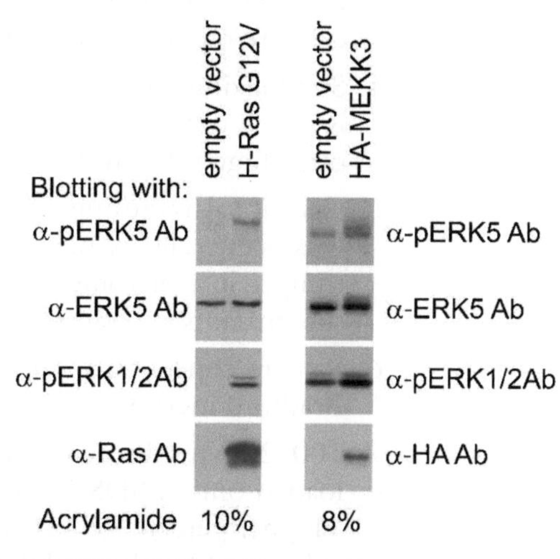

Fig. 1. Detection of ERK5 activation by immunoblotting. *Left panel*: ERK5 activation detected using the anti-phospho-ERK5 antibody. A dominant active H-Ras (H-Ras G12V) in pCMV5 was transiently transfected into HEK293 cells. The total cell lysates were subjected to 10% acrylamide SDS-PAGE. After transfer of proteins onto a nitrocellulose membrane, the membrane is blotted sequentially with anti-phospho-ERK5, -ERK5, and -Ras antibodies with stripping of the filter between antibody blots. Phosphorylation of ERK1/2 is observed due to the cross reactivity of anti-phospho-ERK5 antibody. *Right panel*: ERK5 activation detected by a mobility shift of the phosphorylated ERK5 band. Lysates of HEK293 cells expressing transfected HA-MEKK3 were subjected to SDS-PAGE using an 8% acrylamide gel and sequentially blotted with the same antibodies used in the *left panel*.

antibody, but an antibody recognizing the ERK5 protein) can be used for immunoblotting. Measurement of the slower migrating ERK5 band correlates with phosphorylation of ^{218}T and ^{220}Y in the ERK5 activation loop. Minigels may not work well for the mobility shift assay because the two ERK5 bands are not resolved sufficiently, and optimizing the time of electrophoresis may be required.

3.1.1. Assay Protocol

1. Cells are generally seeded on tissue culture dishes (100 or 35 mm) 1 day before stimulus treatment.
2. Cells are generally grown to 70–90% confluence, and certain stimuli give a greater ERK5 activation if the cells are first starved of serum for 2–8 h before being challenged with stimulus. The stimulus may be a growth factor, cytokine, drug, or a stress such as heat shock. ERK5 is, for example, strongly activated in response to osmostress. For a hyperosmolar stress a 2 M stock of sorbitol (10×) in serum-free medium is added to the cell culture media giving a final concentration of 0.2 M sorbitol.
3. Cells are stimulated for different time periods. Maximal ERK5 activation generally occurs at 10–30 min after stimulation of most mammalian cells. It is recommended to do a time course from 2 to 120 min to define the ERK5 response in a specific cell type. At specific times, the medium is aspirated and the cells are rapidly rinsed with ice-cold 1× PBS.
4. Solubilizing buffer (1 ml for 100 mm or 0.125 ml for 35 mm) is added to the dishes, and the cells are harvested into microcentrifuge tubes using a plastic cell scraper.
5. Occasional vortexing of the collected cells results in efficient lysis as the cells are maintained on ice for 30 min and then centrifuged in a microfuge at 17,000×g at 4°C for 15 min. Cleared supernatants are transferred into new microfuge tubes.
6. Protein concentration of each sample is quantified using the Bradford method.
7. 100–200 μg Protein sample in Laemmli sample buffer is loaded onto a SDS-PAGE gel. For phospho-ERK5 blotting, a higher percent gel (10% or more) is suitable, whereas an 8% gel is used for the mobility shift assay.
8. Proteins separated by SDS-PAGE are transferred onto a nitrocellulose membrane using a wet transfer apparatus. Transfer buffer and apparatus should be prechilled and the transfer conducted in a cold box or cold room.
9. Filters are incubated with blocking buffer for at least 1 h at room temperature or overnight at 4°C. The membranes are blotted with the appropriate anti-phosho-ERK5 or anti-ERK5

antibody. Membranes should be washed with 0.05% Tween 20 in TBS (TBS-T).

10. Phospho-ERK5 immunoblotting can be performed overnight with 1:1,000 dilution of anti-phospho-ERK5 antibody in blocking buffer at 4°C. Secondary antibody blotting with HRP conjugated anti-rabbit antibody (1:3,000 in 5% nonfat dry milk in TBS) is performed for 2 h at room temperature and bands detected using enhanced chemiluminescence reagent. For ERK5 blotting, the anti-ERK5 antibody from Sigma is generally used at 1:4,000 dilution for 2 h at room temperature, followed by HRP-anti-rabbit antibody (1:8,000 final dilution) incubation for 2 h at room temperature. These immunoblotting conditions give total ERK5 as well as a measure of activation based on the mobility shift of phosphorylated ERK5 if an 8% gel is used for SDS-PAGE. Additional vendors provide primary antibodies derived from different hosts (donkey, goat, etc.), which can be useful when reprobing the membrane with additional antibodies for other proteins (e.g., JNK using a mouse monoclonal antibody and a secondary donkey anti-mouse antibody).

3.2. ERK5 In Vitro Kinase Assay Using Myelin Basic Protein or MEF2C as a Substrate

Several commercially available antibodies are good for ERK5 immunoprecipitation (see Table 1). Myelin basic protein (MBP) (14) or recombinant MEF2C (15) can be used as a substrate for in vitro kinase assays.

1. Lysates can be prepared as described in Subheading 3.1. ERK5 is immunoprecipitated from 1 mg of lysate using ERK5-specific antibody and protein-G (primary antibody, mouse) or protein-A (primary antibody: rabbit) Sepharose beads (see Note 4) in microcentrifuge tubes. Continuous tumbling of the tubes at 4°C for at least 2 h is needed for capture of ERK5 from cell lysates.
2. After the 2 h incubation, the beads are washed with solubilization buffer 4× and then washed 2× with in vitro kinase assay buffer.
3. After the second wash, the kinase buffer is carefully aspirated from the tube, leaving the bead pellet unperturbed.
4. For in vitro kinase assay, the bead-bound ERK5 is suspended in 50 μl kinase buffer containing 5 μCi [γ-^{32}P] ATP and 20 μM ATP with substrate (1 μg) (see Note 5) for 20 min at 30°C. Continuous agitation in a shaker bath is recommended.
5. The reaction is stopped by adding Laemmli sample buffer and placing the tubes on ice. The samples are then used for SDS-PAGE.

Table 1
List of commercially available antibodies for ERK5 and other proteins in the ERK5 regulatory pathway

	Vendor	Catalog#	Isotype	Epitope	Applications	Species
Phospho-ERK5 Ab	Cell signaling	3771	Rabbit poly	Thr 218/Tyr 220 (h)	WB	h, m, r
	Invitrogen	44612G	Rabbit poly	Thr 218/Tyr 220 (h)	WB, IHC, ELISA	h, m
	Millipore	07-507	Rabbit poly		WB	h, m, r
	Santa Cruz	sc-16564	Goat poly	Thr 218/Tyr 220 (h)	WB, IF, ELISA	h, m, r
ERK5 Ab	Cell signaling	3772	Rabbit poly		WB, IP	h, m, r, mk
	Eiptomics	1719-1	Rabbit mono	C-terminus (h)	WB, IF, FCM, IP	h, m, r
	Invitrogen	44688M	Mouse mIgG1	N-terminus (h)	WB	h, m, r, cn
		44688G	Rabbit poly		WB	h, m
	Millipore	07-039	Rabbit poly	783–806 (h)	WB, IP	h, m
	Santa Cruz	sc-81460	Mouse mIgG1	N-terminus (h)	WB, IP	h, m, r, dog
		sc-5626	Rabbit poly	516–815 (h)	WB, IF, IHC, FCM, ELISA	h, m, r
		sc-1285	Goat poly	C-terminus (h)	WB, IP, IF, ELISA	h, m, r
		sc-1286	Goat poly	N-terminus (h)	WB, IP, IF, ELISA	h, m, r
	Sigma	E1523	Rabbit poly	789–802 (h)	WB	h, m
MEK5 Ab	BD Transduction Laboratories	610957	Mouse mIgG1	13–188 (h)	WB, IF	h, m, r, dog
	Eiptomics	1789-1	Rabbit mono		WB, IF	h, m, r
	Santa Cruz	sc-81475	Mouse mIgG1	N-terminus (h)	WB, IP	h
		sc-1287	Goat IgG	C-terminus (h)	WB, IP, IF, ELISA	h
		sc-10795	Rabbit poly	351–444 (h)	WB, IP, IF, ELISA	h, m, r
		sc-9320	Goat IgG	C-terminus (r)	WB, IP, IF, ELISA	h, m, r

Phospho-MEK5 Ab	Invitrogen	480024	Rabbit poly	Ser 311/Ser315 (h)	WB	r
	Santa Cruz	sc-135702	Rabbit poly	Ser 311/Ser315 (h)	WB	h, m. r, cow, dog
		sc-130203	Rabbit poly	Ser 142 (h)	WB, IP, ELISA	h
MEKK2 Ab	Eiptomics	1662-1	Rabbit mono	N-terminus (h)	WB, IHC, IF, FCM, IP	h, m, r
		1714-1	Rabbit mono	C-terminus (h)	WB, IHC, IF, FCM, IP	h
	Santa Cruz	sc-1089	Rabbit poly	C-terminus (m)	WB, IF, ELISA	h, m, r
		sc-28768	Rabbit poly	281–360 (h)	WB, IP, IF, ELISA	h, m, r
		sc-1088	Rabbit poly	N-terminus (h)	WB, IP, IF, FCM, ELISA	h, m, r
	Sigma	M7946	Rabbit poly	241–255 (h)	WB	h
		M8071	Rabbit poly	36–50 (h)	WB	h
		AV49223	Rabbit poly	141–190 (h)	WB	h
MEKK3 Ab	BD Transduction Laboratories	611102	Mouse mIgG1	27–135 (h)	IF, WB	h, m, r, dog
	Eiptomics	1672-1	Rabbit mono	N-terminus (h)	WB, IHC, IF, FCM, IP	h, m, r
		1673-1	Rabbit mono	C-terminus (h)	WB, IHC, IF, FCM, IP	h
	Santa Cruz	sc-6843	Goat IgG	C-terminus (h)	WB, IF, ELISA	h, m, r
		sc-6844	Goat IgG	N-terminus (h)	WB, IF, ELISA	h > m, r
		sc-28769	Rabbit poly	291–360 (h)	WB, IP, IF, ELISA	h, m, r

FCM flow cytometry – intracellular staining, *IF* immunofluorescence, *IHC* immunohistochemical staining, *IP* immunoprecipitation, *WB* Western blotting

6. Resolved proteins in the gel are transferred onto a nitrocellulose membrane using a wet transfer apparatus.
7. The kinase activity is visualized by the incorporation of radioactive phosphate into the substrate or ERK5 itself by autophosphorylation of ERK5.
8. Anti-ERK5 immunoblotting can be used to measure ERK5 in each lane on the filter.

3.3. MEF2C Luciferase Assay

The transcription factor MEF2C is a target for both ERK5 and p38 (15). Therefore, measurement of ERK5 and p38 activation is required to determine specificity of MEF2C activation by the two MAPKs. It is possible to use commercially available p38 small-molecule inhibitors for p38 to discriminate between ERK5 and p38 activation of MEF2C.

To measure MEF2C activation, an MEF2C-responsive luciferase reporter gene is used.

1. Cells are seeded in 12-well plates (1.2×10^5 cells/well) 1 day before transfection.
2. MEF2C luciferase reporter plasmid (140 ng) and the control Renilla luciferase (14 ng) expression plasmid are cotransfected for assay of MEF2C activity. Additional expression plasmids (e.g., activated MEK5) can also be cotransfected (see Note 6).
3. 24 h later, cells are serum-starved for 3–24 h and then challenged with a stimulus such as a growth factor, cytokine, or stress for different times, generally up to 8–24 h. Cells are lysed with 1× Passive Lysis Buffer (Promega) for 15 min at room temperature.
4. Following the manufacturer's protocol, firefly and Renilla luciferase activities are measured by a luminescence microplate reader with autoinjectors (10 s each) (10 μl sample lysate, 50 μl Luciferase Assay Reagent II, and 50 μl Stop and Glo). Total luminescence of the MEF2C driven firefly luciferase is standardized to the Renilla luciferase activity as per the manufacturer's instructions.
5. The ratio of luminescence from the MEF2C reporter gene/Renilla luciferase is the relative luciferase activity (Fig. 2), used as the measure of ERK5 activation.

3.4. ERK5 Nuclear Translocation Assay

Like other MAPKs, ERK5 translocates with activation from the cytoplasm to the nucleus with activation. ERK5 encodes a bipartite nuclear localization signal-dependent nuclear import sequence and a CRM1-dependent nuclear export sequence (16, 17). ERK5 nuclear translocation is easily measured using live-cell imaging (9). In general, a GFP-tagged ERK5 construct is expressed in cells either transiently or stably. A constitutively

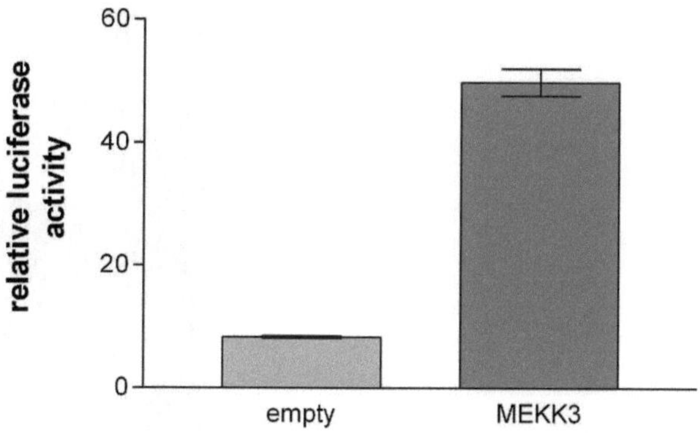

Fig. 2. MEF2C reporter gene activation by MEKK3–MEK5–ERK5 pathway. HA-MEKK3 (4 ng) or empty vector (4 ng) pCMV5 expression plasmids were cotransfected with MEF2C luciferase (140 ng) and Renilla luciferase (14 ng) plasmids in mouse embryo fibroblasts (MEFs). Twenty-four hours later, cells were starved of serum for 8 h. Cells were washed, lysed, and assayed for dual luciferase activity.

activated phospho-mimetic mutant ($^{311}S^{315}T$-DD) of MEK5 is used as a positive control to show ERK5 translocation to the nucleus. Cotransfection in COS cells of eYFP–ERK5 and $^{311}S^{315}T$-DD MEK5 results in ERK5 activation and ERK5 translocation to the nucleus (Fig. 3). Expression of the wild-type MEK5 protein does not induce ERK5 nuclear translocation. Nuclear translocation ERK5 can be assayed following EGF or FGF stimulation of HeLa cells (17).

1. For live-cell imaging, COS7 cells are transfected with eYFP–ERK5 in combination with wild-type or $^{311}S^{315}T$-DD mutant of MEK5 on 25-mm round glass coverslips in 6-well plate (see Note 7).
2. After 24 h, coverslips are used for fluorescence imaging.
3. Imaging analysis is conducted in our laboratory at room temperature using the apparatus described in the item 5 in Subheading 2.1. An objective was coupled with immersion oil to the bottom face of glass coverslips.
4. The images with the filter set in the item 5 in Subheading 2.1 are obtained at 100 ms exposure with 2×2 binning using a multipoint setup function.

3.5. Assay of MEK5 and MEKK2/MEKK3

MEKK2 and MEKK3 are MAPK kinase kinases that phosphorylate and activate MEK5 (9). The simplest method to assay MEK5 activation is to use phospho-specific anti-MEK5 antibody immunoblotting or staining in cells (Invitrogen and Santa Cruz Biotechnology sell antibodies recognizing phospho-$^{311}S^{315}T$ MEK5). In vitro kinase assays can also be performed.

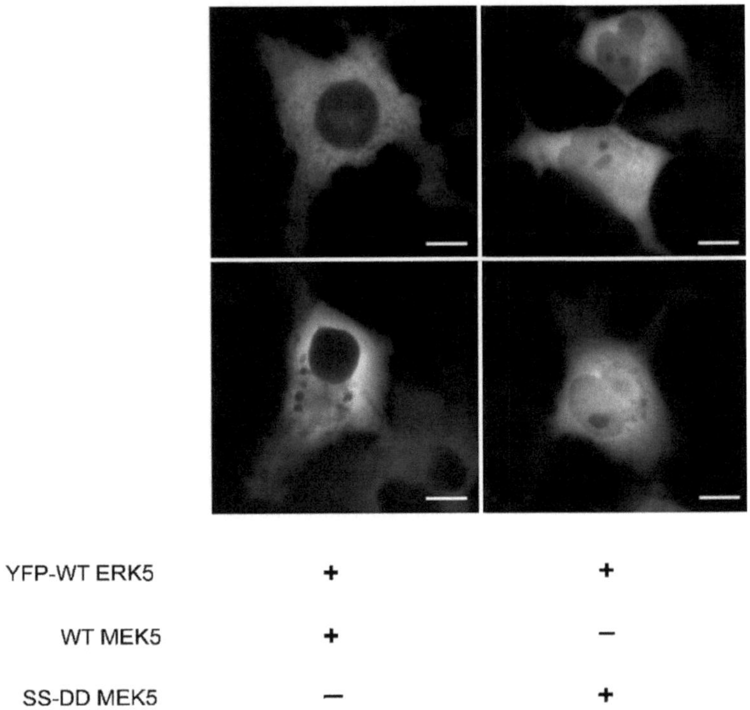

Fig. 3. Nuclear translocation of MEK5-activated ERK5. The YFP-ERK5 expression construct (pEYFP-C1) together with FLAG-MEK5 (WT or $^{311}S^{315}T$-DD) in pCMV5 were expressed by transfection in COS7 cells. Twenty-four hours after transfection, YFP-ERK5 localization in live cells was visualized using a fluorescence microscope. Pictures are two representative images of each condition. *Bar* in *right lower corner* of each picture represents 10 μm.

Methods for MEKK2 and MEKK3 in vitro kinase assay are as follows:

1. Recombinant MEK5 protein is prepared as a substrate to measure kinase activity (see Note 5). Either wild-type or kinase-inactive MEK5 (^{195}K-M) is prepared. The kinase-inactive MEK5 mutant prevents autophosphorylation of the MEK5 protein.
2. MEKK2 and MEKK3 immunoprecipitation and in vitro kinase assay are conducted as described in Subheadings 3.1 and 3.2, except for recombinant MEK5 is used as a substrate.

3.6. Tube Formation Assay Using MEEC Cells

ERK5 is critically involved in the maintenance of vascular integrity and endothelial cell viability (18). In vitro angiogenesis assays using endothelial tube formation as a readout are relatively simple and straightforward methods to assay for functional ERK5 activity. The MAPK kinase kinases, MEKK2 and MEKK3, which activate the ERK5 pathway, bind MEK5 via a PB1–PB1 domain interaction between the MAPK kinase kinase and MEK5. MEKK2, MEKK3, and MEK5 are the only kinases within the MAK signaling network that encode PB1 domains. Expression in cells of the MEKK2 PB1 domain efficiently competes with endogenous MEKK2 or MEKK3 binding to MEK5, leading to the suppression of MEKK2/3-induced MEK5–ERK5 activation (3). Expression of the free MEKK2 PB1 domain in mouse embryonic endothelial cells (MEECs) inhibits tube formation (Fig. 4a top panels). RNAi knockdown of ERK5 gives a similar inhibition (Fig. 4a bottom panels). RNAi knockdown of MEKK3 or MEK5 gives a phenotype similar to that observed with ERK5 knockdown or expression of the MEKK2 PB1 domain (data not shown). Thus, these assays could be used to identify proteins that regulate ERK5-dependent endothelial tube formation.

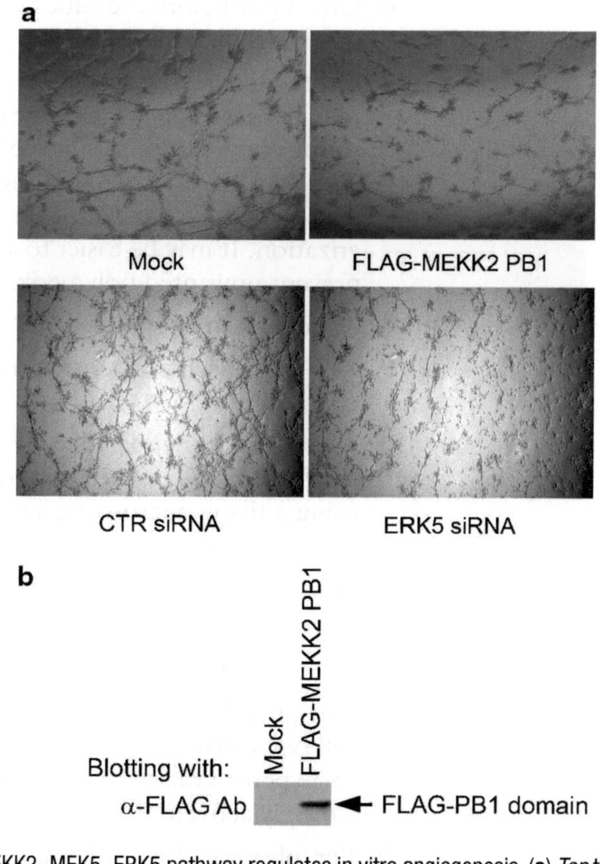

Fig. 4. MEKK2–MEK5–ERK5 pathway regulates in vitro angiogenesis. (**a**) *Top* two panels show the results of the tube formation assay of MEECs stably expressing FLAG-MEKK2 PB1 domain or empty vector (Mock). *Bottom* two panels are the comparison of tube formation assays of transiently expressed siRNA (control (CTR) or ERK5) MEEC cells. (**b**) The expression of FLAG-MEKK2 PB1 domain used in the *top* two panels of A was detected by anti-FLAG antibody blotting.

Methods for introducing siRNA into cells and tube formation assay are as follows:

1. MEECs are trypsinized, suspended as single cells, and counted. 2×10^6 cells are used each per electroporation reaction for introducing siRNA oligonucleotides.
2. 2×10^6 cells are suspended in 150 μl Amaxa Nucleofector solution and mixed with the siRNA (75 pmole/sample).
3. The cell suspension is transferred into an electroporation cuvette and electroporated using an Amaxa program V-001 using Nucleofector (see Note 8).
4. 500 μl MEEC culture medium is added to each cuvette. Electroporated sample is then split into 2×6 cm dishes (325 μl of cell solution per 6 cm dish).
5. The cells are incubated for another 48 h before the tube formation assay.
6. One night before the tube formation assay, 24-well plate and tips are chilled in −20°C.
7. On the day of the assay, BD Matrigel Matrix (growth factor reduced) is kept on ice. Matrigel will rapidly polymerize at room temperature. Using cold pipette tips and keeping the 24-well plate on ice, 350 μl of Matrigel is placed in each well. The plate is left for 30 min at 37°C to allow Matrigel polymerization. It may be easier to aliquot Matrigel in the freezer to prevent unwanted polymerization.
8. MEECs are resuspended at 2×10^5 cells/ml in medium.
9. 845 μl of cell suspension is dispensed in each Matrigel-coated well.
10. The cells are incubated at 37°C in a tissue culture incubator for 8–12 h to see the tube formation that can be visualized using a tissue microscope and camera.

4. Notes

1. Phosphorylation of endogenous ERK5 is detectable in some cell types that express high levels of ERK5. Stress stimuli such as hyperosmolarity and growth factors such as EGF can strongly activate ERK5.
2. For phospho-ERK5 immunoblotting, a 10% acrylamide is recommended for SDS-PAGE.
3. For ERK5 activation assays using the mobility shift of phosphorylated ERK5, an 8% acrylamide gel is recommended for

better resolution of the gel shifted ERK5 from the inactive, nonphosphorylated ERK5.

4. Protein A should be used with rabbit IgG, and protein G with mouse IgG.
5. Substrates for in vitro kinase assay are inserted in the pRSET bacterial expression vector. The proteins are produced in BL-21 bacteria, purified with Ni-NTA beads, dialyzed against the appropriate buffer, and stored at -20°C in buffer with 10% glycerol.
6. For accuracy of the luciferase reporter gene assay, the ratio between firefly and Renilla luciferase constructs should be in the range of 5:1–10:1. It is recommended that the concentration of each cDNA expression construct is optimized in preliminary experiments.
7. Round glass coverslips are treated with 1 M HCl, washed extensively in sterile H_2O and sterilized.
8. Choose an appropriate protocol for each cell type (http://www.amaxa.com/no_cache/cell-datantibodyase).

Acknowledgments

The authors would like to thank Lisa E. S. Crose for ERK5 siRNA experiment. We acknowledge NIH grants DK37871, GM30324, and GM68820 for support of our work on ERK5.

References

1. Kasler, H. G., J. Victoria, O. Duramad, and A. Winoto. 2000. ERK5 is a novel type of mitogen-activated protein kinase containing a transcriptional activation domain. *Mol Cell Biol* 20:8382–8389.
2. Akaike, M., W. Che, N. L. Marmarosh, S. Ohta, M. Osawa, B. Ding, B. C. Berk, C. Yan, and J. Abe. 2004. The hinge-helix 1 region of peroxisome proliferator-activated receptor gamma1 (PPARgamma1) mediates interaction with extracellular signal-regulated kinase 5 and PPARgamma1 transcriptional activation: involvement in flow-induced PPARgamma activation in endothelial cells. *Mol Cell Biol* 24:8691–8704.
3. Nakamura, K., and G. L. Johnson. 2003. PB1 domains of MEKK2 and MEKK3 interact with the MEK5 PB1 domain for activation of the ERK5 pathway. *J Biol Chem* 278:36989–36992.
4. Kato, Y., R. I. Tapping, S. Huang, M. H. Watson, R. J. Ulevitch, and J. D. Lee. 1998. Bmk1/Erk5 is required for cell proliferation induced by epidermal growth factor. *Nature* 395:713–716.
5. Hayashi, M., S. W. Kim, K. Imanaka-Yoshida, T. Yoshida, E. D. Abel, B. Eliceiri, Y. Yang, R. J. Ulevitch, and J. D. Lee. 2004. Targeted deletion of BMK1/ERK5 in adult mice perturbs vascular integrity and leads to endothelial failure. *J Clin Invest* 113:1138–1148.
6. Regan, C. P., W. Li, D. M. Boucher, S. Spatz, M. S. Su, and K. Kuida. 2002. Erk5 null mice display multiple extraembryonic vascular and embryonic cardiovascular defects. *Proc Natl Acad Sci U S A* 99:9248–9253.
7. Sohn, S. J., B. K. Sarvis, D. Cado, and A. Winoto. 2002. ERK5 MAPK regulates embryonic angiogenesis and acts as a hypoxia-sensitive repressor of vascular endothelial

growth factor expression. *J Biol Chem* 277: 43344–43351.

8. Yan, L., J. Carr, P. R. Ashby, V. Murry-Tait, C. Thompson, and J. S. Arthur. 2003. Knockout of ERK5 causes multiple defects in placental and embryonic development. *BMC Dev Biol* 3:11.
9. Nakamura, K., M. T. Uhlik, N. L. Johnson, K. M. Hahn, and G. L. Johnson. 2006. PB1 domain-dependent signaling complex is required for extracellular signal-regulated kinase 5 activation. *Mol Cell Biol* 26:2065–2079.
10. Seyfried, J., X. Wang, G. Kharebava, and C. Tournier. 2005. A novel mitogen-activated protein kinase docking site in the N terminus of MEK5alpha organizes the components of the extracellular signal-regulated kinase 5 signaling pathway. *Mol Cell Biol* 25:9820–9828.
11. Kamakura, S., T. Moriguchi, and E. Nishida. 1999. Activation of the protein kinase ERK5/BMK1 by receptor tyrosine kinases. Identification and characterization of a signaling pathway to the nucleus. *J Biol Chem* 274:26563–26571.
12. Mody, N., J. Leitch, C. Armstrong, J. Dixon, and P. Cohen. 2001. Effects of MAP kinase cascade inhibitors on the MKK5/ERK5 pathway. *FEBS Lett* 502:21–24.
13. Wu, H., F. J. Naya, T. A. McKinsey, B. Mercer, J. M. Shelton, E. R. Chin, A. R. Simard, R. N. Michel, R. Bassel-Duby, E. N. Olson, and R. S. Williams. 2000. MEF2 responds to multiple calcium-regulated signals in the control of skeletal muscle fiber type. *EMBO J* 19:1963–1973.
14. Dinev, D., B. W. Jordan, B. Neufeld, J. D. Lee, D. Lindemann, U. R. Rapp, and S. Ludwig. 2001. Extracellular signal regulated kinase 5 (ERK5) is required for the differentiation of muscle cells. *EMBO Rep* 2:829–834.
15. Kato, Y., V. V. Kravchenko, R. I. Tapping, J. Han, R. J. Ulevitch, and J. D. Lee. 1997. BMK1/ERK5 regulates serum-induced early gene expression through transcription factor MEF2C. *EMBO J* 16:7054–7066.
16. Buschbeck, M., and A. Ullrich. 2005. The unique C-terminal tail of the mitogen-activated protein kinase ERK5 regulates its activation and nuclear shuttling. *J Biol Chem* 280:2659–2667.
17. Kondoh, K., K. Terasawa, H. Morimoto, and E. Nishida. 2006. Regulation of nuclear translocation of extracellular signal-regulated kinase 5 by active nuclear import and export mechanisms. *Mol Cell Biol* 26:1679–1690.
18. Hayashi, M., and J. D. Lee. 2004. Role of the BMK1/ERK5 signaling pathway: lessons from knockout mice. *J Mol Med* 82:800–808.

Chapter 6

Use of Inhibitors in the Study of MAP Kinases

Kimberly Burkhard and Paul Shapiro

Abstract

The mitogen-activated protein (MAP) kinases are ubiquitous intracellular signaling proteins that respond to a variety of extracellular signals and regulate most cellular functions including proliferation, apoptosis, migration, differentiation, and secretion. The four major MAP kinase family members, which include the ERK1/2, JNK, p38, and ERK5 proteins, coordinate cellular responses by phosphorylating and regulating the activity of dozens of substrate proteins involved in transcription, translation, and changes in cellular architecture. Uncontrolled activation of the MAP kinases has been implicated in the initiation and progression of a variety of cancers and inflammatory disorders. As such, the ability to manipulate the activity of MAP kinase proteins with specific pharmacological inhibitors has received much attention as research tools for understanding basic mechanisms of cellular functions and for clinical tools to treat diseases. A variety of pharmacological inhibitors have been developed to selectively block MAP kinases directly or indirectly through targeting upstream regulators. This chapter will provide an overview of some of the current inhibitors that target MAP kinase signaling pathways and provide methodology on how to use selective MAP kinase inhibitors and immunoblotting techniques to monitor and quantify phosphorylation of MAP kinase substrates.

Key words: Mitogen-activated protein kinase, Extracellular signal-regulated kinase, c-Jun N-terminal kinase, p38 MAP kinase, U0126, SB203580

1. Introduction

The mitogen-activated protein (MAP) kinases are ubiquitous regulators of many cellular functions including cell growth, proliferation, differentiation, and inflammatory responses to stress signals (1). The MAP kinase family consists of four major members; the extracellular signal-regulated kinases-1 and 2 (ERK1/2), the c-Jun N-terminal kinases (JNK), p38 MAP kinases, and Big MAP kinase-1 (BMK1) also known as ERK5. Each of the MAP kinases is activated through highly specific interactions with upstream MAP or ERK kinases (MEKs), which phosphorylate

Rony Seger (ed.), *MAP Kinase Signaling Protocols: Second Edition*, Methods in Molecular Biology, vol. 661,
DOI 10.1007/978-1-60761-795-2_6, © Springer Science+Business Media, LLC 2010

threonine and tyrosine residues within the activation loop. Once activated, MAP kinases, in turn, phosphorylate and regulate a variety of substrates including transcription factors, translation regulators, other kinases, structural proteins, and other signaling proteins. Given the prominent role that constitutive activation of the MAP kinases plays in proliferative diseases like cancer, or inflammatory disorders such as rheumatoid arthritis, a number of pharmacological inhibitors have been developed to block MAP kinase signaling (2–4). These inhibitors target multiple proteins in the signaling cascade starting at the plasma membrane receptors all the way to the specific MAP kinase. The ability to manipulate the MAP kinase signaling cascades have been particularly useful for understanding basic biological mechanisms that regulate cell functions and for clinical therapies to treat disease. Table 1 provides a list of some of the major small molecular weight pharmacological inhibitors and their protein targets within the MAP kinase signaling pathway. Other methods for inhibiting MAP kinase signaling pathways in treating disease include monoclonal antibodies that target extracellular domains or ligands of receptor tyrosine kinases. The use of monoclonal antibodies to block MAP kinase signaling will not be discussed and can be found in other reviews (5).

High throughput screening methods have made it feasible to identify potentially target-specific inhibitor compounds with a desired effect from a large pool of chemical compounds. These types of drug discovery projects first develop the appropriate in vitro and cell-based assays to screen large chemical libraries and assess effects on target kinase activity or a cellular response (6). Once active compounds are identified, chemical modifications and refinement of these lead molecules are made to reach greater inhibition in both the in vitro and cell-based models. Drug development efforts also take advantage of the three-dimensional structures of the MAP kinases that have been solved by X-ray crystallography (7). A detailed understanding of the structure–function relationship for MAP kinases allows the design of inhibitor compounds that bind to specific regions on the MAP kinases including the ATP-binding domain or noncatalytic substrate binding domains (8–10). This approach, in combination with testing in biological assays and high throughput screening, provides an opportunity to identify highly specific compounds with better information on their mechanism of action.

Some of the first high throughput screening of chemical libraries aimed at developing target-selective inhibitors of MAP kinase signaling identified the compound PD98059 to be an allosteric inhibitor of MEK1 (11). Since the MEK1/2 proteins are the only known activators of ERK1/2, MEK-selective compounds are effective inhibitors of ERK1/2 activation. Subsequent studies

Table 1
Pharmacological inhibitors of MAP kinases and proteins that regulate MAP kinase signaling pathways

Target family	Specific target	Inhibitor	Vendor (catalog#)
Receptor tyrosine kinase	EGFR	Gefitinib (Iressa®)	American Custom Chemicals (ACC) Corporation (184475-35-2)
	VEGFR	Erlotinib (Tarceva®)	ACC (183321-74-6) Axon Medchem (1128)
		Lapatinib (Tykerb®)	SELLECK (S1028) ACC (231277-92-2)
	PDGFR	Sunitinib (Sutent®)	ACC (557795-19-4) Axon Medchem (1398)
		Sorafenib (Nexavar®)	ACC (284461-73-8) Axon Medchem (1397)
Non-receptor and receptor tyrosine kinases	Bcr–Abl,	Nilotinib (Tasigna®)	SELLECK (S1033) ACC (64157-10-0)
	Bcr–Abl, c-Src	Dasatinib (Sprycel®)	ACC (302962-49-8) Axon Medchem (1392)
	Bcr–Abl, c-SCT, c-Kit, PDGFR	Imatinib (Gleevec®)	ACC (220127-57-1) Axon Medchem (1394)
G-proteins	Ras	Tipifarnib (Zarnestra™)	Onicon Pharmachemie (192185-7201)
MAPKKK	Raf	Sorafenib (Nexavar®)	ACC(284461-73-8) Axon Medchem (1397)
MAPKK	MEK1/2	U0126	EMD Biosciences (662005) SELLECK (S1102)
		PD184352	Axon Medchem (1368) SELLECK (S1020)
		AZD6244	SELLECK (S1008)
	MEK5	BIX02188, BIX02189	Boehringer Ingelheim (*not commercially available*)
MAPK	p38	SB203580	EMD Biosciences (55389) Axon Medchem (1364)
		SB202190	EMD Biosciences (559388) Axon Medchem (1363)
		BIRB-796	Axon Medchem (1353)

The MAP kinases (MAPK) are regulated sequentially through receptor and non-receptor tyrosine kinases, G-proteins, MAP kinase kinase kinases (MAPKKK), and MAP kinase kinases (MAPKK)

developed more potent inhibitors of the MEK1/2 proteins including the small molecules U0126 (12), PD184352 and structurally similar PD0325901 (13), and AZD6244 (ARRY-142886) (14). These pharmacological inhibitors of MEK1/2 have been instrumental in understanding basic functions of ERK1/2 signaling and for clinical testing (2, 4). A variety of potent inhibitors of the p38 MAP kinases have also been developed and include SB203580 (15), SB202190 (16), and BIRB-796 (8). Many of the MEK1/2 and p38 MAP kinase pharmacological inhibitors have been shown to be quite specific in their kinase inhibition profiles (17). Other MAP kinase inhibitors such as SP600125 have been shown to inhibit JNK isoforms by competing with the ATP-binding site (18). However, SP600125 has also been shown to inhibit a number of other kinases (19), which must be considered when using this compound for evaluating JNK pathway regulation. Recent studies have identified pharmacological inhibitors that are reasonably selective for MEK5 causing inhibition of the ERK5 pathway without affecting ERK1/2 signaling (20). Moreover, new approaches are identifying new small molecular weight compounds that are designed to block protein–protein interactions between the ERK1/2 MAP kinases and a selective number of substrates (10). This approach may be advantageous for inhibiting some, but not all, of the substrate proteins that are regulated by a particular MAP kinase.

The ultimate goal in identifying selective MAP kinase inhibitors is to use them to treat human diseases. Several pharmacological inhibitors that target MAP kinase signaling pathways have been approved by the US Food and Drug Administration (FDA) for clinical applications (Table 1). These include drugs that inhibit plasma membrane receptor tyrosine kinases (RTK) that activate MAP kinases and are often overactivated in cancer cells (21, 22). The RTK inhibitors include sunitinib (Sutent®), which targets the PDGF, VEGF, and c-Kit receptors and are approved to treat renal cell carcinoma and gastrointestinal stromal tumors. In addition, gefitinib (Iressa®) and erlotinib (Tarceva®) are small molecular weight inhibitors that target the EGF receptor and are approved to treat non-small cell lung cancer and pancreatic cancer. A relatively nonspecific kinase inhibitor, sorafenib (Nexavar®), targets VEGF receptor and other kinases and is used to treat renal cell and hepatocellular carcinomas. Farnesyl transferase inhibitors (FTIs) were developed to block activation of Ras-G proteins, which are mutated and active in nearly 25% of all human cancers (23). While the FTI class of MAP kinase signaling pathway inhibitors has been less successful in the clinics due to problems with toxicity and lack of efficacy to block Ras, FTIs, such as tipifarnib (Zarnestra™), are still being tested in clinical trials for treating a variety of solid tumors and hematologic disorders (24).

Additional pharmacological inhibitors that target MAP kinase signaling include drugs that were intended to inhibit nonreceptor tyrosine kinases. The most successful example has been imatinib mesylate (Gleevec®), which was developed to inhibit the oncogenic Bcr–Abl fusion protein found in almost every case of chronic myelogenous leukemia (CML) (25). Imatinib mesylate has subsequently been shown to also inhibit other tyrosine kinases such as Src, c-Kit, and PDGF receptors. Several other inhibitors of Bcr–Abl and tyrosine kinases, including dasatinib (Sprycel®) and nilotinib (Tasigna®), have emerged from the successes of imatinib mesylate and are used to treat imatinib-resistant CML and other hematologic disorders (26).

This chapter will focus on the use of some more common pharmacological inhibitors that target-specific MAP kinases or the direct activation of MAP kinases and the evaluation and quantification of substrate phosphorylation in the context of various stimuli. With the availability of phosphorylation state-specific antibodies, methods for analyzing protein phosphorylation can be readily accomplished by immunoblotting techniques without the need for radioisotopes. Semiquantitative methods using densitometry will be discussed to determine the relative amount of protein phosphorylation following immunoblotting.

2. Materials

2.1. Cell Culture Supplies

1. Dulbecco's Modified Eagles Medium (DMEM) (GIBCO® Invitrogen; Carlsbad, CA) supplemented with 10% (v/v) fetal bovine serum (FBS) (Atlanta Biologicals; Lawrenceville, GA) and 1% (v/v) Penicillin/Streptomycin (PS) (GIBCO®).
2. Buffers for washing and passing cells include Hanks' balanced salt solution (HBSS), 0.25% trypsin-EDTA, and phosphate buffered saline (PBS).
3. Six well culture dishes (Becton, Dickinson and Company, BD; Franklin Lake, NJ), teflon cell scrapers (Fisher Scientific; Pittsburgh, PA), and a 5% CO_2 incubator set at 37°C.
4. Tissue lysis buffer (2×) (TLB): 0.2 M Tris–HCl (pH 6.8), 4% (w/v) sodium dodecyl sulfate (SDS), 20% (w/v) glycerol, 0.4 M β-mercaptoethanol, and 0.1% bromophenol blue.

2.2. Chemicals and Reagents

1. The p38 MAPK inhibitor, SB203580 (Calbiochem/EMD Chemicals, Inc.; Gibbstown, NJ), is reconstituted in autoclaved water to 20 mM and stored in aliquots at –20°C.
2. The MEK1/2 inhibitor, U0126 (Calbiochem/EMD Chemicals, Inc.), is stored in 100% DMSO at –20°C in 20 μL aliquots.
3. Epidermal growth factor (EGF) (Sigma; St. Louis, MO) is reconstituted in autoclaved water to a concentration of 50 μg/ml, aliquoted, and stored at –20°C.
4. Anisomycin (Sigma) is dissolved in 100% ethanol to 25 mg/ml, aliquoted, and stored at 4°C.

5. Enhance chemiluminescent (ECL) reagents are from GE Healthcare (Piscataway, NJ).
6. General chemicals: methanol, sodium azide (NaN_3).

2.3. SDS-Polyacrylamide Gel Electrophoresis

1. Stock solutions include 3 M Tris–HCl (pH 8.8), 1 M Tris–HCl (pH 6.8), and 10% (w/v) SDS.
2. 30% (w/v) Acrylamide (National Diagnostics; Atlanta, GA) is stored at 4°C. 1% (w/v) Bis-acrylamide (Fisher Scientific) is filtered through a 0.2-μm filter (Millipore; Billerica, MA) and stored at 4°C.
3. *N,N,N,N′*-Tetramethyl-ethylenediamine (TEMED) is stored at room temperature. A 10% ammonium persulfate solution (APS) is made fresh and stored at 4°C for up to 1 week.
4. Running buffer (10×): 0.25 M Tris–HCl, 2 M glycine, and 1% SDS is stored at room temperature.
5. Full-Range Rainbow™ Molecular Weight Marker (GE Healthcare).

2.4. Antibodies

1. The phosphorylation-specific anti-ppERK1/2 monoclonal mouse antibody is stored at −20°C in 20 μL aliquots. See Table 2 for a list of many of the antibodies used to study the phosphorylation status of the major MAP kinases and their substrates.
2. The total p38 MAPK and p-p38 MAPK antibodies are stored at −20°C.
3. ERK2 substrate phosphorylation-specific antibody: phosphorylated p90Rsk-1 (pRsk-1, Thr573) antibody is stored at −20°C.
4. Anti-α-tubulin monoclonal mouse antibody used as a loading control is stored at −20°C in 20 μL aliquots.
5. Secondary antibodies: anti-mouse and anti-rabbit IgG conjugated to HRP (Sigma) are stored at 4°C.

2.5. Immunoblotting for MAP Kinases and Substrate Phosphorylation

1. Polyacrylamide gel electrophoresis apparatus (C.B.S. Scientific Company, Inc.; Del Mar, CA).
2. MagicMark™ western protein standard (Invitrogen; Carlsbad, CA).
3. Electro-blotter semidry transfer system (Ellard Instrumentation Ltd; Monroe, WA) and slot blotter (Schleicher & Schuell BioScience; Keene, New Hampshire) used for phosphorylation quantification.
4. Transfer solutions include 0.25 M Tris base containing 0.4 M Aminocaproic acid, 1.25 M Tris base, and isopropyl alcohol (IPA). Blotting paper (VWR; West Chester, PA).

Table 2
List of antibodies for the phosphorylated and phosphorylation-independent (total) forms of MAP kinases and phosphorylated forms of MAP kinase substrate proteins

Protein targets	Antibody (phosphorylation site recognized)	Vendor (catalog#)
ERK1/2	Activated ERK1/2 (Thr185/Tyr187)	Sigma-Aldrich (M9692)
		Santa Cruz Biotechnology (sc-16982)
		EMD Biosciences (442706)
		Epitomics (148101)
	Total ERK2	Epitomics (1586-1)
		Santa Cruz Biotechnology (sc-154)
		EMD Bioscience (442685)
	Total ERK1	Santa Cruz Biotechnology (sc-94)
p38α, β, γ	Activated p38 MAP kinase (Thr180/Tyr182)	Cell Signaling Technology (CST) (9211)
		Epitomics (1229-1)
	Total p38α, β, γ MAP kinase	CST (9212)
		EMD Biosciences (506123)
		Epitomics (1544-1)
JNK1/2	Activated JNK1/2 MAP kinase (Thr183/Tyr185)	Santa Cruz Biotechnology (sc-6254)
		CST (9251)
	Total JNK1	Santa Cruz Biotechnology (sc-1648)
	Total JNK2	Santa Cruz Biotechnology (sc-572)
ERK5	Activated pERK5 MAP kinase (Thr218/Tyr220)	CST (2271)
	Total ERK5	Epitomics (1719-1)
		Millipore (07-039)
		Sigma-Aldrich (E1523)
Phosphorylated ERK1/2 substrates	p90RSK-1 (Thr573)	CST (9346)
		BD Biosciences (610225)
	ELK-1 (S383)	CST (9186)
	c-Myc (T58/S62)	CST (9401)
	MNK-1 (T197/T202)	CST (2111)
	PPAR-γ (S112)	Millipore (05-816)
	Connexin-43 (S255)	Santa Cruz Biotechnology (sc-12899-R)
	Tyrosine Hydroxylase (S31)	Millipore (AB5423)
	Estrogen receptor-α (S118)	Epitomics (1091-1)
	Tau (S199/S202)	Biosource (44-768G)
		Epitomics (1242-1)
	eEF2 (T56/T58)	Epitomics (1853-1)
	eIF-2a (S51)	Epitomics (1090-1)
	eIF-4B (S504)	Epitomics (2260-1)
	eIF-4E (S209)	Epitomics (2227-1)
	ATF-2 (T71)	CST (9221)
		Epitomics (1268-1)

(continued)

Table 2 (continued)

Protein targets	Antibody (phosphorylation site recognized)	Vendor (catalog#)
Phosphorylated p38 substrates	MAPKAPK-2 (T334)	CST (3041)
	MNK-1 (T197/T202)	CST (2111)
	Stat-1 (S727)	CST (9177)
	MSK-1 (S369/S376)	CST (9591)
Phosphorylated JNK substrates	c-Jun (S63)	CST (9261)
		Epitomics (1527-1)
	c-Jun (S73)	CST (9264)
		Epitomics (1107-1)
	c-Jun (S63/S73)	Santa Cruz Biotechnology (sc-16312-R)
	p53 (T81)	CST (2676)

Note that not all antibodies against each MAP kinase isoform are listed

5. Tris-buffered saline with Tween (TBS-tween) used for western blotting: 20 mM Tris–HCl (pH 7.4), 150 mM NaCl, and 0.1% Tween-20. Blocking buffer; 5% (w/v) nonfat dry milk in TBS-tween.
6. Polyvinylidene difluoride (PVDF) transfer membrane (PerkinElmer; Waltham, MA).
7. Bio-Max ML autoradiography film (Kodak; Rochester, NY).
8. Quantification by densitometry of films was done using the FLOURCHEM® SP imager (Alpha Innotech; San Leandro, CA) and AlphaEase FC™ software (Alpha Innotech).

3. Methods

Pharmacological inhibitors can be used to help determine the relevance of MAP kinase signaling pathways and their biological responses to extracellular signals. The MAP kinases can phosphorylate and regulate dozens of substrate proteins. With the development of specific antibodies that can distinguish a protein's phosphorylation status, the evaluation of MAP kinase activity can be readily achieved by measuring the phosphorylation of MAP kinase substrate proteins. The following protocol describes immunoblotting methods using phosphorylation-specific antibodies for visualizing, quantifying, and analyzing changes in MAP kinase activity and substrate phosphorylation in the presence of pharmacological inhibitors. The methods shown utilize common MAP kinase activators and pharmacological inhibitors in isolated cell cultures. However, the approach can be adapted for use in the context of different agonists or antagonists as well as assessment of MAP kinase signaling in tissue samples. Lastly, the advantages and disadvantages of the quantitative analysis of protein phosphorylation will be discussed along with the use of appropriate analytical controls.

3.1. Preparation of Cultured Cells for Evaluating MAP Kinase Inhibitors

1. HeLa S3 cells (American Type Culture Collection, catalog #CCL-2.2) at ~80% confluence are washed twice with HBSS buffer and trypsin-EDTA (0.25%) is added to the cells and incubated 2–5 min at 37°C or until most of the cells detach from the plate.
2. Complete DMEM media with 10% FBS and 1% PS is added and cells are seeded in 6-well tissue culture plates at 5×10^5 cells per well. The cells are incubated for an additional 24 h at 37°C with 5% CO_2.
3. Cells are pretreated with 1–10 μM U0126 for 15 min (see Note 1). EGF (50 ng/ml) is added to stimulate the ERK pathway, and the cells are incubated for an additional 5 min. Controls include unstimulated and EGF only treated samples.
4. Cells are pretreated for 10 min with 10 μM SB203580 at 37°C and then stimulated with 25 μg/ml anisomycin to activate the p38 MAP kinase pathway for 20 min at 37°C. Controls include unstimulated and anisomycin only stimulated samples.
5. Immediately after incubating with anisomycin or EGF, the cells are placed on ice and washed twice with cold PBS. 300 μl of TLB is added to each well and the cells are removed from the plate using Teflon cell scrapers. The samples are then transferred to 1.5 ml microcentrifuge tubes and heated at 100°C for 5 min before protein separation by gel electrophoresis and detection by immunoblotting.

3.2. SDS-Polyacrylamide Gel Electrophoresis

1. A 15% gel for an ASG-250 gel apparatus is made by first pouring the separating portion of the gel. The separating gel is made by combining 1.9 ml of 3 M Tris base (pH 8.8), 7.5 ml of 30% acrylamide, 1.3 ml 1% bis-acrylamide, 4.3 ml water, 150 μl 10% SDS. Add 10 μl TEMED and 50 μl APS immediately before pouring gel. Polymerization of the gel is usually complete in 30–45 min.
2. The stacking portion of the gel is made by mixing 625 μl 1 M Tris base (pH 6.8), 835 μl 30% acrylamide, 650 μl 1% bis-acrylamide, 2.8 ml water, and 75 μl 10% SDS. Immediately before pouring the stacking gel, 5 μl TEMED and 25 μl 10% APS is added. The gel is poured, and a comb is inserted

avoiding the introduction of air bubbles. Polymerization is usually complete in 15 min.

3. After the gel has polymerized, any unpolymerized acrylamide remaining in the wells can be removed by rinsing with 1× running buffer using a syringe. Running buffer (1×) is added to the top (cathode) and bottom (anode) chambers of the gel apparatus and 5–20 μl (~50 μg) of the samples are loaded into each well with a Hamilton syringe. The proteins are separated by applying a constant 35 mA to the gel for 1.5–2 h.

3.3. Immunoblotting for MAP Kinases and Substrate Proteins

1. This method is used for a semidry electro-blotter transfer system. PVDF membrane is first soaked in methanol for 30–60 s and then wash with distilled water. Fifteen pieces of blotting paper are cut to the size of the gel that contains the proteins of interest. Six pieces of blotting paper are soaked with Solution A (12.5 ml of 0.25 M Tris base with 0.4 M aminocaproic acid, 25 ml IPA, and 87.5 ml of water), three pieces are soaked in Solution B (2.5 ml 1.25 M Tris base, 25 ml IPA, and 100 ml water), and six pieces are soaked in Solution C (25 ml 1.25 M Tris base, 25 ml IPA, and 75 ml of water) (see Note 2).
2. The transfer assembly is set up on a plastic tray in the following sequential order; blotting paper soaked in Solution A, SDS-polyacrylamide gel electrophoresis (SDS-PAGE) gel, PVDF membrane, blotting paper soaked in Solution B, and blotting paper soaked in Solution C. The transfer assembly is flipped with this electro-blotter system such that the current transfers the protein from the gel closest to the cathode (black) to the PVDF membrane closest to the anode (red). The protein transfer is run at 25 V and 90 mA for 1 h (see Note 3).
3. After the completion of the protein transfer, the PVDF membrane is placed in blocking solution and gently agitated on a rocking platform at room temperature for at least 1 h.
4. The blocking buffer is then removed and the membrane is rinsed briefly with TBS-tween followed by incubation with one of the selected primary antibodies diluted 1:250–1,000 in a sterile 15 or 50 ml conical tube using TBS-tween with 1% (w/v) bovine serum albumin (BSA). For antibody dilutions that will be stored at 4°C for extended periods of time and reused, 0.1% (v/v) sodium azide can be added from a 10% (w/v) stock solution as a preservative. Depending on the antibody, incubations on a rocking platform can range from 1 h at room temperature to overnight at 4°C (see Note 4).

5. Following incubation with primary antibody, the primary antibody can be saved and often reused several times; however, the number of times an antibody can be reused must be determine empirically for each antibody. The membrane is washed three times for 10 min with TBS-tween and then incubated with the appropriate secondary antibody of the appropriate species at a dilution of 1:10,000 in TBS-tween with 1% (w/v) BSA. The membrane is gently rocked for 1 h at room temperature and then washed 3–5 times for 15 min with TBS-tween.
6. The TBS-tween is removed and equal amounts (2.5 ml/blot of each) of the ECL reagents are mixed together and immediately poured onto the membrane taking care to ensure that the entire surface of the membrane is exposed to the ECL reagents. After 1 min incubation, the ECL reagents are removed and the membrane is wrapped in plastic wrap.
7. In a dark room, a piece of autoradiography film (Kodak) is placed on top of the membrane with firm and even pressure for 5–300 s depending on the amount of protein of interest and the specificity of the primary antibody. Exposures of greater than 60 s can be done using an autoradiography film cassette (see Note 5).
8. After exposure, the membrane is rinsed briefly with TBS-tween and reprobed with an antibody that can be used as a protein loading control using the protocol above. Common protein loading controls include α-tubulin or β-actin. An example of an immunoblot for activated ERK1/2, the ERK1/2 substrate Rsk-1, or p38 MAP kinases is shown in Fig. 1.

3.4. Quantification of Protein Levels and Phosphorylation by Densitometry

1. The intensity of the proteins detected in the immunoblot can be semiquantified using densitometry. We use a FluorChem SP imaging system (Alpha Innotech; San Leandro, CA) to create a digital image of the autoradiography film of interest. The images collected are manipulated by AlphaEase FC software. However, densitometry can also be performed with a standard desktop scanner and free software such as ImageJ available through the National Institutes of Health (see Note 6).
2. A digital image of the autoradiography film is generated using the desired scanning device.
3. Using the object function within the various software programs, a square or circle can be drawn around the protein band of interest and the average pixel intensity of the area within the region of interest can be determined (see Note 7).

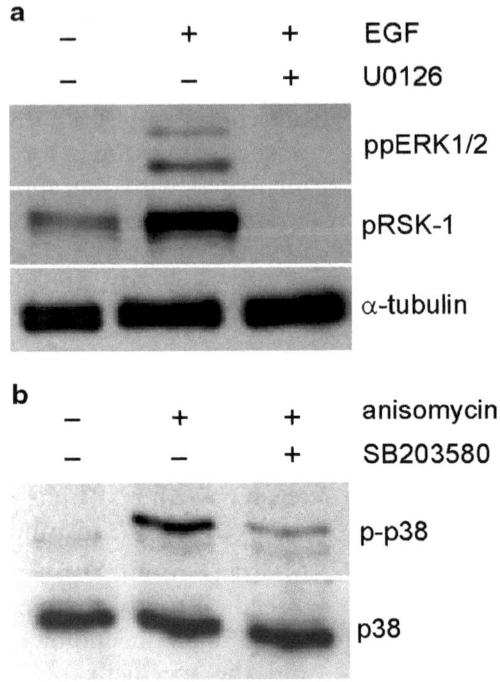

Fig. 1. (**a**) Immunoblots of active ERK1/2 (ppERK1/2) and phosphorylated p90Rsk-1 (pRsk-1) in HeLa cell lysates following treatment with EGF for 5 min in the absence or presence of the MEK1/2 inhibitor U0126. The expression of α-tubulin was used as a protein loading control. (**b**) Immunoblots of active p38 MAP kinase (p-p38) in HeLa cell lysates following treatment with anisomycin in the absence or presence of SB203580. Total p38 MAP kinase expression was used as a protein loading control.

4. The background for each sample on the autoradiography film must be taken into account. This is usually done by determining the average pixel intensity for region of interest that is of the same size as the region drawn around the protein of interest. The background region is usually in an open area of the autoradiography film just above or below the protein band of interest. The background value is subtracted from the value of the sample to get the net intensity of the protein of interest. However, given the variability of the size of the bands to quantify as indicated in Fig. 1, it is often advantageous to have the region of interest be of constant size for each of the samples. This can be accomplished by using a spot or dot blotter as described in the following section.

3.5. Quantification of Phosphorylation by Densitometry Using a Spot Blotter

1. If the primary antibody is specific for the target protein, spot/dot blot systems can be used. Described here is a protocol for the Minifold®I spot blot system (Schleicher & Schuell BioScience; Keene, NH), which allows for a more rapid estimate of protein expression as it eliminates the gel electrophoresis and protein transfer steps (see Note 8).

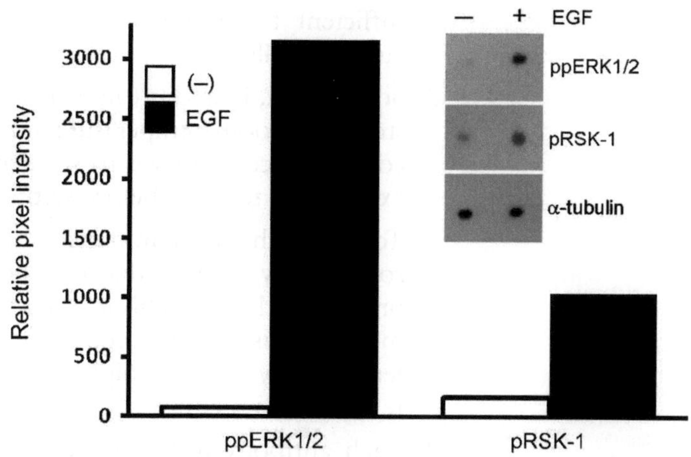

Fig. 2. Quantification of ERK1/2 activation and substrate phosphorylation by spot blotter. The untreated and EGF-treated samples from Fig. 1. were immunoblotted for active ERK1/2 (ppERK1/2), phosphorylated Rsk-1 (pRsk-1), and α-tubulin as a loading control using a slot blotter (*inset*). The graph shows an approximately 40-fold and 8-fold increase in active ERK1/2 and phosphorylated Rsk-1, respectively, when the spot blotter data was quantified by densitometry. These relative changes in protein phosphorylation correlates well with the immunoblots shown in Fig. 1.

2. The filter support plate is placed on the vacuum manifold. Two pieces of pre-wet blotting paper are then placed on the filter support plate. The PVDF membrane soaked in methanol, rinsed with TBS-tween, and placed directly on top of the blotting paper. The sample well plate is carefully placed on top of the membrane and the clamps are securely fastened.
3. Cell lysate samples are then applied to the membrane, taking care not to create air bubbles (see Note 9).
4. Applying a vacuum to the system will aspirate the samples onto the PVDF membrane in 3–5 min.
5. Once the samples have been aspirated through the PVDF membrane, the system is dismantled and the membrane is immunoblotted as described in Subheading 3.3.
6. The spot blotter can provide a more consistent size of signal for each protein of interest in the samples for quantification as described in Subheading 3.4. An example of the use of the slot blotter to quantify ERK1/2 pathway activation following stimulation with EGF is shown in Fig. 2.

4. Notes

1. The MEK1/2 inhibitors have been reported to inhibit the MEK5/ERK5 pathway at concentrations of greater than or equal to 10 μM, although concentrations less than 2 μM are

sufficient to inhibit MEK1/2 and ERK1/2 activation in cultured cells (27).

2. Solutions A, B, and C can be made up in 500 ml volumes and stored at room temperature until needed. Separate plastic containers can be used to soak blotting paper. Gently allow excess fluid to drain before setting up transfer.
3. To ensure that protein transfer is complete, it is important to roll out any bubbles that may be between the PVDF membrane and the gel. This is done twice, once when the PVDF membrane is placed on top of the gel and once after the transfer assembly has been flipped and placed on the electro-blotter transfer system.
4. Each antibody needs to be evaluated for specificity. Many antibodies only recognize the intended target protein, whereas other antibodies react nonspecifically with other proteins. If nonspecific interactions are suspected, controls that increase or decrease the expression of the target protein should be used to validate recognition by the antibody.
5. It is important that the immunoblots be exposed to the autoradiography film for various times so that the protein levels can be accurately quantified by densitometry scanning. Overexposure of autoradiography film can result in very dark bands corresponding to the protein of interest that may exceed the limits of detection for the densitometry scanner and misrepresent the data.
6. There are a number of gel documentation systems that can perform chemiluminescence analysis and substitute for autoradiography film. We have found that in cases where sensitivity is an issue, as is the case with some phosphorylated proteins; autoradiography film still offers an advantage for protein detection sensitivity. Each investigator will need to determine empirically whether their gel documentation system has the sensitivity to image and quantify protein phosphorylation.
7. It should be noted that while densitometry is a convenient way to quantify protein expression, it does have limitations. It is recommended that a densitometry standard curve be established for the specific gel documentation system being used to determine the linear range of sensitivity before quantifying protein expression. This can be done by performing serial dilutions of a known protein that has a highly specific antibody.
8. It is important to note that this method of quantification can only be applied to target proteins that are specifically recognized by their respective antibodies. Nonspecific antibodies interactions should be determined by gel electrophoresis and immunoblotting prior to using the slot blotter.

9. The amount of sample that will be added is dependent upon the number of cells or size of tissue and amount of lysis buffer used. If the number of cells in each condition is approximately equal, then serial dilutions of the positive and negative control samples can be done, if they are available. The dilution that gives the highest signal (positive control) to background (negative control) should be used for the other samples in the experiment. In our hands, the optimum signal to background ratio for several phosphorylation-specific antibodies correspond to ~1–2 μg of total protein loaded per sample. However, the optimal amount of protein loaded should be determined in each laboratory setting.

Acknowledgements

The authors would like to thank Kimberly Still for technical assistance. This work was supported by the National Institutes of Health (CA120215).

References

1. Lewis, T. S., Shapiro, P. S., and Ahn, N. G. (1998) Signal tranduction through MAP kinase cascades. *Adv Can Res 74*, 49–139.
2. McCubrey, J. A., Milella, M., Tafuri, A., Martelli, A. M., Lunghi, P., Bonati, A., Cervello, M., Lee, J. T., and Steelman, L. S. (2008) Targeting the Raf/MEK/ERK pathway with small-molecule inhibitors. *Curr Opin Investig Drugs 9*, 614–630.
3. Cohen, P. (2009) Targeting protein kinases for the development of anti-inflammatory drugs. *Curr Opin Cell Biol 21*, 317–324.
4. Friday, B. B., and Adjei, A. A. (2008) Advances in targeting the Ras/Raf/MEK/Erk mitogen-activated protein kinase cascade with MEK inhibitors for cancer therapy. *Clin Cancer Res 14*, 342–346.
5. Kalyn, R. (2007) Overview of targeted therapies in oncology. *J Oncol Pharm Pract 13*, 199–205.
6. von Ahsen, O., and Bomer, U. (2005) High-throughput screening for kinase inhibitors. *Chembiochem 6*, 481–490.
7. Wang, Z., Canagarajah, B. J., Boehm, J. C., Kassisa, S., Cobb, M. H., Young, P. R., Abdel-Meguid, S., Adams, J. L., and Goldsmith, E. J. (1998) Structural basis of inhibitor selectivity in MAP kinases. *Structure 6*, 1117–1128.
8. Pargellis, C., Tong, L., Churchill, L., Cirillo, P. F., Gilmore, T., Graham, A. G., Grob, P. M., Hickey, E. R., Moss, N., Pav, S., and Regan, J. (2002) Inhibition of p38 MAP kinase by utilizing a novel allosteric binding site. *Nat Struct Biol 9*, 268–272.
9. Regan, J., Breitfelder, S., Cirillo, P., Gilmore, T., Graham, A. G., Hickey, E., Klaus, B., Madwed, J., Moriak, M., Moss, N., Pargellis, C., Pav, S., Proto, A., Swinamer, A., Tong, L., and Torcellini, C. (2002) Pyrazole urea-based inhibitors of p38 MAP kinase: from lead compound to clinical candidate. *J Med Chem 45*, 2994–3008.
10. Hancock, C. N., Macias, A., Lee, E. K., Yu, S. Y., Mackerell, A. D., Jr., and Shapiro, P. (2005) Identification of novel extracellular signal-regulated kinase docking domain inhibitors. *J Med Chem 48*, 4586–4595.
11. Dudley, D. T., Pang, L., Decker, S. J., Bridges, A. J., and Saltiel, A. R. (1995) A synthetic inhibitor of the mitogen-activated protein kinase cascade. *Proc Natl Acad Sci U S A 92*, 7686–7689.
12. Favata, M. F., Horiuchi, K. Y., Manos, E. J., Daulerio, A. J., Stradley, D. A., Feeser, W. S., Van Dyk, D. E., Pitts, W. J., Earl, R. A., Hobbs, F., Copeland, R. A., Magolda, R. L., Scherle, P. A., and Trzaskos, J. M. (1998) Identification of a novel inhibitor of

mitogen-activated protein kinase kinase. *J Biol Chem 273*, 18623–18632.

13. Sebolt-Leopold, J. S., Dudley, D. T., Herrera, R., Van Becelaere, K., Wiland, A., Gowan, R. C., Tecle, H., Barrett, S. D., Bridges, A., Przybranowski, S., Leopold, W. R., and Saltiel, A. R. (1999) Blockade of the MAP kinase pathway suppresses growth of colon tumors in vivo. *Nat Med 5*, 810–816.
14. Kohno, M., and Pouyssegur, J. (2006) Targeting the ERK signaling pathway in cancer therapy. *Ann Med 38*, 200–211.
15. Cuenda, A., Rouse, J., Doza, Y. N., Meier, R., Cohen, P., Gallagher, T. F., Young, P. R., and Lee, J. C. (1995) SB 203580 is a specific inhibitor of a MAP kinase homologue which is stimulated by cellular stresses and interleukin-1. *FEBS Lett 364*, 229–233.
16. Lee, J. C., Laydon, J. T., McDonnell, P. C., Gallagher, T. F., Kumar, S., Green, D., McNulty, D., Blumenthal, M. J., Heys, J. R., Landvatter, S. W., and et al. (1994) A protein kinase involved in the regulation of inflammatory cytokine biosynthesis. *Nature 372*, 739–746.
17. Davies, S. P., Reddy, H., Caivano, M., and Cohen, P. (2000) Specificity and mechanism of action of some commonly used protein kinase inhibitors. *Biochem J 351*, 95–105.
18. Bennett, B. L., Sasaki, D. T., Murray, B. W., O'Leary, E. C., Sakata, S. T., Xu, W., Leisten, J. C., Motiwala, A., Pierce, S., Satoh, Y., Bhagwat, S. S., Manning, A. M., and Anderson, D. W. (2001) SP600125, an anthrapyrazolone inhibitor of Jun N-terminal kinase. *Proc Natl Acad Sci U S A 98*, 13681–13686.
19. Bain, J., McLauchlan, H., Elliott, M., and Cohen, P. (2003) The specificities of protein kinase inhibitors: an update. *Biochem J 371*, 199–204.
20. Tatake, R. J., O'Neill, M. M., Kennedy, C. A., Wayne, A. L., Jakes, S., Wu, D., Kugler, S. Z., Jr., Kashem, M. A., Kaplita, P., and Snow, R. J. (2008) Identification of pharmacological inhibitors of the MEK5/ERK5 pathway. *Biochem Biophys Res Commun 377*, 120–125.
21. Arora, A., and Scholar, E. M. (2005) Role of tyrosine kinase inhibitors in cancer therapy. *J Pharmacol Exp Ther 315(3)*, 971–979.
22. Mendelsohn, J., and Baselga, J. (2006) Epidermal growth factor receptor targeting in cancer. *Semin Oncol 33*, 369–385.
23. Hahn, S. M., Bernhard, E., and McKenna, W. G. (2001) Farnesyltransferase inhibitors. *Semin Oncol 28*, 86–93.
24. Mesa, R. A. (2006) Tipifarnib: farnesyl transferase inhibition at a crossroads. *Expert Rev Anticancer Ther 6*, 313–319.
25. Wong, S., and Witte, O. N. (2004) The BCR-ABL story: bench to bedside and back. *Annu Rev Immunol 22*, 247–306.
26. Bocchia, M., Forconi, F., and Lauria, F. (2006) Emerging drugs in chronic myelogenous leukaemia. *Expert Opin Emerg Drugs 11*, 651–664.
27. Mody, N., Leitch, J., Armstrong, C., Dixon, J., and Cohen, P. (2001) Effects of MAP kinase cascade inhibitors on the MKK5/ERK5 pathway. *FEBS Lett 502*, 21–24.

Part II

Study of MAP Kinase Cascades as Transmitters of Membranal Receptor Signals

Chapter 7

MAP Kinase Activation by Receptor Tyrosine Kinases: In Control of Cell Migration

Gabi Tarcic and Yosef Yarden

Abstract

A myriad of cellular processes instigated by growth factors are mediated by cell surface-associated receptor tyrosine kinases (RTKs). Subsequent downstream activation of signaling cascades, as well as their crosstalk, endows specificity in terms of the phenotypic outcome, e.g., cellular proliferation, migration, or differentiation. Such signaling diversity is exemplified by the ability of the epidermal growth factor receptor (EGFR) to stimulate different MAPK cascades, especially the ERK1/2 cascade. It has been shown that the ability of the ERK1/2 cascade to specify cell fate, such as cell migration, is dependent on signal duration governed by feedback control. Here we focus on one experimental system, MCF10A human mammary cells, and a phenotypic outcome of cell migration. We present methods to identify key components of underlying cascades and their effects on the migratory phenotype. We focus on profiling activation of signaling modules, as well as transcriptional regulation, emphasizing the high-throughput potential of such approaches.

Key words: MAP kinase, ERK1/2, EGFR, Flow cytometry, Real-time PCR, Cell migration

1. Introduction

Animal cells constantly exchange information with their tissue environment by means of signaling molecules (e.g., growth factors; GFs) and structural components (e.g., extracellular matrix; ECM). These molecules harbor essential information, which enables orchestration of key cellular functions leading to proliferation, differentiation or migration. One important class of environment-sensing molecules are receptor tyrosine kinases (RTKs) (1). RTKs are type I transmembrane proteins with an extracellular ligand-binding domain, a kinase domain, and multiple tyrosine phosphorylation sites with regulatory functions. Upon ligand binding, receptors dimerize, thus activating their kinase domains and forming a signaling complex with auto- and trans-phosphorylation capabilities, thereby allowing recruitment of Src homology 2 (SH2)

Rony Seger (ed.), *MAP Kinase Signaling Protocols: Second Edition*, Methods in Molecular Biology, vol. 661,
DOI 10.1007/978-1-60761-795-2_7, © Springer Science+Business Media, LLC 2010

and phosphotyrosine-binding (PTB) containing signaling adaptors (2). One of the most extensively studied subfamily of RTKs is the ErbB group, composed of four family members, each with its own binding partners and signaling capabilities (3, 4).

The prototype of the ErbB family is the epidermal growth factor receptor (EGFR or ErbB-1). EGFR binds not only with EGF, but also with six more ligand growth factors. Upon ligand binding, EGFR is capable of recruiting effector molecules of a variety of signaling cascades, including the mitogen-activated protein kinase cascades (MAPK), phosphatidylinositol 3 kinase (PI3K), signal transducers and activators of transcription (STAT) and phospholipase Cγ (PLCγ) pathways (reviewed in (5)). The activation of these pathways defines the nature of the cellular response, as well as permits signal amplification. One striking feature of this configuration is the ability of a canonical linear pathway to generate several distinct cellular outcomes. A well-defined system exemplifying the ability of a linear cascade to generate multiple phenotypes is the proliferation vs. neurite outgrowth example, first exemplified using rat adrenal pheochromocytoma (PC-12) cells (6). In this system, proliferation can be induced by EGF or insulin, while neurite outgrowth is induced by the nerve growth factor (NGF), both utilizing the ERK1/2 cascade. Among other parameters, outcome specificity is encoded by the duration of ERK1/2 phosphorylation: transient activation leads to cell proliferation, whereas sustained activation results in neuronal differentiation (6). It was later shown that the topology of the MAPK network enables this dichotomy; only the transient mode of activation can induce negative feedback (7). It has long been realized that equally important in shaping the cellular outcome of GF stimulation is the transcriptional response elicited downstream to MAPK and other pathways (8). Recently it has been shown that the transcriptional activation of a module of negative feedback regulators is able to attenuate growth factor signaling, consequently drive robust cellular outcome (9).

In this chapter, we present several methods to explore the cascaded layers of signal propagation. Potentially, this enables reconstruction of a signaling pathway stemming from GF stimulation, through cytoplasmic signaling pathways, nuclear transcription, and eventually functional output. The methods described can serve as a basis for high-throughput screening strategies of a system of choice, as well as unraveling the regulatory hubs and novel feedback loops.

2. Materials

2.1. Cell Culture and Stimulation

1. Dulbecco's Modified Eagle's Medium (DMEM) supplemented with 10% fetal bovine serum (FBS, Gibco/BRL, Bethesda, MD), 1 mM sodium pyruvate (Biological Industries,

Beit Haemek, Israel) for HeLa cells. DMEM/F-12(HAM) 1:1 supplemented with 10 μg/ml insulin (Biological industries), 5% horse serum (Gibco), 1 μg/ml cholera toxin, 1 μg/ml hydrocortisone, and 10 ng/ml EGF (Sigma, St. Louis, MO) for MCF10A cells (see Note 1).

2. Epidermal growth factor (EGF, Sigma) dissolved at 100 ng/ml in PBS and stored at −20°C. MEK1/2 inhibitor U0126 dissolved at 5 mM in dimethylsulfoxide (DMSO) and stored in single use aliquots at −20°C.
3. Solution of trypsin (0.25%) and ethylenediamine tetraacetic acid (EDTA) (1 mM) from Biological Industries.

2.2. Intracellular Flow Cytometry

1. 3% paraformaldehyde (PFA) dissolved in PBS, stored at −20°C.
2. 100% methanol stored at −20°C.
3. Staining medium: 0.5% bovine serum albumin (BSA), 0.02 NaN_3 in PBS, stored at 4°C.
4. Primary antibodies: anti-doubly phosphorylated ERK1/2 conjugated to Alexa-488 (Molecular Probes; Leiden, The Netherlands), anti-phosphorylated p38 antibody conjugated to Alexa-647, anti human HLA-A,B,C conjugated to PE (BD, Franklin Lakes, NJ), anti-EGFR (clone 111.6, Thermo-Scientific, UK).
5. Secondary antibody: FITC-conjugated goat-anti-mouse IgG (FITC-GαM, Jackson ImmunoResearch Laboratories, West Grove, PA).

2.3. RNA Extraction and cDNA Synthesis and qPCR

1. RNA extraction: PerfectPure RNA Cultured Cell Kit (5 Prime, Gaithersburg, MD).
2. cDNA synthesis: High Capacity cDNA Reverse Transcription Kit (Applied Biosystems, Foster City, CA).
3. Real-time PCR: Power SYBR Green PCR Master Mix (Applied Biosystems).

2.4. Transwell Migration Assay

1. Transwell permeable support with 8 μm polycarbonate membrane insert (Corning Incorporated, Corning, NY).
2. Lysis solution: 0.5% Triton X-100 in PBS.
3. Staining solution: 0.3% methyl-violet dissolved in PBS and stored at 22°C.

3. Methods

Multiple signaling cascades are simultaneously activated upon RTK stimulation to achieve correct signal transduction, leading to various cellular responses. The probing of these activation events, mainly exemplified by protein phosphorylation, may be assayed using various techniques, each with its own advantages and drawbacks. There are several main parameters that need to be controlled to attain reliable results. First, each sample analyzed might contain several subpopulations of cells that differ in the activation state of different signaling pathways. This phenomenon becomes more important when dealing with cells that have undergone treatments, such as ectopic expression or silencing of specific genes. Second, biological samples sometimes contain very small number of cells, inappropriate for certain biochemical analyses. Third, simultaneous measurements of several protein active states in the same sample are advantageous, as it allows in-depth analysis of the regulation of signaling cascades. One such technique is intracellular flow cytometry developed by Garry Nolan's lab at Stanford (10, 11). This technique allows simultaneous measurements of multiple protein states at a single cell resolution, offering useful tools for reconstructing signaling networks (12, 13).

After activation of cytoplasmic signaling pathways, the signal is relayed into the nucleus where it is translated into a transcriptional response (9). The identification of such transcriptional events is enabled by using high-throughput platforms, such as DNA microarrays. After initial characterization of specific genes that underwent alterations, it is possible to quantify and verify the events (up- or downregulation) using quantitative real-time PCR (RT-PCR). This method offers both precise determination of transcript levels and a confirmation of transcript identity. The design of the primer pair for each gene is a key step to accomplishing efficient and specific PCR reactions. In general, the primers are designed such that they all contain similar properties (with emphasis on annealing temperature and product length), offering a uniform PCR protocol.

Ultimately, the importance of the different signaling event needs to be tested in respect to the phenotypic changes they evoke. The involvement of RTK stimulation, via MAPKs, in cell motility is well established (5), and one such cell system is EGF stimulation of MCF10A cells (14). MCF10A, an immortalized mammary epithelial cell line, was shown to migrate in response to activation of various RTKs such as the insulin-like growth factor-I receptor (IGF-IR) (14), or the EGFR (9). While the process of cell migration is executed by an intricate machinery involving many regulatory and mechanical steps, the final endpoint of cell motility can easily be measured. One of the widely used assays is the Transwell chamber migration assay in which cells are placed on the top part of a polycarbonate membrane. If the cells are motile they can actively move through the pores to the bottom part of membrane. This simple assay is suitable for many cell types and for small numbers of cells. Further, it can also serve as a chemotactic assay if different media are placed at the two sides of the membrane.

3.1. Preparation of Samples for Intracellular Flow Cytometry Assaying Active MAPK and EGFR Levels (See Fig. 1)

1. HeLa cells are grown in 10-cm tissue culture dishes to 80% confluence. At this point, cells are treated with trypsin and plated onto 6-well plates (0.1×10^6 per well) in 2 ml growth medium. Twenty-four hours later the growth medium is replaced with 2 ml DMEM without FBS (starvation medium) for a starvation period of 16 h (see Note 2).
2. On next day: change the starvation medium to fresh starvation medium, thaw 3% PFA and prepare fresh staining medium.
3. Prepare stimulation medium: dilute EGF in starvation medium (2 ml/sample; preheated to 37°C) to 20 ng/ml.

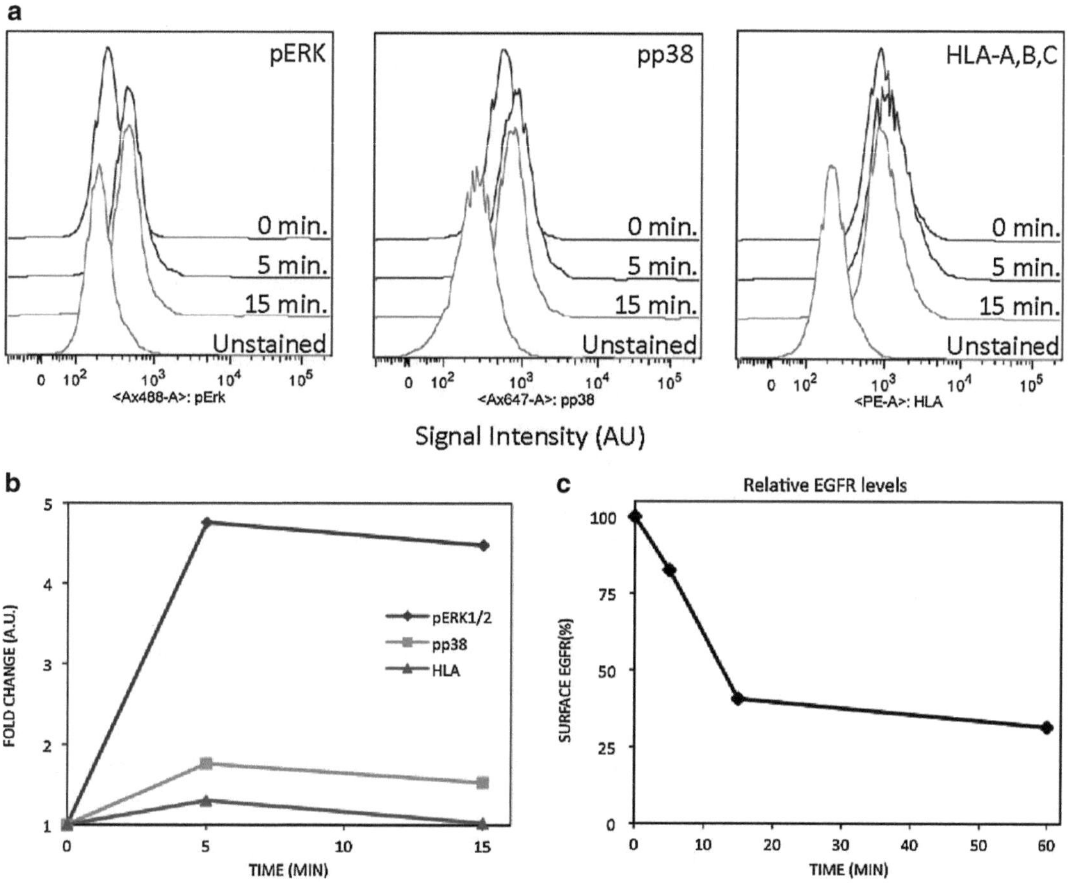

Fig. 1. Time-dependant activation of MAPKs and internalization of EGFR upon stimulation of HeLa cells with EGF. (**a**) HeLa cells were serum starved for 16 h, then stimulated with EGF (20 ng/ml) for the indicated time intervals. Cells were then fixed, permeabilized and stained with a fluorescently labeled Alexa-488-pERK1/2 antibody, Alexa-647-p-p38 antibody or PE-HLA-A,B,C antibody (control). Fluorescence intensity was measured using a flow cytometer. (**b**) Quantification of the results presented in (**a**). Activation of EGFR is followed by strong activation of the ERK1/2 cascade, as reflected by ERK1/2 phosphorylation, and relatively weak activation of p38. The activation of ERK is rapid, yet transient; peaking at 5 min, then slowly decreasing. (**c**) HeLa cells were serum starved for 16 h, then stimulated with EGF (20 ng/ml) for the indicated time intervals. Thereafter, cells were surface labeled with antibodies to EGFR. The remaining surface fraction of EGFR was quantified by flow cytometry and plotted as a function of time of incubation with EGF.

4. The cells are treated with the ligands according to the experimental setup. For each primary antibody used, an unstained sample should be included as a background sample for each secondary antibody.
5. Immediately after the appropriate stimulation, cells are placed on ice, washed once with 1 ml ice cold PBS, and incubated for 5 min, on ice, in 1 ml trypsin solution (see Note 3).
6. For each sample prepare a labeled tube containing 500 μl of 3% PFA (see Note 4).
7. Collect suspended cells to the labeled tubes. Incubate 10 min at room temperature (RT).
8. Pellet the cells by centrifugation (5 min, 500 *g*), then aspirate the medium.
9. Place cells on ice, then resuspend by vigorous vortexing, and add 1 ml 100% methanol drop wise, to avoid clumping. Incubate 10 min at 4°C (see Note 5).
10. Add 3 ml staining medium, let stand for 2–3 min for proper rehydration. Pellet the cells by centrifugation (as aforementioned) and aspirate the medium.
11. Wash once with 3 ml staining medium, centrifuge and aspirate.
12. Resuspend samples in 100 μl staining medium and remove an aliquot (100 μl) to a new tube.
13. Add primary antibody: anti-pERK-488, anti-p-p38-647, anti-HLA-A,B,C-PE (1:100), or anti-EGFR (1:100), then incubate for 30 min at RT
14. Wash with 3 ml staining medium; centrifuge, aspirate the medium and resuspend in 100 μl staining medium.
15. Remove an aliquot (100 μl) to a new tube. Add secondary antibody to the anti-EGFR sample: FITC-GαM (1:5,000), incubate for 30 min at RT
16. Wash with 3 ml staining medium, centrifuge and resuspend in 300 μl staining medium. Samples are ready to be analyzed on a FACS instrument.

3.2. Quantitative Real-Time PCR for Validation of Gene Expression Events Regulated upon EGFR Activation (See Fig. 2)

1. HeLa cells are plated and treated as described above (see Note 6).
2. In the morning of the experiment, change the medium to fresh starvation medium and prepare stimulation medium: dilute EGF to 20 ng/ml and U0126 to 5 μM.
3. Treat the cells according to the experimental design. Cool the cells on ice following the appropriate time intervals.
4. Wash once with ice cold PBS and add 500 μl lysis buffer, supplied with the RNA extraction kit.

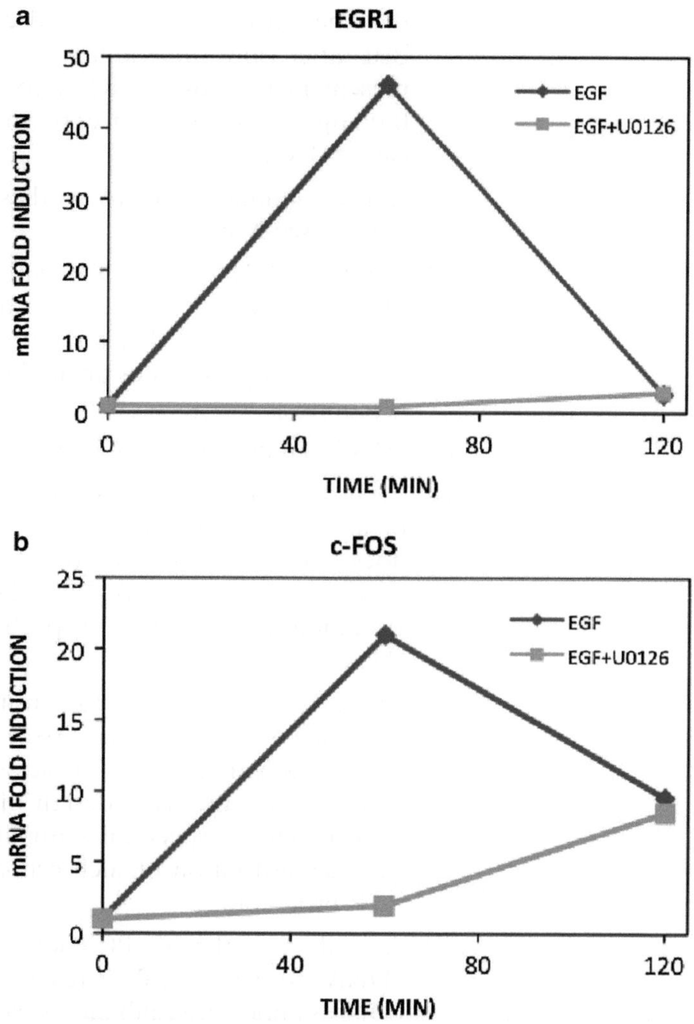

Fig. 2. The transcriptional response induced by EGF depends on MAPK activation. MCF10A (0.1×10^6) cells were seeded in six-well plates. Following a 24-h recovery period, cells were serum starved for 16 h, then stimulated either with EGF alone (10 ng/ml) or with EGF plus a MEK1/2 inhibitor (U0126; 5 μm) for the indicated time intervals. Thereafter, the cells were lysed, total RNA was extracted, and 1 μg was used for cDNA synthesis. (**a**) EGR1 mRNA levels were measured using the synthesized cDNA and real-time PCR, in both EGF- and EGF+U0126-stimulated cells. (**b**) c-FOS mRNA levels were measured, essentially as described in (**a**). As can be seen, both transcription factors (i.e., c-FOS and EGR1), are induced after EGF stimulation and the mRNA levels decrease after 2 h. This induction is mediated by ERK1/2 activation as inhibition of MEK1/2, its direct upstream activator, abolished induction of these genes.

5. Proceed with the RNA purification protocol according to the manufacturer's instructions (see Note 7).
6. After completing the protocol, measure RNA concentration. Make sure that the ratio between 280 and 260 nm is ~2 and

that the ratio between 260 and 230 nm exceeds 1.8. Values that differ from these limits suggest that a contamination is present in the sample. RNA concentration should be over 0.05 μg/μl, as lower concentrations might interfere with efficient cDNA synthesis.

7. Either continue to cDNA synthesis directly or store RNA at −80°C (see Note 8).
8. Use 1 μg of RNA for cDNA synthesis, according to the manufacturer's protocol. Although smaller amounts of RNA may be used, we found that for most purposes 1 μg of RNA is suitable and provides optimal cDNA concentration for PCR reactions.
9. cDNA can be stored at −20°C for long periods of time.
10. Each reaction of RT-PCR should include, besides the genes of interest, a housekeeping gene whose expression pattern does not change between the different experimental conditions. Common genes used are β2-microglobulin or GAPDH. In addition, for each gene probed a non-template control (NTC) is added.
11. Primer design is usually done using online software (such as the Universal ProbeLibrary Assay Design Center provided by Roche Applied Science). The advantage of such software programs is the automatic design of primers to identify exon–exon junctions, specifically amplifying mRNA. Additionally, the standardization of such primers allows usage of a single PCR protocol.
12. According to the RT-PCR kit used, each reaction contains cDNA, primers, and PCR reaction components. Amounts of cDNA samples are calibrated such that the threshold cycle of the control housekeeping gene is between 15 and 20. This allows high sensitivity range for the identification of most transcripts. Primer concentration is 0.5 nM per primer per reaction.

3.3. Transwell Migration Assay to Assess the Contribution of MAPK Activity to EGF-Induced Cell Migration (See Fig. 3)

1. Two hours prior to the beginning of the experiment, incubate an appropriate number of Transwell chambers, depending on the experimental setup, with MCF10A medium without EGF. Add 600 μl of medium to the bottom part of the well and 100 μl to the top part.
2. Before adding cells to the Transwell chambers, aspirate the medium and add the treatment medium. Each experiment needs to include an EGF-free medium as a negative control and an EGF-containing medium as a positive control.
3. MCF10A cells are grown to 70–80% confluence and then transferred onto the preincubated Transwell chambers. Usually, seeding 0.6×10^5 cells per well results in adequate cell

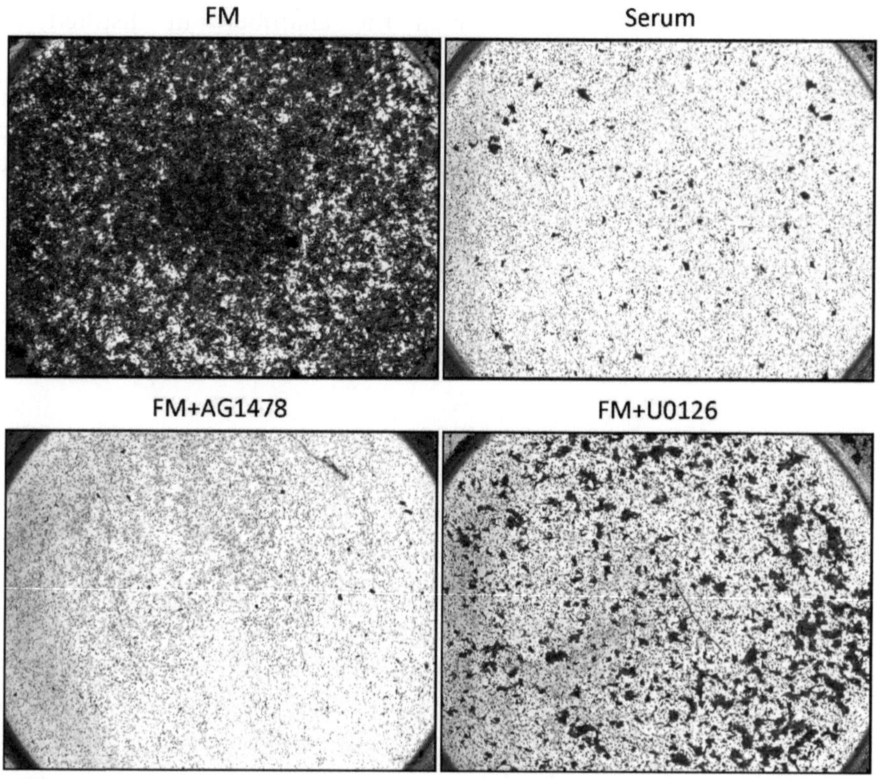

Fig. 3. EGF induces migration of MCF10A cells. MCF10A (0.6×10^5) cells were seeded in Transwell chambers and grown for 24 h in medium containing either serum with EGF (*FM*), same medium without EGF (*Serum*), or with either an EGFR kinase inhibitor (AG1478) or a MEK1/2 kinase inhibitor (U0126). Thereafter, the cells were washed, fixed, permeabilized, and incubated with a crystal violet dye. Photos were taken using a light-microscope connected to a CCD-camera (bar = 500 μm). Evidently, EGF induces robust cell migration, which is dependent on EGFR activation and signaling through the ERK1/2 cascade.

densities. Make sure to avoid adding more that 100 μl of medium-containing cells as this might cause the medium to spill over the sides of the Transwell to its bottom part. Incubate the cells for 16 h in the Transwell chamber (see Note 9).

4. On next day: Thaw 3% PFA and prepare fresh lysis solution.
5. Aliquot 1 ml of PFA, lysis solution or staining solution per sample in a 24-well plate.
6. Wash Transwell chambers in PBS (three times) by lifting each chamber and gently immersing in a vessel containing PBS.
7. Incubate chambers for 15 min at RT in PFA.
8. Wash chambers in PBS three times, as described.
9. Incubate the chambers in lysis solution (15 min at RT).
10. Wash the chambers in distilled water, as described.
11. Incubate the chambers in staining solution (5 min at RT).

12. Thoroughly wash the chambers in distilled water as described.
13. Using a cotton swab, gently scrape the cells on the upper part of the chamber. It is important to make sure all the cells are removed from the top part of the chamber, as these cells are the nonmigrating cells.
14. Photograph the Transwell chambers.

4. Notes

1. When preparing growth medium for MCF10A cells, add all components to the medium, except for EGF. The medium is then filtered through a 0.45-μm filter, and EGF added.
2. The numbers of cells indicated here are the minimal numbers necessary for analysis; higher cell numbers can be used.
3. This step is applicable only for adherent cells. If using nonadherent cells skip this step.
4. It is important to use polystyrene tubes in this experiment as fixed cells adhere to polypropylene tubes.
5. If conducting surface staining of cells, such as staining for EGFR, do not perform this step, as permeabilizing the cells is not necessary.
6. The number of cells may vary according to the purpose of the experiment and cell type. Usually smaller numbers of cells also yield adequate amounts of RNA, enough for cDNA synthesis.
7. There are numerous RNA purification kits that can be used. In this protocol we use a kit supplied by 5 Prime (Gaithersburg, MD); however, other kits able to produce high enough amounts of RNA and with high purity, can be used.
8. Given that RNA is liable to degradation, it is best to proceed immediately with cDNA synthesis. Alternatively, RNA samples may be stored for several months at −80°C.
9. To make sure that an excess volume of medium is not added to the top part of the Transwell, count the cells and adjust to roughly 0.6×10^6 per ml. Moreover, when comparing different cell lines or cells undergoing different treatments, plate in a separate 24-well plate the same amount of cells as that added to the Transwell. This will enable quantification of the number of cells in each Transwell chamber, and allow proper comparison of migrating cells.

Acknowledgments

Our laboratory is supported by research grants from the National Cancer Institute (grant CA72981), the M.D. Moross Institute for Cancer Research and the Willner Family Center for Vascular Biology. Y.Y. is the incumbent of the Harold and Zelda Goldenberg Professorial Chair.

References

1. Hunter T (2000) Signaling–2000 and beyond. *Cell* 100(1):113–127.
2. Pawson T (2004) Specificity in signal transduction: from phosphotyrosine-SH2 domain interactions to complex cellular systems. *Cell* 116(2):191–203.
3. Citri A & Yarden Y (2006) EGF-ERBB signalling: towards the systems level. *Nat Rev Mol Cell Biol* 7(7):505–516.
4. Jones RB, Gordus A, Krall JA, & MacBeath G (2006) A quantitative protein interaction network for the ErbB receptors using protein microarrays. *Nature* 439(7073):168–174.
5. Katz M, Amit I, & Yarden Y (2007) Regulation of MAPKs by growth factors and receptor tyrosine kinases. *Biochim Biophys Acta* 1773(8):1161–1176.
6. Marshall CJ (1995) Specificity of receptor tyrosine kinase signaling: transient versus sustained extracellular signal-regulated kinase activation. *Cell* 80(2):179–185.
7. Santos SD, Verveer PJ, & Bastiaens PI (2007) Growth factor-induced MAPK network topology shapes Erk response determining PC-12 cell fate. *Nat Cell Biol* 9(3):324–330.
8. Treisman R (1996) Regulation of transcription by MAP kinase cascades. *Curr Opin Cell Biol* 8(2):205–215.
9. Amit I, *et al.* (2007) A module of negative feedback regulators defines growth factor signaling. *Nat Genet* 39(4):503–512.
10. Perez OD & Nolan GP (2002) Simultaneous measurement of multiple active kinase states using polychromatic flow cytometry. *Nat Biotechnol* 20(2):155–162.
11. Krutzik PO, Irish JM, Nolan GP, & Perez OD (2004) Analysis of protein phosphorylation and cellular signaling events by flow cytometry: techniques and clinical applications. *Clin Immunol* 110(3):206–221.
12. Irish JM, *et al.* (2004) Single cell profiling of potentiated phospho-protein networks in cancer cells. *Cell* 118(2):217–228.
13. Sachs K, Perez O, Pe'er D, Lauffenburger DA, & Nolan GP (2005) Causal protein-signaling networks derived from multiparameter single-cell data. *Science* 308(5721):523–529.
14. Irie HY, *et al.* (2005) Distinct roles of Akt1 and Akt2 in regulating cell migration and epithelial-mesenchymal transition. *J Cell Biol* 171(6):1023–1034.

Chapter 8

Activation of Ras and Rho GTPases and MAP Kinases by G-Protein-Coupled Receptors

Mario Chiariello, Jose P. Vaqué, Piero Crespo, and J. Silvio Gutkind

Abstract

A complex intracellular signaling network mediates the multiple biological activities of G-protein-coupled receptors (GPCRs). Among them, monomeric GTPases and a family of closely related proline-targeted serine–threonine kinases, collectively known as Mitogen-Activated Protein Kinases (MAPKs), appears to play central roles in orchestrating the proliferative responses to multiple mitogens that act on GPCRs. Upon GDP/GTP exchange, monomeric GTPases control the phosphorylation of conserved threonine and tyrosine residues in MAPKs by their immediate upstream kinases, increasing their enzymatic activity and inducing their translocation to the nucleus where they phosphorylate transcription factors, thereby regulating the expression of genes playing a key role in normal and aberrant cell growth. Recently, a number of GPCRs have been engineered to provide exclusive activation by synthetic drug-like compounds while becoming insensitive to endogenous ligands. These engineered receptors, named Receptors Activated Solely by Synthetic Ligands (RASSLs), promise better understanding of GPCRs signaling in vitro and in vivo, thus representing ideal tools to selectively modulate MAPK signaling routes controlling a wide range of biological functions, from proliferation to differentiation, migration, invasion, and cell survival or death by apoptosis.

Key words: Kinase assays, Western blot, Phosphorylation, Synthetic ligands, RASSLs, ERKs, GTP, Ras

1. Introduction

G-protein-coupled receptors (GPCRs) represent by far the largest family of proteins involved in signal transmission accounting for more than 2% of the total genes encoded by the human genome. These receptors control key physiological functions, including neurotransmission, hormone, and enzyme release from endocrine and exocrine glands, immune responses, cardiac- and smooth-muscle contraction, and blood pressure regulation, to name but a few. Consequently, their dysfunctions contributes

Rony Seger (ed.), *MAP Kinase Signaling Protocols: Second Edition*, Methods in Molecular Biology, vol. 661,
DOI 10.1007/978-1-60761-795-2_8, © Springer Science+Business Media, LLC 2010

to the pathogenesis of a large number of human diseases, as reflected by the fact that GPCRs represent the target of more than 50% of the current therapeutic agents on the market.

GPCRs are coupled to heterotrimeric GTPases that consist of Gα, Gβ, and Gγ subunits that in an inactive state are bound to GDP through the Gα subunit (1). Upon ligand binding, a conformational change in the receptor provokes Gα to release GDP and incorporate GTP. As a consequence, GTP-bound Gα and the Gβγ subunits dissociate, and Gα-GTP and free Gβγ initiate the activation of a multitude of effector molecules. GPCRs are best known by their ability to activate or inhibit the production of a variety of second messengers such as cAMP, cGMP, diacylglycerol, IP_3, PIP_3, arachidonic, and phosphatidic acid, and promoting (Ca^{2+}) elevation and the opening or closing of a variety of ion channels (2, 3). However, cellular responses mediated by GPCRs do not involve the sole stimulation of conventional second messenger-generating systems, but also result from the functional integration of an intricate network of intracellular signaling pathways. Among others, GPCRs also induce the activation of monomeric GTPases of the Ras superfamily which, in turn, stimulate the activation of several members of a family of closely related proline-targeted serine–threonine kinases, collectively known as Mitogen-Activated Protein Kinases (MAPKs) (4). Indeed, both small GTPases such as Ras, RhoA, Rac1, and Cdc42 and MAPKs such as ERK1/2, JNKs, p38α (HOG1), p38γ (ERK6), p38δ (SAPK4), and ERK5 are all activated by GPCRs and control their proliferative effects (Fig. 1).

Following a very complex and not fully elucidated series of events originating from stimulation of GPCRs by appropriate ligands (3), GPCRs activate small GTPases of the Ras and Rho family, which ultimately lead to the phosphorylation and activation of MAPKs on conserved tyrosine and threonine residues by their immediately upstream MAPK kinases (or MEKs) (5). Active MAPKs then can phosphorylate cytosolic targets and translocate to the nucleus. Here, MAPKs phosphorylate transcription factors, thereby regulating the expression of genes that play a key role in normal and aberrant cell growth (5). Thus, efforts to understand the basic molecular processes by which GPCRs regulate the enzymatic activity of small GTPases and MAPKs have yielded fundamental clues about the biochemical routes used by these receptors to exert a wide range of biological functions.

1.1. Use of Receptors Activated by Synthetic Ligands to Build GPCR-Regulated Signaling Networks

Despite the variety and importance of the many GPCRs physiological and pathological functions, the study of these proteins in vivo is often hampered by their expression in multiple tissues. For example, the use of exogenous ligands such as neurotransmitters, lipid mediators, vasoactive peptides, chemokines, vasoconstrictors, and relaxants, etc., would activate not only the receptor

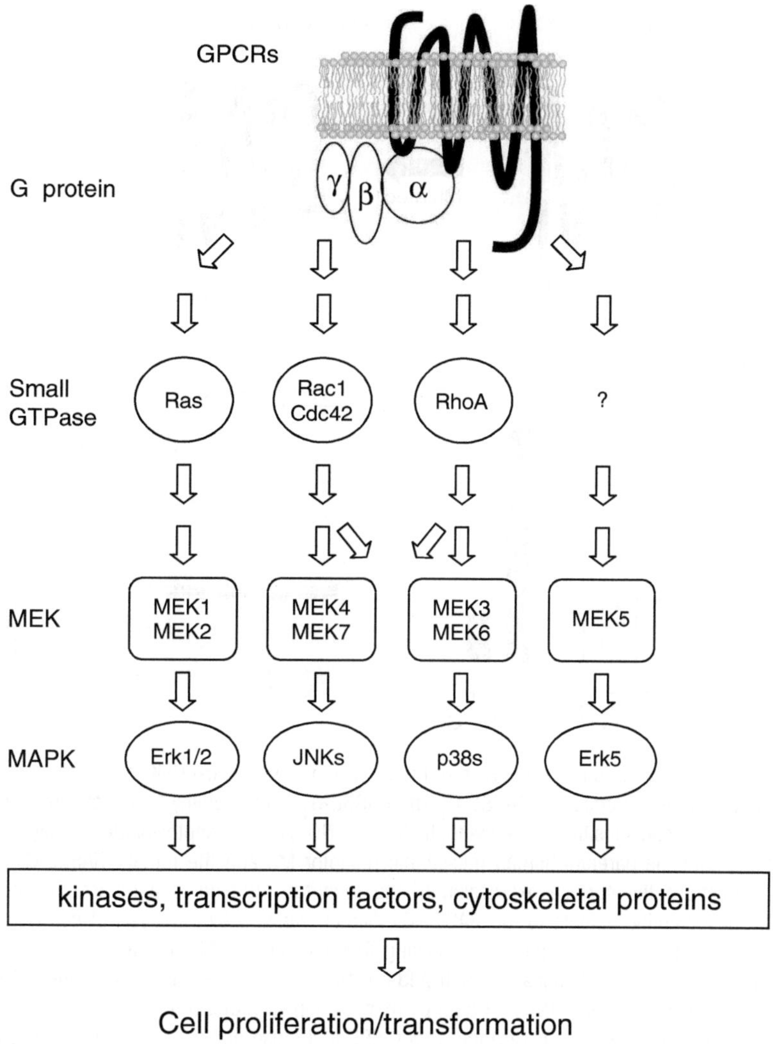

Fig. 1. Schematic representation of the classical three-modular organization of linear MAPK cascades in which receptor-activated small GTPases stimulate a MAPK kinase kinase (omitted in the figure) followed by MAP kinase kinase (MEK) activation, which subsequently activates a MAPK, resulting in phosphorylation and activation of downstream targets.

of interest in the tissue or cell type under investigation, but it would likely elicit a variety of responses that may compromise the ability to interpret the emerging results. This may apply to both, endogenously expressed GPCRs as well as those expressed in a tissue-specific fashion by transgenic approaches. In this regard, an engineered family of GPCRs derived from the human muscarinic acetylcholine receptors has been developed in yeast by random mutagenesis (6). These mutant molecules have lost the ability to be activated by their endogenous ligand (acetylcholine), but gained the ability to be activated by an inert small molecule such

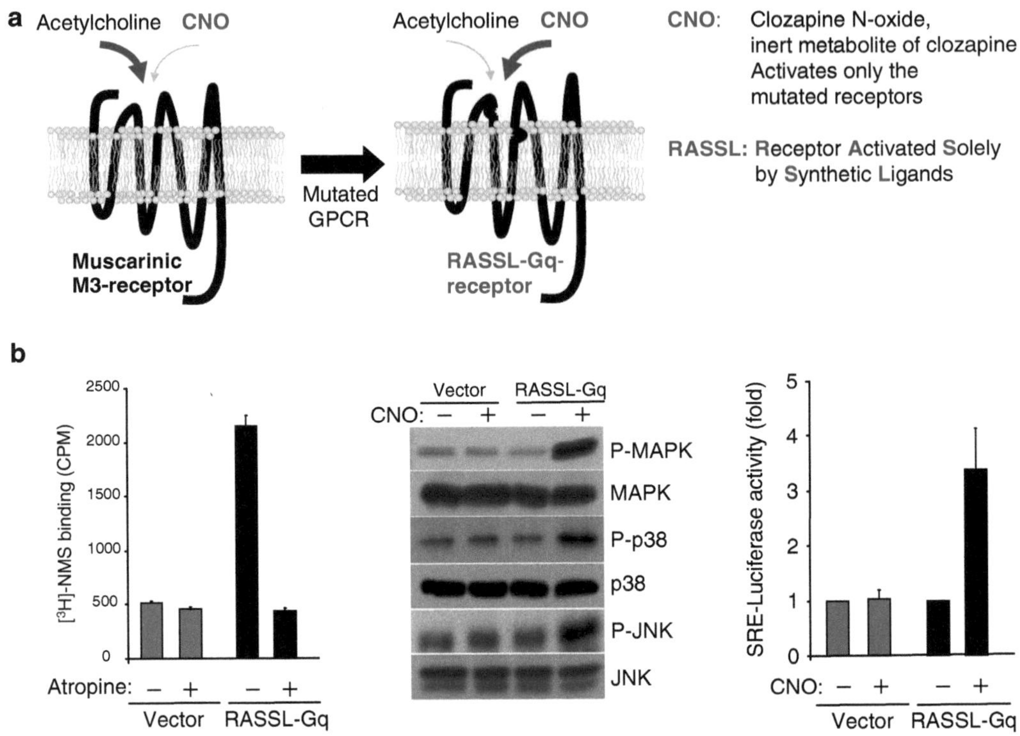

Fig. 2. (**a**) Schematic representation of the activity of RASSLs. (**b**) HEK-293T cells were transiently transfected with an expression vector encoding RASSL-G_q. The RASSL-G_q lost the ability to be activated by its natural ligand, acetylcholine, but it is instead sensitive to the synthetic ligand CNO. *Left*, a receptor-binding assay using (^{3}H)-scopolamine (NMS) that binds both the parental human muscarinic receptor M3 and the mutant RASSL-G_q was performed in these cells to demonstrate the surface expression of the receptor. Atropine is a muscarinic receptor antagonist that prevents (^{3}H)-scopolamine binding to the RASSL-G_q. Center, western blot analysis of cell lysates using anti-P-MAPK antibody (T202/Y204 rabbit polyclonal, Cell Signaling Technology), anti-MAPK (ERK2) (C14 Santa Cruz), anti-P-p38 (T180/Y182 rabbit monoclonal, Cell Signaling), anti-p38 (rabbit polyclonal, Cell Signaling), anti-P-JNK (T183/Y185 rabbit monoclonal, Cell Signaling), and anti-JNK (56G8 rabbit monoclonal), in vector or RASSL-G_q transfected cells stimulated (+) or not (–) with CNO (100 nM) was performed. *Right*, SRE-luciferase activity elicited by RASSL activation after addition of CNO (100 nM) is shown. Cells were co-transfected with an SRE-driven firefly luciferase reporter and a constitutively expressed Renilla luciferase vector. Luciferase activities were measured in a luminometer as reported (8), and firefly luciferase values were normalized based on renilla luciferase values. Data are represented as fold increase with respect to nonstimulated vector-transfected controls. SRE-luciferase activity elicited by the activation of the RASSL after addition of CNO (100 nM). Error Bar: SEM ($n=3$).

as clozapine-N-oxide (CNO) (Fig. 2). This family of mutant receptors, known as RASSLs for Receptors Activated Solely by Synthetic Ligands, can signal through different heterotrimeric G-proteins (reviewed in ref. (7)). Among the receptors, three main RASSLs have been developed: H-M3D or RASSL-G_q that signals through $G\alpha_q$, H-M4D or RASSL-G_i, and a chimeric receptor termed RASSL-G_s. In Fig. 2, we illustrate a scheme of the use of RASSL to modulate downstream signaling pathways, using RASSL-G_q as an example. The CNO-induced activation of MAPKs (ERK, p38, and JNK), as judged by western blotting with specific anti-phospho MAPKs (see below) and transcription from the serum-response element (SRE) by the use of a reporter luciferase assay, are shown in HEK-293T cells transiently transfected with RASSL-G_q. The use of RASSLs to investigate the temporal activation of MAPKs and their regulated transcriptional and signaling networks in the tissues in which GPCRs elicit their numerous biological functions may soon shed key information regarding the contribution of MAPKs to the physiological and pathological roles of GPCRs.

In this chapter, we describe protocols to study, upon stimulation of different GPCRs, the activation of endogenous small GTPases and to assess the activity of endogenously or ectopically expressed MAPKs. In the context of MAPK activity, in vitro kinase assays still represent the most specific and sensitive approach to evaluate the activity of both endogenous and transfected MAPK family members. The established correspondence between phosphorylation of conserved Thr-X-Tyr motifs on MAPKs and their enzymatic activity led to the development of specific anti-phospho MAPK antibodies as valuable tools to score MAPKs activation by GPCRs and other upstream stimuli. Thanks to an always-improving sensitivity, western blots using these antibodies are able to integrate or substitute the in vitro kinase assays in most applications, eliminating the need for ^{32}P radioactive labeling and allowing analysis of limited amounts of samples that are not suitable for immunoprecipitation-based techniques.

2. Materials

2.1. Assaying the Activation of Endogenous Small GTPases by Affinity Precipitation of Their GTP-Bound Forms

1. Examples of antibodies against endogenous GTPases. Anti-Rac1, R56220, BD Transduction Laboratories; anti-RhoA, sc-26C4, Santa Cruz Biotechnology; anti-Pan-Ras, Oncogene Science; anti Cdc42, #2462, Cell Signaling Technology.
2. Affinity precipitation (AP) lysis buffer: 20 mM HEPES, pH 7.5, 1% Triton X-100, 100 mM NaCl, 20 mM $MgCl_2$ (see Note 1), 10 mM EGTA, pH 8.0, 1 mM Dithiothreitol (DTT, add immediately before use), 1 mM sodium orthovanadate (Na_3VO_4, prepare fresh and add immediately before use), 40 mM β-glycerophosphate (prepare fresh and add immediately before use), 10 μg/ml aprotinin/leupeptin (add immediately before use), 1 mM phenylmethylsulfonyl fluoride (PMSF, add immediately before use from a 100 mM stock solution in ethanol). Store at 4°C.
3. Protein-loading buffer: 2% (w/v) SDS, 50 mM Tris–HCl, pH 6.8, 0.2 mg/ml bromophenol blue, 0.1 M DTT; 50% glycerol.

4. PVDF transfer membrane: Immobilon™-P (Millipore) or similar.
5. Tris-Tween-Buffered Saline (TTBS): 50 mM Tris-HCl, pH 7.4, 500 mM NaCl, 0.05% Tween 20.
6. Glutathione-agarose beads (GE Healthcare).

2.2. Assaying the Activation of Endogenous and Ectopically Expressed MAPKs by In Vitro Kinase Assays

1. Examples of antibodies against endogenous MAPKs. Anti-ERK2, C-14, Santa Cruz Biotechnology; anti-JNK1, Cat. no. 15701A, Pharmingen; anti-p38; C-20 and N-20, Santa Cruz Biotechnology; p38γ/SAPK3, Millipore; and ERK5, Cell Signaling Technology.
2. Antibodies against epitopes. Anti-HA.11, Cat. no. MMS-101R, Covance (see Note 2).
3. MAPK lysis buffer: 20 mM Hepes, pH 7.5, 10 mM EGTA, pH 8.0, 1% NP-40, 2.5 mM $MgCl_2$, 1 mM DTT (add immediately before use), 2 mM Na_3VO_4 (prepare fresh and add immediately before use), 20 mM β-glycerophosphate (prepare fresh and add immediately before use), 20 μg/ml aprotinin/leupeptin (add immediately before use), 1 mM PMSF (add immediately before use from a 100 mM stock solution in ethanol) (see Note 3). Store at 4°C.
4. PBS/NP-40 Wash Buffer: PBS 1×, 1% NP-40, 2 mM Na_3VO_4 (prepare fresh and add immediately before use). Store at 4°C.
5. Lithium/Tris Wash Buffer: 0.5 M LiCl, 100 mM Tris–HCl, pH 7.5. Store at 4°C.
6. Kinase reaction buffer: 12.5 mM MOPS, pH 7.5, 7.5 mM $MgCl_2$, 0.5 mM EGTA, 0.5 mM Na_3VO_4, 12.5 mM β-glycero-phosphate, 0.5 mM Sodium Fluoride (NaF). Store at 4°C.
7. Kinase reaction mix (per sample, always prepare for one extra sample): 30 μl kinase reaction buffer, 1 μCi (γ ^{32}P) ATP, 20 μM of unlabeled ATP (Stock 1 mM; store at −20°C), 3 μM DTT (Stock 100 mM; store at −20°C), appropriate amount of specific substrate (5 μg myelin basic protein; 1 μg of purified, bacterially expressed GST-ATF2 or GST-MEF2C). Radiation hazard.

2.3. Assaying GPCR Activation of MAP Kinases by Western Blot Analysis, Using Anti-phospho-specific MAPK Antibodies

1. Phospho-specific antibodies: Phospho-p42/44 MAP Kinase (Thr202/Tyr204) rabbit monoclonal antibody, Cell Signaling Technology; Phospho-p38 MAP Kinase (Thr180/Tyr182), Cell Signaling Technology, Phospho-p38 and Phospho-ERK5 from QCB-Biosource International; Phospho-ERK5 (T218/Y220) rabbit polyclonal or total ERK5 rabbit polyclonal (see ERK5 shifted band) from Cell Signaling Technology; and Phospho-JNK (pTPpY) polyclonal antibody from Promega, and Phospho-JNK (T183/Y185) rabbit monoclonal antibody, from Cell Signaling Technology.

2. Protein-loading buffer: 2% (w/v) SDS; 50 mM Tris–HCl, pH 6.8, 0.2 mg/ml bromophenol blue, 0.1 M DTT, 50% glycerol.
3. PVDF transfer membrane: Immobilon™-P (Millipore) or similar.
4. Tris-Tween-Buffered Saline (TTBS): 50 mM Tris-HCl, pH 7.4, 500 mM NaCl, 0.05% Tween 20.

3. Methods

3.1. Assaying the Activation of Endogenous Small GTPases by Affinity Precipitation of Their GTP-Bound Forms

Members of the Ras superfamily of monomeric GTP-binding proteins serve as molecular switches in regulating a wide range of essential biochemical pathways. In particular, they play a key regulatory function by controlling the activation of MAPKs by different membrane receptors, including a large variety of GPCRs (2, 4). Indeed, among these small GTPases, Ras, Rac1/Cdc42, and RhoA participate in the activation of ERK1/2, JNKs and p38, respectively (reviewed in ref. (5)).

Like other G-proteins, monomeric GTPases cycle between an inactive, GDP-bound state and an active, GTP-bound state. The protocol described in this section allows the isolation, from the total lysate, of different GTPases (Ras, Rac1, Cdc42, and RhoA) in their active conformation based on the specific ability of GTPases effector proteins to interact with their GTP-bound form. The respective binding domains of the downstream effector for each small GTPase (c-Raf for Ras, PAK1 for Rac1 and Cdc42, and Rhotekin for RhoA) are expressed as a GST-fusion protein which, when immobilized on a resin, are used to pull down the active, GTP-bound, GTPase. The purified GTPase is then detected by western blot using specific antibodies/antisera.

1. Prepare bacterially expressed and glutathione-agarose beads-bound GST-Rhotekin-RBD (Rho-Binding Domain) to assay RhoA activity, GST-PAK-N to assay Rac1 and Cdc42 activities or GST-Raf-RBD (Ras-Binding Domain) to assay Ras activity (see Note 4).
2. Aliquot GST-fusion proteins bound to agarose beads in pre-chilled tubes on ice (30 μl/sample) (see Note 5).
3. As small GTPases are activated by serum components, starve cells before appropriate stimulation, to avoid high background signals (see Note 6).
4. After appropriate cellular stimulation (see Note 7), wash cells once with cold PBS. Aspirate the PBS and then immediately add 900 μl of cold AP lysis buffer per 100 mm plate (500 μl per 60 mm plate; 250 μl per 35 mm plate).

5. Quickly scrape plates and clarify lysates in a refrigerated centrifuge at ~10,000 × *g* for 5 min, to remove insoluble debris.
6. Save 20–40 μg of the lysates for western blot analysis of the GTPase of interest, which can be used to normalize the activity of each GTPase by their expression level in the corresponding samples.
7. Immediately incubate remaining lysates with specific GST-fusion proteins bound to beads for 30 min at 4°C.
8. Wash once protein complexes bound to glutathione-agarose beads in AP lysis buffer (500 μl).
9. Boil samples 5 min in 1× protein-loading buffer (50 μl).
10. Load samples onto a 15% (it should be appropriate for all considered small GTPases) denaturing polyacrylamide gel.
11. Run the gel until the dye has reached nearly the bottom of the gel.
12. Blot 1 h at 350 mA (or 100 mA overnight) at 4°C, on a PVDF transfer membrane using a 20% of methanol in the transfer buffer.
13. Incubate PVDF membrane 1 h in blocking buffer, at room temperature (or overnight at 4°C).
14. Discard the blocking buffer and add the specific antibody/antisera recognizing the GTPase of interest (see Subheading 2.1, item 1 in this chapter), diluted 1:1,000 in TTBS + 1% BSA. Incubate 2 h at RT (or overnight at 4°C).
15. Remove the primary antibody (see Note 8) and wash the membrane with abundant TTBS buffer for 30 min, at RT, changing buffer every 10 min.
16. Freshly prepare secondary antibody (horseradish peroxidase-conjugated anti-mouse/rabbit secondary antibody) as a 1:10,000–20,000 dilution (Southern Biotech; Capel) and incubate 1 h at RT
17. Discard secondary antibody and wash with abundant TTBS buffer for 30 min. at RT, changing buffer every 10 min.
18. Detect immunocomplexes by chemo-luminescence by using ECL™ Western blotting detection reagents (GE Healthcare) or equivalent products, exposing the membrane to an autoradiography film (see Note 9).

3.2. Assaying the Activation of Endogenous and Ectopically Expressed MAPKs by In Vitro Kinase Assays

Activity of endogenous MAPKs often represents the most specific and sensitive approach to evaluate the effect of GPCR agonists on the in vivo activity of each MAPK pathway. Several laboratories, including ours, have also used extensively the transient expression of epitope-tagged forms of different MAPK, in readily transfectable cell lines. The advantage of this system is that several proteins can be transiently coexpressed at high levels, avoiding the influence of biological changes that might be manifested during prolonged culturing of stably expressing cells. In addition, the use of stable or transiently expressed epitope-tagged MAPKs allows the evaluation of the regulation of each kinase by GPCRs by in vitro kinase assays, circumventing the need for specific anti-MAPK antibodies. In this context, although the source of enzyme can be as different as crude cell lysate, cellular fractions, immunoprecipitates, partially purified proteins or purified enzymes, this section will specifically deal with in vitro kinase assays performed on endogenous or transfected, epitope-tagged, MAPKs, isolated by immunoprecipitation with specific antibodies.

The following protocol is based on our experience with different cell lines. It can be applied to evaluate the activity of endogenous as well as stably or transiently transfected MAP kinases. Furthermore, it can be successfully used for in vitro kinase assays for the measurement of the activity of a number of MAPK, including ERK1/2, JNK, p38α, p38γ, p38δ, and ERK5.

1. As most MAPKs are activated by serum components, starve cells before appropriate stimulation, to avoid high background signals (see Note 10).
2. After appropriate cellular stimulation (see Note 11), wash cells once with cold PBS. Aspirate the PBS and then immediately add 900 μl of cold MAPK lysis buffer per 100 mm plate (500 μl per 60 mm plate; 250 μl per 35 mm plate).
3. Place plates on ice for 20 min with occasional shaking. Scrape the plates with a cell lifter to collect the cellular lysates and transfer them to 1.5 ml microcentrifuge tubes (Eppendorf).
4. Clarify lysates in a refrigerated centrifuge at ~10,000 × *g* for 10 min, to remove insoluble debris and then transfer the clarified lysates into new microcentrifuge tubes.
5. Determine the protein concentration of the lysates, for example, by BCA Protein Assay (PIERCE). The lysates are now ready for immunoprecipitation.
6. Save 20–40 μg of the lysate for western blot analysis of the kinase of interest, which can be used to normalize the enzymatic activity of each MAPK by their expression level in the corresponding samples.
7. Aliquot the desired amount of lysates into new microcentrifuge tubes, based on the protein concentration of the samples (see Note 12).
8. Add the specific anti-MAPK or anti-epitope antibody to the samples and rotate at 4°C for 2 h to overnight (see Note 13).
9. Add 20 μl of prewashed GammaBind™ G Sepharose™ (GE Healthcare or equivalent) to the samples. Rotate at 4°C for at least 1 h.

10. Wash 3× with cold PBS/NP-40 Wash Buffer. For each wash, centrifuge the samples in a refrigerated microcentrifuge at 10,000 × *g* for 25 s. Aspirate the supernatant with a flat gel-loading tip connected to a vacuum, being careful not to aspirate the beads. Add 400 μl of PBS/NP40 Wash Buffer, mix and centrifuge again the samples, repeating the washing procedure for as many times as needed (usually three times).
11. Wash 1× with cold Lithium/Tris Wash Buffer.
12. Wash 1× with cold kinase reaction buffer. You are now ready to proceed to the in vitro kinase reaction.
13. Prepare the kinase reaction mix, including appropriate kinase substrate (see Note 14). Add 35 μl of the kinase reaction mix to each sample and incubate the microcentrifuge tubes for 30 min at 30°C in a water-bath designated for radioactive samples.
14. Stop the reactions by adding 10 μl of 5× protein-loading buffer to each sample and boil samples for 5 min.
15. Load half of the samples onto a 10–15% (according to the molecular weight of the substrate) denaturing polyacrylamide gel. The remaining material can be stored at –20°C should it be needed. Run the gel until the dye has reached nearly the bottom of the gel.
16. Cut just above the dye and discard the lower part of the gel to eliminate the interference of free-labeled ATP. Dry the upper part of the gel 2 h at 65°C and then expose it either on a phosphoimager plate or on an X-ray film, with an intensifier screen. The length of the exposure depends on many different factors and thus needs to be empirically determined.

3.3. Assaying GPCR Activation of MAP Kinases by Western Blot Analysis, Using Anti-phospho-Specific MAPK Antibodies

Although grouped in at least four different subfamilies (ERKs, JNKs, p38s, and ERK5), all MAPKs require the contemporary phosphorylation of a tyrosine and threonine residue within their activation loop to be enzymatically active. Specific phosphorylation of these residues is usually performed by MAPK kinases, also known as MEKs. To date, phosphorylation of the Thr-X-Tyr motif is the best known mechanism of MAPK regulation, and the amount of the dual phosphorylated form of these proteins is usually considered a good estimate of their enzymatic activity. The use of anti-phospho-MAPK-specific antibodies is therefore a valuable tool to monitor the activity of endogenous MAPKs in response to different GPCR ligands, especially under conditions in which the amount of cellular lysate is not enough to perform immunoprecipitations. In addition, these reagents eliminate the need for ^{32}P radioactive labeling. The success in the use of such reagents is dependent on the quality of the phospho-specific antisera or monoclonal antibodies, which, due to the efforts of a number of companies, has improved dramatically over the last few years.

Ideally, parallel membranes should be probed with antisera or monoclonal antibodies reacting with both the phosphorylated and unphosphorylated species of these MAPKs, thus facilitating the evaluation of the fractional increase in the levels of the activated forms of these kinases.

1. Lyse cells as described in Subheading 3.2, steps 1–5.
2. Aliquot the desired amount of lysates in a new microcentrifuge tube, based on their protein concentration (see Note 15).
3. Add required 5× protein-loading buffer to the different samples and boil them for 5 min.
4. Load the samples onto a denaturing polyacrylamide gel, accordingly to molecular weight of the kinase of interest (see Note 16). Run until the dye front has reached the bottom of the gel.
5. Blot 1 h at 350 mA (or 100 mA overnight) at 4°C, on a PVDF transfer membrane using a 10–20% of methanol in the transfer buffer (see Note 17).
6. Incubate PVDF membrane 1 h in blocking buffer, at room temperature (or overnight at 4°C).
7. Discard the blocking buffer and add the anti-phospho-MAPK antibody (see Subheading 2.3, item 1 in this chapter), diluted in TTBS + 1% BSA or in 5% non-fat dry-milk in TTBS. Incubate 2 h at RT (or overnight at 4°C) (see Note 18).
8. Remove the primary antibody (see Note 8) and wash the membrane with abundant TTBS buffer for 30 min, at RT, changing buffer every 10 min.
9. Freshly prepare secondary antibody (horseradish peroxidase-conjugated anti-mouse/rabbit secondary antibody) as a 1:10,000–20,000 dilution (Southern Biotech; Capel) and incubate 1 h at RT
10. Discard secondary antibody and wash with abundant TTBS buffer for 30 min. at RT, changing buffer every 10 min.
11. Detect immunocomplexes by chemo-luminescence by using ECL™ Western blotting detection reagents (GE Healthcare) or equivalent products, exposing the membrane to an autoradiography film (see Note 9).

4. Notes

1. It is of paramount importance to always keep lysates containing GTP-bound small GTPases in buffer containing $MgCl_2$, to avoid spontaneous dissociation of the nucleotide.

2. In our laboratories, the HA tag (nine amino acids derived from the influenza hemagglutinin HA1 protein), has extensively proved its efficacy to immunoprecipitate epitope-tagged active MAPKs, although epitopes such as MYC, FLAG, AU5, and AU1 can be equally used.
3. MAPKs are relatively stable enzymes in the presence of protease inhibitors such as phenylmethylsulfonyl fluoride (PMSF), leupeptin, and aprotinin. Calcium chelators (EDTA, EGTA) can similarly help to reduce the activity of calcium-activated proteases. Equally important, their activating tyrosine and threonine phosphorylation can be easily preserved for the duration of the assay by keeping them constantly at 4°C, in buffers containing common phosphatase inhibitors.
4. Bacterially purified GST-Rhotekin RBD, GST-PAK-N, and GST-Raf-RBD can be easily prepared by the researcher, if expression vectors for the different fusion proteins are available. Otherwise, ready-to-use stocks of some of these proteins can be purchased as part of all-inclusive kits from a number of companies (for example, Thermo Scientific or Cell Biolabs).
5. It is critical to perform each step on ice. Also centrifuge rotors should be prechilled at 4°C.
6. The length of the starvation period depends on the specific GTPase and can be empirically determined. In our experience, the best results are usually obtained with 12 h serum-starvation for Ras while 2–3 h serum-starvation is sufficient to assay RhoA, Rac1, and Cdc42 activities.
7. To control the proper technical execution of the GTPase activation assays, it is important to include an internal positive control for the experiment, treating the cells with a stimulus that is expected to increase the activity of the considered GTPase. An example of a commonly used positive control is the epidermal growth factor (EGF, 100 ng/ml).
8. The primary antibody can be saved for subsequent experiments by addition of 0.02% final concentration sodium azide (highly toxic) and storage at 4°C. Different antibodies, stored in this way, can be reused (number of times will depend on the specific antibodies used).
9. The optimal time of exposure is dependent on many different factors and therefore should be determined empirically.
10. The length of the starvation period depends on the specific MAPK and can be empirically determined. In our experience, the best results are usually obtained with 12 h serum-starvation for ERK1/2 and ERK5 kinase assays while 2–3 h serum-starvation is sufficient to assay JNK, p38α, p38γ, and p38δ activities. It is also important to remember that most of these

kinases are very sensitive to stress conditions such as those caused by changes in temperature and/or pH, or even prolonged serum-starvation.

11. To control the proper technical execution of the kinase assays, it is important to include an internal positive control for the experiment, treating the cells with a stimulus that is expected to increase the activity of the considered kinase. Examples of commonly used positive controls are: LPA (5–10 μM), TPA (25–100 nM), or serum (10%) for ERK1/2; anisomycin (10 μg/ml) or NaCl (150 mM) for JNK, p38α, p38γ, p38δ; and H_2O_2 (200 μM) for ERK5.
12. The amount of lysate needed to detect kinase activity is dependent on several factors (amount of the MAPKs in each cell type, affinity of the antibody used for immunoprecipitation, efficiency of transfection for assays involving transiently transfected cells), and therefore should be empirically determined.
13. The amount of antibodies needed to immunoprecipitate each MAPK activity is antibody-dependent, and therefore the reader should consult the manufacturer's recommended protocol. In general, 1 μg of antibodies or antiserum is sufficient to immunoprecipitate the majority of the MAP kinases from lysates containing 1 mg of total cellular proteins.
14. In vitro MAPK assays are based on the ability of these proteins to act as phosphotransferase enzymes between a labeled "donor substrate," (γ-^{32}P) ATP, and a protein acting as "acceptor substrate" (the kinase-specific substrate). Selecting a substrate normally depends on the specific kinase under evaluation. Myelin basic protein (MBP) serves as a very good substrate for different MAPKs such as ERKs, p38α, and p38γ. Bacterially purified glutathione S-transferase (GST)-tagged ATF2 is the substrate of choice for JNK, even if GST-c-Jun is an equally good alternative for this kinase. GST-ATF2 also works as a good substrate for p38α. Bacterially purified GST-MEF2C is the best substrate to assay ERK5 kinase activity. Nevertheless, if not available, MBP has been also successfully used in in vitro ERK5 kinase assays. While MBP can be usually purchased from different companies (SIGMA, for example), bacterially purified GST-ATF2, GST-c-Jun, and GST-MEF2C can be easily prepared by the researcher, if expression vectors for the different fusion proteins are available. Otherwise, ready-to-use stocks of some of these proteins (GST-ATF2 and GST-c-Jun) can be purchased from a number of companies (for example, Santa Cruz Biotechnology and Agilent Technologies).
15. The amount of lysates needed for the detection of phosphorylated kinases is dependent on different factors (expression levels of each MAPK in each cell type, affinity of the antibody

used), and therefore should be empirically determined based, if possible, based on manufacturer's recommendations. The remaining part of the cellular lysates may be stored at –80°C, should they be needed.

16. As a reference, in our laboratory we use 10% gels for ERK1/2, JNK1/2, and p38s, and 8% gels for ERK5.
17. We strongly recommend ERK5 to be transferred at 100 mA overnight, at 4°C.
18. The amount of antibodies needed to detect each MAPK by western blot analysis is antibody-dependent, and therefore the reader should consult the manufacturer's recommended protocol. In general, 1:1,000–1:2,000 dilutions are sufficient for most antibodies to detect MAPKs in lysates containing 20–40 μg of total cellular proteins.

References

1. Hepler, J. R., and Gilman, A. G. (1992) G proteins, *Trends Biochem Sci* **17**, 383–387.
2. Gutkind, J. S. (1998) Cell growth control by G protein-coupled receptors: from signal transduction to signal integration, *Oncogene* **17**, 1331–1342.
3. Pierce, K. L., Premont, R. T., and Lefkowitz, R. J. (2002) Seven-transmembrane receptors, *Nat Rev Mol Cell Biol* **3**, 639–650.
4. Gutkind, J. S. (1998) The pathways connecting G protein-coupled receptors to the nucleus through divergent mitogen-activated protein kinase cascades, *J Biol Chem* **273**, 1839–1842.
5. Dorsam, R. T., and Gutkind, J. S. (2007) G-protein-coupled receptors and cancer, *Nat Rev Cancer* **7**, 79–94.
6. Armbruster, B. N., Li, X., Pausch, M. H., Herlitze, S., and Roth, B. L. (2007) Evolving the lock to fit the key to create a family of G protein-coupled receptors potently activated by an inert ligand, *Proc Natl Acad Sci USA* **104**, 5163–5168.
7. Conklin, B. R., Hsiao, E. C., Claeysen, S., Dumuis, A., Srinivasan, S., Forsayeth, J. R., Guettier, J. M., Chang, W. C., Pei, Y., McCarthy, K. D., Nissenson, R. A., Wess, J., Bockaert, J., and Roth, B. L. (2008) Engineering GPCR signaling pathways with RASSLs, *Nat Methods* **5**, 673–678.
8. Marinissen, M. J., Chiariello, M., Pallante, M., and Gutkind, J. S. (1999) A network of mitogen-activated protein kinases links G protein-coupled receptors to the c-jun promoter: a role for c-Jun NH2-terminal kinase, p38s, and extracellular signal-regulated kinase 5, *Mol Cell Biol* **19**, 4289–4301.

Chapter 9

Regulation of MAP Kinase Signaling by Calcium

Colin D. White and David B. Sacks

Abstract

Mitogen-activated protein kinase (MAPK) signaling influences a variety of cellular responses, ranging from stimulation of cell proliferation to induction of senescence and/or apoptosis. Ca^{2+} is a ubiquitous intracellular signaling molecule that controls multiple processes in cells. Published evidence has identified both direct and indirect interactions between the Ca^{2+} and MAPK signaling pathways. Here, we describe assays to accurately determine the effect of changes in intracellular Ca^{2+} concentration on MAPK activation.

Key words: A23187, BAPTA-AM, Ca^{2+}, Confocal microscopy, MAPK signaling, Western blotting

1. Introduction

Mitogen-activated protein kinases (MAPKs) are ubiquitously expressed enzymes that regulate a wide variety of functions in virtually all cell types (1). The term "MAPK" usually refers to the terminal kinase in a three-tier cascade, in which MAPKs are phosphorylated and activated by MAPK kinases (MAPKK or MEK), which themselves are phosphorylated and activated by MAPK kinase kinases (MAPKKK or MEKK). Of the major MAPK pathways, the Ras/Raf/MEK/ERK cascade is the most widely studied and is the focus of this chapter. Engagement of cell-surface receptors by extracellular signaling molecules, such as growth factors, results in activation of the intracellular small G-protein Ras. The resultant change in Ras conformation facilitates its direct interaction with Raf isoforms, namely A-Raf, B-Raf, and C-Raf (also termed Raf-1) (2). The Raf proteins are serine/threonine kinases, which phosphorylate and activate MEK1 and MEK2. In turn, MEK1 and MEK2 catalyze the phosphorylation of the extracellular signal-regulated kinases, ERK1 and ERK2.

Rony Seger (ed.), *MAP Kinase Signaling Protocols: Second Edition*, Methods in Molecular Biology, vol. 661,
DOI 10.1007/978-1-60761-795-2_9,

Once active, ERKs either dimerize and remain in the cytosol where they catalyze the phosphorylation of a variety of substrates, or, as monomers, translocate to the nucleus where they phosphorylate transcription factors (3).

MAPK function is influenced by several pathways, including Ca^{2+} (4). For example, an increase in intracellular free Ca^{2+} concentration ($[Ca^{2+}]_i$) positively regulates Ras signaling in PC12 cells leading to increased ERK phosphorylation (5). Conversely, treating keratinocytes with Ca^{2+} inhibits activation of ERK by epidermal growth factor (EGF) (6). The reasons for these discrepant data are not known, but differences between the cell types are likely to contribute. The ability to manipulate $[Ca^{2+}]_i$ and accurately measure active MAPK is a welcome addition to the researchers' toolbox. In this chapter, we describe straightforward assays for evaluating the effect of Ca^{2+} on growth factor-induced MAPK signaling using Western blotting and confocal immunofluorescence.

1.1. Manipulation of $[Ca^{2+}]_i$

The protocol described in this chapter represents probably the most widely used methods to manipulate $[Ca^{2+}]_i$. A23187 is a Ca^{2+} ionophore that causes a rapid and sustained increase in $[Ca^{2+}]_i$ by permitting entry into the cell of extracellular Ca^{2+}. BAPTA-AM enters cells where it chelates intracellular Ca^{2+}, markedly reducing $[Ca^{2+}]_i$. These reagents therefore allow the investigator to elucidate the regulatory effect of Ca^{2+} on intracellular signaling.

In order to identify the source of the Ca^{2+} responsible for a specific effect, pharmacological compounds are available (which selectively modulate individual Ca^{2+} channels or pumps (Table 1)). Each compound can be broadly characterized as inducing either an "on" or an "off" signal (Fig. 1). On signals increase $[Ca^{2+}]_i$, while off signals reduce it. These are discussed in more detail below.

1.1.1. "On" Signals

Certain extracellular stimuli induce a rise in $[Ca^{2+}]_i$. The increase in $[Ca^{2+}]_i$ is mediated either by Ca^{2+} entering the cell from the outside (across the plasma membrane) or by release from intracellular stores. There are three classes of channels in the plasma membrane which facilitate Ca^{2+} influx from the outside (7). Voltage-activated Ca^{2+} channels respond to changes in the membrane potential of the cell, while ligand-activated Ca^{2+} channels are opened in response to the binding of a specific ligand. Store-activated Ca^{2+} channel opening is stimulated by the emptying of intracellular Ca^{2+} stores.

Ca^{2+} is released from the endoplasmic reticulum, an organelle that acts as an intracellular Ca^{2+} store. The mechanism underlying this release is similar to that of ligand-activated Ca^{2+} channels, but the activating ligands differ. The best studied examples are the inositol triphosphate (IP_3) and ryanodine receptors, which may be modulated by binding of their cognate ligands, IP_3 and ryanodine, respectively. Interestingly, the most important regulator of Ca^{2+}

Table 1
Pharmacological agents used to selectively manipulate $[Ca^{2+}]_i$

Site of manipulation	Compound	Effect on $[Ca^{2+}]_i$	Mode of action	Solubility	Notes
Voltage-activated Ca^{2+} channels	ω-Agatoxin	–	Inhibits P-type Ca^{2+} channels	H_2O	
	(±)-Bay K 8644	+	Activates L-type Ca^{2+} channels	$EtOH/H_2O$	Induces p44/42 MAPK activation in Jurkat cells
	ω-Conotoxin	–	Inhibits N-type Ca^{2+} channels	H_2O	
	(+)-*cis*-Diltiazem hydrochloride	–	Inhibits L-type Ca^{2+} channels	$MeOH/H_2O$	Causes Ca^{2+} release from intracellular stores in neutrophils
	Mibefradil dihydrochloride	–	Inhibits T-type Ca^{2+} channels	H_2O	Inhibits L-type Ca^{2+} channels at high concentrations
	Neomycin trisulfate	–	Inhibits voltage-activated Ca^{2+} channels	H_2O	No effect on Na^+/Ca^{2+} antiporter in neurons
	Nifedipine	–	L-type Ca^{2+} channel antagonist	DMSO/EtOH	Induces apoptosis in human glioblastoma cells
	Nimodipine	–	L-type Ca^{2+} channel antagonist	MeOH	Photosensitive
Receptor-activated Ca^{2+} channels	Ruthenium red	–	Inhibits RyR-activated Ca^{2+} channels	H_2O	May also inhibit voltage-activated Ca^{2+} channels
	Ryanodine	+/–	Locks RyR-activated Ca^{2+} channels in a half open state at nM concentrations. Fully closes them in the μM range	DMSO/EtOH	RyRs are expressed primarily in skeletal and cardiac muscle, and the brain
	Xestospongin C	–	Inhibits IP_3 receptor-activated Ca^{2+} channels	DMSO/EtOH	May also inhibit voltage-activated Ca^{2+} channels
Ca^{2+}-ATPases	Cyclopiazonic acid	+	Inhibits ER/SR Ca^{2+}-ATPase	DMSO	Toxic at high concentrations
	Thapsigargin	–	Inhibits ER/SR Ca^{2+}-ATPase	DMSO/EtOH	Widely used, potent, cell-permeable inhibitor

+, increases $[Ca^{2+}]_i$; –, decreases $[Ca^{2+}]_i$
SR sarcoplasmic reticulum; *ER* endoplasmic reticulum; *RyR* ryanodine receptor

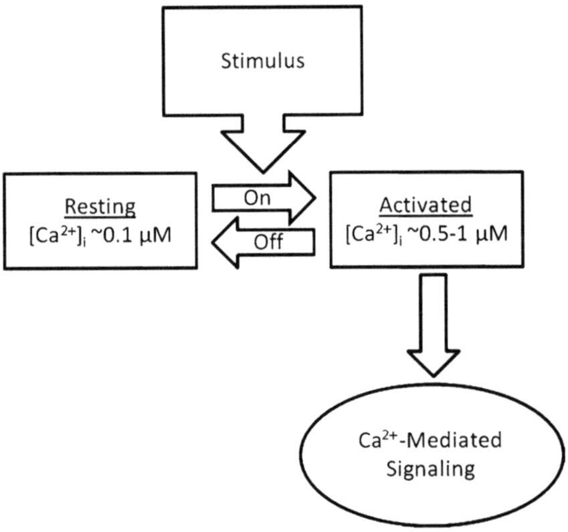

Fig. 1. The basic Ca^{2+} signaling network. A stimulus activates various "on" or "off" signals. "On" signals trigger an increase in $[Ca^{2+}]_i$ which, in turn, induces Ca^{2+}-mediated signaling events. "Off" signals restore $[Ca^{2+}]_i$ to its resting level.

channels on intracellular stores is Ca^{2+} itself. This observation forms the basis of the concept of Ca^{2+}-induced Ca^{2+} release (8, 9).

1.1.2. "Off" Signals

Off signals involve the rapid removal of intracellular free Ca^{2+} from the cytoplasm by a variety of pumps and exchangers. Ca^{2+} can be pumped out of the cell by Ca^{2+}-ATPases or Na^{+}/Ca^{2+} exchangers located on the plasma membrane. Alternatively, Ca^{2+} can be moved into intracellular storage compartments by Ca^{2+}-ATPases on the endoplasmic reticulum or through Ca^{2+} uniporters on the inner mitochondrial membrane.

2. Materials

2.1. Cell Culture, Treatment, and Lysis

Unless otherwise stated, all reagents are stored at room temperature (~22°C).

1. Dulbecco's Modified Eagle's Medium (DMEM) supplemented with 10% fetal bovine serum (FBS) and 1% penicillin/streptomycin/glutamine (PSG). Store at 4°C.
2. DMEM supplemented with 1% PSG and 1 mM 4-(2-hydroxyethyl)-1-piperazineethanesulfonic acid (HEPES). Store at 4°C.
3. 0.05% Trypsin/ethylenediamine tetraacetic acid (EDTA). Store for up to 1 month at 4°C.
4. Sterile phosphate-buffered saline (PBS).

5. A23187 (Sigma, St. Louis, MO) (50 mg/ml in dimethyl sulfoxide (DMSO)). Store in single use aliquots at −80°C. See Notes 1 and 2.
6. 1,2-Bis(2-aminophenoxy)ethane-*N*,*N*,*N*′,*N*′-tetraacetic acid (tetra(acetoxymethyl) ester) (BAPTA-AM) (Sigma, St. Louis, MO) (15 mg/ml in DMSO). Store in single use aliquots at −80°C. See Notes 3 and 4.
7. EGF (Gibco, Carlsbad, CA) (1 mg/ml in sterile PBS). Store in single use aliquots at −80°C. See Note 5.
8. PBS: 150 mM NaCl, 2.7 mM KCl, 10 mM Na_2HPO_4, 1.8 mM KH_2PO_4 (pH 7.4). Store at 4°C.
9. Lysis buffer: 50 mM Tris–HCl, 150 mM NaCl, 1% Triton X-100. Store at 4°C. Prior to use, add 1% 0.1 M phenylmethanesulfonyl fluoride (PMSF), 0.1% Protease Inhibitor Cocktail (Sigma, St. Louis, MO), 10 μg/ml leupeptin, 1% Phosphatase Inhibitor Cocktail 1 (Sigma, St. Louis, MO) and 1% Phosphatase Inhibitor Cocktail 2 (Sigma, St. Louis, MO).
10. Disposable cell lifters.
11. 6× Sample buffer: 180 mM Tris (pH 6.8), 12% (w/v) sodium dodecyl sulfate (SDS), 50% glycerol, 10% (w/v) dithiothreitol, 0.006% (w/v) bromophenol blue. Store at 4°C.

2.2. SDS-Polyacrylamide Gel Electrophoresis

1. 4× Protogel Separating Buffer (National Diagnostics, Atlanta, GA).
2. 4× Protogel Stacking Buffer (National Diagnostics, Atlanta, GA).
3. Protogel (National Diagnostics, Atlanta, GA).
4. 10% (w/v) Ammonium persulfate (APS).
5. *N*,*N*,*N*′,*N*′-tetra-methyl-ethylenediamine (TEMED).
6. Isobutanol: decant 25 ml into a 500 ml spray bottle and use vapor.
7. Running buffer: 50 mM Tris, 0.4 M glycine, 0.1% (w/v) SDS.
8. All Blue Precision Plus Protein Standards (Bio-Rad, Hercules, CA). Store at −20°C.

2.3. Western Blotting for Active MAPK

1. Transfer buffer: 30 mM Tris, 0.25 M glycine. Prior to use, to 800 ml transfer buffer add 200 ml MeOH and 2 ml 10% (w/v) SDS.
2. Prefrozen ice container.
3. Immobilon-P Transfer Membrane (0.45 μm pore) (Millipore, Bedford, MA).
4. MeOH.

5. Tris-buffered saline with Tween (TBS-T): 10 mM Tris, 150 mM NaCl, 0.2% Tween-20 (pH 8.0).
6. Blocking buffer: 4% (w/v) bovine serum albumin (BSA) in TBS-T. Store at 4°C.
7. 10% (w/v) sodium azide.
8. Primary antibody: Anti-phospho-p44/42 MAPK rabbit monoclonal antibody (Cell Signaling Technology, Danvers, MA (catalog no. 4370)). Store at −20°C.
9. Secondary antibody: Horseradish peroxidase (HRP)-linked anti-rabbit immunoglobulin G (GE Healthcare, Buckinghamshire, UK). Store at 4°C.
10. Immobilon Western Chemiluminescent HRP Substrate Kit (Millipore, Bedford, MA). Store at 4°C.
11. Kodak BioMax XAR X-ray Film (Carestream Health, Rochester, NY).

2.4. Stripping Blots and Reprobing for Total MAPK

1. Stripping buffer: 62.5 mM Tris–HCl (pH 6.8), 2% (w/v) SDS, 0.7% β-mercaptoethanol. Make fresh as required.
2. Primary antibody: Anti-p44/42 MAPK mouse monoclonal antibody (Cell Signaling Technology, Danvers, MA (catalog no. 4696)). Store at −20°C.
3. Secondary antibody: HRP-linked anti-mouse immunoglobulin G (GE Healthcare, Buckinghamshire, UK). Store at 4°C.
4. Immobilon Western Chemiluminescent HRP Substrate Kit. Store at 4°C.
5. Kodak BioMax XAR X-ray Film.

2.5. Confocal Immunofluorescence for Active and Total MAPK

1. Microscope Cover Glass.
2. Lab-Tek four-well Glass Chamber Slides.
3. PBS. Store at 4°C.
4. 4% (w/v) Paraformaldehyde (PFA) in PBS. Store at 4°C. See Note 6.
5. Blocking and permeabilization buffer: 0.2% Triton X-100, 3% BSA in PBS. Store at 4°C.
6. Antibody diluent: 0.2% Triton X-100, 1% BSA in PBS. Store at 4°C.
7. Primary antibodies: Anti-phospho-p44/42 MAPK rabbit monoclonal antibody and anti-p44/42 MAPK mouse monoclonal antibody. Store at −20°C.
8. Secondary antibodies: Alexa-Fluor 488-labeled anti-rabbit immunoglobulin G (Molecular Probes, Carlsbad, CA) and Alexa-Fluor 488-labeled anti-mouse immunoglobulin G (Molecular Probes, Carlsbad, CA). Store both in light-protected single use aliquots at −20°C.

9. Nuclear stain: 4,6-diamidino-2-phenylindole (DAPI) (Molecular Probes, Carlsbad, CA). Store in light-protected single use aliquots at −20°C. See Notes 7 and 8.
10. Mounting medium: PermaFluor Aqueous Mounting Medium (Fisher, Pittsburgh, PA). Store at 4°C.

3. Methods

3.1. Cell Culture, Treatment, and Lysis

1. Passage cells when approaching confluence by washing with sterile PBS and detaching with prewarmed 0.05% trypsin/EDTA. One 100 mm dish is required for each data point (each dish holds a volume of ~5–10 ml). Allow cells to attach and approach confluence in prewarmed DMEM supplemented with 10% FBS and 1% PSG.
2. At 90–100% confluence, rinse cells twice with sterile PBS. Starve cells of serum by incubating in prewarmed (37°C) DMEM supplemented with 1% PSG and 1 mM HEPES for 16 h at 37°C.
3. Prepare all materials for cell treatment and lysis (see Notes 1–5). Other materials also required at this stage include three prechilled and labeled microcentrifuge tubes per dish, PBS and lysis buffer (both at 4°C), disposable cell lifters, 70% EtOH and 6× sample buffer.
4. Aspirate growth medium from each 100 mm dish and replace with medium containing either vehicle (DMSO), A23187 or BAPTA-AM as appropriate. Incubate for 20 min at 37°C. See Notes 1–4 and 9.
5. Treat each experimental culture with either vehicle (0.01% BSA) or 100 ng/ml EGF as appropriate. Incubate for 5 min at 37°C. See Note 10.
6. Immediately place all 100 mm dishes on ice and aspirate growth medium. Wash rapidly with cold PBS. Aspirate and add 500 μl cold lysis buffer to each dish. Swirl dishes gently to ensure even coverage. See Note 11.
7. Using a disposable cell lifter, scrape the contents of each 100 mm dish into an appropriately labeled prechilled microcentrifuge tube. Rinse the disposable cell lifter in 70% EtOH between samples. Sonicate twice at high power for 5–10 s and clarify by high speed centrifugation (~15,000 × *g*) for 10 min at 4°C. See Note 12.
8. Carefully aspirate supernatant and transfer into a separate appropriately labeled prechilled microcentrifuge tube. Discard pellet. If desired, protein concentration in an aliquot of the supernatant may be measured using the Modified Bradford Assay (Bio-Rad, Hercules, CA).

9. To 10 μl 6× sample buffer, add 40 μl lysate (or an appropriate normalized volume). Mix well and boil at 100°C for 5 min. Centrifuge briefly, cool to 22°C and proceed to Subheading 3.2. See Notes 13–15.

3.2. SDS-PAGE

This protocol describes the use of the Bio-Rad Mini-Blot Gel System (Bio-Rad, Hercules, CA). Nevertheless, it is easily adaptable to other formats.

1. Prior to (and following) each use, clean each glass plate with 70% EtOH and rinse well with ddH_2O.
2. Prepare a 1.5 mm thick separating gel of the appropriate percentage (Table 2). After addition of TEMED, proceed immediately to step 3.
3. Pour the gel ensuring that ~1.5 cm of space is left at the top for the stacking gel. Use isobutanol vapor to remove any air bubbles. Polymerization should take place in 30–45 min.
4. Prepare the stacking gel by mixing 330 μl Protogel, 630 μl 4× Protogel Stacking Buffer, 1.53 ml ddH_2O, 12.5 μl 10% (w/v) APS and 2.5 μl TEMED. Pour the stack, use isobutanol vapor to remove any air bubbles and carefully insert the comb. Polymerization should take place in 30–45 min. See Note 16.
5. Once the stacking gel has set, carefully remove the comb and use a 5 ml syringe fitted with a 22-gauge needle to wash the wells with running buffer.
6. Assemble the gel unit and fill each chamber with running buffer. Load 10 μl All Blue Precision Plus Protein Standard in well 1. Each sample should be added carefully to a separate well.

Table 2
Separating gel components for different % acrylamide gels

Component	6%	8%	10%	12%	15%
Protogel (ml)	2.0	2.7	3.4	4.0	5.0
4× Protogel separating buffer (ml)	2.5	2.5	2.5	2.5	2.5
ddH_2O (ml)	5.4	4.7	4.0	3.4	2.4
10% (w/v) APS (μl)	100	100	100	100	100
TEMED (μl)	10	10	10	10	10
Typical protein size resolved (kDa)[a]	60–300	40–300	20–300	20–200	10–150

[a]The range of proteins resolved using different % gels is based on our experience using the reagents in this protocol. The use of other reagents may substantially alter these values

7. Complete the assembly of the gel unit and connect to a power supply. Run at 50 mA for ~60 min or until the dye front reaches the bottom of the gel.

3.3. Western Blotting for Active MAPK

1. At this stage it is necessary to prepare for gel transfer. This protocol assumes the use of a "wet" transfer system but is easily adaptable to the "semi-dry" equivalent. Cut a piece of Immobilon-P Transfer Membrane approximately 7 cm × 5 cm in size and soak thoroughly for ~5 min in MeOH. After soaking, rinse thoroughly with ddH_2O and soak in transfer buffer until SDS-PAGE (SDS-Polyacrylamide Gel Electrophoresis) is complete. Both foam pads of the transfer cassette should also be soaked thoroughly in transfer buffer for at least 30 min prior to use (see Note 17).
2. Disconnect the gel unit from the power supply and disassemble. Using a clean razor blade, cut away the stacking gel and discard. Similarly, if still present, cut away anything below the dye front on the separating gel. Carefully submerge the remaining separating gel in transfer buffer.
3. Assemble the transfer cassette as follows: open the cassette and place one soaked foam pad on each side. Place the separating gel on a foam pad and carefully lay the Immobilon-P Transfer Membrane on top. Gently remove any air bubbles in the stack by rolling with a clean test tube, then place the other foam pad on top. Gently remove any air bubbles again and close the transfer cassette (see Note 18).
4. Place the transfer cassette into the transfer tank such that the separating gel is closest to the negative cathode and the Immobilon-P Transfer Membrane to the positive anode. This orientation is critical or the proteins will be lost. Fill the transfer tank with transfer buffer and drop in a small magnetic stir bar. Slot a prefrozen ice container into place.
5. Put the lid on the transfer tank and connect to a power supply. Place the apparatus on a magnetic stirrer and switch on. Transfer at 100 V for 1 h (see Notes 19 and 20).
6. Disconnect the transfer tank from the power supply and remove the transfer cassette. Discard the separating gel and place the Immobilon-P Transfer Membrane in a clean plastic container. If the transfer was successful, the All Blue Precision Plus Protein Standards should be clearly visible.
7. Incubate the Immobilon-P Transfer Membrane in 10 ml blocking buffer for 1 h at 22°C or overnight at 4°C.
8. Prepare the primary antibody solution as follows: To 10 ml blocking buffer add 100 μl 10% sodium azide and 10 μl anti-phospho-p44/42 MAPK. Store at 4°C.

9. Remove the blocking buffer and incubate the Immobilon-P Transfer Membrane in 10 ml primary antibody solution for 1 h at 22°C or overnight at 4°C (see Note 21).
10. Remove the primary antibody solution and wash the Immobilon-P Transfer Membrane three times for 10 min each with 20 ml TBS-T.
11. The secondary antibody is freshly prepared for each experiment. To 10 ml blocking buffer add 2 μl HRP-linked anti-rabbit immunoglobulin G. After washing is complete, add the secondary antibody and incubate for 1 h at 22°C.
12. Remove the secondary antibody and wash the Immobilon-P Transfer Membrane three times for 10 min each with 20 ml TBS-T.
13. Once the final wash is finished, mix together 1 ml of each reagent in the Immobilon Western Chemiluminescent HRP Substrate Kit and pour on to the Immobilon-P Transfer Membrane. Rotate using forceps for 1.5 min to ensure even coverage and place between the leaves of a lightweight sheet protector that has been pre-cut to the same size as an X-ray film cassette.
14. Place the sheet protector in the X-ray film cassette and proceed immediately to a dark room. Delays at this stage in the protocol will result in loss of the chemiluminescent signal.
15. Under safe light conditions, place a sheet of Kodak BioMax XAR X-ray film into the cassette and expose for a suitable time. For most proteins, including phospho-p44/42 MAPK, typical exposure times range between 1 s and 1 min.

3.4. Stripping Blots and Reprobing for Total MAPK

1. Upon satisfactory exposure of active phosphorylated MAPK, it is necessary to strip the Immobilon-P Transfer Membrane and reprobe with an antibody that recognizes both phosphorylated and nonphosphorylated MAPK. This provides a loading control and allows quantification of the various EGF-stimulated responses (see Note 22).
2. Using a preheated waterbath, incubate the Immobilon-P Transfer Membrane in 50 ml stripping buffer for 30 min at 55°C (see Note 23).
3. Remove the stripping buffer and wash the Immobilon-P Transfer Membrane six times for 5 min each with 20 ml TBS-T.
4. Incubate the Immobilon-P Transfer Membrane in 10 ml blocking buffer for 1 h at 22°C or overnight at 4°C.
5. Prepare the primary antibody solution as follows: To 10 ml blocking buffer add 100 μl 10% sodium azide and 10 μl anti-p44/42 MAPK. Store at 4°C.

6. Remove the blocking buffer and incubate the Immobilon-P Transfer Membrane in 10 ml primary antibody solution for 1 h at 22°C or overnight at 4°C (see Note 21).
7. Remove the primary antibody solution and wash the Immobilon-P Transfer Membrane three times for 10 min each with 20 ml TBS-T.
8. As before, the secondary antibody is freshly prepared for each experiment. To 10 ml blocking buffer add 2 μl HRP-linked anti-mouse immunoglobulin G. After washing is complete, add the secondary antibody and incubate for 1 h at 22°C.
9. Remove the secondary antibody and wash the Immobilon-P Transfer Membrane three times for 10 min each with 20 ml TBS-T.
10. Repeat steps 13–15 in Subheading 3.3. Typical exposure times for p44/42 MAPK range from between 1 and 20 s.

3.5. Confocal Immunofluorescence for Active and Total MAPK

1. Passage cells when approaching confluence by washing with sterile PBS and detaching with prewarmed 0.05% trypsin/EDTA. One well of a Lab-Tek 4-well Glass Chamber Slide is required for each data point (each well holds a volume of ~500 μl). Allow cells to attach and approach confluence in prewarmed DMEM supplemented with 10% FBS and 1% PSG.
2. At 70–80% confluence, rinse cells twice with sterile PBS. Starve cells of serum by incubating in prewarmed (37°C) DMEM supplemented with 1% PSG and 1 mM HEPES for 16 h at 37°C (see Note 24).
3. Prepare all materials for cell treatment and permeabilization (see Notes 1–5). Other materials also required at this stage include PBS, PFA, and blocking and permeabilization buffer (all at 4°C).
4. Aspirate growth medium from each well and replace with medium containing either vehicle (DMSO), A23187 or BAPTA-AM as appropriate. Incubate for 20 min at 37°C (see Notes 1–4 and 9).
5. Treat each experimental culture with either vehicle (0.01% BSA) or 100 ng/ml EGF as appropriate. Incubate for 5 min at 37°C (see Note 10).
6. Immediately place all Lab-Tek four-well Glass Chamber Slides on ice and aspirate growth medium. Wash rapidly with cold PBS. Aspirate and add 500 μl cold PFA to each well. Leave for 20 min at 22°C.
7. Wash twice with cold PBS. Aspirate and add 500 μl cold blocking and permeabilization buffer to each well. Leave for 1 h at 22°C.

8. The primary antibody solution is prepared freshly for each experiment. To 500 μl antibody diluent add 5 μl anti-phospho-p44/42 MAPK or 5 μl anti-p44/42 MAPK.
9. Remove the blocking and permeabilization buffer and incubate the experimental cultures in 500 μl primary antibody solution overnight at 4°C.
10. Remove the primary antibody solution and wash three times with cold PBS. The experimental cultures are protected from light for all subsequent steps.
11. The secondary antibody is freshly prepared for each experiment. To 500 μl antibody diluent add 1 μl Alexa-Fluor 488-labeled anti-rabbit immunoglobulin G or 1 μl Alexa-Fluor 488-labeled anti-mouse immunoglobulin G. Add the secondary antibody and incubate for 1 h at 22°C.
12. Remove the secondary antibody and wash three times with cold PBS. Incubate the experimental cultures in 500 μl DAPI for 5 min at 22°C (see Notes 7 and 8).
13. Remove the DAPI and wash three times with cold PBS. Aspirate all the liquid and carefully remove the wells using the supplied tool. Apply ~2–3 ml PermaFluor Aqueous Mounting Medium and a Microscope Cover Glass. Leave in light-protected conditions for 24 h at 4°C.
14. View the slides using phase-contrast microscopy to locate the cells and identify the focal plane. Under confocal conditions, excitation at 488 nm induces green fluorescence for either phospho-p44/42 MAPK or p44/42 MAPK. Excitation at 364 nm induces blue fluorescence for DAPI (see Note 25).

4. Notes

1. Working solutions of A23187 are prepared by diluting to 50 μg/ml in DMSO and subsequent dilution to 5 ng/ml in DMEM supplemented with 1% PSG and 1 mM HEPES.
2. A23187 is a selective Ca^{2+} ionophore (10). It greatly increases the ability of divalent ions to cross biological membranes by forming stable 2:1 complexes with them, thus rendering them cell-permeable. A23187 is commonly used to increase $[Ca^{2+}]_i$ in intact cells. A less Ca^{2+}-selective alternative is Ionomycin (Sigma, St. Louis, MO).
3. Working solutions of BAPTA-AM are prepared by diluting to 30 μg/ml in DMSO and subsequent dilution to 30 ng/ml in DMEM supplemented with 1% PSG and 1 mM HEPES.

4. BAPTA-AM is a Ca^{2+} chelator with 105-fold greater affinity for Ca^{2+} than for Mg^{2+} (10). Once inside the cell, the acetoxymethyl moiety is hydrolyzed by cytosolic esterases and BAPTA, which is unable to cross the plasma membrane, is trapped intracellularly.
5. Working solutions of EGF are prepared by diluting to 100 μg/ml in 0.01% BSA.
6. To dissolve PFA, heat to ~50°C with constant stirring. Precipitation after long term storage indicates that the solution should be discarded.
7. Working solutions of DAPI are prepared by dilution to 200 ng/ml in PBS.
8. DAPI is a known carcinogen. Always wear gloves.
9. It is our experience that the concentrations and incubation times of A23187 and BAPTA-AM we have suggested are sufficient to elicit effects on EGF-induced MAPK activation. Nevertheless, incubation of each reagent at different concentrations for different times should be performed in order to optimize the protocol for each cell type.
10. EGF typically induces maximal p44/42 MAPK activation ~2–5 min poststimulation. Nevertheless, stimulation at different concentrations for different times should be performed to optimize the protocol for each cell type.
11. 500 μl is the recommended initial lysis volume. It can be decreased in order to concentrate protein should a satisfactory phospho-p44/42 MAPK signal not be obtained.
12. Ear protection should be worn when using a sonicator.
13. 6× Sample buffer should be warmed to 22°C before use to allow accurate pipetting.
14. Microcentrifuge tube caps should be "locked" shut in order to prevent them springing open during boiling which may result in loss of some of the sample. If using conventional 1.5 ml microcentrifuge tubes, Microtube Lid Locks (Fisher, Pittsburgh, PA) provide an inexpensive way to achieve this.
15. If required, the protocol may be stopped at this point and the samples stored at −80°C.
16. We use 1.5 mm thick 10-well combs. Both 12- and 15-well models are also available, but limit the sample volume that may be loaded in each well to ~30 μl and ~10 μl, respectively.
17. The Immobilon-P Transfer Membrane is extremely hydrophobic and will not wet in aqueous solutions unless prewet in methanol. After prewetting, do not let the membrane dry. In the event it does dry, it should again be wet in MeOH.

18. Air bubbles should be carefully rolled out to avoid disturbing the flow of current from the negative cathode to the positive anode and thus the transfer of proteins from the separating gel to the Immobilon-P Transfer Membrane.
19. Stir at low speed to avoid the introduction of air bubbles.
20. Coomassie Blue staining can be used to evaluate transfer efficiency. After gel transfer, remove the Immobilon-P Transfer Membrane and incubate the separating gel in Coomassie Blue stain (50% MeOH, 10% acetic acid, 40% ddH_2O, 0.2% (w/v) Coomassie Blue) for 1 h at 22°C. After staining, wash with ddH_2O and incubate in Gel Destain Buffer (10% MeOH, 10% acetic acid, 80% ddH_2O) for ~16 h at 22°C.
21. It is our experience that both anti-phospho-p44/42 MAPK and anti-p44/42 MAPK may be reused ~20 times after which fresh primary antibody solutions should be prepared.
22. Densitometry should be performed to quantify the effect of EGF on MAPK activation. Scan the exposed Kodak BioMax XAR X-ray film into a computer and analyze using a suitable quantification program. Several software packages are available. We recommend ImageJ (available free from http://rsb.info.nih.gov/ij/index.html) as it is both accurate and easy to use. The densitometrical value of each sample when probed with anti-phospho-p44/42 MAPK should be corrected for the value of the same sample when probed with anti-p44/42 MAPK.
23. Incubation of the Immobilon-P Transfer Membrane in stripping buffer may not remove all of the protein-bound primary antibody. When reprobing for a protein of size similar to that already imaged, it is advisable to verify that all of the primary antibody has been removed. After completing step 4 in Subheading 3.4, add the secondary antibody and incubate for 1 h at 22°C without first adding the primary antibody solution. Remove the secondary antibody and wash the Immobilon-P Transfer Membrane three times for 10 min each with 20 ml TBS-T. Repeat steps 13–15 in Subheading 3.3. A positive signal indicates that not all of the primary antibody has been removed during the stripping process. Stripping again, or increasing the temperature at which the Immobilon-P Transfer Membrane is incubated in stripping buffer to 80°C, may solve this problem.
24. Seventy to eighty percent confluence is recommended for microscopy studies in order to ensure that individual cells are clearly visible under the microscope.
25. Confocal laser scanning microscopy allows high-resolution optical images to be obtained. The defining feature is the ability to optically section a sample and thus effectively produce a

three-dimensional image. Confocal microscopes are commonly used in immunofluorescence studies as they generally obtain much higher quality images than would be afforded by a fluorescent microscope.

Acknowledgments

We thank Zhigang Li for critically reviewing the text prior to submission and other members of the Sacks laboratory, past and present, for insightful discussions. Work in the authors' laboratory is funded by the National Institutes of Health (to D.B.S) and the Department of Defense Breast Cancer Research Program (to C.D.W).

References

1. Cuevas, B. D., Abell, A. N. and Johnson, G. L. (2007). Role of mitogen-activated protein kinase kinase kinases in signal integration. *Oncogene* **26,** 3159–71.
2. McKay, M. M. and Morrison, D. K. (2007). Integrating signals from RTKs to ERK/MAPK. *Oncogene* **26,** 3113–21.
3. Casar, B., Pinto, A. and Crespo, P. (2008). Essential role of ERK dimers in the activation of cytoplasmic but not nuclear substrates by ERK-scaffold complexes. *Mol Cell* **31,** 708–21.
4. Agell, N., Bachs, O., Rocamora, N. and Villalonga, P. (2002). Modulation of the Ras/Raf/MEK/ERK pathway by Ca^{2+} and calmodulin. *Cell Signal* **14,** 649–54.
5. Rosen, L. B., Ginty, D. D., Weber, M. J. and Greenberg, M. E. (1994). Membrane depolarization and calcium influx stimulate MEK and MAP kinase via activation of Ras. *Neuron* **12,** 1207–21.
6. Medema, J. P., Sark, M. W., Backendorf, C. and Bos, J. L. (1994). Calcium inhibits epidermal growth factor-induced activation of $p21^{ras}$ in human primary keratinocytes. *Mol Cell Biol* **14,** 7078–85.
7. Berridge, M. J., Lipp, P. and Bootman, M. D. (2000). The versatility and universality of calcium signaling. *Nat Rev Mol Cell Biol* **1,** 11–21.
8. Bardo, S., Cavazzini, M. G. and Emptage, N. (2006). The role of the endoplasmic reticulum Ca^{2+} store in the plasticity of central neurons. *Trends Pharmacol Sci* **27,** 78–84.
9. Endo, M. (2006). Calcium ion as a second messenger with special reference to excitation-contraction coupling. *J Pharmacol Sci* **100,** 519–24.
10. Pressman, B. C. (1976). Biological applications of ionophores. *Annu Rev Biochem* **45,** 501–30.

Chapter 10

Identification of Novel Substrates of MAP Kinase Cascades Using Bioengineered Kinases that Uniquely Utilize Analogs of ATP to Phosphorylate Substrates

Hui Zheng, Adnan Al-Ayoubi, and Scott T. Eblen

Abstract

The Mitogen-Activated Protein Kinase (MAPK) family of signaling molecules regulates a number of cellular processes through the direct phosphorylation and regulation of a plethora of cellular proteins. Identifying the direct substrates of the MAPK pathway proteins is important for determining how the effects of MAPK activation have such profound effects on cell biology. In this chapter, we describe one method for specific labeling and identification of direct MAPK substrates. A single or double point mutation is generated within the ATP binding domain at a particular residue(s) termed the "gatekeeper" that comes into close contact with the N6 position of ATP. Most kinases contain an amino acid larger than alanine at this position. Mutation of the residue(s) to glycine or alanine generates a "pocket" that allows the mutant kinase to bind and uniquely utilize an analog of ATP that contains a chemical substituent at the N6 position. When radiolabeled analog is added to the mutant kinase and a complex mixture of cellular proteins, the only proteins that become radiolabeled are direct substrates of the mutant kinase. To label biologically relevant substrates, we take advantage of the direct binding of MAPKs to their substrates. An epitope tagged mutant kinase is expressed in cells and immunoprecipitated with associated substrates, which are then radiolabeled in an in vitro kinase reaction using (γ-^{32}P) ATP analog. Larger, unlabeled kinase reactions are run in parallel and used to identify the substrates by mass spectrometry.

Key words: ERK, MAPK, p38, MEK1, ATP, Analog, Substrate, Phosphorylation

1. Introduction

The MAP kinase (MAPK) intracellular signaling cascades are activated by a number of extracellular and intracellular stimuli, including growth factors, cytokines, cell stress, cell adhesion, chemotherapeutic drugs, reactive oxygen species, and irradiation. MAPK activation has been linked to a number of cellular processes, including proliferation, migration, apoptosis, and

Rony Seger (ed.), *MAP Kinase Signaling Protocols: Second Edition*, Methods in Molecular Biology, vol. 661, DOI 10.1007/978-1-60761-795-2_10, © Springer Science+Business Media, LLC 2010

differentiation. In order to understand how MAPKs have so many effects on cellular physiology, one must first identify the specific targets of MAPK pathways and how the temporal phosphorylation of the substrates that they modify controls the diverse array of biological responses that occur in response to MAPK activation. While many MAPK substrates have been identified, identification of novel substrates is a very active area of investigation. In this chapter, we focus on the identification of novel substrates of MAPK signaling by the use of kinases mutated to allow them to utilize analogs of ATP to phosphorylate their direct substrates, a technique that was originally developed by Kevan Shokat and Kavita Shah for use with Src tyrosine kinase (1). We have utilized this technique to modify the MAPKs ERK2 (extracellular regulated kinase 2) (2) and p38α, as well as MAPK kinase MEK1/MKK1. This technique has been utilized by others on Raf-1 (3), Jun N-terminal kinase 1 (JNK1) (4), and p38α (5).

The technique for identifying novel kinase substrates relies on the observation that protein kinases have structurally similar ATP-binding domains, with most containing one or more amino acids with a large side chain at a key conserved position(s) that helps regulate the size of the domain around the N^6 position of ATP (Fig. 1a). This amino acid, termed the "gatekeeper" residue (1), comes into close contact with the N^6 position of bound ATP. Mutation of this residue to a smaller amino acid, such as glycine or alanine, creates extra space in the ATP binding site around the

a

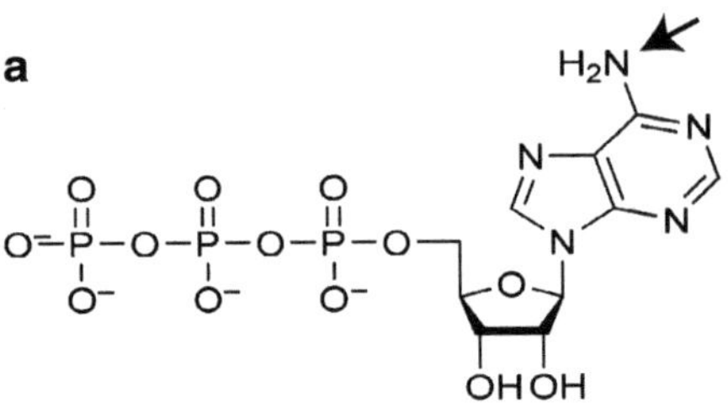

b

ERK2 QMKDVYIV***Q***DLMET-108
p38α EFNDVYLV***T***HLMGA-111
MEK1 SDGEISIC***M***EHMDG-148
JNK EFQDVYIV***M***ELMDA-113--TLKI***L***DFGL-172
Raf1 TKDNLAIV***T***QWCEG-426--NNI***F***LHEGL-480

c

ERK2 Q103G
p38α T106G
MEK1 M143G
JNK M108G/L168A
Raf1 T421A/F475L

Fig. 1. Alignment of MAPK family gatekeeper mutations. (**a**) ATP, with an *arrow* pointing to the N^6 position. (**b**) Alignment of MAPK family proteins, with the gatekeeper residues in *bold italicized*. (**c**) ATP-binding domain mutations in MAPK family proteins.

N^6-position of ATP (1). The increased size of the ATP binding site allows the kinase to still utilize ATP in most cases and also allows the mutant kinase to utilize ATP analogs that have a bulky side group chemically synthesized onto the N^6 position. Depending on the N^6-ATP analog, the wild-type kinase and other cellular protein kinases are unable to efficiently utilize the N^6-ATP analog to phosphorylate substrates due to the large amino acid at the gatekeeper position (1). If the mutant kinase is now used in a kinase reaction with radioactive ATP analog and a simple or complex mixture of proteins, the only proteins that will become radiolabeled are direct substrates of the mutant kinase. The kinase reaction is then run on a gel and labeled proteins visualized by a combination of fluorescent staining and autoradiography. These candidate substrates are then targeted for identification by mass spectrometry.

The methodology that we employ first requires identification and mutation of the "gatekeeper" residue in the kinase, which can be deduced by sequence alignments (Fig. 1) and crystal structure data. The effect of the gatekeeper mutation on kinase activity is then determined. The gatekeeper mutation in subdomain V has a positive effect on ERK2 activity (Fig. 2a) (2,6), partially inhibits p38α activity (Fig. 2a, b), and nearly completely inhibits the kinase activity of Raf (3) and MEK1 (Fig. 3) with normal ATP, although the comutation of subdomain VII partially restores Raf-1 activity (3). Next, N^6-ATP analogs are screened with the mutant kinase for their use as an ATP source to phosphorylate a known, recombinant substrate (2). Commonly used analogs are listed in Subheading 2.4. Surprisingly, MEK1 activity is partially restored by the use of N^6-benzyl-ATP as the ATP source (Fig. 3). Additional tests of the gatekeeper mutant in signaling assays, such as transcriptional reporter assays (2,3), can be performed to determine deleterious effects of the mutation on kinase activity.

Once a mutant kinase and ATP analog pair have been identified (see Note 1), the ADP form of the analog is made radioactive by a simple two-step reaction using nucleotide diphosphate kinase and (γ-^{32}P) ATP. It should be noted that there exists a method to use unlabeled ATP-analog to label potential substrates, where the labeled phosphate is chemically modified and then isolated with an antibody specific to the chemical modification (7). The use of radioactive analog is the method of choice in our laboratory and will be discussed in this protocol.

Certain biological characteristics require that the substrates be labeled in in vitro reactions instead of cells by the mutant kinase and (γ-^{32}P) ATP analog owing to the high (3 mM) concentration of ATP within the cell (8), the impermeability of the plasma membrane to cells to ATP, and the immediate removal of the labeled (γ-^{32}P) from the ATP analog after its introduction into digitonin-permeablized cell (9). To label potential substrates in vitro,

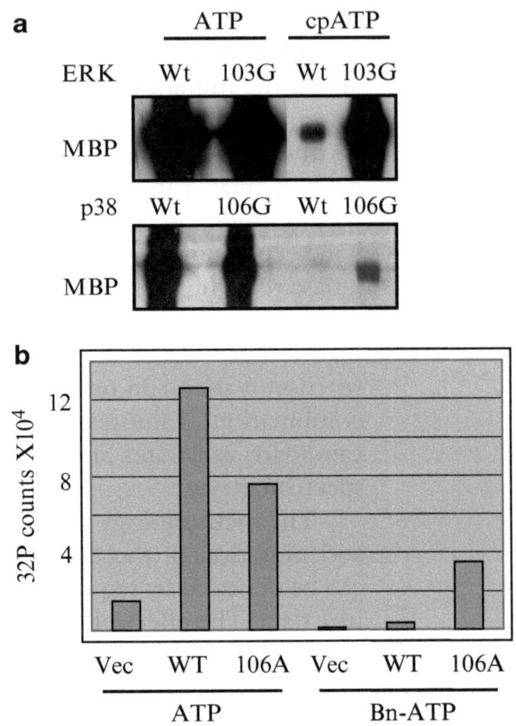

Fig. 2. Utilization of ATP analogs by ERK2 and p38α. (**a**) Plasmids encoding FLAG-ERK2, FLAG-ERK2-Q103G, FLAG-p38α and FLAG-p38α-T106G were individually transfected into cells. FLAG immunoprecipitates from cells stimulated with EGF (for ERK) or H_2O_2 (for p38) were tested for their ability to use [γ-^{32}P]-ATP or [γ-^{32}P]- N^6-cyclopentyl ATP to phosphorylate MBP in an in vitro kinase assay. The samples were run on a gel, transferred to nitrocellulose and exposed for autoradiography. (**b**) Same as in (**a**), except only FLAG-p38α and FLAG-p38α-T106G were tested with [γ-^{32}P]-ATP or [γ-^{32}P]-N^6-benzyl ATP. The labeled MBP bands were excised from the membrane and counted on a scintillation counter.

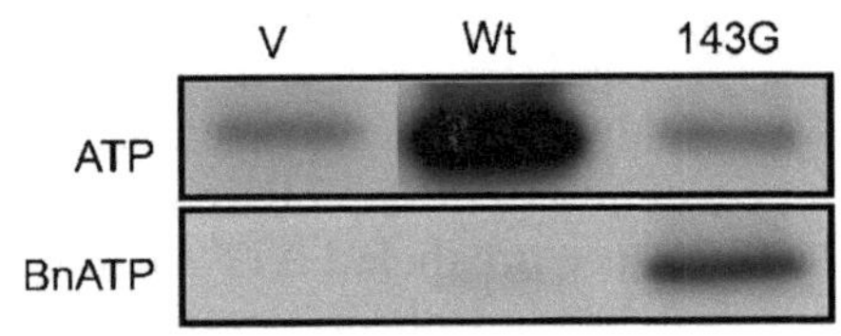

Fig. 3. MEK1 activity is lost with the gatekeeper mutation using [γ-^{32}P]-ATP, but is partially restored when using [γ-^{32}P]-N^6-benzyl ATP. FLAG-MEK1 or FLAG-MEK1-143G were transfected into cells and the cells stimulated with EGF for 2 min. FLAG-immunoprecipitates were tested for their ability to phosphorylate ERK2-K52R with either form of ATP.

recombinant mutant kinase can be added to a dialyzed cellular lysate along with ATP analog. This method will work to label substrates; however, there exists the possibility that the large amount of mutant kinase typically used in this assay phosphorylates proteins that are not biologically relevant, although a candidate substrate identified by any labeling method will have to be independently validated. Our preferred method to label biologically relevant substrates is to take advantage of the direct binding interactions between MAPKs and their substrates, which occur through specific docking domains on the kinases (10). For this method, an epitope-tagged, wild-type or mutant kinase is transiently or stably expressed in mammalian cells, activated, and immunoprecipitated under gentle conditions to retain protein–protein interactions. Nonspecific proteins and cellular ATP are removed by gentle washing, and the radioactive analog is added to the immunoprecipitates to allow phosphorylation of bound substrates by the mutant kinase. The samples are then run on a gel and exposed to autoradiography. Phosphorylated bands that appear in lanes containing the mutant kinase, but not in parallel reactions containing the wild-type kinase, are considered specific and pursued further. Additional parallel reactions using the mutant kinase and unlabelled ATP analogs can be scaled up by using more cells and run on gels alongside the radioactive kinase reactions. The corresponding bands in the unlabelled reactions can then be excised from the gel and identified by mass spectrometry. Overall, this method is effective for the identification of MAP kinase substrates (see Note 2), although other methods have been developed (see Note 3).

2. Materials

2.1. Mutagenesis

1. Quick Change Mutagenesis kit (Stratagene) or another method for gene mutation.

2.2. Cell Culture

1. SKOV-3 ovarian cancer cells (ATCC) are grown in McCoy's 5A (Invitrogen, Carlsbad, CA.) supplemented with 10% (v/v) fetal bovine serum (FBS, Hyclone, Ogden, UT) (see Note 4). Culture the cells on tissue culture dishes (Corning) and incubate them in a humidified incubator at 37°C and 5% CO_2.
2. Solution of trypsin (0.25%) and ethylenediamine tetraacetic acid (EDTA) (1 mM) (Invitrogen).
3. 1X phosphate buffered saline (PBS): 25.6 g Na_2HPO_4-$7H_2O$, 80 g NaCl, 2 g KCl, 2 g KH_2PO_4. Bring to 1 L with distilled, deionized water. Autoclave for 40 min at 121°C.
4. LipofectAMINE 2000 transfection reagent (Invitrogen).
5. Epidermal growth factor (EGF) (R&D Systems, Minneapolis, MN).
6. H_2O_2 (Sigma, St Louis, MO).

2.3. Cell Lysis and Immunoprecipitation

1. M2 lysis buffer (11): 50 mM Tris-base pH 7.4, 150 mM NaCl, 10% glycerol, 1% Triton X100, 0.5 mM EDTA, 0.5 mM ethylene glycol tetraacetic acid (EGTA), 50 mM NaF, 40 mM β-glycerophosphate, 5 mM tetrasodium pyrophosphate, 0.1 mM Na orthovanadate, 10 μg/mL aprotinin, 5 μg/mL leupeptin, and 2 mM phenylmethylsulfonyl fluoride (PMSF) (make fresh in 95% ethanol).
2. M2 anti-FLAG antibody preconjugated to agarose (Sigma).
3. BCA Protein Assay kit (Pierce, Rockville, MD).

2.4. Kinase Reactions

1. 10X kinase buffer: 250 mM HEPES, pH 7.4, 100 mM Mg acetate, and 10 mM dithiothreitol (DTT, make fresh).
2. 10X HEPES buffered saline (HBS): 200 mM HEPES, pH 7.4 and 1.5 M NaCl.
3. [γ-^{32}P]-ATP (6,000 Ci/mmol, Perkin Elmer, Waltham, MA). Store at −70°C. CAUTION: RADIOACTIVE
4. Kinase substrate. Potential substrates to screen with include: Elk1 (2) or myelin basic protein (MBP) (Fig. 2) for ERK, MBP (Fig. 2) and MAPKAP kinase 2 for p38, c-Jun and ATF2 for JNK, MKK1 for Raf-1, and ERK2 K52R for MKK1 (Fig. 3).
5. ATP analogs and ADP analogs (store in aliquots at −20°C and at −70°C when radioactive). Analogs that have been used are: N^6-cyclopentyl ATP for ERK (2), N^6-benzyl ATP for p38 (5) (Fig. 2 and 4), and MEK1 (Fig. 3) and N^6(2-phenethyl) ATP for Jnk1 (4) and Raf-1 (3). ADP and ATP analogs are available from AXXORA, LLC (San Diego, CA), a distributor of products from the Biolog Life Science Institute (Bremen, Germany).
6. 1 M $MgCl_2$.
7. Microspin G50 columns (Amersham Pharmacia Biotech).
8. YM30 columns (Millipore).
9. Nucleotide diphosphate kinase (NDPK), 1,000 units (Sigma).
10. Adenosine triphosphate (Sigma).
11. Acrylic shielding for radioactive work.
12. Personal protective equipment (PPE), including safety glasses.

2.5. SDS-Polyacrylamide Gel Electrophoresis

1. Gel separating buffer. For 30 mL of a 10% polyacrylamide gel: 10 mL of 30% acrylamide/bis (37.5%:1, Bio-Rad, Hercules, CA, stored a 4°C, CAUTION: NEUROTOXIC), 5.6 mL of 2 M Tris base pH 8.8, 14 mL dionized H_2O, 0.1% sodium dodecyl sulfate (SDS), 10 μL *N,N,N,N′*-tetramethyl-ethylenediamine (TEMED, Bio-Rad), and 200 μL of 10% ammonium persulfate (APS) (Bio-Rad; store powder

desiccated at room temperature, prepare fresh, and add last). For a 15% gel, increase the acrylamide to 15 mL and decrease the water to 9 mL.

2. Stacking buffer. For 10 mL: 1.33 mL of 30% acrylamide/bis (37.5%:1), 1.25 mL of 1 M Tris-HCl pH 6.8; 0.1% SDS, 10 μL *N,N,N,N'*-tetramethyl-ethylenediamine (TEMED, Bio-Rad), and 50 μL of 10% APS (add last).
3. SDS running buffer: Make 1 L of a 10X stock containing 30.3 g Tris base, 144 g glycine, and10g SDS. Dilute to 1X before use with deionized water.
4. Hoefer SE 600 and SE 660 gel apparatus.
5. 4X Laemmli sample buffer (LSB). For 10 mL: 2.4 mL 1 M Tris-HCl pH 6.8, 4 mL glycerol, 0.8 g SDS, and 0.01% bromophenol blue. Store at 25°C. Add 1 mL β-mercaptoethanol prior to use.
6. Prestained protein molecular weight markers (Bio-Rad). Store at −20°C.
7. Microcentrifuge tube holders (LabScientific Inc., Livingston, NJ).

2.6. Gel Transfer

1. Transfer buffer: Make 4 L of 1X buffer containing: 12.12 g Tris base, 56.6 g glycine, and 800 mL methanol. Bring to 4 L with deionized water.
2. Bio-Rad gel transfer tank with cassettes and pads.

2.7. Gel Fixation and Staining

1. Gel fixing solution: 40% methanol, 10% acetic acid.
2. SYPRO Ruby gel stain (Bio-Rad). Protect from light.

3. Methods

3.1. Identification and Mutation of the "Gatekeeper" Residue

1. Based on existing crystal structure data, if available, determine the residue(s) in ATP binding domain that comes into close contact with the N-6 position of ATP. These can be in both kinase subdomains V and VII (3,4,12). Additional information on the choice of a residue can come from sequence alignment with kinases that have been successfully engineered to utilize an N^6-ATP analog, as in Fig. 1. Alignment of sequences containing and surrounding subdomain V and VII is more informative than alignments of whole proteins or whole kinase domains. If it is unclear which residue to mutate, it may be necessary to make several mutants and screen them in kinase assays with a known substrate and ATP analogs to see which mutation allows for analog use (see Notes 1 and 5).

2. Perform site-directed mutagenesis to mutate the putative gatekeeper residue to glycine or alanine using the Quick Change Mutagenesis kit.

3.2. Cell Culture and Transfection

1. Plate SKOV-3 cells onto 60 mm dishes at a density of 3×10^5 cells per dish in 5 mL of complete media. The following day, transfect the cells with a mammalian expression vector encoding a cDNA of your epitope-tagged kinase. We prefer to use an N-terminal FLAG epitope tag (amino acid sequence DYKDDDDK). For transfection of SKOV-3 cells, we use a 4:1 ratio of LipofectAMINE 2,000:DNA, with 2 μg of plasmid DNA per 60 mm tissue culture dish. Tranfect cells for 6 h according to the manufacturer's protocol. After transfection, aspirate the media, wash the cells with 1X PBS, and incubate in 10% FCS McCoy's 5A overnight.
2. The following day, aspirate the media and wash the cells twice with 1X PBS. Starve the cells for 12–24 h in serum-free McCoy's 5A. Stimulate the cells with an activator of the MAPK pathway that you are studying. We recommend EGF treatment (10 ng/mL) for 5 min for activation of the ERK pathway, 400 H_2O_2 for 10 min for activation of the p38 pathway, and ultraviolet (UV) irradiation (125 J/m^2) for 10 min for activation of the JNK pathway. After stimulation immediately place the dishes on ice and wash twice with cold 1X PBS.

3.3. Cell Lysis and MAPK Immunoprecipitation

1. Add 1 mL of M2 lysis buffer to the dish, scrape the cells off, and pipette the cells into a microfuge tube on ice. Vortex occasionally over the next 20 min. Centrifuge at $15{,}000 \times g$ for 15 min at 4°C. Transfer the supernatant to a fresh microfuge tube on ice and discard the pellet. Perform a BCA protein assay to determine protein concentration.
2. Immunoprecipitate the kinase from an equal amount of protein (50 μg or more per lysate, brought up to 750 μL with lysis buffer) with M2 anti-FLAG antibody conjugated to agarose for a minimum of 1 h at 4°C with constant rotation. Centrifuge at $13{,}000 \times g$ for 30 s at 4°C. Aspirate the supernatant. Wash the pellet three times, each with 1 mL of M2 lysis buffer, centrifuging the immunoprecipitate between washes. Wash the pellet twice with 1 mL of 1X kinase buffer, with centrifugation between washes.

3.4. Radiolabeling of ADP Analogs with (γ-^{32}P)

This procedure can be performed during the immunoprecipitation.

1. Resuspend lyophilized NDPK (1,000 U) in 200 μL of deionized water and store in 40 μL aliquots at −20°C, until use.

2. Dilute 1 m Ci of (γ-^{32}P)-ATP (6,000 Ci/mmol) with deionized water to 20 μCi/μL (total volume of 50 μL) and store 40 μL aliquots at −70°C, until use. It is recommended that the radioactive ATP be used for generating labeled ATP analog within days of arrival as the half-life of ^{32}P is only 14.5 days. Use shielding and personal safety equipment when working with radioactivity.
3. Combine one aliquot each of NDPK (200 Units) and (γ-^{32}P) ATP (800 μCi) into a 100 μL reaction containing 1X HBS and 5 mM $MgCl_2$. Place the reaction at 25°C for 5 min to allow the NDPK to transfer the radiolabeled γ-phosphate from the ATP onto itself.
4. The [^{32}P]-labeled NDPK is purified from the ADP and unreactive (γ-^{32}P)-ATP using successive centrifugation in Microspin G50 columns according to the manufacturer's instructions. Two columns are needed per reaction per centrifugation due to the 50 μL loading capacity of the columns. Break off and discard the stopper tab at the bottom of each column and place the columns into microfuge tubes. Centrifuge the columns in the tubes for 1 min at 3,000 × *g* to remove excess buffer in the columns. Place the columns into fresh tubes.
5. Pipette 50 μL of the reaction into each of two columns. Centrifuge for 1 min at 3,000 × *g*. Collect the flowthrough and load into the second set of columns. Centrifuge as above. The (^{32}P) NDPK will be in the flowthrough.
6. Combine the purified (^{32}P) NDPK with 1,000 picomoles of ADP analog (be sure to use the ADP form) and 1X HBS in a 150μL reaction volume. Incubate the mixture at 30°C in a water bath for 20 min to allow transfer of the [^{32}P] from the NDPK to the ADP analog.
7. Transfer the reaction to a Microcon YM30 column placed into a microcentrifuge collection tube and centrifuge for 15 min at 15,000 × *g* and 4°C. [γ-^{32}P] ATP analog will be present in the flowthrough, while the NDPK will be retained in the column. Discard the column into a radioactive waste container. We usually obtain 100–200 μCi of purified [γ-^{32}P]-ATP analog. We have determined that the remaining [γ-^{32}P]-ATP contamination after purification constitutes between 0.01–0.05% of the product after these two purification steps. Store the radiolabeled ATP analog in aliquots at −70°C, until use. Discard the column and all previous materials, including tips, in a radioactive waste container.

3.5. Kinase Assay

This general kinase assay protocol can be used to test the effect of the mutation on kinase activity and also to test the ability of the mutant kinase to use ATP analogs (see Note 6). This assay will be adapted in later sections for use to label substrates associated with a MAPK pathway protein.

1. Prepare a kinase reaction mixture (40 μL per reaction) containing: 1X kinase buffer, 100 μM ATP, 20 μg of substrate per reaction, and 10 μCi [γ-^{32}P]-ATP or 10 μ Ci [γ-^{32}P]-ATP analog per reaction. CAUTION: RADIOACTIVE.
2. Aspirate the remaining kinase buffer from the immunoprecipitates using a 27 ga needle at the end of a 1 mL syringe attached to a vacuum flask. Add 40 μL of the kinase reaction mixture to each immunoprecipitate, mix by gently flicking the tube, and incubate in a 30°C water bath for 10 min. Stop the reaction by adding 14 μL of 4X LSB. Vortex and heat the kinase reaction in a 100°C heating block for 3 min (see Note 7). Allow the samples to cool for 2 min and centrifuge at 13,000 × *g* for 30 s.

3.6. SDS-Polyacrylamide Gel Electrophoresis

1. Make a 16 cm 10–15% gradient polyacrylamide SDS gel (see Note 8 for an easy way to make a gradient gel). Pour the mixture between the gel plates immediately after adding APS. Overlay the gel with water-saturated butanol.
2. Allow the gel to polymerize (45 min), pour off the butanol, and rinse the top of the gel with deionized water. Remove excess water by tilting the gel plate to the side and blotting with a Chemwipe. Pour the stacking gel buffer and insert a gel comb. Allow the stacking gel to polymerize completely before removing the gel comb.
3. Load your samples onto the gel and load one lane with prestained molecular weight markers. Run the gel at 50 mA for approximately 2.25 h. Stop the gel when the blue dye front is 2 cm from the bottom of the gel to prevent free [γ-^{32}P]-ATP from running off of the gel and contaminating the buffer. Cut off the stacker and the bottom of the gel just above the dye front with a spacer and discard both pieces into a radioactive waste container.

3.7. Transfer of the Gel to Nitrocellulose Membrane

1. Cut a piece of nitrocellulose membrane and two pieces of 3 M Whatman paper of the size of the gel. Open the transfer cassette and place the black side down in a pan (we use a small autoclave pan that is only used for gel transfers) containing 2 L of transfer buffer, covering the cassette. Place a pad down in the buffer on the cassette. Place one piece of the Whatman paper onto the pad. Gently remove the gel from the glass plate and place on the Whatman paper.
2. Overlay the nitrocellulose onto the gel. Use half of a broken 10 mL pipette to gently smooth the membrane onto the gel, removing any air bubbles. While holding the membrane in

place, place the second piece of Whatman paper on top of the nitrocellulose and smooth out with the pipette. Place the other pad from the cassette on top of the paper. Close the cassette and slide the sealer across to seal.

3. Place a small stir bar into the transfer tank and place the tank into a small autoclave pan on a stir plate. Place the cassette into the transfer tank with the black side facing the black plate electrode. Put the lid on the transfer tank and connect to a power supply. Turn on the stir plate and pack the tank with crushed ice to cool the buffer, which will heat up during the transfer.
4. Turn on the power supply and transfer the gel with constant amperage for a total of 2.5 amps, spread out over at least a 3 h period. If an 8 h transfer is performed at lower voltage, ice and stirring is not required.
5. After transfer is complete, disassemble the transfer apparatus and remove the nitrocellulose membrane. Place the membrane into a glass dish with 100 mL of Ponceau S stain. Stain the gel for 5 min on a rocking platform. Remove the stain (it can be saved and used multiple times) and destain three times for 5 min each with 100 mL of 1% acetic acid.
6. Wrap the membrane in Saran Wrap and expose to autoradiography in a film cassette at –70°C. After autoradiography, cut out the individual substrate bands from the gel, put them into scintillation vials, add 5 mL of liquid scintillation fluid, and count in a scintillation counter. Divide the number of counts from the assay with the mutant kinase by the number in the assay with the wild-type kinase to obtain percent activity.

3.8. Phosphorylation of MAPK-Associated Substrates

1. Transfect SKOV-3 cells with either wild-type (as a negative control, one dish) or mutant MAPK (see below). Transfections can be either transient (24–72 h) or stable. For stable transfections, select cells for 2 weeks with an appropriate drug, based on the drug resistance gene that is in your plasmid. For these experiments, we use as many as twenty 15 cm dishes of cells transfected with the mutant kinase to obtain enough protein for mass spectrometry. Use 15 μg of DNA and 60 μL of LipofectAMINE 2,000 per dish.
2. Stimulate your cells with an agonist to your MAPK pathway. Harvest the cells in M2 lysis buffer, 1.5-mL per dish. Incubate the lysate on ice for 20 min, inverting occasionally; do not vortex.
3. Centrifuge the samples at 15,000 ×*g* for 15 min at 4°C. Transfer 5% of the supernatant to a fresh microfuge tube on ice. Transfer the remainder to a 50-mL conical tube.

4. Immunoprecipitate the FLAG epitope-tagged kinase (wild type and mutant in microfuge tubes, plus mutant in 50-mL conical flask) with M2 anti-FLAG agarose for 1 h at 4°C with constant rotation.
5. Centrifuge the tubes at 6,000 × *g* for 30 s. Transfer the pellet from the 50-mL conical flask to a microfuge tube with 750 μL of kinase buffer. Aspirate the supernatant and gently wash the pellets twice with 1X kinase buffer. Mix by inversion and do not vortex.
6. Aspirate the last of the supernatants with a 27 ga needle attached to a vacuum hose.
7. Add 40 μL of 1X kinase buffer containing 10 μ Ci (γ-^{32}P)-ATP analog to each small immunoprecipitate. Mix gently by flicking the tube. Add 100 μL of 1X kinase buffer containing unlabelled ATP analog to the large immunoprecipitate.
8. Incubate in a 30°C water bath for 3 min (see Note 9). Stop the reactions by adding 4X LSB. Heat the samples to 100°C in a heating block for 3 min. Centrifuge at 6,000 × *g* for 30 s.

3.9. SDS-PAGE, Gel Fixation and Staining

1. Run the samples on a 10% SDS PAGE gel. We prefer to use 24 cm gels for these assays due to the improvement of protein separation on a longer gel. Run the gel at 50 mAmps for approximately 3.5 h. Stop the gel when the blue dye front is 2 cm from the bottom. Cut off the stacker and the bottom of the gel just above the dye front. Discard these pieces into a radioactive waste container.
2. Place the gel in a clean polycarbonate or polypropylene dish on a platform rocker. Fix the gel in 500 mL of fixing solution for 1 h with constant agitation.
3. Stain the gel with Sypro Ruby stain from 3 h to overnight with gentle agitation, according to the manufacturer's protocol. Protect the gel from light during the staining and destaining process.
4. Destain the gel for 1 h, with agitation, at room temperature in destain solution.
5. Visualize the proteins using a fluorescence scanner (Typhoon 9200) set at 532 nm.
6. Cut the gel vertically and expose the radiolabeled lanes for autoradiography at –20°C. Example autoradiographs with differentially radiolabeled proteins are shown in Fig. 4a, b. Compare the autoradiograph to the stained gel.

3.10. Identification of Labeled Substrates by Mass Spectrometry

1. Excise bands from the nonlabeled portion of the gel that correspond to radiolabeled bands present in the lanes from the kinase assay with the mutant, but not the wild-type, kinase (see Notes 6 and 7).

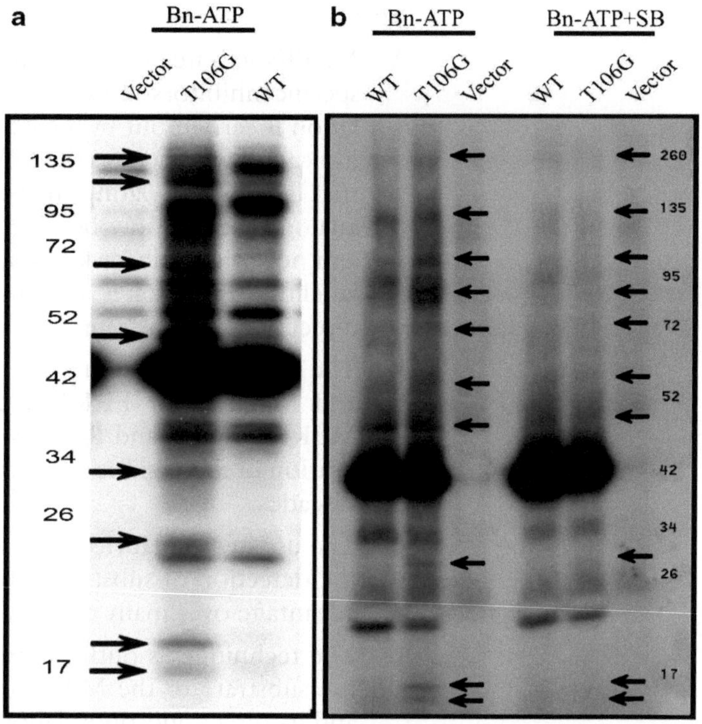

Fig. 4. Specific labeling of FLAG-p38α-T106G with [γ-^{32}P]-N^{6}-benzyl ATP. (**a**) Cells were transfected with vector, FLAG-p38α, or FLAG-p38α-T106G and stimulated with H_2O_2. FLAG immunoprecipitates were washed twice in kinase buffer and incubated with [γ-^{32}P]-N^{6}-benzyl ATP. The reactions were run on a gel, transferred to nitrocellulose and exposed for autoradiography. Phosphoproteins specific to the FLAG-p38α-T106G sample are indicated with *arrows*. (**b**) A similar experiment as in (**a**), with the inclusion of additional immunoprecipitates that were pre-incubated with the p38 inhibitor SB203580 for 15 min prior to the addition of [γ-^{32}P]-N^{6}-benzyl ATP. *The arrows* indicate the loss of FLAG-p38α-T106G specific phosphoproteins.

2. Extract the proteins from the band and perform mass spectrometry to identify the candidate substrates. We typically perform LC-MS/MS in the MUSC Proteomics Facility.

4. Notes

1. Your wild-type kinase should not be able to use the ATP analog, whereas your mutant kinase should have some reactivity towards substrates in the presence of the analog. Chose an analog that is used well by the mutant kinase, but not by the wild-type kinase.
2. There are several pros and cons to the methodology described in this protocol:

Pros:

A. MAPKs activate other protein kinases. The use of pathway-specific inhibitors allows one to determine if a phosphorylation is carried out by a particular pathway, but gives no indication whether it is the MAPK or a downstream kinase that is phosphorylating the substrate. The use of an ATP analog that can only be used by the engineered MAPK, and not other cellular kinases, provides high specificity in labeling of substrates and ensures that the substrate is a direct MAPK substrate.

B. The gatekeeper residue has been successfully identified and mutated for ERK2 (2), JNK (4), p38α (5) and MEK1 (Fig. 3), and Raf-1(3), allowing for easier identification of this residue in other components of the MAPK cascades.

C. The use of radiolabeled ATP allows for high sensitivity in the detection of substrates after SDS PAGE, which is an advantage over many other techniques.

D. The technique as outlined relies on both the binding of the substrate to the MAPK, either directly or indirectly, and its direct phosphorylation by the kinase. This greatly decreases the likelihood of obtaining false positives when identifying substrates, which is an advantage over other techniques that look at global changes in phosphorylation in response to MAPK activation.

Cons:

A. The protocol uses radioactivity to label substrates. Shokat and colleagues (7) have developed a method to chemically modify phosphorylated residues, which can then be isolated with an antibody that recognizes the modified residue on the protein; however, this alternative method is not as sensitive as the isolation step requires a larger amount of the protein to be phosphorylated for the initial detection.

B. The technique requires that enough substrate be co-immunoprecipitated with the MAPK to be identified by mass spectrometry.

3. A number of strategies have been developed for the identification of MAPK substrates. These include two-hybrid assays (13,14), proteomic approaches (15), in vitro phosphorylation of column fractions from cell lysate (16), a solid phase phosphorylation assay (17), and an integrated functional genomic approach (18).
4. This protocol can be performed in any cell type that can be transfected or transduced. The cell type chosen should be one in which you have determined, generally through the use

of MAPK pathway specific inhibitors, that a pathway is involved in regulating a cellular or molecular phenomenon. This can be particularly important in cells in which there is hyperactivation of the pathway, which can be determined by generating lysates from cell cultures in log phase growth and Western blotting with phospho-specific MAPK antibodies. If a starvation–stimulation approach is used on the cells, the starvation conditions and timing of MAPK activation in response to the stimulus will have to be optimized for each cell type. It is particularly useful to generate a cell line that stably expresses the mutant kinase due to the requirement for obtaining enough co-immunoprecipitated substrate for identification by mass spectrometry. We have found that stable cell lines are best generated via viral transduction.

5. Additional control reactions for substrate labeling specificity include pretreatment of the cells with a small molecule inhibitor to an upstream kinase before cell stimulation and treatment of the immunoprecipitate with a small molecule inhibitor to the kinase prior to the kinase reaction for associated proteins, as is shown in Fig. 4b with p38 and SB203580.
6. To screen for potential ATP analogs to use with a mutant kinase, one can use nonradioactive ATP analogs if there is a phospho-specific antibody available for the phosphorylation site(s) on the substrate. Simply add 100 μM unlabelled ATP analog in the kinase reaction in place of the labeled analog. Perform the kinase assay, run and transfer the gel, and then perform a Western blot with the phospho-specific antibody. This is an excellent method to screen several ATP analogs at once for use by a kinase mutant without the requirement for labeling multiple ADP analogs with ^{32}P (2).
7. When heating samples in a heating block at 100°C, place a microfuge tube holder on the cap to keep the cap from popping off due to expansion of gases in the tube.
8. For a gradient gel, dissolve 4 g of sucrose into the higher percentage gel mixture (15%) prior to adding SDS. Pipette up 13 mL of the 10% gel mixture into a 25 mL pipette, followed by 13 mL of the 15% gel/sucrose mixture. Remove the pipette from the solution and draw 15–20 air bubbles up into the pipette to partially mix the two gel solutions. It is important not to overmix the two gel solutions or you will end up with a 12.5% gel instead of a 10–15% gradient. Slowly pipette the gel mixture in between the two gel plates to within two inches of the top.
9. Incubations longer than 3 min for kinase assays for associated proteins do not improve labeling of mutant-specific substrates and actually increase labeling of proteins in the control (wild-type) kinase reactions.

Acknowledgments

We thank our past and present collaborators on various aspects of this project: Michael J. Weber (University of Virginia), Vinay K. Nandicoori (National Institute of Immunology, New Delhi India), Kavita Shah (Purdue University), and Carola Neumann (Medical University of South Carolina). We would particularly like to acknowledge Kevan Shokat (University of California, San Francisco), a former collaborator and the pioneer of this technique. This work was supported by Department of Defense grants W81XWH-04-1-0100 to S.T.E. and W81XWH-07-1-0691 to S.T.E. and Carola Neumann.

References

1. Shah, K., Liu, Y., Deirmengian, C., and Shokat, K. M. (1997) Engineering unnatural nucleotide specificity for Rous sarcoma virus tyrosine kinase to uniquely label its direct substrates. *Proc Natl Acad Sci USA* **94,** 3565–70.
2. Eblen, S. T., Kumar, N. V., Shah, K., Henderson, M. J., Watts, C. K., Shokat, K. M., and Weber, M. J. (2003) Identification of novel ERK2 substrates through use of an engineered kinase and ATP analogs. *J Biol Chem* **278,** 14926–35.
3. Hindley, A. D., Park, S., Wang, L., Shah, K., Wang, Y., Hu, X., Shokat, K. M., Kolch, W., Sedivy, J. M., and Yeung, K. C. (2004) Engineering the serine/threonine protein kinase Raf-1 to utilise an orthogonal analogue of ATP substituted at the N6 position. *FEBS Lett* **556,** 26–34.
4. Habelhah, H., Shah, K., Huang, L., Burlingame, A. L., Shokat, K. M., and Ronai, Z. (2001) Identification of new JNK substrate using ATP pocket mutant JNK and a corresponding ATP analogue. *J Biol Chem* **276,** 18090–5.
5. Ulrich, S. M., Sallee, N. A., and Shokat, K. M. (2002) Conformational restraint is a critical determinant of unnatural nucleotide recognition by protein kinases. *Bioorg Med Chem Lett* **12,** 3223–7.
6. Emrick, M. A., Lee, T., Starkey, P. J., Mumby, M. C., Resing, K. A., and Ahn, N. G. (2006) The gatekeeper residue controls autoactivation of ERK2 via a pathway of intramolecular connectivity. *Proc Natl Acad Sci USA* **103,** 18101–6.
7. Allen, J. J., Lazerwith, S. E., and Shokat, K. M. (2005) Bio-orthogonal affinity purification of direct kinase substrates. *J Am Chem Soc* **127,** 5288–9.
8. Weber, M. J., and Edlin, G. (1971) Phosphate transport, nucleotide pools, and ribonucleic acid synthesis in growing and in density-inhibited 3 T3 cells. *J Biol Chem* **246,** 1828–33.
9. Chaudhary, A., Brugge, J. S., and Cooper, J. A. (2002) Direct phosphorylation of focal adhesion kinase by c-Src: evidence using a modified nucleotide pocket kinase and ATP analog. *Biochem Biophys Res Commun* **294,** 293–300.
10. Tanoue, T., Adachi, M., Moriguchi, T., and Nishida, E. (2000) A conserved docking motif in MAP kinases common to substrates, activators and regulators. *Nat Cell Biol* **2,** 110–6.
11. Eblen, S. T., Catling, A. D., Assanah, M. C., and Weber, M. J. (2001) Biochemical and biological functions of the N-terminal, noncatalytic domain of extracellular signal-regulated kinase 2. *Mol Cell Biol* **21,** 249–59.
12. Hanks, S. K., Quinn, A. M., and Hunter, T. (1988) The protein kinase family: conserved features and deduced phylogeny of the catalytic domains. *Science* **241,** 42–52.
13. Waskiewicz, A. J., Flynn, A., Proud, C. G., and Cooper, J. A. (1997) Mitogen-activated protein kinases activate the serine/threonine kinases Mnk1 and Mnk2. *Embo J* **16,** 1909–20.
14. Maekawa, M., Nishida, E., and Tanoue, T. (2002) Identification of the Anti-proliferative protein Tob as a MAPK substrate. *J Biol Chem* **277,** 37783–7.
15. Lewis, T. S., Hunt, J. B., Aveline, L. D., Jonscher, K. R., Louie, D. F., Yeh, J. M., Nahreini, T. S., Resing, K. A., and Ahn, N. G. (2000) Identification of novel MAP kinase pathway signaling targets by functional proteomics and mass spectrometry. *Mol Cell* **6,** 1343–54.

16. Knebel, A., Morrice, N., and Cohen, P. (2001) A novel method to identify protein kinase substrates: eEF2 kinase is phosphorylated and inhibited by SAPK4/p38delta. *Embo J* **20,** 4360–9.

17. Garcia, J., Ye, Y., Arranz, V., Letourneux, C., Pezeron, G., and Porteu, F. (2002) IEX-1: a new ERK substrate involved in both ERK survival activity and ERK activation. *Embo J* **21,** 5151–63.

18. Arur, S., Ohmachi, M., Nayak, S., Hayes, M., Miranda, A., Hay, A., Golden, A., and Schedl, T. (2009) Multiple ERK substrates execute single biological processes in Caenorhabditis elegans germ-line development. *Proc Natl Acad Sci USA* **106,** 4776–81.

Chapter 11

ERK-MAP Kinase Signaling in the Cytoplasm

Michelle C. Mendoza, Ekrem Emrah Er, and John Blenis

Abstract

ERK-MAPK is activated by dual phosphorylation of its activation loop TEY motif by the MEK-MAPKK. ERK cytoplasmic activity should be measured by assaying both the level of dually phosphorylated ERK and the level of phosphorylated substrate. We describe two complementary methods for quantitatively measuring ERK activity toward the cytoplasmic p90 ribosomal S6 kinase (RSK). The first method is a straightforward immunoblot of endogenous ERK and RSK phosphoepitopes using phospho-specific antibodies. Infrared fluorescent secondary antibodies provide a linear readout that is quantitated using an Odyssey scanner (LI-COR). The second method is an immunoprecipitation of ERK followed by an in vitro immune complex kinase assay with purified GST-RSK as substrate. The level of ERK phosphotransferase activity, or ^{32}P-labeled phosphate transfer, is quantitated using a PhosphorImager.

Key words: Extracellular signal-regulated kinase-mitogen-activated protein kinase, MAPK/ERK-activating kinase, Epidermal growth factor, Protein kinase assay, Odyssey, p90 ribosomal S6 kinase, PhosphorImager, Immunoblot

1. Introduction

The extracellular signal-regulated kinase (ERK)-mitogen-activated protein kinases (MAPKs, ERK-MAPKs) localize to many cellular subcompartments, including the cytoplasm, cell membrane, Golgi apparatus, endosomes, and cytoskeleton. Upon dual phosphorylation and activation by MAPK/ERK-activating kinase (MEK), a detectable fraction of ERK translocates to the nucleus (1–3). Activated ERK is also retained in the extranuclear subcompartments, where it acts upon cytoplasmic substrates (4–7). ERK localization and signaling specificity are controlled by scaffolds, anchoring proteins,

Rony Seger (ed.), *MAP Kinase Signaling Protocols: Second Edition*, Methods in Molecular Biology, vol. 661,
DOI 10.1007/978-1-60761-795-2_11,

interaction domains, and dimerization (8–10). For example, ERK activity toward the cytoplasmic substrates p90 ribosomal S6 kinase (RSK) and $cPLA_2$ requires unique scaffolding proteins and dimerization. Scaffolds and dimerization appear to be disposable for nuclear activity (11).

The availability of antibodies that recognize dually phosphorylated ERK has led the signaling community to adopt the detection of phospho-ERK1/2 as a readout of ERK activity. However, the presence of phosphorylated ERK does not necessarily indicate its activity toward a specific cytoplasmic substrate. We describe two methods for quantitatively measuring ERK activity toward cytoplasmic RSK. There are four RSK isoforms, namely RSK 1–4. Each RSK isoform has two kinase domains, a C-terminal kinase domain that autophosphorylates RSK and an N-terminal kinase domain that phosphorylates RSK substrates. ERK phosphorylates RSK's C-terminal kinase domain on T573 (numbering refers to human RSK1). The activated C-terminal kinase domain then autophosphorylates RSK at the hydrophobic motif S380. Phospho-S380 recruits PDK1, which phosphorylates RSK's N-terminal kinase domain, leading to full RSK activation (12). Inactive RSK1/2 shuttles between the cytoplasm and nucleus but is likely activated in the cytoplasm by cytoplasmic ERK (7, 13). Like ERK, RSK phosphorylates both cytoplasmic and nuclear targets, and ERK-MAPK pathway activation leads to increased RSK translocation and/or retention in the nucleus (1, 14).

We describe two methods to quantitatively measure ERK activity toward RSK in growth factor-stimulated cells. These methods can be easily modified to monitor ERK activity on other cytoplasmic targets, such as $cPLA_2$, paxillin, and others (5, 11). The first method, immunoblotting for endogenous ERK and RSK phospho-epitopes, assesses ERK activation under native conditions. Since RSK is a cytoplasmic target of active ERK (7), this provides a simple assay for monitoring cytoplasmic ERK activation. However, one must be cautious when interpreting data from this assay since it does not directly test whether the observed RSK phosphorylation is due to ERK. In addition, RSK phosphorylation is maintained by multiple signaling pathways at later time points. Incorporating MEK inhibitors (UO126 or PD98059) or ERK RNAi into the experiment will add confirmation that the observed substrate phosphorylations are due to ERK. The second method, an immune complex kinase assay, is advantageous in that it directly measures total, immunoprecipitatable ERK phosphotransferase activity. However, because this assay is in vitro, it does not test whether protein localization and scaffolding are sufficient for activity in intact cells. By performing both the methods, one can quantitatively measure the level of ERK activity toward specific cytoplasmic substrates.

2. Materials

2.1. Cell Culture and Lysis (Suitable for Most Cell Types)

1. Growth Medium: Dulbecco's modified eagle medium (DMEM; Mediatech, Manassas, VA), 10% fetal bovine serum (FBS: Gibco, Invitrogen Carlsbad, CA). Store medium at 4°C and prewarm to 37°C prior to use.
2. Starve medium: DMEM, 20 mM Hepes, pH 7.4. Make up 1 M Hepes, pH to 7.4, and autoclave to sterilize. Add sterile Hepes to DMEM. Store medium at 4°C and prewarm to 37°C prior to use.
3. Solution of phosphate buffered saline (PBS): 137 mM sodium chloride (NaCl), 2.7 mM potassium chloride (KCl), 10 mM sodium phosphate, dibasic, heptahydrate ($Na_2HPO_4 \cdot 7H_2O$), 1.8 mM potassium phosphate monobasic (KH_2PO_4). Adjust to pH 7.4. Autoclave to sterilize (see Note 1).
4. Solution of trypsin: 50 mM Tris–HCl (pH 7.5), 150 mM NaCl, 1 mM ethylenediamine tetraacetic acid (EDTA), 0.25% trypsin. Filter to sterilize. Store trypsin at 4°C and prewarm to 37°C prior to use.
5. Epidermal growth factor (EGF; Peprotech, Rocky Hill, NJ). Dilute in water to 50 μg/mL. Store in aliquots at –20°C.
6. Lysis buffer: 20 mM Tris–HCl (pH 7.5), 150 mM NaCl, 1 mM EDTA, 1% Nonident P-40, 10 mM sodium fluoride (NaF). Store at 4°C. NaF is toxic, so use caution when handling the stock bottle.
7. Protease and phosphatase inhibitors: 200 mM phenylmethanesulfonyl fluoride (PMSF) dissolved in ethanol, 5 mg/mL aprotinin dissolved in water, 5 mg/mL leupeptin dissolved in water, 5 mg/mL pepstatin A dissolved in ethanol, 100 mM sodium pyrophosphate (NaPPi), and 100 mM sodium orthovanadate (Na_3VO_4) dissolved in water. Inhibitors should be stored in aliquots at –20°C.
8. Plastic scrapers (Corning, Corning, NY).
9. Bradford reagent (Bio-Rad, Hercules, CA, see Note 2).
10. Bovine serum albumin (BSA) for protein standard (Sigma): Dilute in water to 1, 0.75, 0.5, 0.25, 0.125, 0.0625, and 0.03125 mg/mL. Store in aliquots at –20°C.
11. Plate reader capable of reading absorbance at 595 nm, such as Victor or Envision plate readers (Perkin Elmer, Waltham, MA).
12. Modified Laemmli sample buffer for protein denaturation (5×): 50% glycerol, 250 mM Tris–HCl (pH 6.8), 10% sodium dodecyl sulfate (SDS), 500 mM dithiothreitol (DTT), 0.5% bromophenol blue. Mix well and store in aliquots at –20°C. Warm up to 37°C just before use to liquify.

2.2. Sodium Dodecyl Sulfate–Polyacrylamide Gel Electrophoresis

1. Separating buffer (4×): 1.5 M Tris–HCl (pH 8.8), 0.4% SDS. Store at room temperature.
2. Stacking buffer (4×): 0.5 M Tris–HCl (pH 6.8), 0.5% SDS. Store at room temperature.
3. 30% acrylamide/0.8% bis-acrylamide solution (37.5:1, National Diagnostics, Atlanta, GA). Acrylamide is a neurotoxin when unpolymerized, so be careful not to inhale or ingest.
4. *N,N,N,N'*-Tetramethylethylenediamine (TEMED; Sigma).
5. Ammonium persulfate (APS; Sigma): Prepare 10% solution in water. Stable for 1–2 months at 4°C or 1–2 years at −20°C.
6. Water-saturated butanol. Mix equal volumes of water and butanol in a glass bottle and allow to separate. Use the top layer. Store at room temperature.
7. Running buffer (5×): 125 mM Tris base, 960 mM glycine, 0.5% (w/v) SDS. Store at room temperature.
8. Prestained molecular weight markers (Invitrogen, Carlsbad, CA).

2.3. Immunoblot for Phosphorylated ERK Substrate

1. Transfer buffer: Prepare 10× stock: 250 mM Tris base, 1.9 M glycine, 0.5% SDS. Store at 4°C. Dilute to 1× and store in transfer apparatus at 4°C. Transfer buffer can be reused five times (see Note 3).
2. Nitrocellulose membrane with 0.45 μm pores and 3MM chromatography paper (Whatman, GE Healthcare, Piscataway, NJ, see Note 4).
3. Tris-buffered saline with Tween (TBST): Prepare 10× stock: 1.5 M NaCl, 100 mM Tris–HCl (pH 7.4), 1% Tween-20. Dilute to 1× with water for use. Store 10× TBST at room temperature and 1× at 4°C.
4. Ponceau S stain: 0.5% ponceau S (w/v), 1% acetic acid in water. Ponceau S stain may be reused.
5. PBS and TBST blocking buffer: 5% nonfat dry milk in PBS and PBST.
6. Mouse monoclonal anti-diphosphorylated ERK-MAPK and rabbit polyclonal anti-ERK (Sigma). Dilute both antibodies to 0.4 μg/mL in TBST blocking buffer (see Notes 5 and 6).
7. Rabbit polyclonal anti-phospho-RSK T573 (R&D Systems, Minneapolis, MN) and mouse monoclonal anti-RSK2 (Santa Cruz, Santa Cruz, CA). Dilute anti-phospho-RSK T573 to 0.5 μg/mL and anti-RSK2 to 0.4 μg/mL in TBST blocking buffer.
8. Secondary antibodies: IRDye 800CW-conjugated goat anti-rabbit IgG (LI-COR) and IRDye 680 goat anti-mouse IgG (LI-COR). These antibodies are light sensitive, so store them

in the dark at 4°C. Dilute antibodies 1:20,000 to 0.05 μg/mL in TBST blocking buffer.

9. Odyssey Infrared Imaging System (LI-COR).

2.4. Purification of Catalytically Inactive RSK C-Terminal Kinase Domain: A Substrate for ERK Phosphotransferase Activity

1. GST-RSK-D2 K/R (GST-RSK, C-terminal kinase domain, inactive) in pGEX2T plasmid (15).
2. Competent BL21 bacteria (Promega, Madison, WI). Store at −80°C. Thaw on ice and keep on ice before use.
3. Luria Broth/Ampicillin (LB/Amp, Sigma): 10 g/L tryptone, 5 g/L yeast extract, 5 g/L NaCl, dissolved in water. Autoclave to sterilize. Add 100 μg/mL ampicillin and store at 4°C. Prewarm to 37°C prior to use.
4. LB/Ampicillin agar plates: 10 g/L tryptone, 5 g/L yeast extract, 5 g/L NaCl, 15 g/L agar, dissolved in water. Make up in Erlenmeyer flask and autoclave to sterilize. When media has cooled to room temperature, add 100 μg/mL ampicillin and pour into sterile plates. Store plates at 4°C and prewarm to 37°C prior to use.
5. 1 L 2× YT media (BD, Franklin Lakes, NJ): 16 g/L tryptone, 10 g/L yeast extract, 5 g/L NaCl, dissolved in water. Make 1 L of media in a 2 L Erlenmeyer flask. Autoclave to sterilize and warm to 37°C before use.
6. Dioxane-free isopropyl-beta-D-thiogalactopyranoside (IPTG; Promega, Madison, WI). Dissolve in water to make a 1 M solution. Filter-sterilize and store in aliquots at −20°C.
7. Dounce homogenizer (Wheaton, VWR, West Chester, PA).
8. Microfluidizer (Microfluidics Corporation, Newton, MA).
9. Lysis buffer G: 1× PBS, pH 7.4 with 50 mM EDTA, pH 8. Store at 4°C.
10. Reducing agent and protease inhibitors: 1 M DTT, 200 mM PMSF, and 5 mg/mL aprotinin, leupeptin, and pepstatin A, as described in Subheading 3.1. Store in aliquots at −20°C.
11. Glutathione Sepharose 4B (GE Healthcare, Piscataway, NJ).
12. Chromatography columns (1.5 cm internal diameter columns and 0.156-in. internal diameter tubing, Kimble Kontes, Vineland, NJ).
13. Elution buffer: Lysis buffer G with10 mM glutathione, pH 7.5.
14. BSA for protein standard: Dilute in water to 1, 0.5, and 0.25 mg/mL. Store in aliquots at −20°C.
15. Coomassie Blue Stain: 45% ethanol, 10% glacial acetic acid, 0.5% (w/v) Coomassie R250.
16. Coomassie Blue Destain: 10% ethanol, 10% glacial acetic acid.

2.5. Transfection of ERK-MAPK to Test ERK Phosphotransferase Activity

1. HA-ERK2 in pcDNA3 vector (Addgene, Cambridge, MA).
2. 2 M Calcium chloride. Filter-sterilize and store at 4°C.
3. 2× Hepes-buffered salt solution (HBSS): 274 mM NaCl, 1.5 mM $Na_2HPO_4{\cdot}7H_2O$, 50 mM HEPES, pH 7.0. Filter-sterilize and store in aliquots at –20°C.

2.6. Immuno-precipitation of ERK-MAPK to Test ERK Phosphotransferase Activity

1. EGF and lysis buffer, as described in Subheading 2.1.
2. Protein A sepharose beads (GE Healthcare). Swell beads in PBS. Preabsorb beads with 1% BSA in PBS for 1 h. Wash in PBS three times and store as a 50% bead slurry in PBS with 0.02% sodium azide at 4°C.
3. Monoclonal anti-HA antibody (homemade or Covance, Princeton, New Jersey).

2.7. In Vitro Kinase Assay to Test ERK Phosphotransferase Activity

1. 10× Kinase Buffer: 250 mM Tris base (pH 7.4), 10 mM DTT, 100 mM $MgCl_2$, 25 mM β-glycerophosphate. Make fresh and keep on ice.
2. One milliCurie (1 mCi) [gamma-^{32}P]ATP in 10 mM Tris buffer (Perkin Elmer, Waltham, MA, see Note 7).
3. 2.5 mM ATP. Store in aliquots at –20°C.
4. HyBlot CL, Autoradiogram Film (Denville Scientific, Metuchen, NJ).
5. Storage Phosphor Screen (Molecular Devices, Sunnyvale, CA).
6. PhosphorImager (GE Healthcare).

2.8. Western Blot for Input Levels of GST-RSK Substrate and Immunoprecipitated ERK-MAPK in the Kinase Assay

1. Transfer buffer, nitrocellulose membrane, TBST, and blocking buffer as described in Subheading 2.3.
2. Rabbit anti-phospho-RSK T573 (R&D Systems) and mouse anti-HA antibodies diluted together in blocking buffer to 0.5 and 1 μg/mL, respectively (see Note 6).
3. Rabbit anti-GST (Sigma) antibody prepared 1:1,000 in blocking buffer.
4. Secondary antibodies as described in Subheading 2.3.
5. Stripping buffer: 0.2 N sodium hydroxide (NaOH).

3. Methods

Protein phosphorylations are inherently labile and sometimes difficult to monitor, due to cellular protein phosphatases and proteases. Therefore, it is necessary to add both protease inhibitors and phosphatase inhibitors in the lysis buffer just before use, process all samples on ice or at 4°C, and work as quickly as possible. We perform a Bradford assay to quantitate the protein concentration of each sample so that equal protein amounts are analyzed. Protein quantitation is particularly important when using prolonged starvations or drug treatments that might cause some cultured cells to grow more slowly and yield less protein than others. Treatment with EGF or phorbol 12-myristate 13-acetate (PMA) provide a positive control for ERK-MAPK activation in most systems.

The Odyssey infrared system uses two solid-state laser diodes to illuminate samples at 685 and 785 nm. The fluorescent secondary antibodies are excited to release fluorophores, which are detected and converted into electrical signals by photodiodes. Infrared detection has low autofluorescence, equal sensitivity, and better signal to noise and linearity than chemiluminescence. The Odyssey method is also very efficient when one wants to probe many different epitopes on the same membrane, as two proteins can be detected simultaneously using two different colors. Two-color imaging requires that the two primary antibodies are derived from different host species, so they can be discriminated by secondary antibodies with different specificities, one labeled with an 800-channel dye and the other labeled with a 700-channel dye. For example, phosphorylated ERK can be detected with the 800-nm spectrum, and total ERK can be detected with the 700-nm spectrum. This abrogates the need for stripping and reprobing.

The wide linear range of Odyssey detection mirrors that of a phosphorimager. Thus, when assaying ERK phosphotransferase activity in vitro, one can use either [gamma-^{32}P]ATP and phosphorimager detection or cold ATP and Odyssey detection of an immunoblot with fluorophore-conjugated secondary antibodies (see Fig. 2, phospho-RSK T573). When assaying ERK activity on a novel substrate or when a phospho-specific antibody is not available, one must use [gamma-^{32}P]ATP and a phosphorimager.

3.1. Cell Culture and Lysis (Suitable for Most Cell Types)

1. Plate cells: 293 T cells are grown in DMEM with 10% FBS. Cells should be passaged just prior to reaching confluency. Plate 293 T cells for the experiment by washing one time with pre-warmed PBS, then trypsinizing and diluting cells 1:8 in 60 mm dishes, about 5×10^5 cells/dish. Grow cells for 48 h at 37°C, 5% CO_2.
2. Starve cells: Starve cells at 70–80% confluency by washing the cells once with starve medium, then adding 2 mL starve medium. Starve for 18–24 h at 37°C, 5% CO_2.
3. Add protease and phosphatase inhibitors to lysis buffer just before use. Dilute aprotinin, leupeptin, and pepstatin A

1:2,000 so that they are 2.5 μg/mL final. Dilute PMSF, NaPPi, and Na_3VO_4 1:100 so that they are 2, 1, and 1 mM, respectively. Prepare a tray of ice for lysing cells.

4. Stimulate cells with EGF: Add 2 μL of EGF into 2 mL of medium so that the final growth factor concentration is 50 ng/mL EGF. Rock to mix. Incubate for desired time at 37°C, 5% CO_2.
5. Place tissue culture (TC) dish on ice and wash cells once with cold PBS. Aspirate off all remaining liquid.
6. Lyse cells in TC dish: Add 100 μL lysis buffer and use scraper to pool lysate into the bottom corner of plate. Pipet lysate into pre-chilled microcentrifuge tubes on ice.
7. Centrifuge 10 min at maximum speed in microcentrifuge (15,000 × *g*) at 4°C.
8. Transfer extract (supernatant) to pre-chilled microcentrifuge tubes on ice at 4°C.
9. Quantitate proteins using Bradford reagent: In a 96-well plate, load 10 μL blank (water) and BSA standards in duplicate. Dilute samples 1:10 in water and load 10 μL of diluted samples, in duplicate, into individual well of the plate. Dilute Bradford reagent 1:5 in water and add 200 μL into each well containing standard or sample. Incubate for 10 min at room temperature. Measure absorbance at 595 nm. Calculate the mean for each duplicated standard and sample. Using Excel or another graphing program, determine the equation for the BSA standard curve. Solve the equation for each sample's absorbance reading to calculate their protein concentrations. Calculate the amount of sample needed for 40 μg of protein and the amount of lysis buffer needed to bring each sample up to 32 μL (see Note 2).
10. In a second set of microcentrifuge tubes, add 40 μg of each sample, 8 μL of 5× modified Laemmli sample buffer, and lysis buffer to bring the total volume up to 40 μL. Heat the samples at 95–100°C for 5 min.

3.2. Sodium Dodecyl Sulfate–Polyacrylamide Gel Electrophoresis

1. These instructions are for using homemade glass plates and spacers but can be adapted for most other gel systems, including minigels. Rinse the glass plates with detergent and at least twice with distilled water. Rinse with 70% ethanol and dry.
2. Using petroleum jelly, line three sides each of a long (18.5 cm × 16 cm) plate and short (16.5 cm × 16 cm) plate with grease and then add spacers to the long plate. Make sure spacers are tight and square next to each other. Add the second shorter plate on top of the spaces and clamp with clips.

3. Prepare a 10% gel (good for visualizing p44, p42 ERK1/2, and larger proteins). Mix 10 mL acrylamide/bis-acrylamide solution with 7.5 mL 4× separating buffer and 12.5 mL water. Add 100 μL of 10% APS solution and 20 μL TEMED. Try to minimize bubbles. Pour the gel, leaving space for the stacking gel, and overlay the top with water-saturated butanol. The gel should polymerize in about 30 min.
4. Pour off the butanol and wash the top of the gel twice with water. Dry the glass with Whatman paper.
5. Prepare the stacking gel: Mix 1.3 mL acrylamide/bis-acrylamide solution with 2.5 mL 4× stacking buffer and 6.1 mL water. Add 50 μL of 10% APS solution and 20 μL TEMED. Pour the stacking gel on top of the polymerized separating gel and insert the comb. The stacking gel should polymerize in about 30 min.
6. Carefully remove the comb under running water. Take out the bottom spacer of the gel and assemble the gel in the gel unit. Prepare 1× running buffer by diluting 5× running buffer in water. Fill the top and bottom chambers with 1× running buffer. Use a syringe and needle to blow out any bubbles that are underneath the gel, between the two glass plates. Bubbles will disrupt the current flow.
7. Using a Hamilton syringe or gel-loading tips, load one well with 10 μL prestained molecular weight marker and the sample wells with 40 μL (40 μg) of each protein sample.
8. Run the samples through the gel at 50 V overnight or 100 V/30 mA through the stacking and 150 V/50 mA through the separating gel. The dye fronts should run off the gel (about 5 h).

3.3. Immunoblot for Phosphorylated ERK Substrates

1. The proteins separated by SDS–polyacrylamide gel electrophoresis (SDS–PAGE) need to be transferred to a nitrocellulose membrane electrophoretically. These instructions use the Amersham Hoefer transfer system but can be adapted to most transfer devices. Dilute 10× transfer buffer to 1× with water and pour into a tray. Place one side of the transfer cassette in the buffer, followed by a piece of foam, and one sheet of Whatman paper.
2. Disassemble the gel unit and pry open the glass plates. Using a razor blade, cut off the stacking gel and any unused lanes and discard them. Wet the gel with some transfer buffer. Pick up the gel and place on top of the Whatman paper in the tray. Cut a piece of nitrocellulose membrane just larger than the gel and wet it in a tray with distilled water. Place the wet nitrocellulose membrane on top of the gel. Place one sheet of Whatman paper on top of the membrane and roll a pipet

over the sandwich to release any bubbles. Place second sponge (presoaked in transfer buffer) over the sandwich and close the cassette.

3. Place the cassette in the transfer tank such that the membrane is closest to the anode (red) and the gel is closest to the cathode (black). The proteins are negatively charged and so will flow from the gel to the membrane, toward the positively charged anode.
4. Transfer the proteins to the membrane at 4°C, with a magnetic stir bar stirring in the tank. Transfer at 30 V overnight or 70 V for 2 h.
5. Disassemble the transfer apparatus and open the cassette. Discard the gel and used Whatman papers. Remove the nitrocellulose membrane and place in a clean box.
6. Stain the membrane with Ponceau S: Cover the membrane with Ponceau S stain and incubate for 2 min. Wash off the stain with distilled water three to four times. Using the protein bands as a guide, cut the membrane so that ERK (p42, p44) and the substrate of interest can the probed for separately. For RSK (p85), we cut the membrane at the 60 kDa marker.
7. Incubate the membranes for 1 h in PBS blocking buffer at room temperature.
8. Discard the blocking buffer. Add the diluted primary antibodies to each membrane (ERK to the bottom half and RSK to the top half). Incubate the primary antibodies for 1 h at room temperature (or overnight at 4°C).
9. Remove the primary antibodies and store at 4°C with 0.02% sodium azide. Wash the membranes three times for 5 min each with TBST.
10. Add the secondary antibodies to the membranes (in an opaque container) and incubate 30 min at room temperature.
11. Discard the secondary antibodies and wash the membranes three times for 5 min at room temperature.
12. Place the membranes between Whatman paper to dry (see Note 8).
13. Scan membranes: Clean the Odyssey scanning surface with water to remove dust and smudges. Dry the surface with Kimwipes (Kimberly-Clark, Neenah, WI), which are low lint. Place the membranes face down on an Odyssey scanner at the 0,0 (x, y) coordinate. Place the silicon mat or a glass plate on top of the membrane to flatten it. Open the Odyssey software. Under the Odyssey "File" menu, select "New Project," assign a name to your project, and direct the file path for saving your data. Click "Scan." Draw a rectangle around the region that

your membrane covers. Set the scanning "Preset" to "Membrane" and select the following scanning parameters: 169-μm resolution, medium quality (amount of photons detector gathers for a given area to form a pixel on the image), and 0 mm focus offset. Set the 700-channel intensity to 3 and 800-channel intensity to 5. Click "Start Scan" and scanning will start at the lower left hand corner of the boxed region. The molecular weight marker will be visible in the 700 channel. If your bands appear white, the fluorescent signal is saturated and the scanning intensity needs to be decreased. The image can be rotated, cropped, and adjusted before saving. These adjustments do not alter the fluorescent signal intensities used for quantitation. Export the image as a tiff file under "File" → "Export image" → "Export Image View."

14. Quantitate the signal: Add a box to the phospho-RSK T573 image by selecting "Add Feature" → "Rectangle" under the "Analyze" menu bar. Use the "Show Details" feature to magnify the box and be sure it is big enough to encompass each band. Copy the box using the "Add Multiple Features" tool or by copying and pasting the highlighted box. Move the boxes over their respective bands. Under "Analyze," select "Background Method" and set the background to "Pixel Boundary Method" with a 1–2 pixel boundary (depending on how close other background bands are to the phospho-RSK T573 signal). Highlight the boxes and click on "Report" → "Report View" to acquire the background-corrected integrated intensity for each box. These values can be exported as a text file. To normalize the phosphorylated T573 signal to that of total RSK2, copy the phospho-RSK T573 boxes, paste them on the RSK2 image, and acquire the RSK2 Integrated Intensities. Manually divide the phospho-RSK T573 Integrated Intensities by those of their respective RSK2 bands and normalize the 0 time point to 1 by dividing each band's intensity by the 0-min time point intensity. An example of the results produced with EGF is shown in Fig. 1.

3.4. Purification of Catalytically Inactive RSK C-Terminal Kinase Domain: A Substrate for ERK Phosphotransferase Activity

1. In a pre-chilled microcentrifuge tube on ice, add 1 μL of 100 ng/μL GST-RSK-D2 K/R construct to 10 μL of BL21 cells. Heat shock the bacteria at 42°C for 45 s; then recover the cells for 30 min on ice. Using a glass spreader, spread the bacteria on an LB/Ampicillin agar plate. Turn the plate upside down (agar on top) and incubate overnight at 37°C.
2. Pick an individual bacterial colony (see Note 9) with a sterile pipet tip and inoculate in 10 mL of prewarmed LB media with 100 μg/mL ampicillin (LB/Amp). Grow overnight at 37°C, 250 RPM.

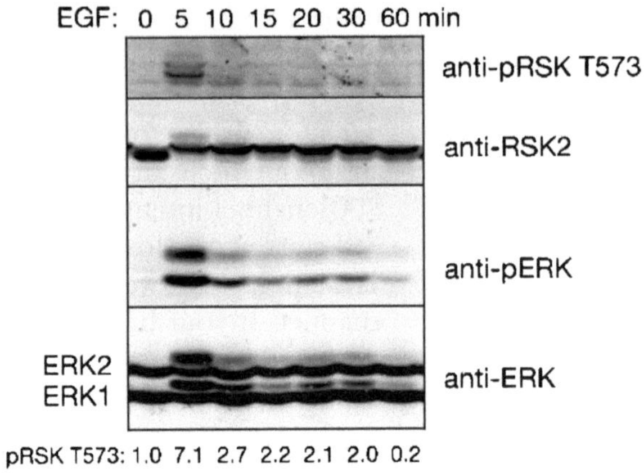

Fig. 1. Time course of 293 T cells stimulated with 50 ng/mL EGF. Cells were treated for the indicated time points and processed for immunoblotting. The intensity of the pRSK T573 signal was quantitated using the Odyssey software and normalized to that of RSK2.

3. The next morning, decant the starter into the flask of 2× YT media with 100 μg/mL ampicillin. Grow the bacteria at 37°C, 250 RPM. After 1 h, begin taking OD_{600} measurements. When the OD_{600} reaches 0.8, induce protein production by adding dioxane-free IPTG to a final concentration of 0.2 mM (see Note 10). Grow the induced bacteria for 6 h at 30°C. Pellet the induced bacteria by centrifuging at 4°C, 1000 *g* for 10 min. Proceed with lysis or snap-freeze the bacteria pellets in liquid nitrogen and store them at –80°C.
4. Lyse the bacteria: Add DTT and protease inhibitors to lysis buffer G (1 mM DTT and PMSF and 2.5 μg/mL aprotinin, leupeptin, and pepstatin A final concentration). Resuspend the bacteria pellet with 40 mL of lysis buffer G on ice. Carry out all the following steps on ice. Homogenize the suspension with a Dounce homogenizer and pass the bacteria through a pre-chilled microfluidizer at 20,000 PSI, twice. Pellet the lysate in an ultracentrifuge, at 10,000 × *g*, 4°C, for 30 min (see Note 10). During the centrifugation, wash 5 mL of Glutathione sepharose beads twice with lysis buffer G and make a 50% slurry in lysis buffer G.
5. GST Pull down: Transfer the supernatant (cell lysate) to a 50 mL conical tube. Add 4 mL of 50% glutathione sepharose beads to the lysates and incubate on an end-over-end rotator for 1 h at 4°C. During the incubation, prepare elution buffer.
6. Wash and elute GST-RSK: In a cold room or refrigerated cold box, decant the slurry into a chromatography column and let it drip (see Note 10). When there is only residual lysis buffer left in the column, drip 500 mL lysis buffer G over the column by gravity force. Run tubing from the buffer G (on a high

shelf) into the column. Washing the beads should take about 15 min. Elute the column with 5 mL of elution buffer and collect the GST-RSK eluate in 1 mL batches in five microcentrifuge tubes.

7. Add 5 μL of modified Laemmli sample buffer to 20 μL of each elution and 20 μL of each BSA standard. Heat at 95–100°C for 5 min. Load molecular weight marker, standard, and samples on an SDS–PAGE gel, as described in Subheading 3.2.
8. Stain the SDS–PAGE gel with Coomassie Brilliant Blue for 1–2 h. Destain the gel for 2–6 h until protein bands are highly visible and gel background is almost clear. GST-RSK should be visible at just below the 60 kDa marker. Approximate the amount of GST-RSK in each elution by comparing it to the BSA samples. Snap-freeze 25 μL aliquots of the elutions with the highest GST-RSK concentrations in liquid nitrogen and store at −80°C.

3.5. Transfection of ERK-MAPK to Test ERK Phosphotransferase Activity

1. Passage a 100 mm plate of 90% confluent 293 T cells 1:10, as described in Subheading 3.1. Grow for 48 h at 37°C, 5% CO_2.
2. Transfect cells with calcium phosphate (see Note 11): for each plate to be transfected, add 500 μL of 2× HBSS in a 5 mL tube. In another 5 mL tube, add 433 μL of sterile water, 61 μL of $CaCl_2$, and 6 μL of 1 μg/μL of pcDNA3/HA-ERK2. Add the $CaCl_2$–DNA solution dropwise to 2× HBSS while gently vortexing the tube. Incubate $CaCl_2$–DNA–HBSS at room temperature for 30 min.
3. During the incubation, change the media on 293 T cells with fresh 8 mL of pre-warmed DMEM with 10% FBS.
4. Add the calcium phosphate–DNA solution to the cells, dropwise, and grow cells for 24 h at 37°C, 5% CO_2.
5. Change media and starve cells for 24 h as described in Subheading 3.1.

3.6. Immuno-precipitation of ERK-MAPK to Test ERK Phosphotransferase Activity

1. Make up lysis buffer with fresh protease and phosphatase inhibitors.
2. Stimulate cells with EGF, lyse cells, and quantitate protein with Bradford assay as described in Subheading 3.1. Transfer 3 mg of protein from the cell lysates to a fresh pre-chilled microcentrifuge tube on ice.
3. Wash and dilute protein A sepharose: Dilute 20 μL of 50% Protein A sepharose (per sample) in 1 mL of lysis buffer. Centrifuge the beads at 5,000 RPM for 2 min at 4°C. Pipet off and discard the buffer without disturbing the bead pellet.

Repeat this wash step two more times. Dilute beads to a 10% slurry by adding 90 μL of lysis buffer (per sample) to the beads. Pipeting larger quantities of diluted beads help prevent pipeting error so that equivalent amounts of ERK2 are immunoprecipitated from each treatment condition.

4. HA Immunoprecipitation: Transfer the supernatant to a fresh microcentrifuge tube and discard the bead pellet. Add 1 μL (2 μg) of anti-HA antibodies per mg of protein in the lysates to the supernatant. Incubate on a nutator for 30 min at 4°C. Add 80 μL of 10% Protein A sepharose washed in step 3 and rotate for an additional 1 h. Centrifuge samples at 5,000 RPM, 4°C for 2 min. Pipet off and discard the supernatant. Wash the beads two times with 1 mL of lysis buffer, centrifuging samples at 5,000 RPM, 4°C for 2 min.
5. Make up fresh 1× kinase assay buffer with cold water. Wash the beads with 1 mL of 1× kinase buffer by centrifuging samples at 5,000 RPM, 4°C for 2 min and pipeting off and discarding the supernatant. Remove as much supernatant as possible by attaching a 27 gauge (or smaller) needle to a vacuum line and aspirating the buffer. Keep the beads on ice. Active ERK2 is bound to the beads and will be used as the kinase in the in vitro kinase assay. Proceed directly to Subheading 3.7 for kinase assay.

3.7. In Vitro Kinase Assay to ERK Phosphotransferase Activity

1. Make a master mix for the kinase reactions: In a microcentrifuge tube, add 2 μL of 10× Kinase buffer, 1 μL of 2.5 mM ATP (125 μM final concentration), 5 μCi of [gamma-^{32}P] ATP, and 2–4 μg of GST-RSK per reaction. Add water to bring the total volume up to 20 μL per reaction. Prepare master mix enough for each sample and one extra reaction (see Note 12).
2. Kinase assay: Consecutively aliquot 20 μL of master mix to each microcentrifuge tube containing HA-ERK2-bound beads. Incubate at 30°C for 10 min. Use a timer and space apart the start of each sample reaction by 30–60 s. Consecutively stop the kinase reactions by adding 20 μL of 2× modified Laemmli sample buffer (5× diluted to 2×), vortexing briefly, and heating for 5 min at 100°C. Load the samples on an SDS–PAGE gel, as described in Subheading 3.2. Stop the radioactive gel when the dye front is still 1 in. from the bottom of the gel. The free, unincorporated ^{32}P-ATP runs in front of the dye and should be cut off of the gel and disposed of in a designated radioactive waste container.
3. Transfer proteins from the SDS–PAGE gel to a nitrocellulose membrane as described in Subheading 3.3. Blot the nitrocellulose membrane dry and cover with Saran Wrap before placing face up, in a film cassette. In a dark room, place HyBlot

EC film on top of the protected nitrocellulose membrane in the film cassette. Expose the film for at least 30 min and develop the film in a film processor.

4. For quantitative analysis, the nitrocellulose membrane should be exposed to a storage phosphor screen. Using a phosphorimager scanner and Quantity One Image software, scan the phosphor storage screen: under the Quantity One "File" menu, select "PMI." Draw a rectangle around the region of the storage phosphor screen exposed to the nitrocellulose membrane. Click "Acquire." Save the image document and open it using "File" → "Open." Under the top menu, select "Volume" → "Volume Rect Tool." Draw a very tight rectangle around the largest gel band. Copy the rectangle and move the copies over the other bands. These equal-sized rectangles are the boundaries for the density measurements of samples. Copy all of these rectangles and move the new rectangles together to a region with no radiation readings. These new rectangles encapsulate the region for measuring the background density. Select "Volume" → "Volume Analysis Report." Under the "Data to Display" menu, select "Name," "Volume," and "Density." Click "Done." On the "Volume Report" panel, choose "Export the Report to Disc." Exporting the data to "Clipboard" in a "Tab" separated format is useful for analyzing the data on a spreadsheet. Subtract the background density from corresponding density readings of the samples. Normalize the density readings to the 0-min time point (sample with no EGF or PMA stimulation) by dividing each band's density by the 0-min time point's density reading. Density readings should also be normalized to total amount of GST-RSK in each sample. An example of the results produced with EGF is shown in Fig. 2.

3.8. Immunoblot for Input Levels of GST-RSK Substrate and Immunoprecipitated ERK-MAPK in the Kinase Assay

1. Incubate the nitrocellulose membrane in blocking buffer for 1 h at room temperature.
2. Incubate the membrane with the diluted anti-pT573RSK and anti-HA antibodies for 1 h at room temperature.
3. Wash the membrane with TBST three times for 5 min at room temperature.
4. Incubate with secondary antibodies for 30 min at room temperature in an opaque container.
5. Wash the membrane with TBST three times for 5 min at room temperature.
6. Scan the membrane and quantify the phospho-RSK T573 signal using the Odyssey, as described in Subheading 3.3.
7. Strip the phospho-RSK T573 antibody: Incubate the membrane in 0.2 N NaOH at room temperature until the

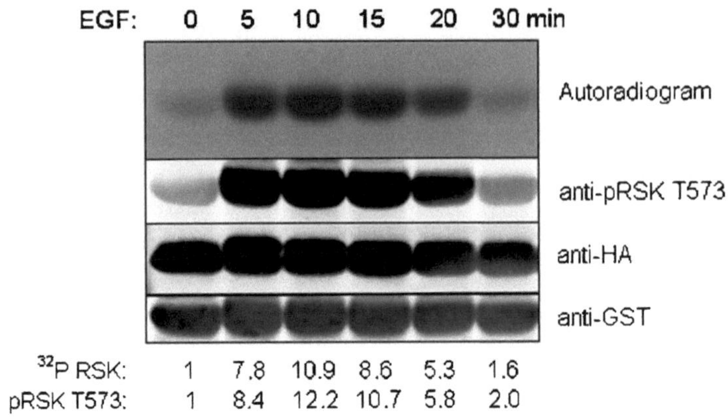

Fig. 2. Time course of 293 T cells stimulated with 50 ng/mL EGF. Cells were treated for the indicated number of minutes. HA-ERK2 was immunoprecipitated and incubated with GST-RSK. The intensity of transferred ^{32}P was quantitated using Quantity One Image software. The intensity of phospho-RSK T573 was quantitated using the Odyssey software and normalized to that of GST. Note the comparable results with phosphorimager detection of ^{32}P incorporation and Odyssey detection of the phospho-RSK T573 immunoblot.

phospho-RSK T573 signal is no longer detectable by the Odyssey scanner (5–10 min). Wash the membrane with water three times.

8. Incubate the membrane with diluted anti-GST antibodies for 1 h at room temperature.
9. Wash the membrane with TBST three times for 5 min at room temperature.
10. Incubate with secondary antibodies for 30 min at room temperature in an opaque container.
11. Wash the membrane with TBST three times for 5 min at room temperature.
12. Dry the membrane between two Whatman papers. Scan the membrane and quantify the GST signal as described in Subheading 3.3.

4. Notes

1. Prepare all solutions with Milli-Q (Millipore) or an equivalent ultrapure deionized water.
2. When complexed with protein, the absorption maximum of Bradford dye shifts from 465 to 595 nm. The amount of absorption is directly proportional to the protein present for protein concentrations in the range of 0.1–1.4 mg/mL.

If your protein concentration is out of this range, you will need to make a series of dilutions to find the dilution that yields an O.D within the linear range. Other protein concentration assays with larger linear ranges (BCA, Pierce, Rockford, IL) can also be used.

3. For smaller proteins (less than 30 kDa), using nitrocellulose with 0.2-μm pores and adding 20% ethanol to the 1× transfer buffer will increase their binding to and retention on the membrane.
4. PVDF (Immobilon-FL, Millipore) membrane may also be used, although PVDF membrane must be briefly activated in ethanol (1 min) before use. When incubating the secondary antibodies, adding 0.1% SDS will reduce background on PVDF membranes. A membrane with 0.2-μm pore size will increase the binding of smaller proteins (less than 30 kDa).
5. The blocking buffer can affect antibody sensitivity and background binding. Blocking the membrane without detergent helps decrease the background. During the wash steps and primary and secondary antibody incubations, including Tween (in TBST) further decreases the background. For less robust antibody signals, the optimal antibody dilution and dilution buffer needs to be empirically determined. Other suitable antibody buffers include: 1–3% nonfat dry milk, 3% BSA, or 30–50% LI-COR blocking buffer (LI-COR, Lincoln, NE), diluted in TBST.
6. When mixing two primary antibodies, it is critical that they are from different species. This allows the primary antibodies to be differentiated with species-specific secondary antibodies, conjugated to distinct fluorophores. The secondary antibodies also need to be compatible in that they should not cross-react with each other. For example, goat anti-mouse IgG and goat anti-rabbit IgG can be incubated together, but goat anti-mouse IgG cannot be incubated with donkey anti-goat IgG. The 800 spectrum tends to have less background, so we use that channel for less robust antibodies. Diluted primary antibodies can be stored at 4°C for multiple uses, provided that 0.01% sodium azide is added to the solution to prevent bacterial growth. The diluted secondary antibodies should be discarded.
7. [gamma-^{32}P]ATP purchased in a Tris buffer can be used directly from the stock vial. We dilute our [gamma-^{32}P]ATP in buffer of 50 mM Tris–HCl (pH 7.2), 2.5 mM EDTA, 65% ethanol and store it at –20°C. The ethanol prevents the stock ATP solution from freezing and aids in long-term stability (2–4 weeks). However, the ethanol needs to be dried off before use, either in a refrigerated vacuum centrifuge or by evaporation in a chemical hood overnight.

8. If you want to later strip the membrane and reprobe with antibodies that you could not co-incubate, it is imperative that you do not allow the membrane to dry out. Dry membranes generally yield brighter and more uniform fluorescent signals, but they cannot be stripped. Using a wet membrane will also yield quantitative results on the Odyssey scanner.
9. While ampicillin selection should yield BL21 colonies containing GST-RSK, it is good practice to pick at least two colonies for protein purification.
10. When purifying a protein for the first time, it is informative to save 200 μL of bacterial cultures (1) before and (2) after induction, (3) 200 μL of lysate supernatant and (4) the pellet, and (5) the flow through after loading the bound GST sepharose beads onto the column. These samples should be run alongside the BSA standard and sample eluates on the SDS–PAGE gel to ensure that the each step worked as expected.
11. ERK1 activity can also be determined by transfecting T7-ERK1 (Addgene). To measure the activity of endogenous ERK1 or ERK2, omit the transfection step and use antibodies against ERK1 (Cell Signaling) or ERK2 (Santa Cruz) to immunoprecipitate either isoform in step 5 of Subheading 3.5. When using a novel substrate in the in vitro kinase assay, one should include a negative control to be sure that the observed phosphotransferase activity is indeed due to ERK and not a nonspecific kinase that binds to the protein A sepharose or antibodies. Suitable controls include transfection of empty vector or immunoprecipitation with nonspecific rabbit IgG in Subheading 3.6.
12. For an in vitro kinase reaction to yield quantitative results, the reaction should be carried out within the linear range of substrate phosphorylation and at saturating levels of ATP. The described protocol is an example of a linear kinase assay where phosphorylation of GST-RSK has not reached saturation. All procedures containing radioactive isotopes should be conducted according to the facility's environmental and radiation safety regulations.

Acknowledgments

The authors would like to thank Greg Hoffman for assistance with the GST-RSK2 purification protocol.

References

1. Chen RH, Sarnecki C, Blenis J. (1992) Nuclear localization and regulation of erk- and rsk-encoded protein kinases *Mol Cell Biol* **12**(3), 915–27.
2. Gonzalez FA, Seth A, Raden DL, Bowman DS, Fay FS, Davis RJ. (1993) Serum-induced translocation of mitogen-activated protein kinase to the cell surface ruffling membrane and the nucleus *J Cell Biol* **122**(5), 1089–101.
3. Lenormand P, Sardet C, Pages G, L'Allemain G, Brunet A, Pouyssegur J. (1993) Growth factors induce nuclear translocation of MAP kinases (p42mapk and p44mapk) but not of their activator MAP kinase kinase (p45mapkk) in fibroblasts *J Cell Biol* **122**(5), 1079–88.
4. Formstecher E, Ramos JW, Fauquet M, et al. (2001) PEA-15 mediates cytoplasmic sequestration of ERK MAP kinase *Dev Cell* **1**(2), 239–50.
5. Ishibe S, Joly D, Zhu X, Cantley LG. (2003) Phosphorylation-dependent paxillin-ERK association mediates hepatocyte growth factor-stimulated epithelial morphogenesis *Mol Cell* **12**(5), 1275–85.
6. Teis D, Wunderlich W, Huber LA. (2002) Localization of the MP1-MAPK scaffold complex to endosomes is mediated by p14 and required for signal transduction *Dev Cell* **3**(6), 803–14.
7. Torii S, Kusakabe M, Yamamoto T, Maekawa M, Nishida E. (2004) Sef is a spatial regulator for Ras/MAP kinase signaling *Dev Cell* **7**(1), 33–44.
8. Casar B, Pinto A, Crespo P. (2009) ERK dimers and scaffold proteins: unexpected partners for a forgotten (cytoplasmic) task *Cell Cycle* **8**(7), 1007–13.
9. Ramos JW. (2008) The regulation of extracellular signal-regulated kinase (ERK) in mammalian cells *Int J Biochem Cell Biol* **40**(12), 2707–19.
10. Shaul YD, Seger R. (2007) The MEK/ERK cascade: from signaling specificity to diverse functions *Biochim Biophys Acta* **1773**(8), 1213–26.
11. Casar B, Pinto A, Crespo P. (2008) Essential role of ERK dimers in the activation of cytoplasmic but not nuclear substrates by ERK-scaffold complexes *Mol Cell* **31**(5), 708–21.
12. Anjum R, Blenis J. (2008) The RSK family of kinases: emerging roles in cellular signalling *Nat Rev Mol Cell Biol* **9**(10), 747–58.
13. Richards SA, Dreisbach VC, Murphy LO, Blenis J. (2001) Characterization of regulatory events associated with membrane targeting of p90 ribosomal S6 kinase 1 *Mol Cell Biol* **21**(21), 7470–80.
14. Zhao Y, Bjorbaek C, Weremowicz S, Morton CC, Moller DE. (1995) RSK3 encodes a novel pp90rsk isoform with a unique N-terminal sequence: growth factor-stimulated kinase function and nuclear translocation *Mol Cell Biol* **15**(8), 4353–63.
15. Fisher TL, Blenis J. (1996) Evidence for two catalytically active kinase domains in pp90rsk *Mol Cell Biol* **16**(3), 1212–9.

Chapter 12

Lentiviral Vectors to Study the Differential Function of ERK1 and ERK2 MAP Kinases

Marzia Indrigo, Alessandro Papale, Daniel Orellana, and Riccardo Brambilla

Abstract

Accumulating evidence indicates that p44^{ERK1} and p42^{ERK2} mitogen-activated protein kinases (MAPKs) have distinct quantitative roles in cell signaling. In our recently proposed model of regulation of ERK1 and ERK2, p42 plays a major role in delivering signals from the cell membrane to the nucleus, while p44 acts as a partial agonist of ERK2 toward effectors and downstream activators, thus providing a fine tuning system of the global signaling output. Here, we describe systems to modulate MAPK signaling in vitro and in vivo via lentiviral vector (LV)-mediated gene transfer, using three systems: RNAi with small hairpin RNAs, microRNA-mediated gene knockdown, and expression of signaling-interfering mutants of MEK1. We show, by using proliferation assays in mouse embryo fibroblasts (MEF) and NIH 3T3 cells, that gene knockdown of ERK1 promotes cell proliferation in a manner indistinguishable from a constitutively active MEK1 construct, while ERK2 RNAi causes a significant growth arrest, similar to that observed with the ectopic expression of a dominant negative MEK1 mutant.

Key words: ERK1, ERK2, MEK1, Lentiviral vector, RNAi, microRNA, Gene knockdown, Cell proliferation, Mouse embryo fibroblast, NIH 3T3

1. Introduction

One of the major dilemmas in signal transduction is related to the "redundancy problem": why does a cell need multiple isoforms of a given signal transducer to perform its physiological function? In recent years, the idea that slightly different isoforms may perform a set of nonoverlapping functions has become a central concept to be tested experimentally, to reach a deeper understanding of the complexity of cell signaling mechanisms. Certainly, the availability of genetically modified mice and RNA interference (RNAi) techniques has shed new light on this complexity as it is

Rony Seger (ed.), *MAP Kinase Signaling Protocols: Second Edition*, Methods in Molecular Biology, vol. 661, DOI 10.1007/978-1-60761-795-2_12, © Springer Science+Business Media, LLC 2010

now clear that ablation of different MAP kinases may lead to quantitative and qualitative differences in phenotype. This is particularly evident for the two main ERK isoforms, $p44^{ERK1}$ and $p42^{ERK2}$. A number of groups have independently generated gene-targeted mice for both isoforms and the general consensus is clear: ERK2 ablation has a large impact on animal survival, while ERK1 loss has not. In fact, *Erk2* gene deletion causes an early phenotype, which is incompatible with the completion of the embryonic development (1–6). On the contrary, *erk1* gene ablation is fully compatible with the adult life (7–9).

In 2006, we published a seminal paper in which we demonstrated for the first time that ERK1 and ERK2 MAP kinases play a differential role in the control of cell proliferation (10). That publication was the culmination of an intense research activity performed by our laboratory in the pursue of a molecular explanation for a crucial observation made already by Pagès et al. in 1999 and later confirmed by us in 2002: ERK1-deficient cells, from embryonic fibroblasts to neurons, show an enhanced phosphorylation of the remaining ERK2 isoform, without significant changes in the ERK2 protein level (7, 9).

At first glance, this observation could have been interpreted as a compensatory action put in place by the cell signaling machinery to overcome ERK1 loss. However, using a combination of ERK1 knockout mouse embryonic fibroblasts (MEF), ERK1 and ERK2 knockdown cells obtained via RNAi, and cells overexpressing either epitope tagged ERK1 or ERK2 proteins, we suggested a different scenario. In our view, ERK1 could act, in physiological settings, i.e., without major alterations of the cell signaling machinery and without an extreme, ERK-independent deregulation of cell growth, as a fine regulator of ERK2, which is, in the large majority of tissues, the most abundant of the two MAP kinases. We essentially found, both in MEF and NIH 3T3 cells, that a complete (>90%) ablation of ERK1 protein leads to a significant growth advantage, while ablation of ERK2 is essentially incompatible with cell viability. However, we also found that in NIH 3T3, overexpression of both WT ERK1 and a kinase defective form of ERK1 caused a significant reduction in oncogenic Ras-mediated cell growth while basal activity was unchanged. The ability of the kinase dead form of ERK1 to inhibit oncogenic cell growth is, in our opinion, sufficient and definitive proof that a competition for either upstream regulators (e.g., MEK1/2 kinases) or downstream effectors (e.g., MSK-1 kinases or MKP1/2/3 phosphatases) occurs between ERK1 and ERK2. Moreover, we observed that the same mutation in ERK2 causing the kinase-defective phenotype in ERK1 resulted in an even stronger dominant negative effect on cell proliferation mediated by Ras, confirming that ERK isoforms interact with some limiting factors in the cell cycle. From these experiments, we originally proposed that (1) ERK1 is a weaker signal transducer than ERK2, not necessarily because it bears an intrinsic lower kinase activity but because it has a reduced capability to interact with crucial signaling activators/effectors, (2) ERK1, in appropriate conditions, can antagonize ERK2-mediated signaling by partially limiting the ability of ERK2 to interact with yet to be identified relevant effectors. In this context, the term "appropriate conditions" is crucial since not in all experimental settings and cell types can the effect on cell proliferation be observed and clearly, a nonlinear and nonreciprocal response pattern seems to be in place. For instance, in our NIH 3T3 clone, ERK1 overexpression is capable of attenuating cell growth only when oncogenic Ras is expressed, while ERK1 ablation facilitates cell growth in basal conditions but, paradoxically, causes a minor, yet significant, detrimental effect on Ras-mediated transformation. In addition, in other cells types with a higher proliferative potential, including primary astrocytes (our unpublished data), B and T cells (5, 11, 12), hepatocytes (13–15), or myoblasts (16, 17), the absence of ERK1 alone seems to have no effect at all on either cell proliferation or survival, while ERK2 single depletion severely impacts on cell physiology. However, a recent report indicates that hepatocyte cell survival is increased in either ERK1 knockout or knockdown cells when apoptosis is induced (18), suggesting that the expected potentiated ERK2 activity could protect these cells from cell death.

So far, only one report has indicated that loss of ERK1 alone could negatively impact on cell proliferation, similar to an ERK2 knockdown, but that was observed in HeLa cells, a tumor cell line (19). Surprisingly, keratinocytes established from ERK1-deficient mice show paradoxical effects: they are impaired in their proliferative index, but at the same time they are resistant to apoptosis. This complex response results in a reduced tumor growth in response to carcinogens (20).

Altogether, the large majority of available data seems to support the notion that the main ERK isoform is $p42^{ERK2}$, while $p44^{ERK1}$ appears to play an ancillary role. However, our own data also indicate that ERK1 can antagonize ERK2 activity in some physiological conditions and cell types and that the main direct effect of ERK1 depletion is hyperactivation of ERK2 phosphorylation and stimulation of downstream signaling. In order to gain more information about the mechanism of action of ERK1, we recently identified the structural determinant responsible for most of the functional differences between the two MAPK isoforms. Despite the fact that the overall sequence identity between the two kinases is 85%, a unique N-terminal stretch of 20 aminoacids is exclusively present in ERK1. We found that this domain is responsible for the slow nucleocytoplasmic shuttling of ERK1, a key phenomenon implicated in ERK signal termination. Since the upstream kinases of ERKs, MEK1/2, are mainly localized in the cytoplasm, dephosphorylation prevails in the nucleus. Therefore, signaling starts in the cytoplasm and terminates in the nucleus. The maintenance of a functional level of activated ERKs in the nucleus depends on the inflow of phosphorylated ERKs from the cytoplasm, which is in equilibrium with the efflux of dephosphorylated ERKs. This process of shuttling has been recently demonstrated in living cells (21–23). Furthermore, we recently showed that ERK2 shuttles much faster than ERK1, and thus p44 is more susceptible to nuclear inactivation than p42 (24). The strength of this effect depends critically on the differences between the trafficking of ERK1 and ERK2, which is likely due to differential binding on nuclear import/export molecular components. Importantly, in that study, we demonstrated that the N-terminal domain of ERK1 is not only necessary but also sufficient for determining the trafficking differences between ERK1 and ERK2 and for the inhibitory effect of ERK1 on Ras-mediated cell transformation. Indeed, anchoring of the N-terminal domain of ERK1 to ERK2 confers ERK1-like signaling properties to ERK2, while removal from ERK1 converts it into an ERK2-like kinase.

Unfortunately, our view that different signaling isoforms may play, qualitatively and quantitatively, different cellular functions is not shared by some scientists, namely Pouyssegur and colleagues, who recently published a provocative paper in Molecular and Cellular Biology, followed by a fairly biased perspective article in Cell Cycle (25, 26). Firstly, what Lefloch et al. showed is that gene silencing of ERK1 does not impact on cell proliferation, as measured in a NIH 3T3 cell clone they had in the lab, while silencing of ERK2 severely affects cell growth in the same cell line. Interestingly, after prolonged puromycin selection, ERK2-depleted cells appeared to overactivate ERK1, as an extreme compensatory attempt to rescue p42 loss. Finally, and most importantly, Lefloch et al. showed that when ERK2 protein is reduced to very low levels and total ERK1/2 activity is only down to 70% of the controls, thanks to ERK1 hyperactivation, the silencing of ERK1 would negatively impact on cell proliferation, by further decreasing the inhibitory effect on cell growth caused by ERK2 depletion.

Thus, the main hypothesis proposed by Lefloch et al. in their perspective article, based on their 2008 publication, claims that ERK1 and ERK2 play similar roles in cell signaling and that their apparent differences are exclusively due to their differential expression. In this view, ERK1 could completely vicariate for ERK2 provided that its expression level would be comparable to ERK2. Indeed, Lefloch et al. went further on to propose that ERK1 does play a positive role in controlling cell proliferation but only when ERK2 levels are clamped down to lower than physiological levels, suggesting that ERK1, in extreme conditions in which ERK2 levels are not enough to support cell viability, can compensate for the main ERK isoform loss.

As we stated in a fair and well-balanced letter to Mikhail Blagosklonny, the Cell Cycle Editor, which was deliberately disregarded and surprisingly not accepted for publication, "Lefloch et al. have ignored the fact that their interpretation is likely to be a special case of a rather more complex scenario. The key issue here is not whether ERK1 and ERK2 are qualitatively similar but rather whether they can provide to cells a quantitatively different signaling potential in physiological conditions in which ERK1 expression levels are significantly lower than ERK2." In fact, the major criticism to their data is that they used a NIH 3T3 clone which, by judging from the cell density reached in the proliferation assays, is already transformed and not suitable for such functional studies. Together with the notion that "the absence of evidence is not the evidence of absence," this last point is very relevant for the interpretation of their negative data and their inability to detect a growth advantage in ERK1 ablated cells. Moreover, after plasmid transfection, Lefloch et al. used a selection procedure with puromycin to obtain subclones that are likely to additionally affect the basic proliferative features of the cells, further complicating the interpretation of their data. In our experience, the growth of ERK2-deficient cells is extremely poor and any attempt to generate stable clones is virtually linked to failure, at best leading to a selection of revertants with a highly deregulated growth, as a result of the necessity of sustaining an ERK2-independent cell proliferation. All these considerations lead us to conclude that the data and explanations provided by Lefloch et al. are rather inconclusive and poorly informative.

In conclusion, in our view, ERK1 can be seen as a partial agonist to ERK2 in interacting with signaling partners either upstream or downstream. In pharmacology, a partial agonist is a molecule that binds to a receptor as a full agonist but is less efficient in either causing receptor activation or promoting downstream signaling. Thus, a partial agonist effect can only be observed in a tripartite binding scheme, in which one receptor can bind to at least two ligands. Importantly, a key feature of a partial agonist is that its activity changes depending on its relative expression levels and the intensity of signaling. In general terms, at low levels of receptor occupancy and low signaling intensity, the partial agonist has little effects because the limiting factor is the efficient binding between the full agonist and the receptor. At moderate receptor occupancy, the presence of the partial agonist reduces the overall signaling since it competes with the full agonist in the binding to the receptors. However, at high receptor occupancy, the partial agonists may show a different effect depending on its level: (1) if expressed at low level, it will positively contribute to signaling since it will bind to the receptor quota that is not bound to the full agonist; (2) if expressed at high level, it will compete with the full agonist for binding to the receptor, thus inhibiting signaling.

If one considers ERK2 as a full agonist and ERK1 as a partial agonist while the receptor can be any of their signaling effectors (or upstream activators), our data can be largely explained.

In order to facilitate future research on these crucial topics, we provide methods to study the differential function of ERK1 and ERK2 in physiological conditions, both in vitro and in vivo. Lentiviral vectors (LV) are powerful and efficient tools to modify gene expression in primary cells without the need to use a selection marker, which is a major complication for in vitro study of cell proliferation (10, 27–33). Here, we show that both in MEF and NIH 3T3 cells, ablation of ERK1, using two independent systems (shRNA- and microRNA-mediated gene knockdown), provides a significant growth advantage, while ERK2 deficiency does just the opposite. Remarkably, the ERK1 knockdown phenotype strongly parallels that observed when a constitutive active form of MEK1 is overexpressed via LV. On the contrary, the ERK2 knockdown phenotype is similar to what observed with a dominant negative form of MEK1 (34). Thus, in our opinion, these data unequivocally and conclusively demonstrate that ERK1 and ERK2 can play different and opposing function in the control of cellular functions.

2. Materials

2.1. Cell Culture, Transfection and Viral Vector Production

1. Dulbecco's Modified Eagle Medium with GlutaMax (DMEM GlutaMax, Gibco).
2. Iscove's Modified Dulbecco Medium (IMDM, Sigma).
3. Newborn Calf Serum (BCS, Euroclone).
4. Fetal Bovine Serum (FBS, Euroclone).
5. 100× glutamine: 200 mM glutamine.
6. 100× Pen/Strept: 10,000 U/ml penicillin, 10,000 μg/ml streptomycin (Gibco).
7. Sodium butyrate.
8. Dulbecco's phosphate saline buffer (PBS).
9. Trypsin solution: 0.05% Trypsin, 0.53 mM ethylenediaminetetraacetic acid (EDTA).
10. Tris–EDTA (TE) buffer: 10 mM Tris pH 8.0, 1 mM EDTA.
11. Collagenase (Sigma).
12. 100 mm, 150 mm and 6-well plates.
13. 2× HBS: 281 mM NaCl, 100 mM HEPES (Sigma), 1.5 mM Na_2HPO_4 pH 7.09 (Sigma), filter-sterilized and stored at −20°C.
14. $CaCl_2$ solution: 2.5 M $CaCl_2$ (Sigma), filter (0.22 μm) and store at −20°C.
15. Carrier DNA from salmon sperm (Gibco).

16. Sterile water.
17. Syringe-driven filter units (0.22 μm) MILLEX GP (Millipore).
18. Dimethyl sulfoxide (DMSO, Sigma).

2.2. Proliferation Curve

1. Fixing solution: 2% paraformaldehyde in PBS.
2. Cell Coulter Z1 (Beckman).
3. Saline solution: 0.9% NaCl solution in sterile water.

2.3. Sodium Dodecyl Sulfate–Polyacrylamide Gel Electrophoresis

1. Stacking gel: 125 mM Tris–HCl pH 6.8, 4.5% acrylamide-bis (19:1), 0.1% sodium dodecyl sulfate (SDS), 0.2% ammonium persulfate, 0.2% *N,N,N′,N′*-tetramethylethylenediamine (TEMED, Bio-Rad).
2. Running gel: 375 mM Tris–HCl pH 8.8, 12% acrylamide-bis (19:1), 0.1% SDS, 0.1% ammonium persulfate, 0.1% TEMED.
3. Running buffer: 25 mM Tris–HCl pH 8.8, 190 mM glycine, 0.1% SDS.
4. 5× sample buffer: 250 mM Tris–HCl pH 6.8, 10% SDS, 50% glycerol, 0.02% bromophenol blue, 110 mM dithiotreithol (DTT).
5. Lysis buffer: 125 mM Tris–HCl pH 6.8, 2.5% SDS.

2.4. Western Blot

1. Transfer buffer: 25 mM Tris, 190 mM glycine, 20% methanol.
2. Protran nitrocellulose membrane (Whatman).
3. Tris buffered saline buffer (TBS: 20 mM Tris–HCl pH 7.4, 150 mM NaCl) with 0.1% Tween (TBS-TW).
4. Blocking buffer: 5% bovine serum albumin (BSA, Sigma) in TBS-TW.
5. Antibody buffer: 3% BSA in TBS-TW.
6. Antibodies: Phospho-p44/42 MAPK (Erkl/2) (Thr202/Tyr204) Antibody #9101 (Cell Signaling), p44-42 MAPK #SC-153 (Santa Cruz Biotechnology), GAPDH #SC-25778 (Santa Cruz Biotechnology).

3. Methods

3.1. Mouse Embryonic Fibroblasts Preparation

MEF cultures are prepared from wild-type E13.5 embryos obtained from commercially available C57 B/6 mice (Charles River) as previously described (9, 10) (see Note 1).

1. Dissection is carried out in PBS.
2. Remove the internal organs and the head, reduce the remaining tissues in small pieces, and transfer in a 50-ml tube.
3. Centrifuge at 3,000 RCF for 5 min.

4. Wash in PBS and incubate for 1 h at 37°C in a solution of 0.25% collagenase and 20% FBS in PBS.
5. Dissociate tissues using a syringe.
6. Centrifuge at 3,000 RCF for 5 min at room temperature.
7. Resuspend the cells in DMEM GlutaMax containing 10% FBS, 1× Pen/Strept.
8. Count the cells and seed 5×10^6 cells on 15-mm plates.
9. Freeze the cells at confluence in 95% serum and 5% DMSO.

3.2. Lentiviral Vectors Production

We use all third-generation LV, modifications of the originally described backbone (28) (see Note 2).

1. Seed and incubate 9×10^6 HEK 293 T cells (ATCC, CRL-11268) in 150-mm dishes, approximately 24 h before transfection. The medium used is DMEM GlutaMax containing 10% FBS, 1× Pen/Strept. Use low-passage cells (not more than P12-15) and do not ever let cells grow to confluence.
2. Change medium 2 h before transfection with IMDM supplemented with 10% FBS, 1× Penicillin/Streptomycin and 1× glutamine (22-ml final volume).
3. Prepare the plasmid DNA mix by adding together: 9 μg ENV plasmid (VSV-G), 12.5 μg packaging plasmid (pMDLg/pRRE or CMV R8.74), 6.25 μg of pRSV-REV, and 32 μg of gene transfer plasmid. The plasmid mix solution is made up to a final volume of 1,125 μl with 0.1× TE buffer (1×: 10 mM Tris pH 8.0; 1 mM EDTA pH 8.0 in water). Finally, 125 μl of 2.5 M $CaCl_2$ is added.
4. Leave the mix 15 min at room temperature.
5. The precipitate is formed by dropwise addition of 1,250 μl of 2× HBS (281 mM NaCl, 100 mM HEPES, 1.5 mM Na_2PO_4, pH 7.06–7.12) solution to the 1,250 μl DNA–TE–$CaCl_2$ mixture from step 3 while vortexing at full speed. The precipitate should be added to HEK 293 T cell immediately following the addition of the 2× HBS. High-magnification microscopy of the cells should reveal a very small granular precipitate of $CaPO_4$-precipitated plasmid DNA, initially above the cell monolayer, and after incubation in a 37°C incubator overnight, on the bottom of the plate in the large spaces between the cells.
6. The $CaPO_4$-precipitated plasmid DNA should be allowed to stay on the cells for 14–16 h, after which the media should be replaced with fresh medium (IMDM with 10% FBS, 1× Pen/Strept, 1× glutamine, and 1 M sodium butyrate).
7. Collect the cell supernatants at 36 h after changing the medium; filter (0.22 μm) and centrifuge at 50,000 rcf, at 20°C for 2 h (Beckman Ultracentrifuge, SW32Ti rotor).

8. Discard the supernatant and resuspend the pellet in sterile PBS 1×.
9. Aliquot and store at −80°C.

3.3. Titration of the Lentiviral Vectors

1. Seed and incubate 5×10^4 HEK 293 T cells in 35-mm dishes, 12–14 h before the infection.
2. Make serial dilution of the LV in the growing medium (DMEM GlutaMax, 10% FBS, 1× Pen/Strept) and transduce the cells with the desired dilutions in a final volume of 1 ml.
3. After 5 days, collect the cells, wash in PBS with 1% FBS, and resuspend in 2% paraformaldehyde in PBS 1×.
4. The titer (transforming units (TU)) is determined by using FACScalibur (BD Bioscience) and counting the percentage of GFP-positive cells in each dilution. When the percentage is between 2.5 and 25%, the titer can be determined using the following formula:

$$\text{Titer (TU/ml)} = \frac{\begin{array}{c}\text{Percentage of GFP positive cells} \times \\ \text{cells plated the first day } (5 \times 10^4)\end{array}}{\text{Dilution}}$$

3.4. Transient Transfection of NIH 3T3 Cells with $CaPO_4$

The NIH 3T3 clone used in this work has previously been use with success in performing functional assays (10, 24, 35–40) (see Note 3).

1. Thaw NIH 3T3 cells, at passage 3, by keeping them at 37°C for a few seconds. Immediately after thawing, add 1 ml of fresh prewarmed medium (DMEM GlutaMax supplemented with 10% BCS, 1× Pen/Strept) to each cryovial. Split on 100-mm tissue plates and incubate at 37°C, 5% CO_2.
2. Allow the cells to grow until approaching to confluence (70–90%) and split them with 1% trypsin/EDTA for 5 min at 37°C to new maintenance 100-mm plates and experimental 6-well plates.
3. The day before transfection seed 1.2×10^5 cells on 6-well plates and incubate in a total of 2 ml of medium. Two hours before the transfection change the medium with fresh pre-warmed medium.
4. Set up the plasmid DNA mix: For each well use 5 μg of plasmid DNA of interest, 2 μg of carrier DNA, 7 μl of 2.5 M $CaCl_2$ and TE buffer 0.1× up to a final volume of 62.5 μl.
5. Mix and incubate at room temperature for 15 min. (NOTE: the plasmid DNA mix used for transfection must be adjusted according to the size of the well/dish used for the specific experiment. In this case, all the DNA mixes are for 6-well plates).

6. The precipitate is formed by dropwise addition of 62.5 μl of 2× HBS (pH 7.09) solution, to the previous DNA–$CaCl_2$ mixture, while the mix is vortexed at full speed.
7. The precipitate should be added to the cells immediately following the addition of the 2× HBS. The $CaPO_4$-precipitated plasmid DNA should be allowed to stay on the cells for 14–16 h, after which the medium should be replaced with fresh medium.

3.5. Proliferation Assays

The protocol described here works very well with different cells lines such as NIH 3T3 cells and MEF. Cells are initially plated in the exponential phase and followed for 5 days, till they reach a subconfluent stage (see Note 4).

1. The day before the infection (or the transfection), seed 1.25×10^5 cells/well in 6-well plates. It is better to use low passage cells; in this case, P4-5 cells were used. Prepare one plate for each day of counting for each condition.
2. The day after the infection (day 0), count one plate for each condition. The count is performed following this protocol:
   (a) Detach the cells with the trypsin solution for 5 min at 37°C.
   (b) Collect the cells and wash carefully the wells with PBS 1× with 1% serum to be sure that all the cells have been taken.
   (c) Centrifuge at 1,500 RCF for 5 min at RT.
   (d) Resuspend the cells in 1 ml of fixing solution.
   (e) Dilute 200 μl of cell suspension in 10 ml of saline and determine the cell number with the Coulter counter.
3. Transduce (at multiplicity of infection (MOI) 5, in a volume of 1 ml of fresh medium) or transfect the remaining plates.

Every day count one plate for each condition and replace the others with fresh medium containing 5% serum.

4. At day 1 after transfection, or day 2 after infection, the cells are controlled for GFP expression to check the efficiency of the process.

4. Notes

1. MEF, prepared following this protocol, are cells with a relatively low proliferative potential, and they should be used exclusively at low passages (maximum two times after thawing) since they rapidly undergo senescence. We never refreeze MEF after the initial freezing. Although MEF normally grow at 10% FBS and can be starved at 0.5–1% FBS for inducing

entry into the G_0 phase of the cell cycle, they can also be grown at 5% FBS, as done in our proliferation assays. Since MEF are resistant to plasmid transfection, we use for the experiments in Figs. 1 and 3 LV-assisted infection with the constructs described in Note 2. Using our infection conditions (MOI = 5), transduction efficiency of MEF is very high (<90% of GFP positive cells).

2. All lentiviral vectors used for the MEF proliferation assays (Figs. 1 and 3) were self-inactivating, third-generation vectors (41). The plasmids used for packaging, kindly provided by Luigi Naldini, were pREV (expressing REV protein), pVSVG (expressing the envelope: G glycoprotein of Vesicular Stomatitis Virus), and pRRE (expressing capsid, polymerase, protease, and integrase proteins). The lentiviral vector used as control in the MEF proliferation assay was pCCLsin.PPT.hPGK.eGFP.PRE (28). This vector allows expression of eGFP under the human phosphoglycerate kinase (hPGK) promoter. The two LV used for the knockdown of ERKl and ERK2

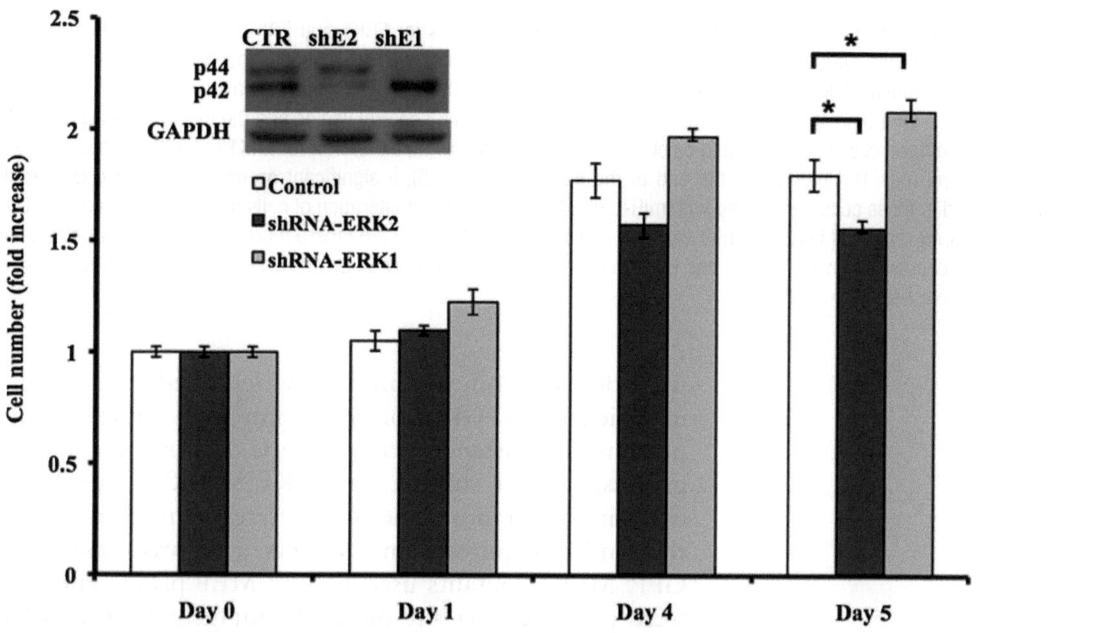

Fig. 1. Proliferation assay on MEF transduced with shRNAs for ERK1 and ERK2. Proliferation was assessed by counting the cells every day using a Coulter counter. 1.25×10^5 cells/well were seeded in 6-well plate and maintained in culture medium (DMEM Glutamax, 1× P/S) supplemented with 10% FBS. The following day (day 0), cells were transduced at MOI = 5 with lentiviral vectors expressing either shRNA-ERK1 or shRNA-ERK2 or GFP as a control, and the medium was changed to 5% FBS. No differences in number could be observed in cells treated with either shRNA for ERK1 or ERK2, when compared with the control (day 1). However, at the end of the experiment (day 5), a significant growth advantage was observed in cells transfected with shRNA-ERK1. Conversely, the proliferation of those cells transfected with shRNA-ERK2 was significantly reduced (day 5). Results are the mean of $n = 6$. In the *upper left inset* a representative Western blot is shown, with the expression levels of ERK1 and ERK2 in the transduced cells (day 5). On day 5, statistical analysis was done using the Student *t*-test: $^*p < 0.05$.

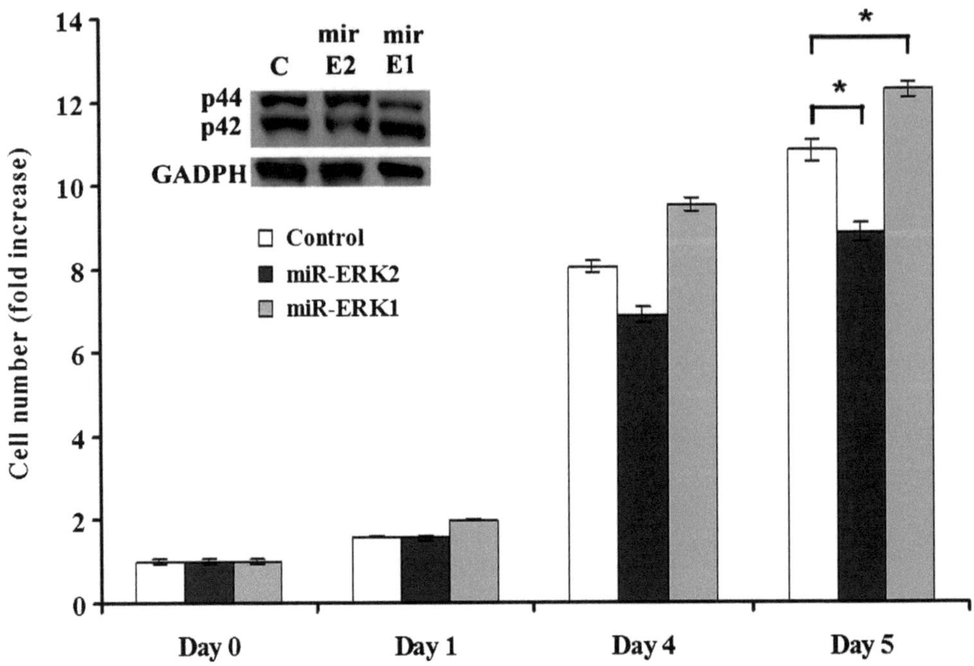

Fig. 2. Proliferation assay on NIH 3T3 cells transfected with miRs for ERK1 and ERK2. Proliferation was assessed by counting the cells every day using a Coulter counter. 1.25×10^5 cells/well were seeded in 6-well plate and maintained in culture medium (DMEM Glutamax, 1× P/S) supplemented with 10% BCS. The following day (day 0), cells were transiently transfected with 5 µg of miR-ERK1, miR-ERK2, or a GFP expressing control vector, and the medium was changed to 5% BCS. No differences in number could be observed in cells treated with either miR for ERK1 or ERK2 when compared with the control (day 1). However, at the end of the experiment (day 5), a significant growth advantage could be clearly observed in those cells transfected with miR-ERK1. Conversely, the proliferation of cells transfected with miR-ERK2 was significantly reduced (day 5). Results are the mean of $n = 6$. In the *upper left inset* a representative Western blot is shown, with the expression levels of ERK1 and ERK2 in the transfected cells (day 5). On day 5, statistical analysis was done using the Student *t*-test: $^*p < 0.05$.

were derived from the previously described constructs (10) in which the hPGK promoter controlled expression of the puromycin resistance gene and the Histone H1 promoter expressed either shRNA for ERK1 or ERK2, subcloned in opposite orientation. The only difference in the new LV was that, in it, the puromycin resistance gene was substituted for GFP. MEK1 mutants used for the MEF proliferation assays were produced via site-directed mutagenesis (QuikChange Site-Directed Mutagenesis Kit, Stratagene) of the WT full length MEK1 cDNA (kindly provided by Rony Seger): after adding at the N-terminus a myc-tag, two different mutations were produced. The two mutations were (Lys97Ala) and (Δ31-52; Ser118Glu; Ser122Glu), already known to generate a dominant negative and a constitutively active form of the enzyme, respectively (34). These MEK1 mutants have been subcloned in a LV (pCCL.sin.cPPT.hPGK.ΔLNGFR.IRES.eGFP.WPRE, kindly provided by Luigi Naldini) that allows expression of the transgene under the hPGK promoter and of the GFP under internal ribosomal entry site (42).The plasmids used in the NIH 3T3 proliferation assays were modified from a microRNA (miR) 30-based shRNA expression system driven by a RNA pol II promoter, pPRIME (kindly provided by Frank Stegmeier) (32). The expression of GFP-miR30-shRNA was under human cytomegalovirus (CMV) promoter. A cotransfection of two plasmids expressing shRNA for ERK1 and two for ERK2 were used.

shRNA sequences for ERK1:

AACGCTACACGCAGCTGCAGTATAGTGAAGC-CACAGATGTATACTGCAGCTGCGTGTAGCGTG, AAGGACCTTAATTGCATCATTATAGTGAAGC-CACAGATGTATAATGATGCAATTAAGGTCCTC.

shRNA sequences for ERK2:

CCGGCATGGTTTGCTCTGCTTATAGTGAAGCC-ACAGATGTATAAGCAGAGCAAACCATGCCGT, CGCTCTGGATTTACTGGATAAATAGTGAAGCC-ACAGATGTATTTATCCAGTAAATCCAGAGCT.

A control vector containing the FF3 hairpin shRNA (which targets firefly luciferase) is used as a negative control. The Western blot assay shown in the inset of Fig. 2 indicates that this transfection protocol achieved approximately 60% of reduced protein levels for both ERK1 and ERK2, without affecting the other isoform (mean of $n = 3$). However, it must be taken into consideration that knockdown efficiency is likely to be underestimated since the efficiency of transfection is much lower than that achieved with LV-mediated infection, allowing approximately only 70% of GFP-positive cells.

3. The NIH 3T3 clones used in this work were originally obtained from the laboratory of Mariano Barbacid (now at the Centro Nacional de Investigaciones Oncológicas, Madrid, Spain). They were exclusively grown in 10% BCS and we carefully selected serum batches in order that the maximum confluence density reached did not exceed 1,000 cells/mm^2, a typical contact inhibition profile for untransformed NIH 3T3. As a matter of comparison, Lefloch et al., in their 2008 paper (25), measured a maximum cell density in a strictly exponential phase as high as 1,600 cells/mm^2. This value, supposedly far below the density reached at confluence, is in fact rather close to what we normally observe in oncogenic Ras transformed clones. To maximize the difference among controls, miR-ERK2 and miR-ERK1 transfected cells, in this work we performed the proliferation assays with 5% BCS. Since NIH 3T3 cells are not readily infected with LV, for the experiments shown in Fig. 2, we used the plasmid versions of the

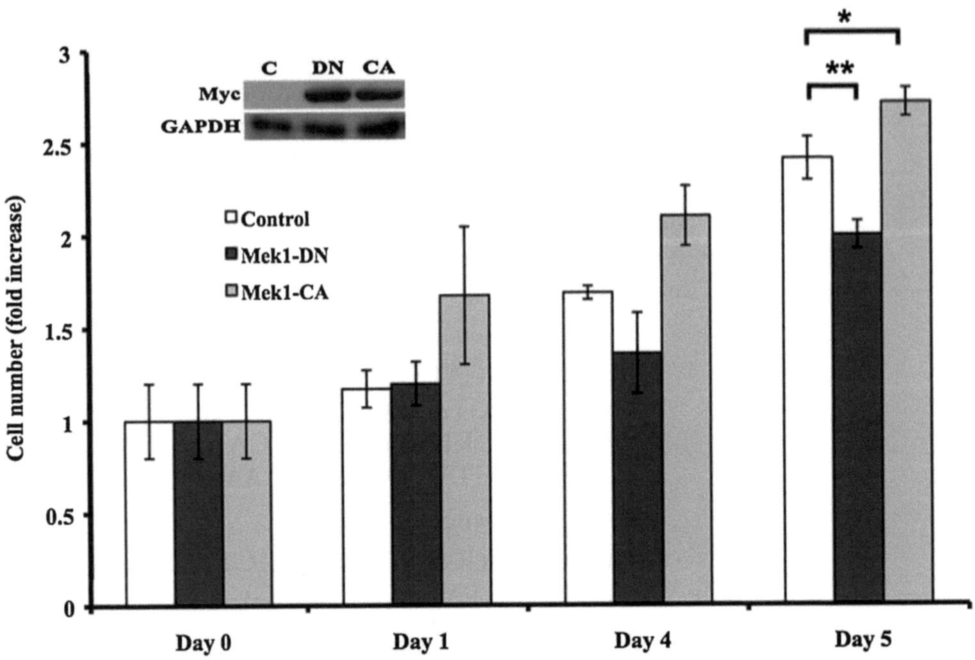

Fig. 3. Proliferation assay on MEF transduced with MEK1 mutants. Proliferation was assessed by counting the cells every day using a Coulter counter. 1.25×10^5 cells/well were seeded in 6-well plate and maintained in culture medium (DMEM Glutamax, 1× P/S) supplemented with 10% FBS. The following day (day 0), the cells were transduced at MOI = 5 with lentiviral vectors expressing either the dominant negative (DN, Lys97Ala), or the constitutively active (CA, Δ33-51; Ser118Glu, Ser122Glu) form of MEK1 or the GFP as a control. At day 0 the medium was changed to 5% FBS. No differences in number were observed in cells treated with the different vectors when compared with the control (day 1). However, a significant advantage in proliferation could be observed in those cells transfected with the CA mutant of MEK1. Conversely, the proliferation of those cells transfected with the DN form of MEK1 was significantly reduced (day 5). Cells transduced with the CA mutant of MEK1 showed an apparently different morphology when observed in a phase contrast optical microscopy, with respect to the other lentiviral vectors, starting from day 2; this may be due to dependency on ERK hyperactivation, but further investigation will be required to clarify this issue. Results are the mean of $n = 6$. In the *upper left inset* a representative Western blot is shown, with the expression levels of the MEK1 transgenes in the transfected cells (day 5). On day 5, statistical analysis was done using the Student *t*-test: *$p < 0.05$, **$p < 0.01$.

constructs described in Note 2. However, the same constructs can be used for generating LV as described in the protocols and tested on MEF or any other primary culture.

4. At day 5 of the proliferation assays, the following cell densities (mean ± SEM) were reached. For Fig. 1: control vector, 245 ± 5 cells/mm^2; shRNA-ERK1, 291 ± 7 cells/mm^2; shRNA-ERK2, 202 ± 5 cells/mm^2. For Fig. 2: control vector, 732 ± 34 cells/mm^2; miR-ERK1, 830 ± 31 cells/mm^2; miR-ERK2, 599 ± 24 cells/mm^2. For Fig. 3: control vector, 426 ± 21 cells/mm^2; Mek1-CA, 479 ± 14 cells/mm^2; Mek1-DN, 353 ± 13 cells/mm^2.

Acknowledgments

This work was supported by the Michael J Fox Foundation for Parkinson's Research and the Parkinson's Disease Society of the UK, as well as by the Italian Ministry of Health (to RB).

References

1. Saba-El-Leil MK, *et al.* (2003) An essential function of the mitogen-activated protein kinase Erk2 in mouse trophoblast development. *EMBO Rep* 4(10):964–968.
2. Yao Y, *et al.* (2003) Extracellular signal-regulated kinase 2 is necessary for mesoderm differentiation. *Proc Natl Acad Sci U S A* 100(22):12759–12764.
3. Hatano N, *et al.* (2003) Essential role for ERK2 mitogen-activated protein kinase in placental development. *Genes Cells* 8(11): 847–856.
4. Meloche S, Vella FD, Voisin L, Ang SL, & Saba-El-Leil M (2004) Erk2 signaling and early embryo stem cell self-renewal. *Cell Cycle* 3(3):241–243.
5. Fischer AM, Katayama CD, Pages G, Pouyssegur J, & Hedrick SM (2005) The role of erk1 and erk2 in multiple stages of T cell development. *Immunity* 23(4):431–443.
6. Satoh Y, *et al.* (2007) Extracellular signal-regulated kinase 2 (ERK2) knockdown mice show deficits in long-term memory; ERK2 has a specific function in learning and memory (Translated from English). *J Neurosci* 27(40):10765–10776 (in English).
7. Pagès G, *et al.* (1999) Defective thymocyte maturation in p44 MAP kinase (Erk 1) knockout mice. *Science* 286(5443):1374–1377.
8. Selcher JC, Nekrasova T, Paylor R, Landreth GE, & Sweatt JD (2001) Mice lacking the ERK1 isoform of MAP kinase are unimpaired in emotional learning. *Learn Mem* 8(1):11–19.
9. Mazzucchelli C, *et al.* (2002) Knockout of ERK1 MAP kinase enhances synaptic plasticity in the striatum and facilitates striatal-mediated learning and memory. *Neuron* 34:807–820.
10. Vantaggiato C, *et al.* (2006) ERK1 and ERK2 mitogen-activated protein kinases affect Ras-dependent cell signaling differentially. *J Biol* 5(5):14.
11. D'Souza WN, Chang CF, Fischer AM, Li M, & Hedrick SM (2008) The Erk2 MAPK regulates CD8 T cell proliferation and survival (Translated from English). *J Immunol* 181(11):7617–7629 (in English).
12. Yasuda T, *et al.* (2008) Erk kinases link pre-B cell receptor signaling to transcriptional events required for early B cell expansion (Translated from English). *Immunity* 28(4):499–508 (in English).
13. Bessard A, Fremin C, Ezan F, Coutant A, & Baffet G (2007) MEK/ERK-dependent uPAR expression is required for motility via phosphorylation of P70S6K in human hepatocarcinoma cells (Translated from English). *J Cell Physiol* 212(2):526–536 (in English).
14. Bessard A, *et al.* (2008) RNAi-mediated ERK2 knockdown inhibits growth of tumor cells in vitro and in vivo (Translated from English). *Oncogene* 27(40):5315–5325 (in English).
15. Fremin C, *et al.* (2007) ERK2 but not ERK1 plays a key role in hepatocyte replication: an RNAi-mediated ERK2 knockdown approach in wild-type and ERK1 null hepatocytes (Translated from English). *Hepatology* 45(4):1035–1045 (in English).
16. Lips DJ, *et al.* (2004) MEK1-ERK2 signaling pathway protects myocardium from ischemic injury in vivo (Translated from English). *Circulation* 109(16):1938–1941 (in English).
17. Li J & Johnson SE (2006) ERK2 is required for efficient terminal differentiation of skeletal myoblasts (Translated from English). *Biochem Biophys Res Commun* 345(4):1425–1433 (in English).
18. Fremin C, *et al.* (2009) Multiple division cycles and long-term survival of hepatocytes are distinctly regulated by extracellular signal-regulated kinases ERK1 and ERK2 (Translated from English). *Hepatology* 49(3):930–939 (in English).

19. Liu X, Yan S, Zhou T, Terada Y, & Erikson RL (2004) The MAP kinase pathway is required for entry into mitosis and cell survival (Translated from English). *Oncogene* 23(3): 763–776 (in English).
20. Bourcier C, *et al.* (2006) p44 mitogen-activated protein kinase (extracellular signal-regulated kinase 1)-dependent signaling contributes to epithelial skin carcinogenesis (Translated from English). *Cancer Res* 66(5):2700–2707 (in English).
21. Live cell imaging of ERK and MEK: simple binding equilibrium explains the regulated nucleocytoplasmic distribution of ERK (Translated from English). *J Biol Chem* 280(5):3832–3837 (in English).
22. Costa M, *et al.* (2006) Dynamic regulation of ERK2 nuclear translocation and mobility in living cells (Translated from English). *J Cell Sci* 119(Pt 23):4952–4963 (in English).
23. Ando R, Mizuno H, & Miyawaki A (2004) Regulated fast nucleocytoplasmic shuttling observed by reversible protein highlighting (Translated from English). *Science* 306(5700):1370–1373 (in English).
24. Marchi M, *et al.* (2008) The N-terminal domain of ERK1 accounts for the functional differences with ERK2 (Translated from English). *PLoS ONE* 3(12):e3873 (in English).
25. Lefloch R, Pouyssegur J, & Lenormand P (2008) Single and combined silencing of ERK1 and ERK2 reveals their positive contribution to growth signaling depending on their expression levels (Translated from English). *Mol Cell Biol* 28(1):511–527 (in English).
26. Lefloch R, Pouyssegur J, & Lenormand P (2009) Total ERK1/2 activity regulates cell proliferation (Translated from English). *Cell Cycle* 8(5):705–711 (in English).
27. Naldini L, *et al.* (1996) In vivo gene delivery and stable transduction of nondividing cells by a lentiviral vector (Translated from English). *Science* 272(5259):263–267 (in English).
28. Follenzi A, Ailles LE, Bakovic S, Geuna M, & Naldini L (2000) Gene transfer by lentiviral vectors is limited by nuclear translocation and rescued by HIV-1 pol sequences. *Nat Genet* 25(2):217–222.
29. Follenzi A & Naldini L (2002) Generation of HIV-1 derived lentiviral vectors. *Methods Enzymol* 346:454–465.
30. Brummelkamp TR, Bernards R, & Agami R (2002) A system for stable expression of short interfering RNAs in mammalian cells. *Science* 296(5567):550–553.
31. Rubinson DA, *et al.* (2003) A lentivirus-based system to functionally silence genes in primary mammalian cells, stem cells and transgenic mice by RNA interference. *Nat Genet* 33(3):401–406.
32. Stegmeier F, Hu G, Rickles RJ, Hannon GJ, & Elledge SJ (2005) A lentiviral microRNA-based system for single-copy polymerase II-regulated RNA interference in mammalian cells (Translated from English). *Proc Natl Acad Sci U S A* 102(37):13212–13217 (in English).
33. Papale A, Cerovic M, & Brambilla R (2009) Viral vector approaches to modify gene expression in the brain. *J Neurosci Methods* 185(1):1–14.
34. Jaaro H, Rubinfeld H, Hanoch T, & Seger R (1997) Nuclear translocation of mitogen-activated protein kinase kinase (MEK1) in response to mitogenic stimulation. *Proc Natl Acad Sci U S A* 94(8):3742–3747.
35. Jainchill JL, Aaronson SA, & Todaro GJ (1969) Murine sarcoma and leukemia viruses: assay using clonal lines of contact-inhibited mouse cells (Translated from English). *J Virol* 4(5):549–553 (in English).
36. Klein R, *et al.* (1991) The trkB tyrosine protein kinase is a receptor for brain-derived neurotrophic factor and neurotrophin-3 (Translated from English). *Cell* 66(2):395–403 (in English).
37. Klein R, Lamballe F, Bryant S, & Barbacid M (1992) The trkB tyrosine protein kinase is a receptor for neurotrophin-4 (Translated from English). *Neuron* 8(5):947–956 (in English).
38. Brambilla R, *et al.* (1995) Membrane-bound LERK2 ligand can signal through three different Eph-related receptor tyrosine kinases. *Embo J* 14(13):3116–3126.
39. Brambilla R, *et al.* (1996) Similarities and differences in the way transmembrane-type ligands interact with the Elk subclass of Eph receptors (Translated from English). *Mol Cell Neurosci* 8(2–3):199-209 (in English).
40. Labrador JP, Brambilla R, & Klein R (1997) The N-terminal globular domain of Eph receptors is sufficient for ligand binding and receptor signaling (Translated from English). *Embo J* 16(13):3889–3897 (in English).
41. Naldini L, Blomer U, Gage FH, Trono D, & Verma IM (1996) Efficient transfer, integration, and sustained long-term expression of the transgene in adult rat brains injected with a lentiviral vector. *Proc Natl Acad Sci U S A* 93(21):11382–11388.
42. Amendola M, Venneri MA, Biffi A, Vigna E, & Naldini L (2005) Coordinate dual-gene transgenesis by lentiviral vectors carrying synthetic bidirectional promoters. *Nat Biotechnol* 23(1):108–116.

Part III

Structure-Function Relationships and Localization of MAP Kinases

Chapter 13

Structural Studies of MAP Kinase Cascade Components

Elizabeth J. Goldsmith, Xiaoshan Min, Haixia He, and Tianjun Zhou

Abstract

MAPK cascade components have been the subject of structural analysis, advancing our understanding of how these enzymes are activated and how they interact. A surprising finding has been that unique inactive conformers are adopted by many of these kinases. These inactive conformers are interesting and often require experimental phases to determine their crystal structures because molecular replacement techniques are not successful. Here, we describe the preparation of MAP2K MEK6 and MAP3K TAO2 substituted with selenomethionine (SeMet) for de novo phasing. TAO2 and SeMet TAO2 were expressed in insect cells.

Key words: MAPK, MAP2K, MAP3K, MEK6, TAO2, Selenomethionine, Crystallography, Docking, ERK2, p38alpha

1. Introduction

MAPK cascades are central transducers of information from the plasma membrane to intracellular targets and participate in the regulation of diverse cellular processes including proliferation, differentiation, transformation, cell death, and senescence (1, 2). Aberrant signaling through these cascades has been associated with human diseases including inflammation, cancer, and disease of the central nervous system (3–6). Many of these enzymes are being actively pursued as drug targets. p38alpha MAP kinase is under study for the treatment of rheumatoid arthritis, pain, dementia, and stroke (7–9). In the ERK1 and ERK2 pathway, kinases at all three levels are under consideration for treatment of various kinds of cancer and other diseases (3, 4, 10–12). Good recent reviews of the role of MAP kinases in disease and their potential as drug targets are available (3, 5).

Rony Seger (ed.), *MAP Kinase Signaling Protocols: Second Edition*, Methods in Molecular Biology, vol. 661,
DOI 10.1007/978-1-60761-795-2_13,

MAPK pathways are comprised of three enzymes acting in concert, an MAP3K (for MAP kinase kinase kinase), an MAP2K (for MAP kinase kinase), and an MAPK (MAP kinase). Several distinct pathways have been identified, the best studied of which leads to activation of the MAPK ERK2, activation of the MAPK p38, and activation of the MAPK JNK (13). Thus, mechanisms must be in place to assure pathway specificity (14). One such mechanism is the use of docking interactions (mediated at loci outside the active site). MAP2Ks utilize docking interactions to bind both MAPKs (15) and MAP3Ks (16). MAPKs utilize docking interactions to bind substrates and phosphatases as well (reviewed, for example, in (17)). A second mechanism is the use of inactive conformers, as has been observed in the MAP2K MEK6 (18) (discussed below).

Given the medical and biochemical interest in these proteins, it is perhaps not surprising that there are over 200 Protein Data Bank entries for MAPKs and their activators. These structures include contributions from academic labs, numerous drug industry labs, and structural genomics consortia. Here, we summarize the primary crystallographic studies to date (not including inhibitor complexes). To obtain these structures, most often the protein of interest has been crystallized and molecular replacement from the original structure of ERK2 (19) has been used to generate and refine electron density maps . In contrast, kinases at the MAP2K and MAP3K levels have proved to be unexpectedly unique and have required de novo phasing of crystallographic data. Incorporation of selenomethionine (SeMet) in place of methionine into proteins during expression has proved to be a relatively straightforward and robust way to modify the electron content of a crystal without changing the lattice constants or space group, thus making single anomalous dispersion phasing straightforward (20). Here, we offer protocols for expressing SeMet-containing MAP2Ks and MAP3Ks, one protocol for expression of MEK6 in bacteria and one for expression of the MAP3K TAO2 in insect cells. Methods for crystallizing MAPKs (21, 22), expressing phosphorylated MAPKs (23), and an article devoted to p38alpha expression and crystallization (24) have been published.

1.1. Structures of MAPK Cascade Components

The vast majority of MAP kinase pathway crystallographic studies have been on MAPKs. Structures are available for the extracellular signal-regulated kinases ERK1, ERK2, and ERK3 (Table 1). In the MAPK p38 subfamily, structures include those of p38alpha, p38gamma, and p38delta. Structures of Jnk1 and Jnk3 have also been solved. The structures of ERK2 (2ERK) and p38gamma (1CM8) are of the doubly phosphorylated ERK2, ERK2/T*EY* and doubly phosphorylated p38gamma, p38gamma/T*GY*, building the active structure. The structure of ERK1 (2ZOQ)

Table 1
Protein Data Bank entries for the lead crystallographic studies of MAPKs, MAP2Ks, and MAP3Ks

Level	Protein	PDB		Res. (Å)	Ligand	PHOS	Species	Phasing	References
MAPK	ERK2	1ERK	ERK2	2.3	Apo	U	Rat	HA	(19)
		2ERK	ERK2	2.4	Apo	T*EY*	Rat	MR	(43)
		2GPH	ERK2	1.9	HePTP peptide	U	Rat	MR	(28)
		2FYS	ERK2	2.5	MKP-3 peptide	U	Rat	MR	(44)
		1GOL	ERK2 K52R	3.0	ATP	U	Rat	MR	(45)
		1PME	ERK2 Penta	2.0	Inhibitor	U	Human	MR	(46)
	ERK1	2ZOQ	ERK1	2.4	Inhibitor	TEY*	Human	MR	(47)
	ERK3	2I6L	ERK3	2.3	Apo	U	Human	MR	
	P38α	1P38	p38a	2.1	Apo	U	Mouse	MR	(48)
		1LEW	p38a	2.3	MEF2A peptide	U	Mouse	MR	(49)
		1LEZ	p38a	2.3	MKK3b peptide	U	Mouse	MR	(49)
		1WFC	p38a	2.3	Apo	U	Human	n/a	(50)
		1R39	p38a C162S	2.3	Apo	U	Human	MR	(51)
	P38γ x	1CM8	p38g	2.4	AMPPNP	T*GY*	Human	MIR	(52)
	P38δ x	3COI	p38d	2.1	Apo	U	Human	MR	
	JNK3	2OK1	JNK3	2.4	Inhibitor	U	Human	MR	(53)
	FUS3	2B9F	Fus3	1.8	ADP	U	*Saccharomyces cerevisiae*	MR	(54)
		2F49	Fus3	1.9	Ste5 peptide	U	*S. cerevisiae*	MR	(55)
		2B9I	Fus3	2.5	Msg5 peptide	U	*S. cerevisiae*	MR	(54)
		2B9J	Fus3	2.3	Far1 peptide	U	*S. cerevisiae*	MR	(54)
		2FA2	Fus3	2.9	Apo	VEF	*S. cerevisiae*	MR	(55)
		2B9H	Fus3	1.6	Ste7 peptide	U	*S. cerevisiae*	MR	(55)

(continued)

Table 1 (continued)

Level	Protein	PDB		Res. (Å)	Ligand	PHOS	Species	Phasing	References
MAP2K	MEK2	1S9I	MEK2	3.2	ATP + Inhibitor	U	Human	MR	(32)
	MEK1	1S9J	MEK1	2.4	ATP + Inhibitor	U	Human	(I) SAD	(32)
		3EQD	MEK1	2.1	ATP-γS	U	Human	MR	(33)
		3EQI	MEK1	1.9	ADP	U	Human	MR	(33)
	MEK6	3ENM	MEK6/DeltaN40/DD	2.3	Apo	DD	Human	Se (SAD)	(18)
		3FME	MEK6	2.3	Staurosporine	U	Human	MR	
	MEK7	2DYL	MEK7/DeltaN	2.5	n/a	EE	Human	MIR	
MAP3K	B-RAF	1UWH	B-RAF	3.0	Inhibitor	U	Human	MR	(56)
	ASK1	2CLQ	ASK1		Staurosporine	U	Human	Se (SAD)	(56)
	TAO2	1U5R	TAO2/1-320		ATP	T*	*Rattus norvegicus*	MR	(57)
	Other	2C60	MAP3K3 PHOX domain	1.3	Glycerol	n/a	Human	Se (SAD)	
	Domains	2O2V	MAP2K5 PHOX domain	1.8	n/a	n/a	h	MR	
		2JRH	MEKK3 PB1 domain	1.8	n/a	n/a	Human	NMR	(58)
		2NPT	MAP2K5 PHOX domain	1	n/a	n/a	Human	MR	

Res resolution, *PHOS* phosphorylation state, *U* unphosphorylated, *D* or *E* mutation to aspartic acid or glutamic acid; method of determining the phases of the X-ray diffraction pattern reveals the inactive Tyr monophosphorylated intermediate in the activation of MAPKs.

Docking interactions of MAPKs have also been studied crystallographically. The best studied and most ubiquitous are the D-motif sequences (25, 26). Structural data of ERK2, p38alpha, and JNK1 (Table 1) show that each form interacts with a hydrophobic groove near helices D and E of the kinase domain and an acidic patch in the C-terminus called "CD" for common docking (27). The most extensive and specific interactions were observed in ERK2 bound to a peptide derived from hematopoietic tyrosine phosphatase (HePTP) (28). Good electron density was obtained only after introducing a disulfide bridge between the peptide and ERK2 (in a locus dictated by the low resolution wild-type complex). As noted above for ERK2 and ERK2/T*EY*, the structure was solved for rat ERK2. Similarly, the work on p38alpha was done for the same sequence from mouse. Thus, detailed comparison of the structures can be made without interference from sequence variation. We observed in both ERK2 and p38alpha that docking interactions induced long-range changes that formed a distinct third conformation, representing neither the inactive structure nor the active structure. MAP2K-derived and phosphatase-derived peptides induced long-range conformational changes that affected the active site and activation loop. In p38alpha, we demonstrated that the activation loop became more susceptible to proteolysis on binding docking motif peptides. Since the conformational changes resulted in either activation loop disorder (p38alpha) or made the activation loop more accessible to solvent (ERK2), we conclude that the induced conformation may be for "processing," facilitating the binding of the activation loop into the active site of another enzyme (MAP2K or phosphatase). The conformational change also absorbs intrinsic binding energy, solving the conundrum in signal transduction of modest affinities between interacting components but very high levels of specificity (29). Substrate-derived peptides also induce conformational changes and perhaps also reduce affinity for substrate by the conformational change energy. In ERK2, the peptide-induced structure is closer to the active [root mean square deviation (rmsd) = 0.3 Å] than the inactive (rmsd = 0.8 Å) ERK2 structure, consistent with substrate peptide binding active form.

Another docking motif that has been extensively characterized contains an "FXFP" motif. There is no crystallographic study so far of FXFP docking motifs. Deuterium exchange protection experiments (30) have been used to map out the binding sites to the MAP kinase insert and helix G. Also, electron density appeared in this region in the structure of ERK2 complexed with a phosphatase-derived D-motif peptide, suggestive of part of the D-motif peptide being bound there nonspecifically (31).

At the MAP2K level, Parke-Davis provided the first structural insights with the structures of MEK1 and MEK2 in complex with ATP and inhibitors (32). They made an N-terminal truncation mutant eliminating the docking motif for ERK2. They found that the inhibitors touched ATP and that much of the activation loop was formed into a helix, a clearly inactive configuration. A recent Protein Data Bank entry of an inactive MEK1 (3EQD) (33) is also interesting because the entire activation loop is visible. Following the lead from Parke-Davis, we made a similar truncation mutant in MEK6, an activator of p38alpha, and introduced aspartic acid mimetics in the phosphorylation sites MEK6/αN/DD. The aspartic acids confer activity toward p38alpha in the context of the full-length protein, although both the full length and truncated MEK6 do not phosphorylate myelin basic protein (unpublished data). To our surprise, MEK6 adopts a completely inactive configuration despite the activating point mutations (Fig. 1) (18). The dimer was confirmed in low angle X-ray scattering experiments (18). The activation loop is in the dimer interface of MEK6. Thus, the processing of MAP2K phosphorylation sites may also be controlled by interactions with the MAP3K and phosphatases.

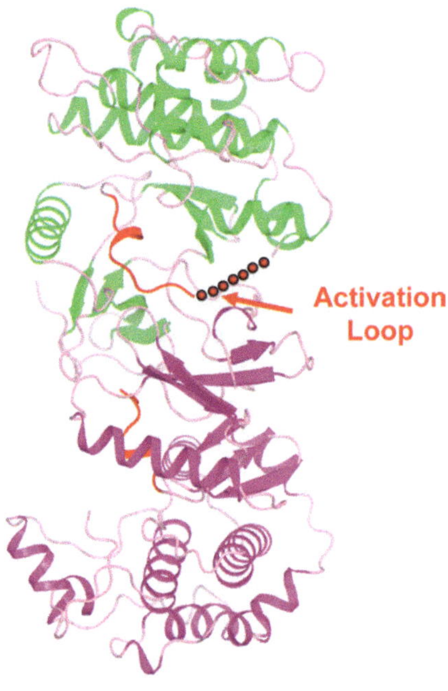

Fig. 1. The structure of MEK6/DeltaN/DD. MEK6 with an N-terminal deletion (residues 45-344 included) and the two phosphorylation site residues replaced with aspartic acid. Subunits are *magenta* and *green*, activation loops are *red*, and *dots* indicate the disordered part of the activation loop in the *green subunit*. Note that both active sites are blocked in the dimer. The dimer was verified by low angle X-ray scattering and was also found to be a feature of the full-length unphosphorylated MEK6 (18).

Relatively few structures have been reported for MAP3Ks. B-RAF is a major activator of MEK1/2. It is a known oncogene (34), which undergoes activating point mutations and deletions in various kinds of melanomas and tumors (35). Drug discovery efforts have dominated structural studies of this enzyme. In addition to the entry in Table 1 [a complex with the Bayer inhibitor (Bay439006)], the Protein Data Bank possesses the entries 1UWJ, 2FB8, 3C4C, and 3C4D, which are all bound by ATP competitive inhibitors. The structures invariably adopt an inactive configuration. We determined the structure of TAO2, which is an activator of MEK3 and MEK6 (36, 37). The kinase domain became phosphorylated as expressed in insect cells and was active, revealing an active MAP3K kinase domain. The paucity of data on MAP3Ks probably stems from their poor expression in bacteria. We used insect cells and were able to crystallize TAO2/1-320, which is the standard kinase domain (38) with two additional helices at the C-terminus.

Even with the numerous Ser/Thr protein kinases available, it is not always possible to determine a new kinase structure using molecular replacement. We found this to be the case for both MEK6 and TAO2. The structure of MEK6 (Fig. 1) clearly showed why molecular replacement had failed. To determine the structures of MEK6 and TAO2, we expressed proteins containing SeMet, for Single Anomalous Dispersion phasing (20). The following protocols were used for bacterial expression of SeMet-derivatized MEK6 and insect cell expression of SeMet-derivatized TAO2.

2. Materials

2.1. Reagents for Making the MAP2K MEK6 with and Without SeMet

2.1.1. Reagents for Cell Culture and Lysis of MAP2K MEK6 in Bacterial Cells

1. Protein expression: Rosetta 2 (DE3) pLysS competent cells (Novagen).
2. LB medium (QbioGene).
3. Lysis buffer [50 mM Tris–HCl (pH 8.0), 300 mM NaCl, and 5 mM imidazole].
4. Antibiotics ampicillin and chloramphenicol (Research Products International).
5. For SeMet incorporation, the *met* auxotrophic strain B834 (Novagen) is used.
6. For SeMet incorporation, SelenoMet medium complete (Molecular Dimensions) is used.

2.1.2. Reagents for Purification of MAP2K MEK6s

1. Ni^{2+}-nitrilotriacetic acid agarose (Ni-NTA; Qiagen).
2. Ni-NTA elution buffer A: 50 mM Tris–HCl (pH 8.0), 300 mM NaCl, 5 mM imidazole.

3. Ni buffer B: 50 mM Tris–HCl (pH 8.0), 300 mM NaCl, 250 mM imidazole.
4. MonoQ buffer A: 50 mM Tris–HCl (pH 8.0), 50 mM NaCl, 1 mM dithiothreitol (DTT), 10% glycerol.
5. MonoQ buffer B: 50 mM Tris–HCl (pH 8.0), 1 M NaCl, 1 mM DTT, 10% glycerol.
6. Gel filtration buffer: 25 mM HEPES (pH 7.4), 50 mM NaCl, 1 mM EDTA, 1 mM DTT.
7. Protein inhibitor cocktail: Sigma Catalog Number: P2714.

2.2. Reagents Making the MAP3K TAO2 with and Without SeMet

2.2.1. Reagents for Cell Culture and Lysis of MAP3K TAO2 in Insect Cells with and Without SeMet

1. Recombinant virus production: SF9 cells (Invitrogen), grown in IPL-41 Insect Medium, supplemented with 10% fetal bovine serum, 10% Yeastolate Ultrafiltrate, 5% Pluronic F-68 (all from Invitrogen), and 10% Tryptose Phosphate Broth (MP Biochemicals).
2. Protein expression: High Five cells (Invitrogen), grown in EX-CELL 405 Serum-Free Medium (JRH Biosciences).
3. For SeMet incorporation, EX-CELL 405 Serum-Free Medium without L-Met (JRH Biosciences) is used.
4. For SeMet incorporation, L-SeMet (Acros) is used.
5. 27°C shaking incubator.
6. Cell lysis buffer: 10 mM HEPES–NaOH, 10 mM NaCl (pH 8.0), 1 mM β-mercaptoethanol (β-ME), 0.2 mM Phenylmethanesulfonyl fluoride (Sigma-Aldrich), supplemented with Protein Inhibitor Cocktail P2714 (Sigma-Aldrich) as per manufacturer's instructions.
7. Dounce homogenizer (Wheaton).

2.2.2. Reagents for Protein Purification of MAP3K TAO2

1. Ni-NTA (Qiagen).
2. Ni-NTA elution buffer A: 5 mM imidazole, 20 mM HEPES–NaOH (pH 8.0), 0.5 M NaCl, 1 mM β-ME.
3. Ni-NTA elution buffer B: 250 mM imidazole, 20 mM HEPES–NaOH (pH 8.0), 0.5 M NaCl, 1 mM β-ME.
4. MonoQ 5/50 column (Pharmacia).
5. MonoQ buffer A: 50 mM Tris–HCl (pH 8.0), 1 mM DTT.
6. MonoQ buffer B: 50 mM Tris–HCl (pH 8.0), 1.0 M NaCl, 1 mM DTT.
7. Crystallization buffer: 30 mM HEPES–NaCl, 0.1 M NaCl, 1 mM EDTA (pH 7.5), 1 mM DTT.

3. Methods

3.1. *Escherichia coli* Expression and Purification of MAP2K MEK6 with and Without SeMet

3.1.1. Expression of the MAP2K MEK6 Without SeMet

1. Transform pHis-Mek6 into Rosetta 2 (DE3) pLysS competent cells.
2. Inoculate 50 ml of LB starter culture containing ampicillin and chloramphenicol with a single colony from a freshly grown selective plate. Grow overnight in 37°C shaker, at 250 RPM.
3. Transfer 5–10 ml overnight culture to 1-l LB medium with antibiotics and grow at 37°C, 250 RPM for about 4 h until $OD_{600} = 0.7$.
4. Add IPTG at 0.5 mM final concentration.
5. After induction with IPTG, grow 12 h at 18°C, 200 RPM.
6. Harvest cells by spinning for 30 min at 800 *g*, 4°C in 1 l bottles. Remove the supernatant, resuspend the pellet in 10–15 ml lysis buffer, and add Protease Inhibitor Cocktail (see Note 1).
7. The SeMet-incorporated protein was expressed in the *met* auxotrophic strain B834 (Novagen) grown in minimal media supplemented with SeMet and other nutrients, and purified using the same protocol as for the native protein (39) [*key step for selenomethionine (SeMet) – labeled MEK6*].

3.1.2. Expression of SeMet-Incorporated MEK6

The SeMet-incorporated protein was expressed in the *met* auxotrophic strain B834 (Novagen) grown in minimal media supplemented with SeMet and other nutrients (39, 40).

1. Transform pHis-Mek6 into B834 (Novagen) competent cells.
2. Inoculate 50 ml of LB starter culture containing ampicillin and chloramphenicol with a single colony from a freshly grown selective plate. Grow overnight in 37°C shaker, at 250 RPM.
3. Centrifuge the overnight culture and discard the medium. Resuspend the cell pellet in minimal medium. Repeat this step once.
4. Inoculate 1 l of SelenoMet complete medium with 10–20 ml of overnight culture. Grow at 37°C, 250 RPM until $OD_{600} = 0.7$ (see Note 2).
5. Add IPTG at 0.5 mM final concentration.
6. After induction with IPTG, grow 12–24 h at 18°C, 200 RPM.

7. Harvest cells by spinning 30 min at 4,000 RPM, 4°C in 1l bottles. Remove the supernatant, resuspend the pellet in 10–15 ml lysis buffer, and add Protease Inhibitor Cocktail.

3.1.3. Purification of MEK6

All steps are to be conducted on ice or at 4°C, unless otherwise stated.

1. Lyse cells using an Avestin cell disruptor. Centrifuge the lysate in Ti45 rotor (Beckman) 1 h at 35,000 × *g*.
2. Discard the pellet and pass the supernatant through a 0.45 μM cellulose acetate membrane filter.
3. Load the filtered supernatant onto a Ni-NTA Sepharose column.
4. Wash the Ni column with at least five column volumes (CV) of Ni buffer A.
5. Elute the target protein stepwise with Ni buffer B.
6. Remove small samples of all eluted fractions and determine which ones contain the target protein by PAGE/Coomassie blue stain.
7. Pool fractions containing the target protein and perform buffer exchange to MonoQ buffer A by dialysis.
8. Load MEK6 in MonoQ buffer A onto a MonoQ HR 5/5 column (GE Life Sciences).
9. Elute the target protein with MonoQ buffer B by a linear gradient.
10. Determine which fractions contain the target protein by PAGE/Coomassie blue stain.
11. Pool these fractions and run size-exclusion chromatography on a Superdex 75 16/60 column (GE Life Sciences).
12. Buffer exchange to 30 mM HEPES, 100 mM NaCl, 1 mM EDTA, 1 mM DTT, 1 mM TCEP, and concentrate MEK6 to 12 mg/ml using an Amicon concentrator.

3.2. Expression and Purification of MAP3K TAO2 (1-320) in Insect Cells with and Without SeMet Incorporation

3.2.1. Expression of TAO2 Without SeMet

Rat TAO2 kinase domain (1-320) was expressed in the *Baculovirus*/insect cell expression system (Invitrogen). The TAO2 kinase domain gene was cloned into pRSETB (Invitrogen) to incorporate a $MRGSH_6$ tag and subsequently transferred into the baculoviral shuttle vector pVL1393 (Invitrogen). Recombinant viruses were selected, and were expressed and harvested from Sf9 cells as described by Hutchison et al. (41).

1. Infect High Five cells at 2.0×10^6 cell/ml density with recombinant TAO2 (1-320) virus.
2. Grow cells in 27°C shaking incubator at 125 RPM for 48 h.

3. Harvest cells by sedimentation at 2,000 × *g* for 30 min at 4°C.
4. Resuspend cells in lysis buffer and homogenize with a Dounce homogenizer.

3.2.2. Expression of SeMet-Incorporated TAO2 (1-320)

The TAO2 kinase domain (1-320), containing SeMet was expressed in High Five (Hi5) cells according to the protocol described by Bellizzi et al. (42) with substantial modifications.

1. Infect High Five cells at 2.0×10^6 cell/ml density with recombinant TAO2 (1-320) virus.
2. Grow cells in 27°C shaking incubator at 125 RPM for 20–24 h.
3. Remove the growth medium by aseptic sedimentation at 500 × *g* for 10 min at 4°C.
4. Resuspend cells in L-methionine-deficient medium.
5. Grow cells for 4–6 h to deplete them of the remaining L-methionine.
6. Repeat aseptic sedimentation.
7. Resuspend cells in L-methionine-deficient medium, supplemented with 100 mg/l of SeMet.
8. Grow cells for additional 24 h.
9. Harvest cells and purify the protein the same way, as the unlabeled TAO2 (1-320).

3.2.3. Purification of TAO2

1. Clarify the lysate by spinning at 100,000 × *g* for 1 h at 4°C.
2. Further clarify the supernatant by filtering through a 0.45 μM filter.
3. Load the clarified supernatant onto a Ni-NTA column, preequilibrated with buffer A.
4. Elute TAO2 (1-320) with a 20–250 mM imidazole gradient.
5. Pool the fractions, containing TAO2 (1-320), and dilute fivefold in MonoQ buffer A.
6. Apply the diluted pool to a preequilibrated MonoQ 5/50 column.
7. Elute TAO2 (1-320) with a 0–1.0 M NaCl gradient.
8. Pool the fractions, containing TAO2 (1-320), as assessed by SDS gel.
9. Concentrate the pool in Amicon Ultra protein concentrator.
10. Dialyze the pool against crystallization buffer.

4. Notes

1. MEK6: MEK6 is prone to form aggregates when expressed at higher temperatures. A good practice in our lab is to first grow the bacterial culture at 37°C to OD ~ 0.5 and then adjust the temperature of the shaker to 18°C and decrease the shaker speed. In about 30 min, the culture will be cooled down to the right temperature, and OD will reach around 0.7. Add IPTG only after the temperature of the culture has dropped to the right temperature. Decreasing the shaker speed after adding IPTG also decreases the aggregate formation. The bacteria grow more slowly in SeMet minimal medium. For this reason, we usually opt to use LB to grow a starter culture for large-scale inoculation. It is important to wash the starter culture with the minimal medium twice before inoculation. This will decrease carryover of the methionine to the minimal medium. When both Met and SeMet are present, *E. coli* will utilize Met first. Alternatively, one can also use minimal medium with 1–5% LB to grow starter. The cell density of SeMet culture is usually lower, and the protein yield is also lower.
2. TAO2: As with the bacterial expression, the insect cells will utilize Met before SeMet. However, the cells must be grown in normal medium to get started . The critical step is to grow the cells for several hours in a completely Met-deficient medium to clear out all of the remaining methionine, and then switch to SeMet-containing medium. Also, the EX-CELL medium is important for cell growth.

Acknowledgments

Our thanks go to Thomas Moon and Saurabh Mendiratta for assistance in preparing Table 1. Our work in this area has been supported by an NIH grant DK46003 and a grant from the Welch Foundation, I1128.

References

1. Deng, Q., Liao, R., Wu, B. L., and Sun, P. (2004) High intensity ras signaling induces premature senescence by activating p38 pathway in primary human fibroblasts, *J Biol Chem 279*, 1050–1059.
2. Johnson, G. L., and Lapadat, R. (2002) Mitogen-activated protein kinase pathways mediated by ERK, JNK, and p38 protein kinases, *Science 298*, 1911–1912.
3. Lawrence, M. C., Jivan, A., Shao, C., Duan, L., Goad, D., Zaganjor, E., Osborne, J., McGlynn, K., Stippec, S., Earnest, S., Chen, W., and Cobb, M. H. (2008) The roles of MAPKs in disease, *Cell Res 18*, 436–442.

4. Roberts, P. J., and Der, C. J. (2007) Targeting the Raf-MEK-ERK mitogen-activated protein kinase cascade for the treatment of cancer, *Oncogene 26*, 3291–3310.
5. Sebolt-Leopold, J. S., and English, J. M. (2006) Mechanisms of drug inhibition of signalling molecules, *Nature 441*, 457–462.
6. Whitmarsh, A. J., and Davis, R. J. (2007) Role of mitogen-activated protein kinase kinase 4 in cancer, *Oncogene 26*, 3172–3184.
7. Schindler, J. F., Monahan, J. B., and Smith, W. G. (2007) p38 pathway kinases as anti-inflammatory drug targets, *J Dent Res 86*, 800–811.
8. Borders, A. S., de Almeida, L., Van Eldik, L. J., and Watterson, D. M. (2008) The p38alpha mitogen-activated protein kinase as a central nervous system drug discovery target, *BMC Neurosci 9 Suppl 2*, S12.
9. Loeser, R. F., Erickson, E. A., and Long, D. L. (2008) Mitogen-activated protein kinases as therapeutic targets in osteoarthritis, *Curr Opin Rheumatol 20*, 581–586.
10. Wang, Y. (2007) Mitogen-activated protein kinases in heart development and diseases, *Circulation 116*, 1413–1423.
11. Muslin, A. J. (2008) MAPK signalling in cardiovascular health and disease: molecular mechanisms and therapeutic targets, *Clin Sci (Lond) 115*, 203–218.
12. Miloso, M., Scuteri, A., Foudah, D., and Tredici, G. (2008) MAPKs as mediators of cell fate determination: an approach to neurodegenerative diseases, *Curr Med Chem 15*, 538–548.
13. Raman, M., Chen, W., and Cobb, M. H. (2007) Differential regulation and properties of MAPKs, *Oncogene 26*, 3100–3112.
14. Bardwell, L. (2006) Mechanisms of MAPK signalling specificity, *Biochem Soc Trans 34*, 837–841.
15. Bardwell, A. J., Abdollahi, M., and Bardwell, L. (2003) Docking sites on mitogen-activated protein kinase (MAPK) kinases, MAPK phosphatases and the Elk-1 transcription factor compete for MAPK binding and are crucial for enzymic activity, *Biochem J 370*, 1077–1085.
16. Takekawa, M., Tatebayashi, K., and Saito, H. (2005) Conserved docking site is essential for activation of mammalian MAP kinase kinases by specific MAP kinase kinase kinases, *Mol Cell 18*, 295–306.
17. Sharrocks, A. D., Yang, S. H., and Galanis, A. (2000) Docking domains and substrate-specificity determination for MAP kinases, *Trends Biochem Sci 25*, 448–453.
18. Min, X., Akella, R., He, H., Humphreys, J. M., Tsutakawa, S. E., Lee, S. J., Tainer, J. A., Cobb, M. H., and Goldsmith, E. J. (2009) The structure of the MAP2K MEK6 reveals an autoinhibitory dimer, *Structure 17*, 96–104.
19. Zhang, F., Strand, A., Robbins, D., Cobb, M. H., and Goldsmith, E. J. (1994) Atomic structure of the MAP kinase ERK2 at 2.3Å resolution, *Nature 367*, 704–711.
20. Dauter, Z., Dauter, M., and Dodson, E. (2002) Jolly SAD, *Acta Crystallogr D Biol Crystallogr 58*, 494–506.
21. Goldsmith, E. J., Cobb, M. H., and Chang, C. I. (2004) Structure of MAPKs, *Methods Mol Biol 250*, 127–144.
22. Lee, S. J., Zhou, T., and Goldsmith, E. J. (2006) Crystallization of MAP kinases, *Methods 40*, 224–233.
23. Wilsbacher, J. L., and Cobb, M. H. (2001) Bacterial expression of activated mitogen-activated protein kinases, *Methods Enzymol 332*, 387–400.
24. Bukhtiyarova, M., Northrop, K., Chai, X., Casper, D., Karpusas, M., and Springman, E. (2004) Improved expression, purification, and crystallization of p38alpha MAP kinase, *Protein Expr Purif 37*, 154–161.
25. Kallunki, T., Su, B., Tsigelny, I., Sluss, H. K., Derijard, B., Moore, G., Davis, R., and Karin, M. (1994) JNK2 contains a specificity-determining region responsible for efficient c-Jun binding and phosphorylation, *Genes Dev 8*, 2996–3007.
26. Yang, S. H., Whitmarsh, A. J., Davis, R. J., and Sharrocks, A. D. (1998) Differential targeting of MAP kinases to the ETS-domain transcription factor Elk-1, *EMBO J 17*, 1740–1749.
27. Tanoue, T., and Nishida, E. (2002) Docking interactions in the mitogen-activated protein kinase cascades, *Pharmacol Ther 93*, 193–202.
28. Zhou, T., Sun, L., Humphreys, J., and Goldsmith, E. J. (2006) Docking interactions induce exposure of activation loop in the MAP kinase ERK2, *Structure 14*, 1011–1019.
29. Zhou, T., Raman, M., Gao, Y., Earnest, S., Chen, Z., Machius, M., Cobb, M. H., and Goldsmith, E. J. (2004) Crystal structure of the TAO2 kinase domain: activation and specificity of a Ste20p MAP3K, *Structure 12*, 1891–1900.
30. Lee, T., Hoofnagle, A. N., Kabuyama, Y., Stroud, J., Min, X., Goldsmith, E. J., Chen, L., Resing, K. A., and Ahn, N. G. (2004) Docking motif interactions in MAP kinases revealed by hydrogen exchange mass spectrometry, *Mol Cell 14*, 43–55.

31. Akella, R., Moon, T. M., and Goldsmith, E. J. (2008) Unique MAP kinase binding sites, *Biochim Biophys Acta 1784*, 48–55.
32. Ohren, J. F., Chen, H., Pavlovsky, A., Whitehead, C., Zhang, E., Kuffa, P., Yan, C., McConnell, P., Spessard, C., Banotai, C., Mueller, W. T., Delaney, A., Omer, C., Sebolt-Leopold, J., Dudley, D. T., Leung, I. K., Flamme, C., Warmus, J., Kaufman, M., Barrett, S., Tecle, H., and Hasemann, C. A. (2004) Structures of human MAP kinase kinase 1 (MEK1) and MEK2 describe novel noncompetitive kinase inhibition, *Nature Struct Mol Biol 11*, 1192–1197.
33. Fischmann, T., Smith, C., Mayhood, T., Myers, J., Reichert, P., Mannarino, A., Carr, D., Zhu, H., Wong, J., Yang, R. S., Le, H., and Madison, V. (2009) Crystal structures of MEK1 binary and ternary complexes with nucleotides and inhibitors, *Biochemistry 48*, 2661–2674.
34. Garnett, M. J., and Marais, R. (2004) Guilty as charged: B-RAF is a human oncogene, *Cancer Cell 6*, 313–319.
35. Li, N., Batt, D., and Warmuth, M. (2007) B-Raf kinase inhibitors for cancer treatment, *Curr Opin Investig Drugs 8*, 452–456.
36. Chen, Z., and Cobb, M. H. (2001) Regulation of stress-responsive mitogen-activated protein (MAP) kinase pathways by TAO2, *J Biol Chem 276*, 16070–16075.
37. Chen, Z., Hutchison, M., and Cobb, M. H. (1999) Isolation of the protein kinase TAO2 and identification of its mitogen-activated protein kinase/extracellular signal-regulated kinase kinase binding domain, *J Biol Chem 274*, 28803–28807.
38. Hanks, S. K., and Hunter, T. (1995) Protein kinases 6. The eukaryotic protein kinase superfamily: kinase (catalytic) domain structure and classification, *FASEB J 9*, 576–596.
39. Doublie, S. (1997) Preparation of selenomethionyl proteins for phase determination, *Methods Enzymol 276*, 523–530.
40. Doublie, S. (2007) Production of selenomethionyl proteins in prokaryotic and eukaryotic expression systems, *Methods Mol Biol 363*, 91–108.
41. Hutchison, M., Berman, K. S., and Cobb, M. H. (1998) Isolation of TAO1, a protein kinase that activates MEKs in stress-activated protein kinase cascades, *J Biol Chem 273*, 28625–28632.
42. Bellizzi, J. J., Widom, J., Kemp, C. W., and Clardy, J. (1999) Producing selenomethionine-labeled proteins with a baculovirus expression vector system, *Structure 7*, R263–R267.
43. Canagarajah, B. J., Khokhlatchev, A., Cobb, M. H., and Goldsmith, E. J. (1997) Activation mechanism of the MAP kinase ERK2 by dual phosphorylation, *Cell 90*, 859–869.
44. Liu, S., Sun, J. P., Zhou, B., and Zhang, Z. Y. (2006) Structural basis of docking interactions between ERK2 and MAP kinase phosphatase 3, *Proc Natl Acad Sci U S A 103*, 5326–5331.
45. Robinson, M. J., Harkins, P. C., Zhang, J., Baer, R., Haycock, J. W., Cobb, M. H., and Goldsmith, E. J. (1996) Mutation of position 52 in ERK2 creates a nonproductive binding mode for adenosine 5 -triphosphate, *Biochemistry 35*, 5641–5646.
46. Fox, T., Coll, J. T., Xie, X., Ford, P. J., Germann, U. A., Porter, M. D., Pazhanisamy, S., Fleming, M. A., Galullo, V., Su, M. S., and Wilson, K. P. (1998) A single amino acid substitution makes ERK2 susceptible to pyridinyl imidazole inhibitors of p38 MAP kinase, *Protein Sci 7*, 2249–2255.
47. Kinoshita, T., Yoshida, I., Nakae, S., Okita, K., Gouda, M., Matsubara, M., Yokota, K., Ishiguro, H., and Tada, T. (2008) Crystal structure of human mono-phosphorylated ERK1 at Tyr204, *Biochem Biophys Res Commun 377*, 1123–1127.
48. Wang, Z., Harkins, P. C., Ulevitch, R. J., Han, J., Cobb, M. H., and Goldsmith, E. J. (1997) The structure of mitogen-activated protein kinase p38 at 2.1-Å resolution, *Proc Natl Acad Sci U S A 94*, 2327–2332.
49. Chang, C. I., Xu, B. E., Akella, R., Cobb, M. H., and Goldsmith, E. J. (2002) Crystal structures of MAP kinase p38 complexed to the docking sites on its nuclear substrate MEF2A and activator MKK3b, *Mol Cell 9*, 1241–1249.
50. Wilson, K. P., Fitzgibbon, M. J., Caron, P. R., Griffith, J. P., Chen, W., McCaffrey, P. G., Chambers, S. P., and Su, M. S. (1996) Crystal structure of p38 mitogen-activated protein kinase, *J Biol Chem 271*, 27696–27700.
51. Patel, S. B., Cameron, P. M., Frantz-Wattley, B., O'Neill, E., Becker, J. W., and Scapin, G. (2004) Lattice stabilization and enhanced diffraction in human p38 alpha crystals by protein engineering, *Biochim Biophys Acta 1696*, 67–73.
52. Bellon, S., Fitzgibbon, M. J., Fox, T., Hsiao, H. M., and Wilson, K. P. (1999) The structure of phosphorylated p38gamma is monomeric and reveals a conserved activation-loop conformation, *Structure 7*, 1057–1065.
53. Aronov, A. M., Baker, C., Bemis, G. W., Cao, J., Chen, G., Ford, P. J., Germann, U. A., Green, J., Hale, M. R., Jacobs, M., Janetka, J. W., Maltais, F., Martinez-Botella, G., Namchuk,

M. N., Straub, J., Tang, Q., and Xie, X. (2007) Flipped out: structure-guided design of selective pyrazolylpyrrole ERK inhibitors, *J Med Chem 50*, 1280–1287.

54. Remenyi, A., Good, M. C., Bhattacharyya, R. P., and Lim, W. A. (2005) The role of docking interactions in mediating signaling input, output, and discrimination in the yeast MAPK network, *Mol Cell 20*, 951–962.

55. Bhattacharyya, R. P., Remenyi, A., Good, M. C., Bashor, C. J., Falick, A. M., and Lim, W. A. (2006) The Ste5 scaffold allosterically modulates signaling output of the yeast mating pathway, *Science 311*, 822–826.

56. Wan, P. T., Garnett, M. J., Roe, S. M., Lee, S., Niculescu-Duvaz, D., Good, V. M., Jones, C. M., Marshall, C. J., Springer, C. J., Barford, D., and Marais, R. (2004) Mechanism of activation of the RAF-ERK signaling pathway by oncogenic mutations of B-RAF, *Cell 116*, 855–867.

57. Zhou, T. J., Sun, L. G., Gao, Y., and Goldsmith, E. J. (2006) Crystal structure of the MAP3K TAO2 kinase domain bound by an inhibitor staurosporine, *Acta Biochim Biophys Sin (Shanghai) 38*, 385–392.

58. Hu, Q., Shen, W., Huang, H., Liu, J., Zhang, J., Huang, X., Wu, J., and Shi, Y. (2007) Insight into the binding properties of MEKK3 PB1 to MEK5 PB1 from its solution structure, *Biochemistry 46*, 13478–13489.

Chapter 14

Analysis of MAP Kinases by Hydrogen Exchange Mass Spectrometry

Kevin M. Sours and Natalie G. Ahn

Abstract

Hydrogen exchange mass spectrometry (HX-MS) is an experimental technique that can be used to examine solvent accessibility and conformational mobility in biological macromolecules. This chapter summarizes studies using HX-MS to examine the regulation of conformation, protein mobility, and ligand binding to MAP kinases. We describe the planning and design of HX-MS experiments, strategies for data analysis and interpretation, and available software.

Key words: Extracellular signal-regulated kinase, Hydrogen–deuterium exchange, Mass spectrometry, Conformational mobility

1. Introduction

Hydrogen–deuterium exchange methods have provided useful approaches for exploring conformational and dynamic properties of biomolecules. Although often performed by nuclear magnetic resonance (NMR), hydrogen–deuterium exchange coupled with mass spectrometry has increased in popularity over the last 15 years (1, 2). Advances in the resolution and speed of mass spectrometry instrumentation allow experiments to be performed readily on macromolecules. Early publications provided the basic strategy for HX-MS of proteins, which is still prevalent among current studies of protein folding, macromolecular interactions, ligand binding, enzyme activation, and conformational mobility (3, 4).

A sizeable body of work has used HX-MS to examine properties of enzyme activation, regulated conformational mobility, and substrate binding in MAP kinases, including ERK2, p38 MAPK, and MKK1 (5–11). Studies of ERK2 showed distinct differences

Rony Seger (ed.), *MAP Kinase Signaling Protocols: Second Edition*, Methods in Molecular Biology, vol. 661,
DOI 10.1007/978-1-60761-795-2_14, © Springer Science+Business Media, LLC 2010

in regional hydrogen exchange rates between inactive (unphosphorylated) and active (diphosphorylated) states of the enzyme (5). For example, altered HX was observed in the activation lip, as expected from the dramatic conformational reorganization which follows lip phosphorylation. Also observed were increased HX within the glycine-rich ATP-binding loop and the hinge between N- and C-terminal domains. These could not readily be accounted for by differences in X-ray structure between active and inactive ERK2, suggesting that phosphorylation and activation of this enzyme regulates backbone conformational mobility in key regions of this enzyme. For example, increased conformational mobility at the hinge might enable domain closure needed for catalysis.

Follow-up experiments were conducted to test this hypothesis. First, site-directed spin label electron paramagnetic resonance (SDSL-EPR) spectroscopy was carried out on ERK2, in which cysteine residues were individually engineered at different regions of the hinge onto which nitroxide spin label groups were added (12). Residues at the hinge showed significant changes in correlation rate without evidence for conformational perturbations, consistent with altered side-chain dynamics at the hinge upon ERK2 activation. Second, HX-MS was used to monitor binding of a nonhydrolyzable nucleotide analog between active and inactive ERK2 (8). In inactive ERK2, Mg^{2+}-AMP-PNP binding caused steric protection from deuteration in regions within the N-terminal domain that were known to interact with nucleotide, including the glycine-rich loop, sheet β3/helix αC, and the hinge region, but little protection within the C-terminal domain. In contrast, nucleotide binding to active ERK2 led to comparable protection in N-terminal regions and the hinge, but increased protection around the DFG motif in the C-terminal domain. Both forms of ERK2 bound Mg^{2+}-AMP-PNP with similar affinity as measured by isothermal titration calorimetry. The findings led to novel insight into the solution conformation of ERK2, by demonstrating that the inactive enzyme is constrained from forming a closed domain solution conformation needed for catalysis, which the active state is able to adopt (Fig. 1). This result implies that closure of domains to form a competent catalytic site represents a barrier to enzyme activation, which may be overcome by regulating conformational mobility at the hinge.

HX-MS has also been used to understand binding interactions between ERK2 and substrate, as well as with MAP kinase phosphatases (MKP) (11, 13, 14). One study used HX-MS to identify binding sites for docking motifs, sequences within substrates and scaffolds, phosphatases and MAP kinase kinases, which confer high affinity binding to MAP kinases (11). Binding of a DEJL docking motif (consensus sequence Arg/Lys-X_{2-4}-φ-X-φ) interacted with a hydrophobic groove previously identified in X-ray cocrystal structures, while a DEF docking motif (consensus Phe-X-Phe) interacted

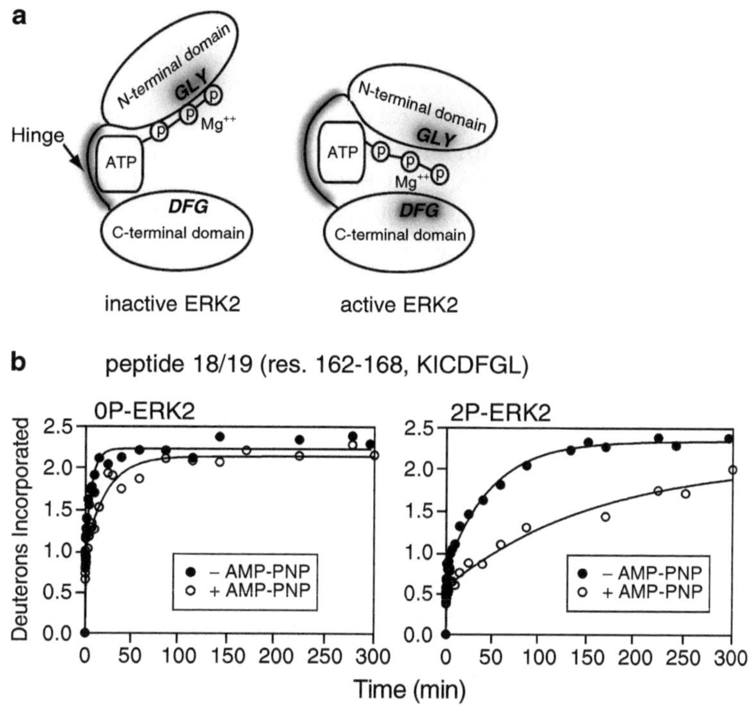

Fig. 1. (**a**) A proposed model for domain closure as revealed by differential binding of Mg^{2+}-AMP-PNP to inactive unphosphorylated (0P) ERK2 and active phosphorylated (2P) ERK2. Both forms bind Mg^{2+}-AMP-PNP comparably within the N-terminal domain and the hinge, but 2P-ERK2 shows increased protection of the C-terminal domain, suggesting a closed domain solution conformation. (**b**) HX-MS time courses for the C-terminal domain peptide, KICDFGL, used in this interpretation. The lower deuterium incorporation in 2P-ERK2, upon Mg^{2+}-AMP-PNP binding, reveals greater protection from solvent in the C-terminal domain, compared to 0P-ERK2. Reproduced with permission from ref. 8.

with a novel binding pocket formed upon rearrangement of the activation lip following phosphorylation. In studies of the binding interface between ERK2 and MKP3 (13, 14), HX-MS was used to map regions within kinase and phosphatase that were protected from exchange in the heterodimeric complex. Site-directed mutagenesis confirmed that binding of MKP3 involved the DEJL and DEF docking motif binding sites and the activation lip of ERK2, while binding of ERK2 involved corresponding docking motif sequences in MKP3, revealing that two distinct domains in the MKP3 interact with distally spaced regions in ERK2. These studies highlighted the importance of structural changes in ERK2 following activation and how they direct binding of substrates and phosphatase regulators.

These studies illustrate how HX-MS can be effective for understanding the regulation of protein mobility and its importance for MAP kinase activation, as well as protein interactions involved in MAP kinase enzymatic function and regulation. Additional studies showed that activation of p38 MAPK and MKK1 also caused changes in HX that were consistent with regulated conformational mobility (7, 9). Such changes occur within distinctive regions of each enzyme, suggesting that regulated motions vary significantly across related MAP kinases. Further investigations are needed to understand how changes in protein motions control kinase activation and how backbone flexibility is controlled by phosphorylation. Undoubtedly, HX-MS will play a valuable role in future experiments which address these questions.

2. Materials

Studies of MAP kinases (5–9) provide useful examples describing the design of HX-MS experiments for kinases and other proteins. A basic strategy compares proteins in two or more states (e.g., active vs. inactive kinase, kinase bound vs. unbound to nucleotide or substrate). In order to obtain information about localized changes in hydrogen exchange rate, proteins are incubated in D_2O for varying times, then proteolyzed, and the resulting deuterated peptides are analyzed by mass spectrometry (Fig. 2).

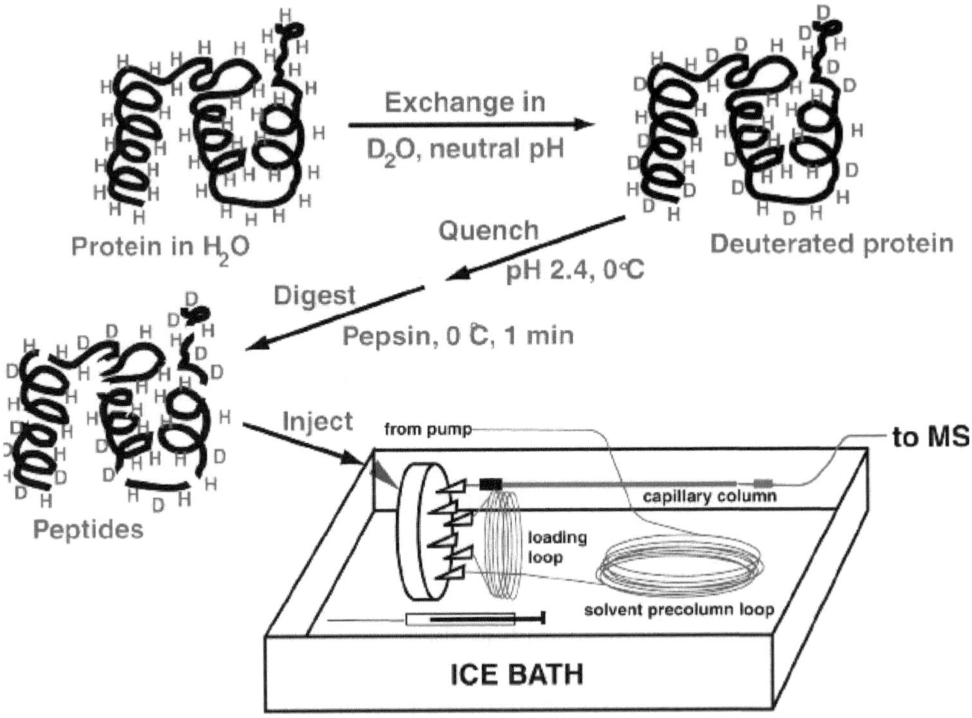

Fig. 2. Schematic of the HX-MS experiment. Proteins are incubated in D_2O for varying times; in-exchange is quenched by lowering temperature and pH, and proteins are rapidly proteolyzed by pepsin addition. Peptides are then resolved and masses measured by LC–MS.

In this way, the degree of regional hydrogen–deuterium exchange is reported by the increased mass of peptides following deuteration of the intact protein. An effort is made to achieve maximal sequence coverage, in order to monitor HX behavior over the entire protien (see Notes 1, 2, and 3). Different mass spectrometry strategies have been used successfully, including electrospray ionization mass spectrometry coupled to liquid chromatography (LC-ESI-MS) and matrix-assisted laser desorption ionization mass spectrometry (MALDI-MS) (15, 16). We favor LC-ESI-MS due to its advantages in separating highly complex peptide mixtures by liquid chromatography and its reduced back-exchange of deuterated residues in water. Here, we discuss steps useful for the design and implementation of a successful HX-MS experiment by LC-ESI-MS.

2.1. Equipment and Instrumentation

Materials listed are those used in our lab, but can vary according to instrument availability, sample type, and experimental design.

1. Ice-water bath able to fit a reverse-phase HPLC column, a 6-port injector valve, polyether ether ketone (PEEK) solvent cooling lines and sample loop. Constant column temperature at 0°C is needed to maintain low back exchange during chromatography.
2. Water bath for deuteration reactions. We generally set the incubation temperature to 10°C, in order to measure increases in hydrogen exchange over the timescale of minutes to hours.
3. An ethanol/water bath (1:2 v/v) is maintained at −10°C by adding dry ice.
4. Blunt-end HPLC injection syringes (50 μL, 250 μL, 1 mL, from Hamilton or Scientific Glass Engineering). The syringes are wrapped with Parafilm to 5-mm thickness, in order to minimize heat transfer to the solution when handling the syringe.
5. PEEK HPLC injector 6-port valve (Rheodyne), outfitted with a PEEK sample loop (1 mL) for loading sample and forming the acetonitrile gradient in the loop.
6. PEEK tubing (we use a 3-mL sample loading loop) for precooling the solvent from the pump entering the 6-port valve.
7. Agilent 1100 capillary HPLC pump. Any HPLC capable of delivering solvent at 20–40 μL/min will suffice.
8. HPLC columns (~150-mm length, 0.5 mm i.d.). We pack these manually with POROS 20 R1 resin (Applied Biosystems). The columns are made from fused silica tubing (Polymicro Technologies), with zero or low dead volume fittings (Upchurch Scientific), Teflon or PEEK sleeves, and PEEK or stainless steel frits (e.g., Upchurch Scientific). This allows for smooth control of flow and pressure at the column outlet.

9. A quadrupole time-of-flight mass spectrometer. Any instrument with a capillary electrospray source will work, but TOF instruments provide high mass resolution which facilitates data analysis. We currently use a QSTAR Pulsar QqTOF (Applied Biosystems) with standard electrospray source.

2.2. Reagents

1. Deuterium oxide (D_2O).
2. Nucleotide analog, AMP-PNP (Roche).
3. Acid quench buffer containing 25 mM citric acid, 25 mM succinic acid, titrated to pH 2.40, passed through a 0.22-µm syringe filter, and stored in 5-mL aliquots at −20°C. All experiments are carried out using the same solution of quench buffer.
4. Pepsin (0.4 mg/mL) dissolved in water, passed through a 0.22-µm syringe filter, and stored in 50 µL aliquots at −80°C. Each day, a new aliquot is thawed and clarified by microcentrifugation at 10,000 × *g* for 10 min at 4°C.
5. HPLC buffers:
   (a) Buffer A: 0.05% v/v trifluoroacetic acid (TFA) in HPLC grade water.
   (b) Buffer B: 0.05% TFA v/v in HPLC grade acetonitrile.
   (c) Step gradient buffers: 5, 7.5, 10, 12.5, 15, 17.5, 20, 22.5, 25, 30, 35, 40, and 50% buffer B in buffer A. Buffers are stored at −20°C in 1–2-mL tubes outfitted with O-ring caps.
6. Protein sample: The purity of the protein should be ≥95%. Studies with the ABI QSTAR require protein concentrations that yield 1–5 µM in a 100-µL deuteration reaction mixture. The protein should be stored frozen in aliquots. Buffers which catalyze hydrogen exchange (e.g., amines, Lewis bases) should be avoided because these enhance back exchange.

3. Methods

3.1. ERK2 Identification

Here, we illustrate an experiment designed to identify regions of ERK2 which form interfaces for binding nucleotides or other ligands (8, 11). The experiment involves preincubation of kinase with ligand to achieve equilibrium binding, prior to addition of D_2O. The protocol can be adapted to compare the behavior of active vs. inactive MAP kinases by HX-MS, with minor changes in reaction solution (5, 7, 9).

1. Set up the HPLC to pass a steady flow of 3:97 acetonitrile:water in 0.05% v/v TFA at 40 µL/min. This can be done using

isocratic flow through a single pump, or by mixing 97% buffer A and 3% buffer B using a dual pump. Solvent is precooled to 0°C by inserting a 3-mL sample loading loop between the mixer and injector valve, which is buried in the ice–water mixture. The injector valve is outfitted with a 1-mL sample loading loop and reverse-phase column, the latter of which is initially uncoupled from the mass spectrometer. All components (3-mL precooling loop, 1-mL sample loading loop, reverse-phase column, injector valve) are immersed in the ice–water mixture, leaving only the needle injection port above the ice.

2. Prepare the ethanol/water/dry-ice bath to achieve constant temperature at −10°C. Prepare a separate ice bucket containing stock solutions (protein, ligand, D_2O, pepsin, acid quench buffer, step gradient solvents) and syringes. We also find it useful to chill the necessary P200 and P20 pipettors outfitted with tips by inserting them into 15-mL conical tubes on ice. It is important to prechill the solutions, syringes, and pipettors to minimize back exchange.
3. Place 8 μL (~100 pmol, ~4 μg) of ERK2 in a low protein binding tube (e.g., 1-mL titer tubes) in a water bath at 10°C. Incubate for 30 s.
4. Add 7 μL of ligand solution or control solution to the protein sample (e.g., to achieve final concentrations of 1 mM AMP-PNP, 10 mM $MgCl_2$ in the ligand preincubation, and 10 mM $MgCl_2$ in the control). Preincubate at 10°C for 10 min to achieve equilibrium binding.
5. Initiate the deuteration reaction by adding to the binding reaction 85 μL D_2O ligand (containing 1 mM AMP-PNP, 10 mM $MgCl_2$) or control solution (containing 10 mM $MgCl_2$), which has been preincubated at 10°C. The final volume is 100 μL.
6. Incubate at 10°C for varying times (see Notes 4 and 5).
7. Transfer the tube to the ethanol/water/dry-ice bath for 5 s. The timing should allow rapid cooling of the sample to 0°C, while avoiding freezing.
8. Add 90 μL of 25 mM citrate/25 mM succinate pH 2.4 to the mixture. The final solution should be at pH 2.4.
9. Incubate in ethanol/water/dry-ice bath for an additional 15 s.
10. Add 10 μL of pepsin (~4 μg, approximately 1:1 w/w). The amount of pepsin needed for optimal proteolysis may vary and should be determined for each protein in preliminary trials (see Note 2).
11. Immediately load the sample (200 μL) into the PEEK injector loop using a prechilled 250 μL syringe, wrapped with Parafilm to minimize heat transfer from fingers. Allow the

sample to digest in the sample loading loop for 1 min following pepsin addition.

12. Switch the valve from Load to Inject, to inject the sample onto the column at 40 μL/min. A 200-μL sample requires 5 min for complete injection. After 5 min, switch the injector back to the Load position.
13. Wash the sample on the column for 6 min after returning the valve to the Load position.
14. While the column is washing the sample, rinse the injector loop with buffer A using the 1-mL HPLC syringe (2 mL is needed to wash the 1 mL loop) to remove air bubbles or D_2O crystals that may remain in the loop.
15. Load the step gradient buffers (stored on ice) into the injector loop using the 50-μL HPLC syringe. Begin by adding 40 μL of 100% buffer B, and then load 17.5 μL of each step gradient buffer in descending order of concentration. Because the loop is last on first off, the last solution injected (5% buffer B) is the first to be added onto the column. Because the buffer from the pump is being chilled by flowing through 3 mL of PEEK tubing in an ice bath prior to reaching the column, the gradient must be loaded onto the column manually to deliver a timely, sharp, step gradient. Having the pump deliver the gradient is not practical under these conditions due to mixing of the steps, dilution of the gradient, and length of time each run would require.
16. After washing, connect the column outlet to the electrospray source on the mass spectrometer. Immediately reduce the flow rate to 20 μL/min, and begin data acquisition on the mass spectrometer.
17. After 1 min, switch the valve position from Load to Inject. The gradient flows rapidly onto the column and peptides are eluted to the mass spectrometer over 15 min.
18. At the end of the gradient, switch the injector to Load and apply 200 μL of 100% buffer B, to further wash the column. Allow the column to reequilibrate with 3% buffer B for the next run.

3.2. Data Analysis

Instrument-specific software is used for data collection and analysis. For each ion, intensities are measured for each isotope peak and used to calculate weighted average mass (WAM). After correcting for back-exchange, the level of deuteration can be plotted as a function of incubation time. Observed exchange rate constants are obtained by nonlinear least squares fitting of time courses to a sum of exponentials. Our studies use Analyst QS (version 1.1) for analysis of .wiff data files collected on the ABI QSTAR instrument.

3.2.1. Calculation of Weighted Average Mass

1. Open data files collected at each time point and condition.
2. Display the total ion chromatogram (TIC) and enter the mass range of each ion of interest to display its extract ion chromatogram (XIC). Highlight the extract ion peak and show its corresponding mass spectrum, summed over all scans within the highlighted time window.
3. Zoom in on the desired ion and measure m/z and intensities for each of the isotopic peaks. In some software, these measurements can be automatically determined and saved to a separate text file or spreadsheet, generating lists of all isotope peaks and their corresponding intensities for each. Separately calculate m/z values for each isotope peak, based on the peptide sequence and charge, and match the calculated m/z values to their corresponding intensities.
4. In a spreadsheet, calculate WAM for each ion at each time point, using the measured intensities, and calculate m/z values for each isotope peak in the following equation:

$$\mathrm{WAM} = \left[\frac{i\sum I_i \times \frac{m}{z_i}}{i\sum I_i} \right] \times \mathrm{Z} - \mathrm{Z}$$

   Note that calculated m/z values are used instead of the observed m/z, to avoid variance due to calibration error.
5. Repeat steps 1–4 for calculating WAM for each peptide ion at each HX time point, under each experimental condition.

3.2.2. Correction for Artifactual In-Exchange

Artifactual in-exchange occurs at the $t = 0$ time point, due to deuterium incorporation during proteolysis. The degree of artifactual in-exchange varies with time as the peptide deuteration increases and can be calculated as follows:

$$M_{t,\mathrm{corr(IE)}} = \frac{M_{t,\mathrm{wa}} - LM_{\infty,85}}{1 - L} \quad (1)$$

$$L = \frac{M_{0,\mathrm{wa}} - M_{\mathrm{calc}}}{M_{\infty,85} - M_{\mathrm{calc}}} \quad (2)$$

Where, $M_{t,\mathrm{corr(IE)}}$ is the peptide mass corrected for artifactual in-exchange, $M_{t,\mathrm{wa}}$ is WAM at time t, and $M_{\infty,85}$ is the peptide mass at infinite time in 85% D_2O. L is a correction factor normalizing the fractional of artifactual in-exchange under the deuteration conditions in the experiment. $M_{0,\mathrm{wa}}$ is WAM at time zero, while M_{calc} is the calculated average mass of the peptide.

3.2.3. Correction for Back-Exchange

During proteolysis and LC separation, deuterons will back-exchange with protons in aqueous solution, leading to underestimates of the extent of deuteration. Back-exchange is rapid with deuterated side chains, but slower with deuterated amides; thus, HX-MS measurements mainly report deuteration at backbone amide groups. We measure the back-exchange of peptides using three approaches.

In a first approach experimental measurement of back-exchange is carried out by fully deuterating peptides and then measuring their decrease in mass with exchange back to hydrogen. This is accomplished by modifying the HX protocol outlined earlier. Perform steps 1–18 from Subheading 3, substituting 50 μL of protein and 50 μL of H_2O in the reaction mixture, generating peptides which are loaded onto the reverse-phase column. Instead of eluting peptides with a step gradient, elute with 30 μL 40% buffer B and collect the eluate off-line. Lyophilize and redissolve the sample in 15 μL of the buffer used in the exchange experiment, add 85 μL D_2O, and heat at 90°C for 90 min. This leads to complete deuteration of all peptides, which are then cooled on ice and injected onto the sample loading loop. After 1 min, proceed with step 11 in Subheading 3 and analyze the sample by LC–MS. The back exchange for each peptide is calculated as follows:

$$\mathrm{BE} = \frac{M_{\infty,85} - M_{\mathrm{BE}}}{M_{\infty,85} - M_{\mathrm{calc}}} \tag{3}$$

where $M_{\infty,85}$ is the calculated peptide mass after deuteration in 85% D_2O, M_{BE} is the observed peptide mass following back-exchange during LC–MS, and M_{calc} is the calculated average mass of the peptide in water.

In a second approach Estimated back-exchange for different peptides can be calculated using an empirical equation previously determined by our laboratory, based on back-exchange vs. elution time.

$$\mathrm{BE} = \mathrm{L} \times \left(\frac{\%\,\mathrm{H_2O}}{\%\,\mathrm{D_2O}} \right) + \left\{ \left(\text{peptide elution time in min} + 6\ \text{min} \right) \times \left(\frac{1\%\,\mathrm{min}^{-1}}{100} \right) \right\} \tag{4}$$

L is the same fraction of artifactual in-exchange from Eq. 2, and the equation assumes 1% back-exchange for the peptide per minute of elution.

A third approach estimates back-exchange based on HX rates of model peptides. Amide backbone hydrogen–deuterium exchange rate constants for tri-amino acid peptides in solution were previously determined by NMR (17). The program HXPep (created by Dr. Zhongqi Zhang, Amgen, Thousand Oaks, CA) uses these experimentally measured rate constants to estimate HX rates for amides located between different flanking amino acids. Rates are determined from the peptide sequence, temperature (0°C), and pH/pD (2.4). For an individual backbone amide hydrogen:

$$\mathrm{BE}_{\mathrm{amide}} = k_{\mathrm{HXPep}} \times (\text{elution time} + \text{wash time}) \tag{5}$$

where, BE_{amide} is the estimated back-exchange for a given backbone amide, and k_{HXPep} is the exchange rate estimated by HXPep for a particular backbone amide. The fractional back-exchange for the entire peptide can be calculated by:

$$BE = \sum \frac{BE_{amide}}{M_{\infty} - M_{calc}} \quad (6)$$

where $\Sigma(BE_{amide})$ sums all amide hydrogens in the peptide from Eq. 14.5, M_{∞} is the calculated peptide mass if every amide hydrogen were exchanged to deuterium in 100% D_2O, and M_{calc} is the calculated average mass of the peptide.

3.2.4. Deuteration Corrected for Back-Exchange and Artifactual In-Exchange

Following estimation of the back-exchange, the data are corrected by the following equation:

$$M_{t,corr(BE)} = M_{calc} + (M_{t,corr(IE)} - M_{calc}) / (1 - BE) \quad (7)$$

in which $M_{t,corr(BE)}$ is the corrected mass of the peptide at time t, BE is the fractional back-exchange, $M_{t,corr(IE)}$ is the mass at time t, corrected for artifactual in-exchange as in Eq. 14.1, and M_{calc} is the calculated average mass of the peptide. The number of deuterons incorporated ($N_{deuterons}$) is then calculated as $M_{t,corr(BE)} - M_{calc}$.

3.2.5. Data Fitting

Once the data are collected and corrected, they are fit to a multiple exponential rate curve to estimate apparent rate constants. Typically, time courses can be fit to sums of three exponential terms, which allows the number and rates of fast, intermediate, and slow exchanging amides to be estimated. We use SigmaPlot (Systat Software, v. 9, http://www.sigmaplot.com) for this purpose, but any nonlinear least squares program can be used (Fig. 3).

1. Enter the time vs. number of deuterons ($N_{deuterons}$) into the NLSQ spreadsheet. For the example of nucleotide binding, four columns are used, which report time vs. $N_{deuterons}$ for control and nucleotide-bound samples of OP-ERKZ and ZP-ERKZ.

   Fit the curve to one, two, or three exponentials, using the general equation:

$$N_{deuterons} = N_{max} - Ae^{-k_1 t} - Be^{-k_2 t} - Ce^{-k_3 t} \quad (8)$$

   where, $N_{deuterons}$ is the extent of deuteration at time t, N_{max} is the number of deuterons after maximal in-exchange, and A, B, and C are the number of amides which exchange with rate constants k_1, k_2, and k_3, respectively.

   Input parameters are used to provide starting estimates of numbers of amides and rate constants. N_{max} is estimated as the asymptote of the time course and constrained by

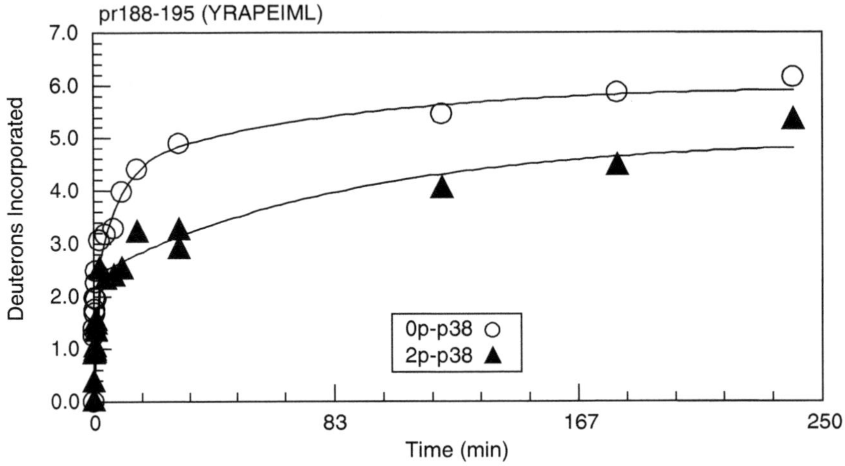

	0P-p38	2P-p38
N_{max}	**6.0**	**5.0**
A	1.97 (0.25)	2.22 (0.21)
B	**2.36 (0.47)**	**0.0**
C	**1.61 (0.51)**	**2.60 (0.16)**
NE	**0.0**	**1.0**
k_1	**5.74 (1.46)**	**1.11 (0.24)**
k_2	0.14 (0.05)	0.14
k_3	0.012 (0.005)	0.011 (.002)
RSS	0.61	1.27

Fig. 3. Time courses comparing HX of a peptide from 0P-p38 and 2P-p38, fit by nonlinear least squares to a sum of exponentials as in Eq. 8. Best fits to parameters N_{max}, A, B, C, k_1, k_2, and k_3 with standard deviations (*parentheses*), and residual sum of squares are indicated. NE indicates the number of amides that are nonexchanging over the time window of the experiment. Significant changes in parameters are indicated in bold. The decreased HX in 2P-p38 relative to 0P-p38 can be explained by conversion of one amide from an intermediate (0.14 per min) to slow (0.01 per min) rate constant, and a second amide from intermediate to nonexchanging (NE) on the experimental time scale. Reproduced with permission from ref. 9.

$N_{max} = A + B + C$. For time courses ranging from 0 to 240 min, reasonable initial estimates for k_1, k_2, and k_3 range from 0.1 min^{-1} to 10 min^{-1} .

3.2.6. Software for Semi-automated Data Analysis

Manual HX-MS data analysis can be tedious; therefore, several computer programs have been developed to automate this process. Errors arise when isotopic peaks are improperly chosen; therefore, most programs display raw data to allow manual inspection and selection of isotope peaks that are then used for the WAM calculations.

1. HXExpress (http://www.hxms.com/HXExpress/) is built on a Microsoft Excel file, using macros to enable semi-automated analysis of datasets (18). Users import peptide isotope *m/z* for each time point from any MS output, and

the software calculates WAM. Additional masses measured under conditions of no deuteration or complete deuteration enable calculation of percent deuteration at each time point. The software operates with MS Excel'97 and higher versions, using Windows 2000 and XP.

2. HX-Analyzer is an interactive tool for identifying isotope peaks and calculating WAM from HX-MS datasets, developed in-house and implemented in Visual Basic. The program requires local installation of Analyzer QS 1.1, Microsoft Office 2003, and Microsoft Windows XP software library modules ((9); request free software from authors). The software inputs a list of LC–MS/MS files (.wiff format) and a user-generated Excel spreadsheet with information about peptide ion mass and approximate elution times. It then links to Analyzer QS modules which display isotopic peaks for each ion for manual inspection to detect anomalous peaks, and compares ions across multiple datasets. Isotopic masses for each peptide ion are captured, WAMs calculated, and output in an Excel file.
3. Hydra (http://www.ucalgary.ca/~dschriem) allows automated data capture and analysis for HX-MS. Different versions are built for .wiff files, requiring local installation of Analyst QS version 1.1, or for mzXML files (19). The software requires an input Excel file summarizing sequence, *m/z*, charge state, and retention time for each peptide, a project tree which associates raw data files with experiments, and a file specifying the data analysis workflow. The software can analyze MS or MS/MS data, enables manual inspection of isotope peaks, outputs calculated WAMs and deuterium incorporation, and displays time courses.
4. HD Desktop (http://deuterator.florida.scripps.edu/) is a web-based platform built on previously described Deuterator software (20), with calculations performed on the HD Desktop server at Scripps. The software inputs mzXML data files and files containing protein sequence and peptide ion information (e.g., sequence, mass, retention time, charge state). The software estimates WAM and calculates deuterium incorporation. The program also consolidates results from peptides with overlapping sequences, for displaying localized HX information onto secondary and tertiary structures.

4. Notes

1. The first step involves proteolyzing undeuterated proteins followed by peptide identification by LC–MS/MS, to determine the cleavage sites of peptides produced by proteolysis.

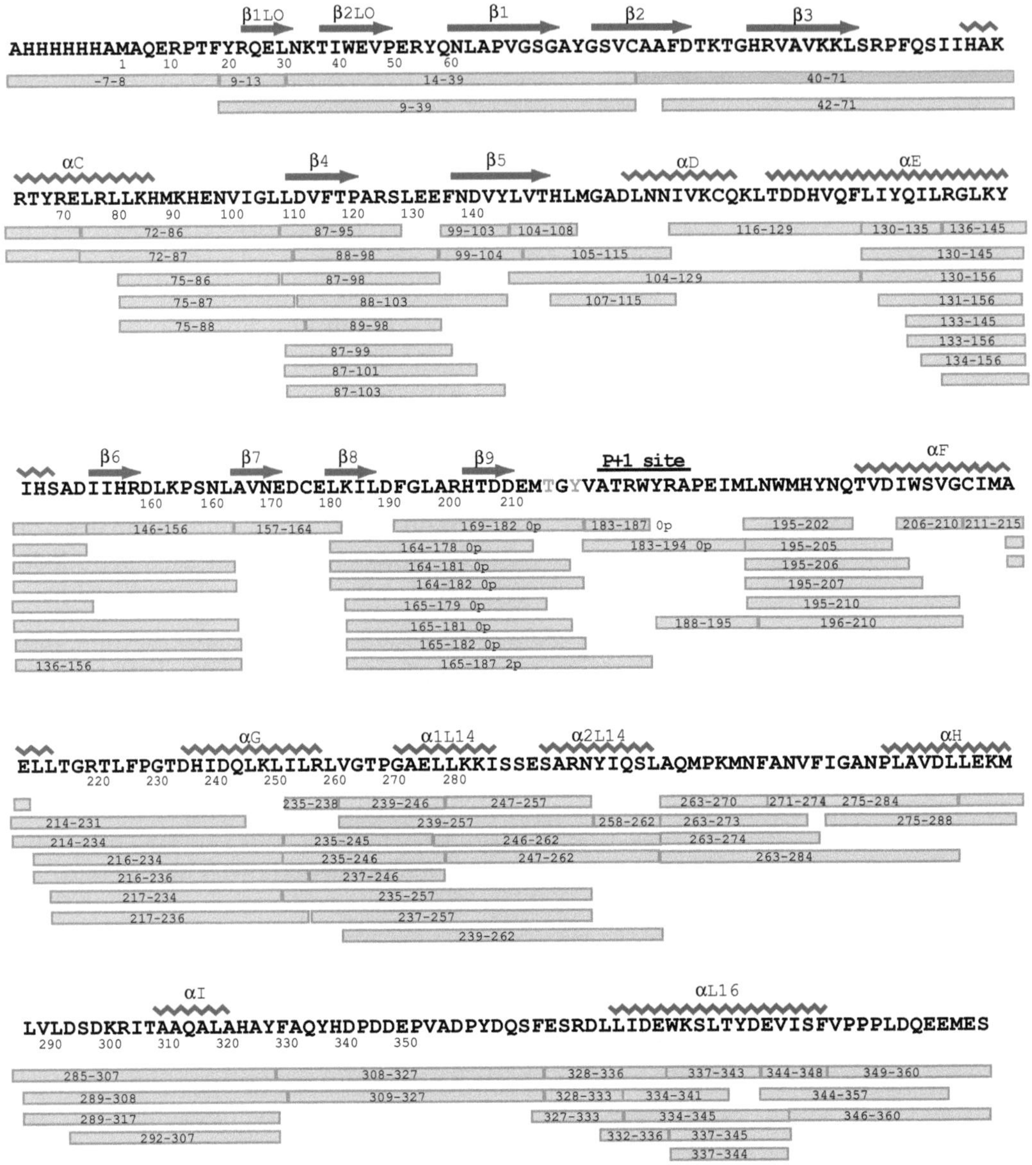

Fig. 4. Sequence coverage map of p38 MAP kinase, indicating peptides that were identified by sequencing and analyzed by HX-MS. Peptide identification and determination of cleavage sites is the first step in a HX-MS experiment to evaluate sequence coverage and experiment feasibility. Reproduced with permission from ref. 9.

The goal is to maximize sequence coverage, measured by the number of amides contained within the observed peptides (Fig. 4). Thus, sequence coverage is optimized with respect to peptide size and recovery, by adjusting proteolysis and LC gradient conditions. We recommend performing six or more replicate LC–MS/MS runs to sample high numbers of identified peptides (9). Not every identified peptide can be analyzed for deuteration in the HX datasets; therefore, the number of analyzable peptides is enhanced by increasing the number of identified peptides.

2. Pepsin is usually preferred for proteolysis because it is active at pH and temperatures which minimize back-exchange (0°C, pH 2.4). Other proteases can be used in parallel with or in place of pepsin and have been shown to enhance peptide recovery and improve sequence coverage (21). Pepsin catalyzes nonspecific and incomplete cleavages, which requires high confidence peptide identifications. It also generates overlapping peptides, which can often be compared to localize deuteration events within narrower areas of sequence.

3. Accurate peptide identification is very important for data analysis and interpretation. We initially match MS/MS spectra to peptide sequences using automated search programs (e.g., Mascot) and applying score thresholds corresponding to low false discovery rates (FDR ≤ 0.01%) (22, 23). Later, every peptide assignment is confirmed by manual analysis, using well-established rules for chemical plausibility. Manual validation is possible for HX datasets which typically include 100–200 peptides. Together, these steps ensure high confidence peptide assignments.

4. Our HX-MS experiments on MAP kinases typically involve 10–20 time points collected between 5 s and 4 h. Incubation of enzymes in D_2O for long times can sometimes lead to partial unfolding and should be monitored for the appearance of bimodal deuteration patterns in certain regions, where unfolded proteins undergo faster HX, yielding higher deuteration and mass. Each day, $t=0$ measurements should be made by adding acid quench buffer to the reaction prior to addition of D_2O and pepsin to correct for artifactual in-exchange of deuterium during proteolysis (see below). Errors in deuterium incorporation largely result from variations in sample handling, back-exchange, and temperature control. We find it useful to quantify errors and assess reproducibility by measuring mass after deuteration for 1 and 10 min, with each time point measured in triplicate or more over the course of several days. This allows standard deviations in mass to be calculated for different peptides at times when deuteration changes rapidly. Typically, we observe standard deviations in peptide mass of 0.1 Da or lower. Changes in mass of 0.5 Da or higher are accepted as true differences if they occur systematically over consecutive time points.

5. Data collection over an HX time course is carried out by randomizing deuteration time points and experimental conditions. Data quality should be monitored by examining two or three ions in each dataset over the days required for data collection. We often choose ions corresponding to peptides

where the exchange behavior can be predicted, for example, peptides from the protein surface might exchange rapidly and completely, and peptides from the protein interior might exchange slowly or not at all. By plotting WAM vs. time for each ion, outliers can be detected thatreflect nonuniform back-exchange or protein denaturation and often systematically affect all peptides within the same dataset. Such time points can then be repeated, if necessary.

References

1. Hoofnagle, A. N., Resing, K. A., and Ahn, N. G. (2003) Protein analysis by hydrogen exchange mass spectrometry. *Annu. Rev. Biophys. Biomol. Struct.* **32,** 1–25.
2. Englander, S. W. (2006) Hydrogen exchange and mass spectrometry: a historical perspective. *J. Am. Soc. Mass Spectrom.* **17,** 1481–1489.
3. Zhang, Z. and Smith, D. L. (1993) Determination of amide hydrogen exchange by mass spectrometry: a new tool for protein structure elucidation. *Protein Sci.* **2,** 522–531.
4. Johnson, R. S. and Walsh, K. A. (1994) Mass-spectrometric measurement of protein amide hydrogen-exchange rates of apo-myoglobin and holo-myoglobin. *Protein Sci.* **3,** 2411–2418.
5. Hoofnagle, A. N., Resing, K. A., Goldsmith, E. J., and Ahn, N. G. (2001) Changes in protein conformational mobility upon activation of extracellular regulated protein kinase-2 as detected by hydrogen exchange. *Proc. Natl. Acad. Sci. U.S.A.* **98,** 956–961.
6. Resing, K. A., Hoofnagle, A. N., and Ahn, N. G. (1999) Modeling deuterium exchange behavior of ERK2 using pepsin mapping to probe secondary structure. *J. Am. Soc. Mass Spectrom.* **10,** 685–702.
7. Resing, K. A. and Ahn, N. G. (1998) Deuterium exchange mass spectrometry as a probe of protein kinase activation: analysis of wild-type and constitutively activate mutants of MAP kinase kinase-1. *Biochemistry* **37,** 463–475.
8. Lee, T., Hoofnagle, A. N., Resing, K. A., and Ahn, N. G. (2005) Hydrogen exchange solvent protection by an ATP analogue reveals conformational changes in ERK2 upon activation. *J. Mol. Biol.* **353,** 600–612.
9. Sours, K. M., Kwok, S. C., Rachidi, T., Lee, T., Ring, A., Hoofnagle, A. N., Resing, K. A., and Ahn, N. G. (2008) Hydrogen-exchange mass spectrometry reveals activation-induced changes in the conformational mobility of p38alpha MAP kinase. *J. Mol. Biol.* **379,** 1075–1093.
10. Hamuro, Y., Coales, S. J., Morrow, J., Griffin, P. R., Southern, M. R., and Weber, P. C. (2007) Application of hydrogen/deuterium-exchange to p38 mitogen-activated protein kinase. *Am. Biotech. Lab.* **25,** 28–30.
11. Lee, T., Hoofnagle, A. N., Kubuyama, Y., Stroud, J., Min, X., Goldsmith, E. J., Chen, L., Resing, K. A., and Ahn, N. G. (2004) Docking motif interactions in MAP kinases revealed by hydrogen exchange mass spectrometry. *Mol. Cell.* **14,** 43–55.
12. Hoofnagle, A. N., Stoner, J. W., Lee, T., Eaton, S. S., and Ahn, N. G. (2004) Phosphorylation-dependent changes in structure and dynamics in ERK2 detected by SDSL and EPR. *Biophys. J.* **86,** 395–403.
13. Zhou, B., Zhang, J., Liu, S., Reddy, S., Wang, F., and Zhang, Z. (2006) Mapping ERK2-MKP3 binding interfaces by hydrogen/deuterium exchange mass spectrometry. *J. Biol. Chem.* **281,** 38834–38844.
14. Zhou, B. and Zhang, Z. (2007) Application of hydrogen/deuterium exchange mass spectrometry to study protein tyrosine phosphatase dynamics, ligand binding, and substrate specificity. *Methods* **42,** 227–233.
15. Truhlar, S. M., Croy, C. H., Torpey, J. W., Koeppe, J. R., and Komives, E. A. (2006) Solvent accessibility of protein surfaces by amide H/2H exchange MALDI-TOF mass spectrometry. *J. Am. Soc. Mass Spectrom.* **17,** 1490–1497.
16. Sabo, T. M., Brasher, P. B., and Maurer, M. C. (2007) Perturbations in factor XIII resulting from activation and inhibition examined by solutions based methods and detected by MALDI-TOF MS. *Biochemistry* **46,** 10089–10101.
17. Bai, Y., Milne, J. S., Mayne, L., and Englander, S. W. (1993) Primary Structure effects on peptide group hydrogen exchange. *Proteins* **17,** 75–86.

18. Weis, D. D., Kass, I. J., and Engen, J. R. (2006). Semi-automated analysis of hydrogen exchange mass spectra using HX-Express. *J. Am. Soc. Mass Spectrom.* **17(12),** 1700–1703.
19. Slysz, G. W., Baker, C. A. H., Bozsa, B. M., Dang, A., Percy, A. J., Bennett, M., and Schriemer, D. C. (2009) Hydra: software for tailored processing of H/D exchange data from MS or tandem MS analyses. *BMC Bioinformatics* **10,** 162.
20. Pascal, B. D., Chalmers, M. J., Busby, S. A., and Griffin, P. R. (2009) HD Desktop: an integrated platform for the analysis and visualization of H/D exchange data. *J. Am. Soc. Mass Spectrom.* **20,** 601–610.
21. Cravello, L., Lascoux, D., and Forest, E. (2003) Use of different proteases in acidic conditions to improve sequence coverage and resolution in hydrogen/deuterium exchange of large proteins. *Rapid Commun. Mass Spectrom.* **17,** 2387–2393.
22. Perkins, D. N., Pappin, D. J., Creasy, D. M., and Cottrell, J. S. (1999) Probability-based protein identification by searching sequence databases using mass spectrometry data. *Electrophoresis* **20,** 3551–3567.
23. Elias, J. E. and Gygi, S. P. (2007) Target-decoy search strategy for increased confidence in large-scale protein identifications by mass spectrometry. *Nat. Methods* **4,** 207–214.

Chapter 15

A "Molecular Evolution" Approach for Isolation of Intrinsically Active (MEK-Independent) MAP Kinases

Vered Levin-Salomon, Oded Livnah, and David Engelberg

Abstract

Mitogen-activated protein (MAP) kinases are a large family of enzymes composed of about four subfamilies, each containing several isoforms and splicing variants. Many MAP kinases are coexpressed in each eukaryotic cell and coactivated in response to various stimuli. It is, therefore, difficult to explore the specific downstream effects of each species of MAPK. Expression of an intrinsically active variant of a MAPK, while other MAPKs are not active, allows for tracking of a specific array of substrates, target genes, and biological/pathological effects corresponding to the expressed molecule. This chapter describes a method for obtaining such intrinsically active MAPKs. Because of the unique mode of MAPK activation, which is absolutely dependent on unconventional phosphorylation (on neighboring Thr + Tyr residues), a rational design of mutations that would render the kinase intrinsically active is currently unfeasible. Our method is based, therefore, on a "Molecular Evolution" approach that uses the power of yeast genetics and is unbiased toward the mutation sites. We describe in detail how to prepare a large population of randomly mutated molecules of the desired MAPK and how to screen this library in a yeast strain lacking the relevant MAPK kinase (MAPKK). The idea is to identify MAPK variants that are fulfilling all MAPK functions and allow growth of this strain – namely, MAPK molecules that function biologically in the complete absence of their upstream activator. We further describe the details of the "plasmid-loss" assay used for distinguishing between true positive and false positive clones. Finally, we report on a new yeast strain lacking four MAPKKs that could serve as a universal target for screening for active MAPK of all subfamilies.

Key words: MAP kinases, Yeast, Active mutants, Molecular evolution, Pbs2, Mkk1, Mkk2, Ste7, MAPKK knockout

1. Introduction

1.1. Prelude

This chapter describes a method for creating a large population of mutated mitogen-activated protein (MAP) kinases and isolating, out of that population, novel, intrinsically active variants. This

Rony Seger (ed.), *MAP Kinase Signaling Protocols: Second Edition*, Methods in Molecular Biology, vol. 661,
DOI 10.1007/978-1-60761-795-2_15,

type of screens that involve creating and isolating molecules with desired properties, which were not bestowed on the parental molecules by nature, is generally known as a "Molecular Evolution" or "Site-directed Evolution" approach. Before sketching the technical details, we shall briefly explain why it is particularly important to attain intrinsically active variants of MAP kinases and why the "Molecular Evolution" approach is the most suitable one (in fact, the only one that works) for this purpose.

1.2. Why Are Intrinsically Active Variants Most Important for MAPK Research?

MAP kinases is a generic name for a large number of enzymes (ERK, JNK, p38, and BMK/ERK5 families) conserved in all eukaryotes. Many isoforms of these enzymes are expressed in each eukaryotic cell. They are involved in numerous processes (1) and abnormal activity of them is associated with inflammatory diseases, degenerative diseases, and cancer (2–5). Although being heavily studied, many aspects of MAPKs biology have not been revealed. Very little is known, for example, about their involvement in the etiology of human diseases.

MAP kinases are similar in sequence and structure and activated via a similar mechanism that requires dual phosphorylation of neighboring Thr and Tyr residues. This dual phosphorylation, catalyzed by MAPK kinases (MAPKKs), is unique to MAPKs (6) and is absolutely critical for their activity. Although highly similar in structure and function (reviewed in (7)), each isoform and splicing variant of MAPKs has its own unique functions, array of substrates, and pharmacological and biochemical properties (8).

MAP kinases are activated following exposure of the cell to relevant stimuli. In response to growth factors, for example, ERKs are strongly activated, and JNKs and p38 are moderately activated. In general, almost all MAP kinases in the cell are concomitantly activated (to different degrees), making it difficult to study the specific downstream biochemical, physiological, or pathological effects of each MAPK molecule. A significant progress in revealing the role of each MAPK was achieved by downregulating their expression or activity using siRNAs, pharmacological inhibitors, or knockout animals (9–14). A complementary approach could be the activation of each MAPK species individually in the cell and tracking the biochemical and physiological consequences. This could be achieved by expressing, in the biological system of interest, an intrinsically active variant of each MAP kinase. Activation of just one MAPK molecule, under conditions in which other MAPKs are not active, will disclose the specific array of substrates, target genes, and physiological effects of this particular MAPK. The goal of the method described here is to develop intrinsically active variants that when introduced to cells will be spontaneously and individually active.

1.3. Intrinsically Active MAP Kinases Have Not Been Obtained via Conventional Approaches

Intrinsically active variants of enzymes, receptors, GTP binding proteins, and transcription factors have been obtained from two major sources. One is human and animal diseases, primarily cancers, in which such mutations are part of the cause of the diseases (15–17). The other is the rational design of mutations on the basis of understanding of the mechanism of action and the structure/function relationships of the protein (18, 19). None of the approaches seems valid for MAP kinases because activating mutations in MAPKs have not been found so far in any disease and because understanding their mechanism of action (1, 6, 20, 21) did not provide ideas on mutations that may render them intrinsically active (reviewed in (22)). Thus, for obtaining bona fide intrinsically active variants, a different approach should be taken. We devised the "Molecular Evolution" approach described in here, which is unbiased towards the mutations' sites and towards their mechanism of action.

1.4. An Unbiased "Molecular Evolution" Approach Was Successfully Applied for Isolation of Proteins with Desired Properties

The desire to change properties of proteins for the benefit of basic and applied research, as well as for the industry, is not restricted of course to MAPKs. Yet, similar to the case of MAPKs, thorough mechanistic and structural knowledge was not useful in many cases for predicting the mutations that would impose the desired property on the protein of interest. This situation pushed forward the approach known as "Molecular Evolution" or "Site-directed Evolution," in which one does not have to envisage the mutations, but should rather prepare a large population of mutants of the protein in question, hoping that one or a few of the mutants within that population would acquire the desired property. Then, one should apply an appropriate selective pressure to select, out of the population, those molecules with the desired property. This approach (reviewed in (23–25)) is useful only if the selection system is powerful and accurate. "Molecular/Site-directed Evolution" approaches have been used successfully for isolation of proteins that are 1,000-fold more stable than their parental molecule, that are catalytically more efficient, and even proteins with modified substrate specificity (reviewed in (23, 24)).

We have applied the principles of the approach for devising a method for the isolation of intrinsically active MAP kinases.

2. Materials

2.1. Media

1. LB medium: 1% tryptone, 0.5% yeast extract, and 1% NaCl (2.4% agar).
2. YPD medium: 1% yeast extract, 2% peptone, and 2% glucose, (3% agar).

3. Y-nb medium: 0.17% y-nb, 0.5% $(NH_4)_2SO_4$, 2% glucose, and 0.004% of essential amino acids or nitrogenous bases (commonly: Trp, Ura, Ade, His, Leu, Met, Lys), (3% agar).

2.2. Maxi-Prep of Plasmid DNA

1. STE: 0.1 M NaCl, 10 mM Tris–Hcl (pH 8.0), andl mM EDTA (pH 8.0).
2. Solution I: 50 mM glucose, 25 mM Tris–Hcl (pH 8.0), and 10 mM EDTA (pH 8.0). The solution can be prepared in batches of ~100 ml, autoclaved, and stored at 4°C.
3. Solution II: 0.2 N NaOH and 1% SDS.
4. Solution III: 60 ml of 5 M potassium acetate, 11.5 ml of glacial acetic acid, and 28.5 ml DDW. The resulting solution is 3 M with respect to potassium and 5 M with respect to acetate.
5. Phenol–Chloroform in LETS: 250 ml phenol, 250 ml chloroform, and 500 ml LETS.
6. LETS: 1 M LiCl, 0.5 M EDTA, 1 M Tris–Hcl pH = 7.4, and 10% SDS.

2.3. High-Efficient Introduction of Plasmid DNA (Library) to Yeast Cells

1. Li-Sorb: 100 mM LiAC in TE, 1 M sorbitol.
2. LiPEG: 40% PEG3550 and 0.1 M LiAC, TEx1.
3. TEx10: 100 mM Tris–Hcl pH 8.0 and 10 mM EDTA pH 8.0.

2.4. Extraction of Plasmid DNA from Yeast: Boiling Prep

1. STET: 8% glucose, 50 mM EDTA pH 8.0, 50 mM Tris–Hcl pH 8.0, and 0.1% Triton X-100.
2. Lysozyme: 10 mg/ml, dissolved in DDW or TE.
3. Glass beads: 425–600 μm, washed several times with nitric acid, then several times with water, and a final wash with TE.

3. Methods

3.1. The Concept: Only an Intrinsically Active Variant of a MAPK Will Activate the Entire Downstream Pathway in the Complete Absence of Its MAPKK

The concept of the approach is depicted in Fig. 1A. The principal idea is simple, suggesting that if the activity of a MAP kinase is absolutely dependent on its MAPKK, eliminating the MAPKK would totally abolish the catalytic activity and the related biological effects of the MAPK (the knockout strain in Fig. 1Aa). Importantly, the dependence of the MAPK on its MAPKK should be absolute, so that even overexpression of the MAPK in cells lacking the relevant MAPKK will not rescue the phenotype of those cells. However, if the MAPK would acquire an intrinsic activity, it will no longer depend on the MAPKK and would rescue the phenotype (Fig. 1A bottom). A MAPK mutant possessing an intrinsic activity could be isolated from a library of mutants (Fig. 1Ab–d). Each of the yeast MAP kinase pathways (Fig. 1B)

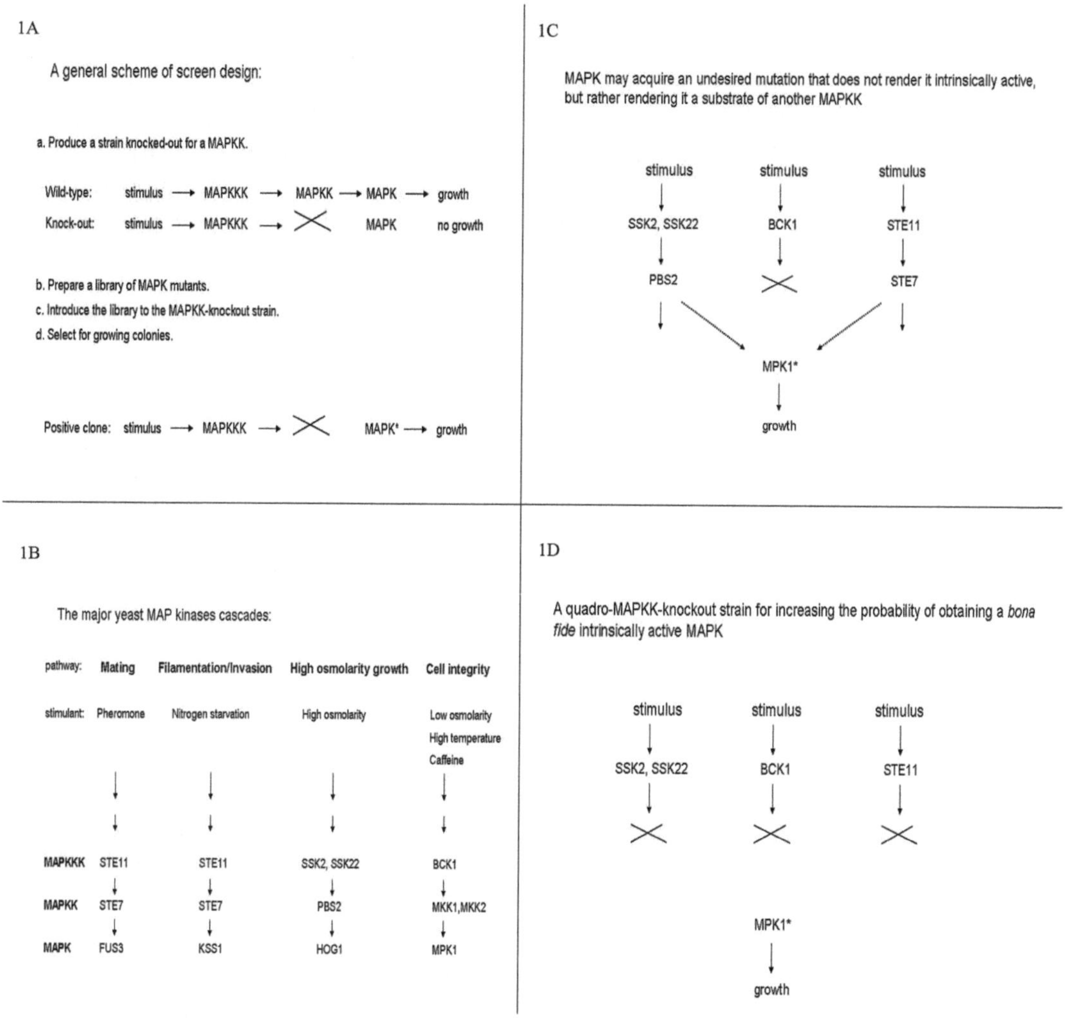

Fig. 1. (**A**) A general scheme of the "Molecular Evolution" screen for isolation of intrinsically active, MAPKK-independent, variants of MAPK. (a) The approach is based on the absolute dependence of a MAPK on its upstream MAPKK. Namely, the MAPK cannot function in a strain knocked out for the relevant MAPKK. (c–d) Only an intrinsically active MAPK variant that is independent of upstream regulation can function in cells lacking the relevant MAPKK. (**B**) Presentation of major MAPK cascades in yeast. In principle, the method described in this chapter can be applied in cells knocked out for each of the yeast MAPKKs. (**C**) A case of undesired mutations: In the "Molecular Evolution" screen, MAPK molecules may acquire undesired mutations that render them a substrate to a nonrelevant MAPKK. Thus, rescue of the *mapkk*Δ phenotype would not be a result of intrinsic activity, but rather of MAPKK-dependent activity. (**D**) In order to avoid obtaining such undesired mutants, it is possible to perform the screen using a quadro-MAPKK-knockout strain, lacking MKK1, MKK2, STE7 and PBS2.

could be used for the screen. Namely, each of the yeast MAPKK could be eliminated and the respective MAPK gene could be mutated in vitro and the resulting library of mutated genes could be screened. A potential problem, however, is that the screen may select for a MAPK variant that is not intrinsically active but acquired a mutation that allows its recognition and activation by a MAPKK that naturally does not activate it. Such a nonnatural phosphorylation of the MAPK by this MAPKK would also rescue the phenotype (Fig. 1C). It is possible to prevent isolation of this undesired type of mutants by using a strain mutated for all MAPKKs (Fig. 1D; quadro-MAPKK-knockout; see below). Because activating mutations are very rare, a crucial parameter for the success of the screen is the ability to produce as large number as possible of mutants of the MAP kinase. It is also important that these mutants will carry a single point mutation (see below). Importantly, because members of MAPK pathways of other organisms are able to replace their yeast orthologs (26), it should be possible to screen not only for active variants of the yeast MAPKs but also for active variants of the relevant mammalian, insect, or plant ortholog.

3.2. Overview of the Experimental Procedure (Brief Steps)

1. Constructing the yeast strain in which the screen will take place, namely, knocking out the relevant MAPKK (Fig. 1A "knockout"). An attractive alternative possibility is to use a strain we recently constructed, which lacks four MAPKKs (a quadro-MAPKK-knockout; see ref. (26), (Fig. 1D) and is suitable for screening for active variants of many MAP kianses.
2. Characterizing and calibrating the relevant phenotype of the strain constructed in step 1 and verifying its reliability. The critical experiment here is to introduce into the *mapkk*Δ strain the native (not mutated) MAPK in question (preferably overexpressed), to obtain a large number of colonies, to expose them to selective pressure and to count the number of false positives. In most cases, the number of revertants should be zero, but a ratio of 1/10,000 is also acceptable and allows for the use of this strain for screening.
3. Producing the library of mutants to be screened.
4. Introducing the library into the relevant strain and applying a selective pressure.
5. Identifying the true positive and the false positive clones via "plasmid-loss" assays.
6. Isolating the plasmids from positive clones, sequence the gene for identification of the activating mutation, and continue with biochemical studies.

3.3. Establishment and Implementation of the Screening System in Detail

3.3.1. Obtaining or Constructing a Yeast Strain Knocked Out for the Relevant MAPKK

Based on the MAPK of interest, for which intrinsically active variants are desired, a yeast strain knocked out for the respective MAPKK should be used. As various "knock-out libraries" are available from commercial or academic sources, such a strain could be purchased or requested. However, it might be preferable, in some cases, to construct the knockout strain in house, using a genetic background known to manifest the phenotype in an optimal way. The strain to be used could be knocked out for a single gene or for several. For example, to isolate active mutants of MAPKs of the Hogl/p38 family, a *pbs2*Δ strain should be constructed (see Fig. 1B), whereas a *mkk1Δ mkk2Δ* strain should be constructed if the MAPK of interest is of the Mpk1/ERK family (Fig. 1B).

Importantly, to prevent isolation of undesired mutants that are not intrinsically active, but rather acquired affinity to another MAPKK (Fig. 1C), the strain used for screening may be knocked out for nonrelevant MAPKKs as well. We suggest the use of the *mkk1Δ mkk2Δ pbs2Δ ste7Δ* strain, which is knocked out for the four major yeast MAPKKs.

3.3.2. Characterization of Screening Conditions

Before implementing a screen for isolation of active mutants of a specific MAPK, it is necessary to characterize and calibrate the conditions under which the screen should be held. The conditions to be chosen are those that activate the particular signal transduction pathway known to activate the MAPK of interest. The yeast strain knocked out for the relevant MAPKK should not survive under these conditions, for example:

1. For isolation of active mutants of MAP kinases of the Hog1/p38 family, that are essential for surviving under osmotic pressure ((27); Fig. 1B), the screen should be done on plates supplemented with 0.9–1.3 M NaCl. *pbs2Δ* cells overexpressing Hog1WT should not grow under these conditions.
2. For isolation of active mutants of MAP kinases of the Mpk1/ERK family, that are critical for maintenance of the yeast cell wall ((26); Fig. 1B), the screen should be performed on hypotonic media or on media supplemented with 10–15 mM caffeine. *mkk1Δ mkk2Δ* cells harboring an empty vector or a vector overexpressing Mpk1WT should not grow under these conditions (see Note 1).

3.3.3. Production of a Library of Randomly Mutated MAPK Molecules

The protocol below has been used in our laboratory for preparing several libraries, including two libraries of mutated MAPKs (27, 28). The principle of this method is to propagate the plasmid carrying the cDNA of the MAPK of interest in an *Escherichia coli* strain that is defective in its DNA repair machinery. The *E. coli* strain we are using, LE30, contains just one mutation (mutD5) in the repair machinery. As a result, mutagenesis is very efficient with respect to the number of plasmids mutated, but is not efficient with respect to the number of mutations each plasmid carries. Namely, mutated plasmids commonly carry just one point mutation. We believe this is a great advantage because most mutations in a MAPK are expected to be deleterious to the kinase's activity. Indeed, our experience shows that the intrinsically active mutants should harbor just one point mutation. The presence of another mutation usually abolishes activity. Thus, although other methods could be used for the random mutagenesis procedure, “too” efficient methods are not recommended. The method we use is based on reference (29).

1. Transform the LE30 *E. coli* strain with a plasmid carrying the MAPK of interest.
   (a) Mix 1 μl of plasmid (from a common mini-prep) with 200 μl of LE30 cells, which were made competent using the $CaCl_2$ method (30).
   (b) Incubate on ice for 20 min.
   (c) Incubate at 42°C for 2 min and return to ice.
   (d) Add 300 μl LB without ampicillin.
   (e) Incubate at 37°C for 1 h without shaking.
   (f) Mix the tube well and plate transformants on five selective plates (100 mm plates of LB + agar + ampicillin), 100 μl per plate.
   (g) Incubate plates overnight at 37°C. The plates should be covered (high density) with colonies, creating a lawn.
2. Add 1 ml LB onto each plate. Using a rubber policeman, scrape all transformants and transfer them from the plates to a flask containing 100 ml LB + ampicillin. Grow the culture for 4–5 h at 37°C with shaking.
3. Transfer the 100 ml culture into a flask containing 2 L of LB + ampicillin. Grow the culture overnight at 37°C with shaking.
4. Harvest the bacterial cells by centrifugation at 2,700 × *g* for 15 min at 4°C. Harvest 4 × 500 ml. Discard the supernatant.
5. Extract DNA plasmid from cells using Maxi-prep protocol.
   (a) Resuspend the pellet from a 500 ml culture in 100 ml of ice-cold STE.
   (b) Centrifuge at 2,700 × *g* for 15 min at 4°C. Discard the supernatant.
   (c) Resuspend the washed bacterial pellet in 10 ml solution I.
   (d) Add 1 ml of a freshly prepared solution of lysozyme (10 mg/ml in 10 mM Tris–HCl pH 8.0) (see Note 2).
   (e) Add 20 ml of freshly prepared solution II. Close the top of the centrifuge bottle and mix the contents thoroughly by gently inverting the bottle several times. Store the bottle at room temperature for 5–10 min.
   (f) Add 15 ml of ice-cold solution III. Close the top of the centrifuge bottle and mix the contents thoroughly by gently inverting the bottle several times. There should no longer be two distinguishable liquid phases. Store the bottle on ice for 10 min. A flocculent white precipitate should form (see Note 3).
   (g) Centrifuge the lysate at 4,200 × *g* for 15 min at 4°C. Allow the rotor to stop without braking (see Note 4).

(h) Filter the supernatant through four layers of cheesecloth into a 250-ml centrifuge bottle.

(i) Add one volume of phenol–chloroform. Mix by vortexing. Centrifuge at 4,200 × g for 10 min at 4°C. Take upper phase to a new bottle.

(j) Add two volumes of 100% ethanol. Recover the nucleic acids by centrifugation at 4,200 × g for 15 min at room temperature (see Note 5). Decant the supernatant.

(k) Rinse the pellet with 70% ethanol at room temperature. Centrifuge at 4,200 × g for 10 min.

(l) Dissolve the pellet in 3 ml of DDW.

3.3.4. Estimation of Mutagenesis Rate

As explained at the beginning of paragraph in Subheading 3.3.3, the mutagenesis rate is kept low in purpose. In addition, the number of relevant mutations in the library (those rendering the MAPK intrinsically active) is probably extremely low. It is difficult therefore to quantify the rate of mutagenesis, but some evaluation that the mutagenesis procedure worked at all could be obtained. Such an estimation could be made by introducing the library of the mutated MAPK into a yeast strain lacking this MAPK. For example, introduction of a library of MPK1 mutants into *mpk1Δ* cells. Transformants should be plated on a medium that selects for the presence of plasmids (e.g., -URA). Each plate should contain only a few hundreds of colonies, and 10–20 plates should be used. After 48–72 h, plates should be replica plated onto plates containing caffeine. If mutagenesis procedure did not work and all MPK1 clones of the library are in fact identical to wild type, all colonies should grow on caffeine. The number of colonies that fail to grow on caffeine should give some idea of the mutagenesis rate. From our experience, if about 1 of a thousand colonies fails to grow (i.e., contains a destructive mutation in the MAPK), the library is valid for screening in a *mapkkΔ* strain as long as a large number of colonies (~2–5 × 10^5) is obtained in this screen.

3.3.5. Introduction of the Mutated Plasmids Library to the Knocked-Out MAPKK Strain for Screening

1. Transform the knocked-out MAPKK yeast strain of choice (or the quadro-MAKK-knockout) with the mutated plasmids library.

   (a) Transfer *mapkkΔ* cells into 10 ml of YPD medium. Incubate the culture overnight at 30°C with shaking.

   (b) Dilute cells to OD_{600} = 0.5 in a volume of 50 ml using YPD medium. Grow for 2–4 h (to log-phase) at 30°C with shaking (see Note 6).

   (c) Centrifuge culture at 2,060 × *g* for 5 min at room temperature. Remove supernatant.

(d) Resuspend pellet with 25 ml Li-Sorb. Centrifuge at 2,060 × *g* for 5 min at room temperature. Remove supernatant.

(e) Resuspend pellet in 1 ml of Li-Sorb.

(f) Prepare transformation mixture in two Eppendorf tubes. Add to each tube: 10 μl boiled salmon-sperm, 3 μl DNA from mutated plasmid library, 100 μl yeast cells, and 700 μl LiPEG. Add components in the order above. Mix by inverting the tube few times.

(g) Incubate tubes for 30 min at 30°C. Invert tubes for mixing every 10 min.

(h) Add 86 μl of DMSO. Mix.

(i) Heat shock cells in a water bath for 10 min at 42°C.

(j) Centrifuge tubes at 17,900 × *g* for 20 s. Remove supernatant.

(k) Resuspend pellet in ~1 ml of 1 M sorbitol.

(l) Plate 100 μl of transformants on selective plates, to select cells harboring plasmid of the library (e.g., -URA plates, if the plasmids library is carrying the URA3 gene). All other growth conditions should be optimal. Namely, selections for phenotype should not be applied at this stage (i.e., media should not be supplemented with caffeine or NaCl). Incubate at 30°C for 2–3 days (until colonies grow). From our experience, a number of $2\text{–}5 \times 10^5$ colonies at this step would provide 10–50 true positive clones (see Note 7).

3.3.6. Transfer the Transformants to Selective Plates to Select for Required Phenotype, Using Replica Plating

Take each of the plates from the previous step and replica plate it, using velvet stripes, to five selective plates (see Note 8).

1. As the number of transformants in the original plate is very high (a lawn), many cells are transferred to the selective plates, making it sometimes difficult to monitor the positive ones over the background of not growing cells. Thus, the first replica plate should be discarded. Namely, its purpose is only to reduce the number of cells transferred to the next plates.
2. Plates 2–4 should contain growth medium that enforces manifestation of the phenotype of choice (e.g., high salt, caffeine). Plates 2–4 can be either identical or may contain different concentrations of the drug or salt.

   This step has two purposes. First, it allows elimination of false positives or contaminants – there is a high probability that a colony that grew on one of the replica plates and not on its twin plates is a contaminant or a revertant. Second, if replica plating of the transformants is done onto different stringencies of selection (achieved by different concentrations of drug or salt), it will be possible to classify the positives into groups of strong positives (grow under all conditions) and weak positives (grow only under mild selective pressure).

3. Plate 5 should contain selection only for cells harboring the plasmids library (e.g., -URA). All cells that grew on the original plate should grow on this plate too. If not all cells were grown on this plate, replica plating should be problematic and hence repeated, perhaps with less than five replicas.

3.3.7. Analyzing the Putative Positive Colonies: Identification of True Positives and False Positives Using Plasmid Loss Assays

1. Positive clones should be picked up using a toothpick and plated as patches on selective plates (i.e., supplemented with salt or caffeine) (see Note 9). In order not to scratch the agar and to "draw" patches easily (see example of patches in Fig. 2B), it is best to use flat toothpicks available at any grocery store (toothpicks should be autoclaved in a glass beaker). The purpose of this step is to enrich positive cells and eliminate other cells that might be transferred on the toothpick together with the positive colony (one should remember that cells that do not grow are not dead and if transferred to nonselective media would multiply). It is expected that following the growth of the patch under selective conditions, nearly 100% of its cells originate from the positive colony.
2. Cells from each of the patches (that originate from independent positive colonies) should be picked up and plated, as a new patch, on a fresh plate in which selection pressure is only for plasmid maintenance (i.e., -URA plates).
3. Once the patches are grown, cells from each patch should be plated again as patches, but on a nonselective YPD plate to encourage loss of plasmid. Following 36–48 h of growth, cells from each of these patches (some cells lose the plasmid and some do not) should be streaked on a fresh YPD plate so that 50–100 individual colonies will be obtained.
4. These 50–100 colonies should be picked up and plated as small patches on a YPD plate, using a grid drawn on a paper that is placed under the plate (Fig. 2A).
5. Once all patches are grown, they are tested for a linkage between rescuing the phenotype and the plasmids library. Namely, if the ability to rescue the MAPKK knockout strain is associated with the mutated MAPK on the plasmid, then colonies that lose the plasmid will also lose the ability to rescue the phenotype. For example, the plates on the left side of Fig. 2B represent a true positive colony. After the original colony was allowed to grow as a patch on YPD plate for ~48 h, cells were streaked on YPD plate and about 50 single colonies were plated as patches. Twenty-four hours later, this

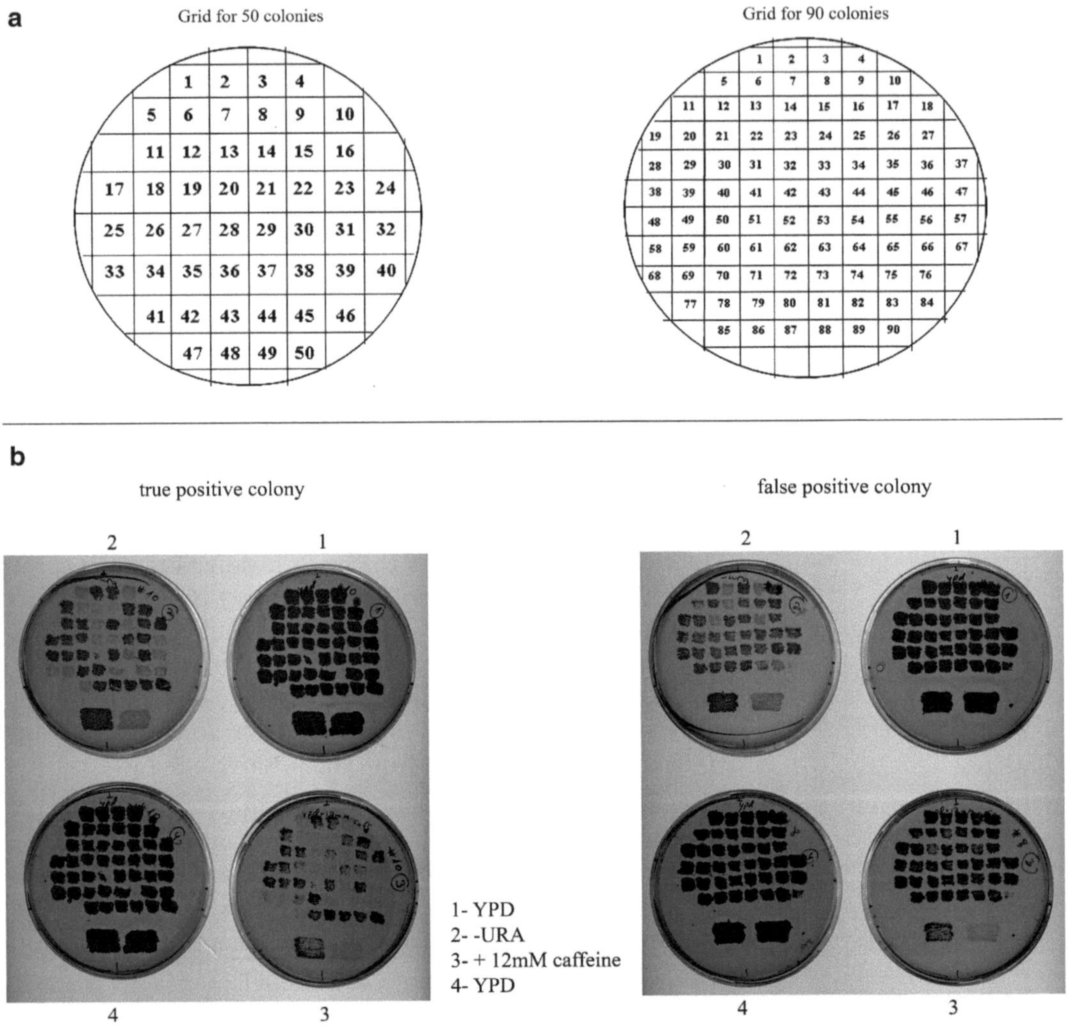

Fig. 2. Example for the results of a plasmid loss assay of a screen for intrinsically active Mpk1/ERK molecules. (**a**) A template for 50 and 90 colonies grids. These templates can be placed under the plate on which one should plate patches of single colonies (see text for details) for plasmid loss assays. (**b**) An example for two different results of plasmid loss assays, one for a false positive colony (*right*) and the other for a true positive colony (*left*). Patches of colonies originated from a clone that was allowed to lose its plasmid, were plated on a YPD plate, grown overnight, and replica plated to four different plates in the indicated order. Large patches on the bottom of the plates are cells of the *mkk1Δmkk2Δ* strain (*right*; negative control) and the *mpk1Δ* strain (*left*; positive control), both expressing Mpk1WT. Note that in the true positive colony (*left panel*), each of the patches that grew on -URA (plate 2) also grew on 12 mM caffeine (plate 3), whereas each patch that did not grow on -URA also did not grow on caffeine. Namely, there is an absolute linkage between the library plasmid (carrying the URA3 marker) and the property to grow on caffeine. In the false positive colony (*right panel*), many patches that lost the plasmid (did not grow on -URA) grew on medium supplemented with caffeine.

plate was replica plated in the following order: (1) To a plate supplemented with YPD (plate 1). (2) To a plate supplemented with -URA medium (plate 2). (3) To a plate supplemented with 12 mM caffeine (plate 3). (4) To another plate supplemented with YPD (plate 4). Comparison of plates 2 and 3 reveals that every patch that did not grow on -URA medium (i.e., lost the plasmid carrying the URA3 marker) did not grow also on medium supplemented with 12 mM caffeine. On the other hand, each patch that grew on -URA also grew on caffeine. These results strongly suggest that the property to grow on caffeine and on -URA are linked and located on the same plasmid. On the other hand, if the ability to grow on caffeine is maintained even if plasmid is lost, (i.e., many patches grow on caffeine, but cannot grow on -URA) then the colony is false positive (Fig. 2B right panel).

In summary, to test for the linkage between ability to grow on caffeine and the library plasmid, patches are replica plated from the YPD plates to four plates in the following order (Fig. 2B):

1. YPD plate. The purpose of this plate is to reduce the number of cells transferred to the next plates.
2. Selective plate for cells harboring plasmid (i.e., -URA plate).
3. Selective plate for manifestation of phenotype (i.e., caffeine, salt).
4. YPD plate. The purpose of this plate is to monitor whether all patches are replicated. Incubate all plated at 30°C for 24–48 h.
5. Monitoring growth of patches. Consider positive clones in those plates in which there is a clear linkage between growth under selective pressure and presence of plasmid (see Fig. 2B).

3.3.8. Isolating Plasmid from Positive Colonies

1. Extract plasmid from true positive yeast cells.
   - (a) Grow yeast colony overnight on 3 ml Y-nb medium that enforces maintenance of plasmid (e.g., -URA) at 30°C.
   - (b) Transfer 1.5 ml to a new Eppendorf tube. Centrifuge at 3,200 × g for 5 min. Discard supernatant.
   - (c) Resuspend yeast cells in 100 μl STET.
   - (d) Add 0.2 g glass beads. Mix by vortexing for 5 min (see Note 10).
   - (e) Boil tubes for 3 min at 100°C. Replace tubes on ice.
   - (f) Centrifuge at 17,900 × g for 10 min at 4°C.
   - (g) Using sterile tip, remove 100 μl of upper phase to a new tube that contains 50 μl 7.5 M ammonium acetate. Mix.
   - (h) Incubate for 1 h at −20°C.
   - (i) Centrifuge at 17,900 × g for 10 min at 4°C.
   - (j) Take upper phase to a new tube. Add 200 μl ice-cold ethanol to every 100 μl of upper phase. Mix by vortexing.

(k) Incubate for 1 h at −20°C.

(l) Centrifuge at 17,900 × *g* for 15 min at 4°C. Discard upper phase.

(m) Rinse the pellet with 70% ethanol. Centrifuge at 17,900 × *g* for 4 min at 4°C.

(n) Air-dry the pellet.

(o) Dissolve the pellet in 10 μl of DDW.

2. In order to increase concentration of the extracted DNA and to separate the plasmid DNA from remains of cellular DNA, it should be transformed to *E. coli*, according to the protocol in Subheading 3.3.3 (step 1) (see Note 11).
3. Plasmid DNA rescued from positive clones and purified from *E. coli* should be used to transform the same MAPKK knocked-out strain, used in the screen, according to the protocol in Subheading 3.3.5.
4. Transformants should be grown under selective conditions to verify the ability of the plasmids selected in the screen to rescue the phenotype (see Note 9).
5. Plasmids that are found to be positive in this test should be sequenced for identifying the point mutation that render the MAPK intrinsically active.

4. Notes

1. It is impossible to autoclave media supplemented with caffeine. Therefore, caffeine in the desired concentration should be prepared separately. Caffeine should be dissolved in pre-autoclaved water, in a minimal volume (about 20 ml), in a sterile Erlenmeyer flask, containing a sterile magnetic stirrer. It is recommended to dissolve the caffeine on a hot plate (not too hot), and keep it on the hot plate until adding it to the media (otherwise it may become solid). If caffeine is not melting, add more water. Alternatively, subtract the volume of water in which you dissolve the caffeine in from the volume of water you add to the media. Only after the media is autoclaved, and cooled (to less than 50°C), the dissolved caffeine should be added. Mix well and pour to plates.
2. Lysozyme will not work efficiently if the pH of the solution is less than 8.0.
3. The precipitate that forms during storage at 0°C consists of chromosomal DNA, high-molecular weight RNA, and potassium/SDS/protein/membrane complexes.

4. If the bacterial debris do not form a tightly packed pellet, recentrifuge at 4,200 × *g* for another 30 min and then transfer as much of the supernatant as possible to a fresh bottle. Discard the viscous liquid remaining in the centrifuge bottle. The failure to form a compact pellet is usually a consequence of inadequate mixing of the bacterial lysate with solution III.
5. Salt may precipitate if centrifugation is carried out at 4°C.
6. Turn on heating block to 42 and 100°C. Boil Salmon sperm for 3 min at 100°C and place it on ice.
7. It is useful to obtain about $2\text{–}5 \times 10^5$ transformants. However, it is important not to plate transformants densely. A maximum of 5,000 transformants should be plated on a 100-mm plate.
8. Number plates by the total number of plates from previous step. For example, if you plated transformants on 25 plates, number plates in this step from 1 to 25. In addition, number plates of each replica series from 1 to 5, according to the sequence of replica plating. For example, plates of the first series will be marked 1.1, 1.2, 1.3, 1.4, and 1.5. Plates of the second series will be marked 2.1, 2.2, 2.3, 2.4, 2.5, etc. Also, mark each plate within a series with an indicative line, located at the same place in each plate, so you can identify the orientation of each plate and compare between cells grown on different plates.
9. In addition to patches originated from putative positive clones, one should also plate on the same plate patches of a wild-type strain (positive control), a knocked-out MAPKK strain harboring an empty vector and same strain harboring the wild-type MAPK (of interest) on a plasmid (two negative controls).
10. In order to prevent heating the samples, do not vortex 5 min continuously. Instead, vortex each sample for 1 min and keep it on ice until you finish to vortex all other samples for 1 min, and then vortex it again for 1 min, and so on.
11. Obtaining colonies following transformation of plasmid DNA extracted from yeast is not an easy task. Therefore, transform 2 and 5 μl of DNA. Plate only half the volume of bacteria onto LB + agar + ampicillin plates. If no colonies appear after overnight incubation, plate the remaining half of bacteria.

References

1. Pearson, G., Robinson, F., Beers Gibson, T., Xu, B. E., Karandikar, M., Berman, K., and Cobb, M. H. (2001) *Endocr Rev* **22**, 153–83.
2. Engelberg, D. (2004) *Semin Cancer Biol* **14**, 271–82.
3. Esteva, F. J., Sahin, A. A., Smith, T. L., Yang, Y., Pusztai, L., Nahta, R., Buchholz, T. A., Buzdar, A. U., Hortobagyi, G. N., and Bacus, S. S. (2004) *Cancer* **100**, 499–506.
4. Kuan, C. Y., and Burke, R. E. (2005) *Curr Drug Targets CNS Neurol Disord* **4**, 63–7.
5. Kumar, S., Boehm, J., and Lee, J. C. (2003) *Nat Rev Drug Discov* **2**, 717–26.
6. Kyriakis, J. M., and Avruch, J. (2001) *Physiol Rev* **81**, 807–69.

7. Cobb, M. H., and Goldsmith, E. J. (1995) *J Biol Chem* **270**, 14843–6.
8. Yoon, S., and Seger, R. (2006) *Growth Factors* **24**, 21–44.
9. Yao, Y., Li, W., Wu, J., Germann, U. A., Su, M. S., Kuida, K., and Boucher, D. M. (2003) *Proc Natl Acad Sci U S A* **100**, 12759–64.
10. Yang, D. D., Conze, D., Whitmarsh, A. J., Barrett, T., Davis, R. J., Rincon, M., and Flavell, R. A. (1998) *Immunity* **9**, 575–85.
11. Tamura, K., Sudo, T., Senftleben, U., Dadak, A. M., Johnson, R., and Karin, M. (2000) *Cell* **102**, 221–31.
12. Sabapathy, K., Hu, Y., Kallunki, T., Schreiber, M., David, J. P., Jochum, W., Wagner, E. F., and Karin, M. (1999) *Curr Biol* **9**, 116–25.
13. Adams, R. H., Porras, A., Alonso, G., Jones, M., Vintersten, K., Panelli, S., Valladares, A., Perez, L., Klein, R., and Nebreda, A. R. (2000) *Mol Cell* **6**, 109–16.
14. Pages, G., Guerin, S., Grall, D., Bonino, F., Smith, A., Anjuere, F., Auberger, P., and Pouyssegur, J. (1999) *Science* **286**, 1374–7
15. Garnett, M. J., and Marais, R. (2004) *Cancer Cell* **6**, 313–9.
16. Thein, S. L., Oscier, D. G., Flint, J., and Wainscoat, J. S. (1986) *Nature* **321**, 84–5.
17. Hayward, N. K., Keegan, R., Nancarrow, D. J., Little, M. H., Smith, P. J., Gardiner, R. A., Seymour, G. J., Kidson, C., and Lavin, M. F. (1988) *Hum Genet* **78**, 115–20.
18. Mansour, S. J., Matten, W. T., Hermann, A. S., Candia, J. M., Rong, S., Fukasawa, K., Vande Woude, G. F., and Ahn, N. G. (1994) *Science* **265**, 966–70.
19. Cowley, S., Paterson, H., Kemp, P., and Marshall, C. J. (1994) *Cell* **77**, 841–52.
20. Marshall, C. J. (1995) *Cell* **80**, 179–85.
21. Canagarajah, B. J., Khokhlatchev, A., Cobb, M. H., and Goldsmith, E. J. (1997) *Cell* **90**, 859–69.
22. Askari, N., Diskin, R., Avitzour, M., Yaakov, G., Livnah, O., and Engelberg, D. (2006) *Mol Cell Endocrinol* **252**, 231–40.
23. Eijsink, V. G., Gaseidnes, S., Borchert, T. V., and van den Burg, B. (2005) *Biomol Eng* **22**, 21–30.
24. Tao, H., and Cornish, V. W. (2002) *Curr Opin Chem Biol* **6**, 858–64.
25. Matsuura, T., and Yomo, T. (2006) *J Biosci Bioeng* **101**, 449–56.
26. Levin-Salomon, V., Maayan, I., Avrahami-Moyal, L., Marbach, I., Livnah, O., and Engelberg, D. (2009) *Biochem J* **417**, 331–40.
27. Bell, M., Capone, R., Pashtan, I., Levitzki, A., and Engelberg, D. (2001) *J Biol Chem* **276**, 25351–8.
28. Levin-Salomon, V., Kogan, K., Ahn, N. G., Livnah, O., and Engelberg, D. (2008) *J Biol Chem* **283**, 34500–10.
29. Silhavy, T. J., Berman, M. L., Enquist, L. W. (1984) *Experiments with Gene Fusion*, Cold Spring Harbor Laboratory, Cold Spring Harbor, NY, pp 75–78.
30. Sambrook, J., Fritsch, E. F., and Maniatis, T. (1989) *Molecular Cloning: A Laboratory Manual*, Cold Spring Harbor Laboratory, Cold Spring Harbor Laboratory, NY.

Chapter 16

Reconstitution of the Nuclear Transport of the MAP Kinase ERK2

Arif Jivan, Aarati Ranganathan, and Melanie H. Cobb

Abstract

The nuclear–cytoplasmic distribution of ERK2 is regulated in response to various stimuli and changes in cell context. Furthermore, the nuclear flux of ERK2 occurs by several energy- and carrier-dependent and -independent mechanisms. ERK2 has been shown to translocate into and out of the nucleus by facilitated diffusion through the nuclear pore, interacting directly with proteins within the nuclear pore complex, as well as by karyopherin-mediated transport. Nuclear export has been suggested to be CRM1- and MEK1/2-dependent. Here, we describe a general nuclear import assay of wild-type ERK2 that can be employed to identify different mechanisms governing nuclear entry of the protein kinase, adapted to evaluate ERK2 mutants that impair nuclear entry to dissect energy- and carrier-dependent and -independent mechanisms, and extended to characterize export mechanisms.

Key words: MAP kinase, MAPK, ERK1/2, Nuclear import, Nuclear export, Ran, Karyopherins

1. Introduction

1.1. Nuclear–Cytoplasmic Distribution of ERK2

The mitogen-activated protein kinase (MAPK) signaling module containing the extracellular signal-regulated kinases 1/2 (ERK1/2) is subject to diverse modes of control within the cell (1). These include modulating activation kinetics of the enzyme, spatial restriction to distinct subcellular compartments and dynamic trafficking to sites of action. One such method of regulation is the nuclear translocation of ERK1/2, in response to various hormones or changes in cell state. Moreover, the mechanisms that dominate the control of ERK1/2 nuclear entry depend upon on cell type, context, activation state, and stimulus specificity. Unphosphorylated, phosphorylated, and active ERK1/2 have distinct but overlapping modes of nuclear entry. Initial studies

Rony Seger (ed.), *MAP Kinase Signaling Protocols: Second Edition*, Methods in Molecular Biology, vol. 661,
DOI 10.1007/978-1-60761-795-2_16,

examining ERK1/2 localization suggested that nuclear translocation occurs upon stimulation with growth factors that cause the phosphorylation and activation of these protein kinases (2). Two groups have shown that unphosphorylated ERK2 can enter the nucleus independently from active import processes that require carrier proteins and the Ran-GTP cycle (3, 4). Instead, the nuclear entry of inactive ERK2 can occur through its direct interactions with the nucleoporins (Nups), proteins of the nuclear pore complex, in a manner independent of transport factors and Ran-GTP. Live-cell imaging studies provide some support for the energy-independent entry mechanism, as well as potentially contradictory views. One study suggested that ERK2 phosphorylation and activation are required for nuclear shuttling because inhibiting the upstream MAPK kinases MEK1/2 with U0126 blocked the nuclear accumulation of ERK2 (5). Fluorescence resonance energy transfer and recovery after photobleaching (FRET, FRAP) showed a steady-state rate of energy-independent entry of ERK2 even in unstimulated cells, confirmed the cytoplasmic association of ERK2 and MEK, and suggested that dissociation from MEK was sufficient to account for growth factor-enhanced nuclear entry (6). Because dissociation from MEK is caused by ERK2 activation, the MEK inhibitor will disfavor their dissociation; this suggests a possible reconciliation of the live-cell findings. Later reconstitution studies indicated that phosphorylated ERK2 can enter by both energy-independent and energy-dependent transport, as the addition of exogenous transport factors and an energy regenerating system enhanced nuclear import of the protein kinase (7). Recently, a Ser–Pro–Ser sequence was identified in ERK1/2 that is phosphorylated and required for importin 7-mediated nuclear translocation upon stimulation with serum or tissue plasminogen activator (8). Several other proteins have been implicated in the nuclear trafficking of ERK1/2.

In addition to the direct facilitated or active nuclear import/export of ERK1/2, nucleocytoplasmic shuttling of ERK1/2 may be functionally coupled to a diverse array of cellular processes, each potentially contributing to a different subset of interactions that facilitate the nuclear entry and exit of ERK1/2. Canonically, ERK1/2 are believed to translocate to the nucleus to participate in the activation of gene transcriptional programs as well as epigenetic regulation events. ERK1/2 signaling has also been implicated in regulating aspects of nuclear transport processes themselves. Notable examples include the nuclear export of certain mRNA transcripts, providing an element of posttranscriptional regulation of gene expression (9), and an indirect effect on the Ran gradient, which is necessary for active import (10). Although inferential, these diverse processes could give rise to distinct binding modalities that regulate ERK1/2 nuclear function as well as trafficking.

1.2. Differential Interactions that Contribute to the Nuclear Import of ERK2

Differential interactions also contribute to the different modes of nuclear import of unphosphorylated, phosphorylated, and/or active ERK1/2. Binding to anchoring proteins in multiple cell compartments contributes substantially to the distribution of ERK1/2. Both MEK and microtubules bind ERK1/2 in the cytoplasm, and their release may facilitate nuclear entry, as suggested above (6, 11, 12). Lamin A/C have been suggested as sites of ERK1/2 sequestration at the nuclear envelope (13). Mxi2, a p38α MAPK splice variant, has been shown to promote the nuclear translocation and accumulation of ERK2 when overexpressed, and is thought to compete with cytoplasmic anchoring proteins (14). Binding directly to nucleoporins may also concentrate ERK2 at the nuclear pore (3, 4, 15). The nuclear import characteristics of a variety of ERK2 mutants defective in known protein interactions, including mutations to the common docking (CD) motif as well as the MAP kinase insert in ERK2, have been tested for impaired nuclear entry using reconstitution experiments such as described here (16). ERK2 CD domain mutants defective in binding to basic/hydrophobic docking (D) motifs were imported into the nucleus normally. However, ERK2 MAP kinase insert mutants that are unable to bind to proteins containing FXF motifs displayed impaired nuclear import kinetics (17). This import defect may be attributed to a decreased ability to bind to the nuclear pore complex proteins, several of which contain FXF motifs.

1.3. Nuclear Import Assay

In vitro nuclear import reconstitution employs digitonin permeabilization of cells to remove the majority of the soluble cytoplasm, leaving the nuclei intact. The protocol calls for REF52 cells, but any cell type may be used. The assay relies on the specificity of digitonin for cholesterol, which being present at a higher concentration in the plasma membrane than in the nuclear envelope (18). Depending on the stringency of washes after permeabilization, soluble components of the cytosol required for nuclear import, including transport factors (i.e., karyopherins/importins), the small GTPase Ran, and nucleotides (i.e., GTP) are largely removed as well. Thus, the protein components must be added back to the reconstitution experiment, either in the form of crude cytosolic extract or purified proteins (19). Additionally, GTP is required for continual nucleocytoplasmic cycling of the Ran GTPase and karyopherins for active nuclear transport to occur (20). Typically, an energy regenerating system is added to the reconstitution reaction to maintain a replenishing source of GTP for the duration of the experiment. Import may be observed in real time or after the desired time has elapsed. The experiment is terminated, and the nuclei are fixed and visualized. If using indirect immunofluorescence, cells will be immunostained and visualized for nuclear signal.

Several different parameters can be measured during the nuclear import assay. Given a user-defined length of time for the reaction, a steady-state analysis can yield relative values of nuclear import between different samples or proteins of interest. The kinetics of nuclear import can be measured by examining a time course, as long as the reaction has not already achieved steady state. Quantification of nuclear import can be performed using any standard image-analysis software that defines a two-dimensional nuclear region/area and measures the intensity of fluorescent signal. This assay can be extended to examine nuclear export. After a fixed time of import under the desired conditions, the import solution may be removed and export is allowed to occur under various test conditions, for example, with and without a source of the export factor, CRM1, which is implicated in export of ERK pathway components (11). Fractional export is determined by comparing to the import detected at the end of the initial incubation period.

2. Materials

2.1. Cell Culture

1. Rat embryonic fibroblasts (Ref) 52 cells are grown on glass coverslips in Dulbecco's Modified Eagle's Medium (DMEM) supplemented with 10% fetal bovine serum (FBS), 1% L-glutamine, 100 units/mL penicillin, and 100 μg/mL streptomycin at 37°C in 10% CO_2.
2. To decrease the amount of endogenous active ERK1/2 in import reconstitution assays, cells can be placed in serum-free DMEM for 2–4 h.

Optional:

3. Epidermal growth factor (EGF; BD Biosciences): 100 μg/mL stock dissolved in H_2O. Aliquot and store at −80°C up to 6 months.
4. Phorbol 12-myristate 13-acetate (PMA; Sigma): 100 μM working stock dissolved in dimethyl sulfoxide (DMSO).

2.2. Nuclear Import Reconstitution

1. 1× Transport buffer (TB): 20 mM HEPES pH 7.3, 110 mM potassium acetate, 2 mM magnesium acetate, 1 mM ethylene glycol-bis(b-aminoethyl ether)-*N*,*N*,*N*′,*N*′-tetraacetic acid (EGTA), and 2 mM dithiothreitol (DTT). A 10× stock of transport buffer (TB; 200 mM HEPES pH 7.3, 1.1 M potassium acetate, 20 mM magnesium acetate, 10 mM EGTA) can be made and filter sterilized using a 0.22 μm filter for long-term storage. Add DTT fresh to a 1× working solution. 1× TB can be made in advance and stored at 4°C for no more than 1 week due to the stability of DTT in solution.

2. Permeabilization solution: 1× TB, 70 μg/mL digitonin.
3. 1× Tris-buffered saline (TBS): 25 mM Tris–HCl pH 7.4, 150 mM NaCl, 2 mM KCl. TBS is typically made as a 10× stock and diluted with deionized H_2O, prior to use.
4. 1× TBST: 1× TBS, 0.1% (v/v) Tween 20. Make fresh, prior to use.
5. Fixation: 3% paraformaldehyde (PFA) (v/v) in TBS.
6. Energy regeneration system (listed as final concentrations in the import reaction mixture): 1 mM ATP, 1 mM GTP, 5 mM phosphocreatine, 20 units/mL creatine phosphokinase.
7. Source of import factors: REF52, HeLa (7) or Xenopus oocyte (19) cytosol. Cytosol is dialyzed in 1× TB prior to use and can be stored at −80°C for up to 1 year. HeLa cytosol is used at a final concentration of 2.5 mg/mL in the import reaction mixture. Alternatively, recombinant import factors may be used as follows (at the indicated final concentrations in the import reaction mixture): 0.5 μM karyopherin-α2, 0.25 mM karyopherin-β1, 2 μM Ran GTPase, 0.4 μM p10/NTF2. Other factors may be substituted as desired.
8. His_6-tagged GFP and GFP-ERK2 wild type and mutants are expressed and purified as described (7, 16). Tetramethylrhodamine B isothiocyanate (TRITC)-NLS-BSA is generated as described (7). Purification of pERK2 is as described (21). All recombinant, purified protein stocks are stored at −80°C.

2.3. Immunofluorescence and Visualization

Import reconstitution assays are visualized using epifluorescence microscopy. If indirect immunofluorescence is employed to visualize nuclear ERK2, the following reagents must be used:

1. Blocking buffer: 1% (w/v) bovine serum albumin (BSA) in 1× TBST. Buffer should be made fresh prior to use and kept at 4°C.
2. Mouse monoclonal pERK1/2 antibody (Sigma). Stored at −20°C. Working aliquots may be kept at 4°C no longer than 2 months.
3. Rabbit polyclonal ERK1/2 antibody (multiple sources). Store at 4°C.
4. 4′,6-diamidino-2-phenylindole (DAPI). Store at 4°C.
5. Alexa Fluor 546 nm anti-mouse or anti-rabbit secondary antibody (Molecular Probes). Light sensitive and stored at 4°C.

The following are required for all assays:

6. Mounting solution: Aqua-Poly/Mount (Polysciences, Inc.).
7. Clear nail polish.

8. To quantify nuclear intensity: Slidebook 4.1 software (Intelligent Imaging Innovations, Inc.) or Image J (NIH; http://rsb.info.nih.gov/ij/).

3. Methods

3.1. Cell Culture

1. Place approximately 6–10 glass coverslips into a 10 cm diameter dish for culturing cells. Dip coverslips in 100% ethanol, wick off excess EtOH on a paper towel, and flame dry prior to placing in dish.
2. Plate REF52 cells onto coverslips 24 h prior to experiment, to yield approximately 60–70% confluent population on the day of the experiment. Culture cells to desired confluency at 37°C under 10% CO_2 atmosphere (see Note 1).
3. To diminish endogenous ERK1/2 activity: aspirate serum-containing medium and replace with serum-free medium for 2–4 h (see Note 2). Cells can be treated with diverse ligands/stimuli at this step to examine the differential import of unphosphorylated ERK1/2 versus phosphorylated, active ERK1/2. Ligands include: 10–100 ng/mL EGF, 10–100 nM PMA, or others as desired.
4. Figure 1 provides an overview of the nuclear import reconstitution assay.

3.2. Cell Permeabilization

1. Examine the coverslips under a brightfield microscope to ensure that cells are attached to the coverslips. Transfer the plate of REF52 cells from tissue culture incubator to ice.
2. Aspirate medium (see Note 3).
3. Wash cells twice briefly with 5 mL cold 1× TB; aspirate solution after each wash.
4. Transfer coverslips into a 24-well plate (one coverslip per well).
5. Add 1 mL of cold permeabilization solution to each well and incubate on ice for 5 min (see Note 4).
6. Aspirate the permeabilization solution.
7. Wash twice briefly with 1× TB; aspirate buffer after each wash.

3.3. Nuclear Import of ERK2

1. Prepare import reaction mixture: 40 μL of protein-containing solution for each coverslip.
   (a) Substrates: wild-type or mutant GFP-ERK2 (or untagged ERK2) is used at a final concentration of 0.8 μM diluted in 1× TB. TRITC-NLS-BSA at 0.14 μM (final concentration) is used as a positive control. GFP-alone at 0.8 μM is used as a negative control (see Note 5).

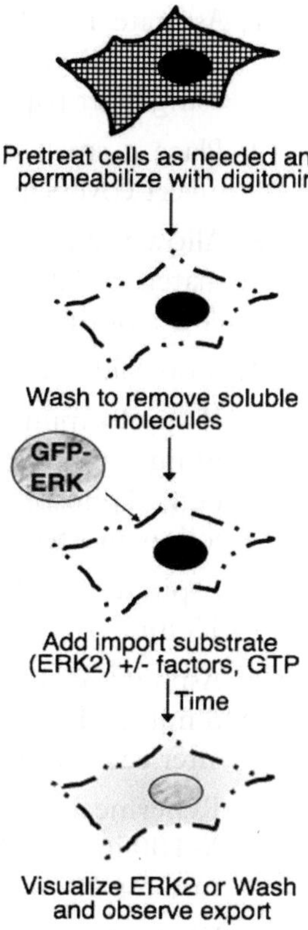

Fig. 1. Schematic of reconstitution of nuclear import and export of GFP-ERK2.

   (b) In samples requiring transport factors and an energy regeneration system, use necessary proteins and other components at the concentrations indicated in Subheading 2.2.

2. Prepare a humidified chamber and pipette 40 μL of the import reaction mixture onto the parafilm at the bottom of the chamber (see Note 6).
3. Place coverslips on top of the import reaction mixture drops, seal the top of the chamber, and incubate at room temperature for 15 min or desired time, protected from light.
4. To terminate the reaction, float the coverslips from the parafilm surface by pipetting 200 μL of cold 1× TB underneath the coverslip.
5. Transfer coverslips back to the 24-well plate (one coverslip per well) containing cold 1× TB in each well, and place the plate on ice.

3.4. Fixation and Mounting

If using GFP-tagged recombinant proteins (otherwise start at step 5 below):

1. Aspirate 1× TB and add 1 mL 3% PFA. Incubate on ice for 15 min. Remove coverslip from fixative and wick off excess using filter paper.
2. Place a small drop of Aqua-Poly/Mount onto a slide and invert coverslip onto the mounting medium.
3. Allow the coverslip to dry at room temperature for approximately 5–10 min, and then apply clear nail polish around the edges of the coverslip.
4. Store slides at 4°C.

   If performing indirect immunofluorescence experiments using antibodies to detect and distinguish nonfluorescently tagged phosphorylated, active ERK2 and/or total ERK2, the following steps should be performed:
5. Aspirate 1× TB and add 1 mL 3% PFA. Incubate on ice for 15 min.
6. Aspirate 3% PFA solution and wash the coverslips three times 5 min each in 1× TBS at room temperature (aspirating buffer after each wash – applicable for each subsequent step).
7. Repermeabilize the fixed cells by adding 1 mL of 0.5% Triton X-100 in 1× TBS to each well and incubate at room temperature for 10 min.
8. Wash twice for 5 min each time with 1× TBS at room temperature (see Note 7).
9. Wash with 1× TBST (note: 1× TBS, 0.1% (v/v) Tween 20) for 5 min at room temperature.
10. Incubate coverslips with blocking buffer for 30 min at room temperature.
11. Dilute primary antibody (pERK1/2 or ERK1/2) 1:300 in blocking buffer.
12. Transfer coverslips to humidified chamber, placing them cell-side up on the parafilm-covered bottom surface.
13. Add 80 μL of diluted primary antibody solution on top of each coverslip (see Note 8).
14. Incubate coverslips for 24 h at 4°C.
15. Transfer coverslips to a new 24-well plate and wash three times 10 min each with 1× TBST at room temperature.
16. Prepare secondary antibody mixture: 1:3,000 Alexa Fluor 546 nm anti-mouse or anti-rabbit secondary antibody (for pERK1/2 and ERK1/2 respectively) in blocking buffer. DAPI can be added at a 1:5,000 dilution to visualize DNA and nuclear boundary.
17. Once again, place coverslips into a humidified chamber and add 80 μL of secondary antibody mixture on top of each coverslip.

18. Incubate the coverslips for 30 min at 37°C or 1 h at room temperature. Wrap the humidified chamber in aluminum foil or store enclosed in a dark environment to prevent bleaching from light (see Note 9).
19. Wash three times 10 min each with 1× TBST.
20. Wash briefly with H_2O prior to mounting coverslips onto slides with Aqua-Poly/Mount. Store slides at 4°C.

3.5. Visualizing and Quantifying Nuclear Import

1. Observe fluorescent signal by epifluorescence microscopy using 63× oil objective (see Note 10).
2. Figure 2 is a representative micrograph of GFP-ERK2 nuclear import in REF52 cells.
3. Figure 3 is a representative micrograph of GFP-ERK2 nuclear export.
4. Quantification of import assays and micrographs using Image J:

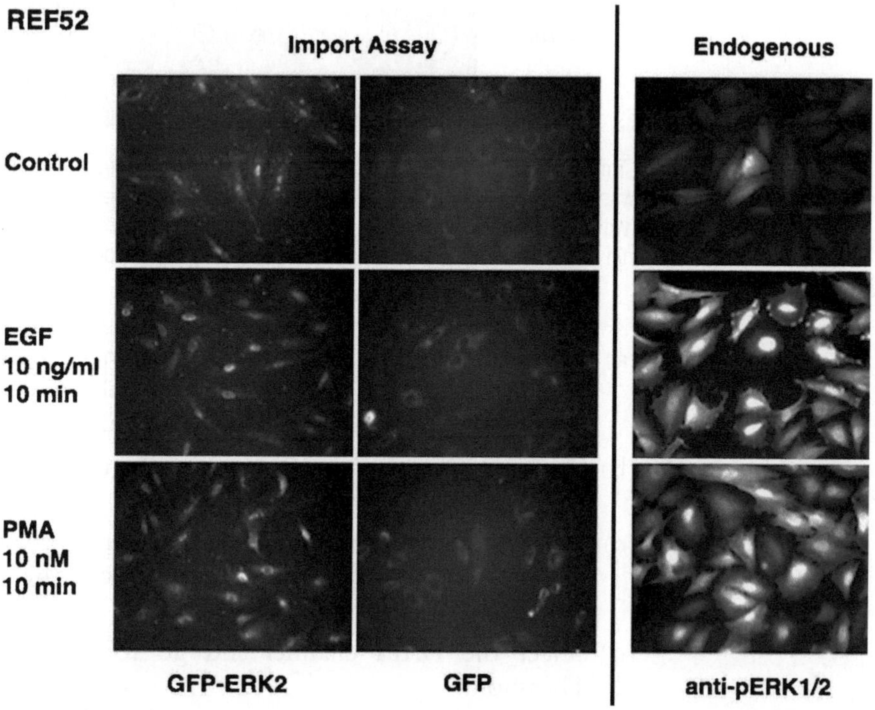

Fig. 2. Nuclear import of GFP-ERK2 compared to localization of endogenous ERK1/2. *Left panels*: Nuclear import of GFP-ERK2 was imaged in REF52 cells that were unstimulated or stimulated for 10 min with 10 ng/mL EGF or 10 nM PMA prior to permeabilization and reconstitution. *Middle panels*: A comparable experiment using GFP to show that it is excluded from the nucleus under all conditions. TRITC-NLS-BSA was used as a positive control and is imported into the nucleus as long as transport factors and GTP are present (not shown). *Right panels*: Immunofluorescence of endogenous phosphorylated ERK1/2 (pERK1/2) in cells that were untreated or stimulated with EGF or PMA as above. No permeabilization or recombinant proteins were used. In some cell types, EGF does not induce nuclear accumulation of pERK1/2.

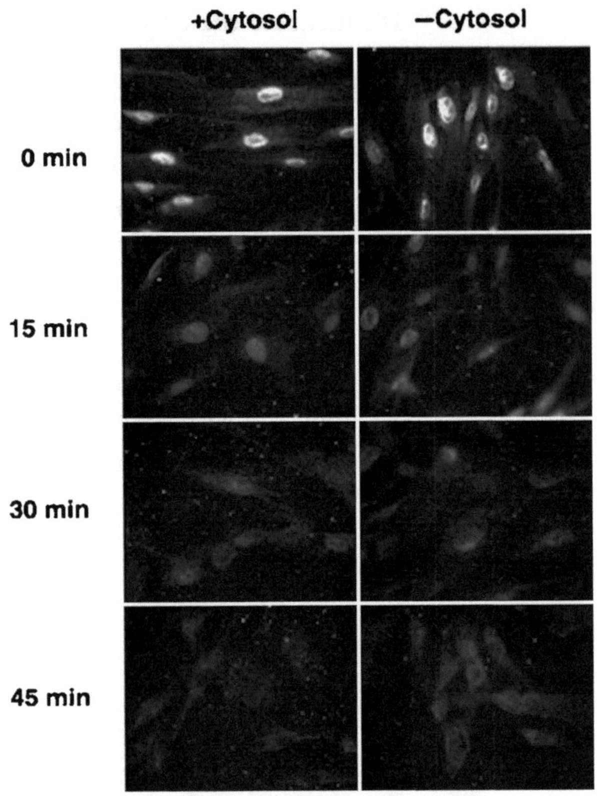

Fig. 3. Time course of GFP-ERK2 nuclear export with and without added cytosol. GFP-ERK2 was imported as in Fig. 2. Cells were quickly washed and placed in fresh transport buffer with or without cytosol as in 7. GFP-ERK2 remaining in nuclei was assessed over a 45-min time course. Cytosol provides a source of Ran and export factors.

(a) Open the captured image file using Image J. A micrograph of the DAPI channel is required along with one of the protein to be quantified (i.e., GFP-ERK2).

(b) Convert image to 8 bit monochrome (Image/Type/8bit).

(c) Auto-adjust the brightness/contrast on both channels (Adjust/Brightness/Contrast and click the "Auto" button).

(d) Select the DAPI channel and go to Image/Adjust/Threshold.

(e) Change the drop down menu from "Red" to "Over/Under". The DAPI image should have a blue mask around the nuclei (which should remain open). Adjust the slider bars in the Threshold window to visualize all possible nuclei in the field.

(f) Select the "magic wand" tool from the Image J toolbar.

(g) While holding down the shift key, select each nucleus in the micrograph, resulting in a yellow outline around each selected nucleus. This defines the nuclei as the regions of interest (ROI).

(h) Next, in the Analyze/Tools/ROI manager menu, click the "Add" button and select the file name corresponding to the micrograph of the protein of interest (i.e., GFP-ERK2 channel).

(i) In the ROI manager window, click on the added data set, and the yellow nuclear outlines will be highlighted on the GFP-ERK2 channel.

(j) To obtain intensity characteristics on the GFP-ERK2 channel, select Analyze/Measure. The resultant window will display area, mean, minimum and maximum intensity measurements for the defined ROI.

4. Notes

1. Populations of isolated single cells as well as cells with multiple cell–cell contacts should be present on coverslips as visualized under a brightfield microscope. The doubling time of cells is variable, depending on their passage number and confluency. Many other cell lines such as HeLa, human embryonic kidney (HEK) 293, and BJ (human foreskin fibroblasts immortalized with h-TERT) have been utilized for reconstitution assays as well. The number of cells plated should be optimized to yield desired confluency on the day of the experiment.
2. Serum starvation is critical if using antiphospho-specific antibodies to examine nuclear transport of a phosphorylated subset of endogenous or exogenous protein population.
3. Make sure to aspirate medium from the side of the well as opposed to directly over the coverslip to prevent aspirating cells.
4. Do not permeabilize cells longer than 5 min because the nuclear envelope may be damaged.
5. TRITC-conjugated BSA coupled to a nuclear localization signal (NLS) is used as a positive control substrate for energy- and transport factor-dependent nuclear import. Purified His_6-GFP is used as a negative control and measured as background in the assay. Wheat germ agglutinin (WGA), which binds to glycosylated proteins in the nuclear pore complex (NPC), thus occluding the nuclear pore, can be used to inhibit nuclear

import/export through the NPC. If cells treated with WGA prior to (or in conjunction with) import reconstitution display no nuclear signal, or if nuclear accumulation is observed when cells are treated with WGA prior to (or in conjunction with) the export reaction, one can expect that transport occurs through the NPC rather than as a result of disruption of the nuclear envelope. Apyrase, a nucleotide hydrolase, can be added to the import/export reaction to further verify the necessity of energy (i.e., nucleotide) to facilitate active nuclear transport.

6. To make a humidified chamber: wet Whatman paper (or a paper towel) and place it at the bottom of a sealable container (Tupperware or other comparable ones). Drain excess water. Cover wet surface with Parafilm.
7. Rocking is not required during the wash steps. Cells may come off of the coverslips.
8. Use a secondary-antibody alone mixture as a negative control for immunostaining. Thus, at the primary antibody incubation step, pipette blocking buffer alone (without primary antibody) onto one coverslip.
9. All subsequent wash steps should be performed in the dark (e.g., by wrapping the plate with aluminum foil).
10. Nuclei should be visualized using epifluorescence microscopy (nonconfocal) as it allows for examining total nuclear fluorescence signal rather than from a particular nuclear plane. Although 63× under oil immersion is recommended, magnification is determined by user preference.

Acknowledgments

We would like to acknowledge grant DK34128 from the National Institutes of Health and I1243 from the Welch Foundation for support of work from the Cobb Laboratory.

References

1. Whitehurst, A. W., Cobb, M. H., and White, M. A. (2004). Stimulus-coupled spatial restriction of extracellular signal-regulated kinase1/2 activity contributes to the specificity of signal-response pathways. *Mol Cell Biol* **24**, 10145–10150.
2. Chen, R. Y., Sarnecki, C., and Blenis, J. (1992). Nuclear localization and regulation of erk and rsk-encoded protein kinases. *Mol Cell Biol* **12**, 915–927.
3. Matsubayashi, Y., Fukuda, M., and Nishida, E. (2001). Evidence for existence of a nuclear pore complex-mediated, cytosol-independent pathway of nuclear translocation of ERK MAP kinase in permeabilized cells. *J Biol Chem* **276**, 41755–41760.
4. Whitehurst, A. W., Wilsbacher, J. L., You, Y., Luby-Phelps, K., Moore, M. S., and Cobb M. H. (2002). ERK2 enters the nucleus by a

carrier-independent mechanism. *Proc Natl Acad Sci U S A* **99**, 7496–7501.

5. Costa, M., Marchi, M., Cardarelli, F., Roy, A., Beltram, F., Maffei, L., and Ratto, G. M. (2006). Dynamic regulation of ERK2 nuclear translocation and mobility in living cells. *J Cell Sci* **119**, 4952–4963.
6. Burack, W. R. and Shaw, A. S. (2005). Live cell imaging of ERK and MEK: Simple binding equilibrium can explain ERK's regulated nucleocytoplasmic distribution. *J Biol Chem* **280**, 3832–3837.
7. Ranganathan, A., Yazicioglu, M. N., and Cobb, M. H. (2006). The nuclear localization of ERK2 occurs by mechanisms both independent of and dependent on energy. *J Biol Chem* **281**, 15645–15652.
8. Chuderland, D., Konson, A., and Seger, R. (2008). Identification and characterization of a general nuclear translocation signal in signaling proteins. *Mol Cell* **31**, 850–861.
9. Phelps, M., Phillips, A., Darley, M., and Blaydes, J. P. (2005). MEK–ERK signaling controls Hdm2 oncoprotein expression by regulating hdm2 mRNA export to the cytoplasm. *J Biol Chem* **280**, 16651–16658.
10. Yoon, S. O., Shin, S., Liu, Y., Ballif, B. A., Woo, M. S., Gygi, S. P., and Blenis, J. (2008). Ran-binding protein 3 phosphorylation links the Ras and PI3-kinase pathways to nucleocytoplasmic transport. *Mol Cell* **29**, 362–375.
11. Adachi, M., Fukuda, M., and Nishida E. (2000). Nuclear export of MAP kinase (ERK) involves a MAP kinase kinase (MEK)-dependent active transport mechanism. *J Cell Biol* **148**, 849–856.
12. Reszka, A. A., Seger, R., Diltz, C. D., Krebs, E. G., and Fischer, E. H. (1995) Association of mitogen-activated protein kinase with the microtubule cytoskeleton. *Proc Natl Acad Sci U S A* **92**, 8881–8885.
13. Gonzalez, J. M., Navarro-Puche, A., Casar, B., Crespo, P., and Andres, V. (2008) Fast regulation of AP-1 activity through interaction of lamin A/C, ERK1/2, and c-Fos at the nuclear envelope. *J Cell Biol* **183**, 653–666.
14. Casar, B., Sanz-Moreno, V., Yazicioglu, M. N., Rodriguez, J., Bercioano, M. T., Lafarga, M., Cobb, M. H., and Crespo, P. (2007). Mxi2 promotes stimulus-independent ERK nuclear translocation. *EMBO J* **26**, 635–646.
15. Vomastek, T., Iwanicki, M. P., Burack, W. R., Tiwari, D., Kumar, D., Parsons, J. T., Weber, M. J., and Nandicoori, V. K. (2008). Extracellular signal-regulated kinase (ERK2) phosphorylation sites and docking domain on the nuclear pore complex protein Tpr cooperatively regulate ERK2–Tpr interaction. *Mol Cell Biol* **28**, 6954–6966.
16. Yazicioglu, M. N., Goad, D. L., Ranganathan, R., Whitehurst, A. W., Goldsmith, E. J., and Cobb, M. H. (2007). Mutations in ERK2 binding sites affect nuclear entry. *J Biol Chem* **282**, 28759–28767.
17. Lee, T., Hoofnagle, A. N., Kabuyama, Y., Stroud, J., Min, X., Goldsmith, E. J., Chen, L., Resing, K. A., and Ahn, N. G. (2004). Docking motif interactions in MAP kinases revealed by hydrogen exchange mass spectrometry. *Mol Cell* **14**, 43–55.
18. Colbeau, A., Nachbaur, J., and Vignais, P.M. (1971). Enzymatic characterization and lipid composition of rat liver subcellular membranes. *Biochim Biophys Acta* **249**, 462–492.
19. Moore, M. S. and Blobel, G. (1992). The two steps of nuclear import, targeting to the nuclear envelope and translocation through the nuclear pore, require different cytosolic factors. *Cell* **69**, 939–950.
20. Melchior, F., Paschal, B., Evans, J., and Gerace, L. (1993). Inhibition of nuclear protein import by nonhydrolyzable analogues of GTP and identification of the small GTPase Ran/TC4 as an essential transport factor. *J Cell Biol* **123**, 1649–1659.
21. Khokhlatchev, A., Canagarajah, B., Wilsbacher, J. L., Robinson, M., Atkinson, M., Goldsmith, E., and Cobb, M. H. (1998). Phosphorylation of the MAP kinase ERK2 promotes its homodimerization and nuclear translocation. *Cell* **93**, 605–615.

Chapter 17

Localization and Trafficking of Fluorescently Tagged ERK1 and ERK2

Matilde Marchi, Riccardo Parra, Mario Costa, and Gian Michele Ratto

Abstract

The action of ERK1 and ERK2 activity on the nuclear substrates requires crossing the nuclear envelope and the localization of phospho-ERK into the nucleus. The nucleo-cytoplasmic trafficking of ERK is therefore crucial for the correct functioning of the pathway. Indeed, this step is necessary for the correct control of gene expression by growth-factors, for morphological transformation of fibroblasts and for neurite extension in PC12. Furthermore, disruption of ERK2 localization in the nucleus severely affects the transduction of ERK2 signaling. This process has now been observed and quantitatively measured by expressing fluorescently tagged ERK1 and ERK2. These experiments provide important insight on the operation of these signaling modules and have revealed an hitherto unknown functional difference between ERK1 and ERK2.

Key words: MAP Kinase, ERK, FRAP, GFP, Nuclear envelope

1. Introduction

The Extracellular Regulated Kinase 1 and 2 (ERK1 and 2) convert a variety of extracellular stimuli into a complex set of cellular responses, regulating processes as diverse as proliferation, differentiation, and synaptic plasticity (1–4). ERK1/2 acts on both cytosolic and nuclear targets, and the list of the downstream partners, nuclear and cytosolic, is ever expanding, underscoring the complexity and pervasiveness of their action. The phosphorylation of nuclear substrates is essential for the induction of specific programs

Matilde Marchi and Riccardo Parra have contributed equally to this chapter

Rony Seger (ed.), *MAP Kinase Signaling Protocols: Second Edition*, Methods in Molecular Biology, vol. 661, DOI 10.1007/978-1-60761-795-2_17, © Springer Science+Business Media, LLC 2010

of gene expression: this signalling requires that ERK1 and 2, upon stimulation, translocate into the nucleus to carry on the activation of the nuclear targets (5). Notwithstanding the importance of this process, little is known about the modalities, time course and regulation of ERK exchange between nucleus and cytoplasm in living cells. In the past few years, it has been demonstrated that this process is rapid and that it is regulated by activity (7–10).

By expressing low levels (<150 nM) of fluorescently tagged ERK2 in living fibroblasts, we have shown that nuclear concentration can change bidirectionally with a time constant of a few minutes and that this process requires continuous MEK activity upstream of ERK. The rate of ERK-GFP shuttling between nucleus and cytoplasm can be determined by Fluorescence Recovery After Photobleaching (FRAP) experiments, and these experiments have determined that shuttling is accelerated after ERK activation, becoming fast enough not to be rate-limiting for translocation (9).

The characterization of ERK1 and 2 trafficking by live cell imaging contributes to clarify another important issue. The large degree of overlap between ERK1 and ERK2 is witnessed by their 85% of amino acid identity and by the identity of the upstream activators and downstream substrates (11–13). Although these data suggest that ERK1 and 2 are functionally equivalent, several studies have shown critical functional differences between these two proteins. Notably, while the genetic ablation of ERK2 in mice results in embryonic lethality, loss of ERK1 only causes subtle effects (14, 15). Significant differences between the two kinases also appear in the control of cell growth, in cultured fibroblasts (16) and hepatocytes, and in liver tumor (17, 18).

One interpretation of these data is that the apparent differences between ERK1 and ERK2 are merely due to their different levels of expression (13): since ERK1 is much less abundant than ERK2, its ablation can be fully compensated by ERK2. The opposite cannot occur, since ERK1 is expressed more weakly. This model is consistent with the fact that ERK1 and ERK2 are interchangeable in vitro, but there is another determinant of nuclear signaling that cannot be replicated in vitro: to act on the nuclear downstream targets, ERK1 and ERK2 must remain phosphorylated during their permanence in the nucleus. Since dephosphorylation prevails in the nucleus, the maintenance of nuclear activity requires continuous shuttling of activated protein from the cytoplasm. Indeed, the speed of nucleo-cytoplasmic trafficking sets the transfer efficiency of the biochemical information carried by the activated ERK to the nuclear compartment. For this reason, different nuclear-cytoplasmic trafficking of ERK1 and 2 would cause a differential signaling capability.

The analysis of the FRAP experiments on cells transfected with either ERK1-GFP or ERK2-GFP demonstrated that ERK1 shuttles through the nuclear membrane far more slowly than ERK2, thus suggesting that ERK1 has less signaling capability than ERK2 toward the nuclear targets (6). Here, we review the basic tools used for these measures in living cells.

2. Materials

2.1. Cell Culture

2.1.1. NIH-3T3: Culture Medium

1. Dulbecco's Modified Eagle's Medium (DMEM, Invitrogen), supplemented with 10% Fetal Bovine Serum (FBS, Sigma) and antibiotics (100 units/mL penicillin – 0.1 mg/mL streptomycin, Sigma).
2. Glass disks 14 mm or Willco dish (Willco Wells) (see Note 1).

2.1.2. NIH-3T3: Medium for Cell Starvation

1. DMEM supplemented with 1% FBS and 100 units/mL penicillin – 0.1 mg/mL streptomycin.

2.1.3. PC12: Culture Medium

DMEM supplemented with 10% Horse Serum and 5% Fetal Bovine Serum and antibiotics (100 units/mL penicillin – 0.1 mg/mL streptomycin, Sigma). Glutamine 2 mM final concentration (Invitrogen). Nerve Growth Factor for differentiation; final concentration 50 nM (Alomone Labs).

2.1.4. Imaging Solution: HEPES Buffer

130 mM NaCl, 3.1 mM KCl, 1.0 mM K_2HPO_4, 4.0 mM $NaHCO_3$, 5.0 mM dextrose, 1.0 mM $MgCl_2$, 2.0 mM $CaCl_2$, 10 mM HEPES, 1.0 mM ascorbic acid, 0.5 mM myo-inositol, 2 mM pyruvic acid, pH 7.3 (see Notes 2 and 3).

3. Cell Transfection

1. Box of ice.
2. Lipofectamine 2000 (Invitrogen) or GenePORTER 2 (Genlantis).
3. Plasmids.
4. OptiMEM (Invitrogen).

4. Methods

4.1. ERK-GFP Fusion Protein

Fluorescently tagged plasmids can be obtained by ATCC or can be produced locally. In brief, ERK RNA was amplified by one-step reverse transcription-polymerase chain reaction (RT-PCR) from total RNA extracts obtained from mouse brain.

The amplification products were ligated to the corresponding restriction site in the vectors of the fluorescent protein of choice to fuse the fluorescent reporter to the *N*-terminus of ERK1 or ERK2. The *C*-terminal fusion proteins were obtained by PCR amplification, removing the ERK stop codon from the reverse primer. The amplification products were ligated to the vector expressing the fluorescent protein (see (6, 9) for further details). Verification of correct sequence and framing was obtained from forward- and reverse-automated fluorescence sequencing. In some cases we used constructs purchased from American Type Culture Collection (ATCC). It is a good practice to repeat the experiments with at least two different constructs, and, if possible, to verify whether the *C* and *N*-fusion proteins return similar results.

4.2. Cell Preparation and Culture

The NIH-3T3 cell line was cultured in modified Dulbecco's medium (see Subheading 2) in 5% CO_2.

Rat pheochromocytoma PC12 cells were cultured in modified Dulbecco's medium in 5% CO_2. PC12 were induced to differentiate for 7–10 days by adding NGF to a final concentration of 50 ng/mL.

The cells were plated on glass disks or on glass-bottomed dish at a density of about 60,000–80,000 cells/mL, until about 60–70% confluence.

4.3. Transfection

1. For each plasmid to transfect two Eppendorf tubes are prepared as follows: 0.5 mL of OptiMEM per well are added in each of the two Eppendorf tubes. For example, to transfect three wells, 1.5 mL (0.5 × 3) are added in each Eppendorf tube. Three coverslips can be placed in each well. If the cells are plated on the Willco dish, each requires 75 μL of Optimem in each Eppendorf tube.
2. In one of the two Eppendorf tubes, 2 μg (per well) (see Note 2) of DNA plasmid is added and mixed. For example, to transfect three wells, 6 μg (2 × 3) are added. Each Willco dish requires 1 μg of DNA.
3. In the second Eppendorf tube, 4 μL (per well) or 2 μL (per dish) of Lipofectamine 2000 is added and mixed (see Note 4). Both Eppendorf tubes are kept for 5 min at RT.
4. Both solutions are thoroughly mixed together and left for 30 min at RT.
5. The conditioned medium is removed from wells (see Note 5) and collected in a Falcon tube (for PC12 only).
6. Rapidly 1 mL (150 μL per dish) of transfecting mix is carefully added to each well.
7. Plates or dishes are then incubated for 30 min at 37°C.

8. The collected conditioned medium is centrifuged for 10 min at 1,500*g* (PC12 only).
9. During medium centrifugation, plates are rinsed twice with 1 mL of OptiMEM at room temperature.
10. In each well OptiMEM is removed and, depending on its dimension, 1.0 mL or 1.5 mL of medium (NIH 3T3) or centrifuged supernatant (PC12) is added.

After transfection, cells are left undisturbed for 24 h before any further experimental manipulation. When required, cells were starved by keeping them in 1% FBS for 24 h. At least 1 h before the experiment, cells are placed in the imaging medium (either saline or minimum medium) and are gently brought to the temperature of the microscope stage.

4.4. Imaging and Measure of Nuclear Trafficking

4.4.1. Imaging of ERK-GFP in Living Cells

The confocal microscope is an ideal tool for time lapse imaging of living cells. Most instruments offer the possibility of executing the described experiments even in the absence of custom software modules; of course, the operative details depend on the specific instrument. Cells must be secured very stably under the microscope and all possible sources of mechanical artifacts have to be removed (see Note 6). If possible cells should be contained within a microscope stage incubator, maintain correct temperature, humidity, and CO_2 concentration. However, if an incubator is not available, cells can be imaged for a few hours at room temperature in an HEPES buffered imaging solution. These experiments can be performed both with upright or inverted microscopes; however, some specific considerations need to be made.

Upright microscope: A water immersion lens must be used in order to be able to immerge the lens in the imaging medium. These lenses are characterized by a white terminal cone that is compatible with immersion in saline solution and that has a low thermal conductibility, which is especially important if the experiment is performed on a heating stage. The numerical aperture of these lenses is generally between 0.8 and 0.9. Cells can be seeded both on Willco dishes or on coverslips, since imaging is not performed through the bottom.

Inverted microscope: Imaging is performed through the coverslip, so it is much better and more practical to use the Willco dish. Oil immersion lenses can be used, offering a better numerical aperture than the water immersion lenses, and this helps in increasing the collection of fluorescence, thereby improving the sensitivity of detection. It is also easier to maintain the cells in a correct environment in an enclosed microscope incubator, since the lens is not immersed in the solution.

Cells must be selected by measuring the fluorescence of ERK-GFP, to avoid cells with excessive overexpression (see next section). Given the weak fluorescence and the need to contain photobleaching, detection sensitivity must be optimized at the expense of resolution by fully opening the confocal aperture and using a wide-emission bandpass. In these conditions it is common to observe fluorescent debris overimposed on the cells. During quantification of fluorescence, these artefacts should be avoided; the possible presence of autofluorescence of the cells have to be carefully measured to estimate the average fluorescence background that must be measured on nontransfected cells and must be subtracted from all measurements. Obviously this requires that all imaging session must be performed in similar conditions.

To evaluate the distribution of ERK-GFP between the nuclear and cytosolic compartments, the average fluorescence of the nucleus (F_{Nuc}) and of a surrounding ring of thickness approximately equal to the nucleus radius (F_{Ring}) can be measured. The parameter defined as Concentration Index (CI) is computed as:

$$\mathrm{CI} = (F_{\mathrm{Nuc}} - \mathrm{BG}) / (F_{\mathrm{Ring}} - \mathrm{BG}), \quad (1)$$

where *BG* is the average background.

4.4.2. Effects of the Expression Level on Localisation and Trafficking: Calibration of Fluorescence

The subcellular localisation of ERK-GFP in bright cells is rather surprising, since it is concentrated in the nucleus (Fig. 1a), independently on the stimulation of the ERK pathway. This ectopic localization is due to the disruption of the relative ratios of MEK and ERK concentration (5, 19, 20). In previous studies on trafficking or localization of tagged ERK, this problem was either ignored (8) or overcomed by coexpressing MEK1 to rescue this unbalance (21, 22), thus introducing another exogenous protein. An alterna-

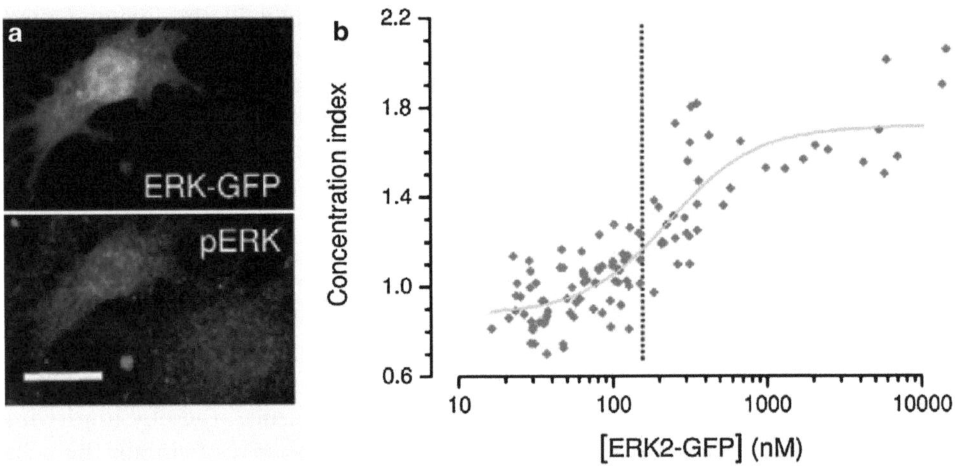

Fig. 1. (**a**) NIH 3 T3 have been transfected with ERK2-GFP, starved in 1% serum and processed for immunofluorescence against pERK. Brightly fluorescent cells show a strong accumulation in the nucleus (*upper panel*), even in the absence of phosphorylation, as shown by the uniform staining of the *lower panel*. (**b**) Cumulative results. Each symbol represents the ERK2-GFP localization in function of the concentration. For values below about 150 nM, the localization is independent on the concentration, and this is the upper limit used in all the following experiments. Modified from reference (9).

tive is to minimally perturb the system by identifying the upper limit of ERK-GFP expression compatible with a normal ERK localization in starved cells. In Fig. 1b, the localization of ERK2-GFP is plotted as a function of the total fluorescence of each cell. The corresponding ERK2-GFP concentration can be computed after a calibration process that is described below. This plot shows that at low concentrations ERK2-GFP is mainly localized in the cytoplasm, as it should be. This allows to estimate the maximum level of fluorescence (and therefore of ERK overexpression) that is admissible for a physiological localization of ERK. This effect was saturable, since we never observed a concentration index CI > 2.2, even in extremely bright cells. It is quite useful to refine the transfection protocol in order to avoid an excessive level of expression, and, to this effect, a plot as in Fig. 1b can result quite useful.

4.4.3. Calibration of the Imaging Setup

Although the above analysis does not require to know the relationship between measured fluorescence and concentration of ERK-GFP, this can be obtained by comparing the cell fluorescence with a known standard. Artificial cells are prepared by dispersing a water solution of recombinant GFP, at known concentration, in mineral oil, and a calibration curve can be obtained relating the measured fluorescence to the GFP concentration (Fig. 2). These artificial cells are imaged exactly in the same conditions of the experiments on living cells. That means objective, filter set, pinhole diameter, laser setting, and photomultiplier settings must be exactly matched. Furthermore, it would be useful

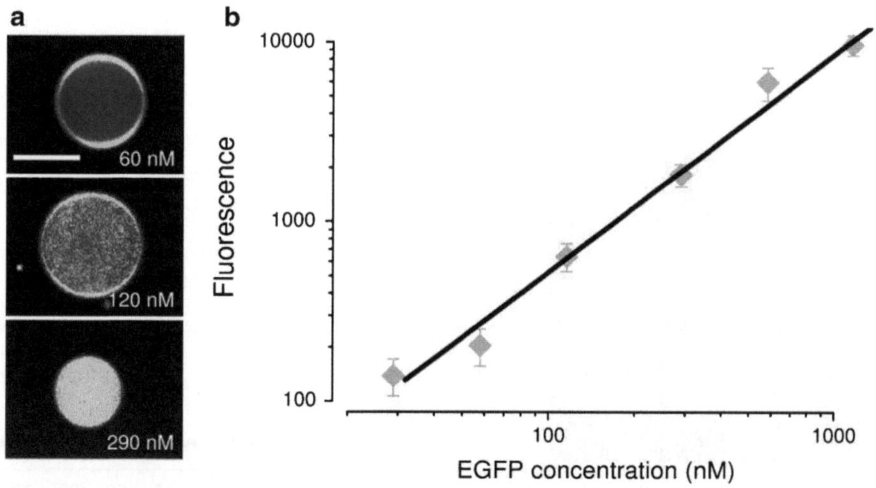

Fig. 2. Calibration of the imaging system. (**a**) Recombinant EGFP was diluted at decreasing concentrations in saline solution. The micelles were prepared by dispersing the solution in mineral oil. Imaging was performed on the confocal microscope in carefully controlled conditions to ensure applicability of the calibration to the live cell experiments. The three panels show micelles at three different concentrations; quantification has been performed in the central part of the droplet to avoid spurious effects due to lensing occurring nearby the edge. Bar 10 μm. (**b**) Calibration data: each point is the average of at least 10 measures (30 at the three lower concentrations) (9).

to monitor the laser power at the focal plane of the lens with an appropriate laser meter. If the laser power changes between experiments, this measure allows a simple linear correction.

4.4.4. Time Lapse Imaging of ERK-GFP

Figure 3a shows the subcellular localization of ERK2-GFP in low-expressing cells in starved conditions and after stimulation of the ERK pathway. In this population of cells, the kinase relocation is not occluded by the ectopic localization present in overexpressing cells. The stimulus-dependent nuclear localization can be temporally resolved by time lapse imaging (Fig. 3b). In these experiments, cells are stimulated by adding either serum or fibroblast growth factor 4 (FGF4, 50 ng/mL) to the medium. Upon stimulation, ERK2-GFP accumulated in the nucleus reaches 90% of translocation within 9 min of the onset of stimulation following

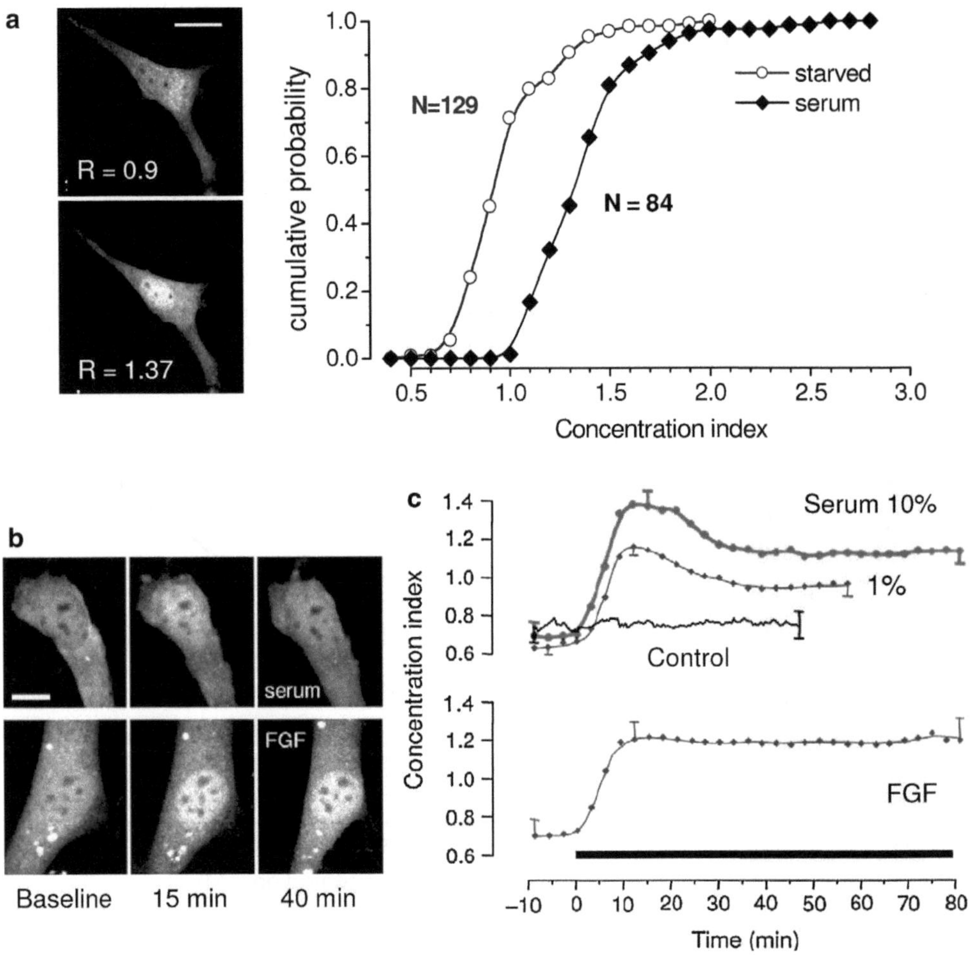

Fig. 3. (**a**) ERK2-GFP concentrates in the nucleus of living cells after stimulation as shown by the change in of the CI. (**b**) Nuclear fluorescence begun increasing within 4 min from stimulation with 10% serum or FGF4. Bar, 10 μm. (**c**) Average time course of the CI after stimulation with two different doses of serum and FGF or in cells transfected with an empty GFP vector.

an exponential time course (9). Interestingly, different stimuli cause distinct temporal dynamics as shown in Fig. 3c.

4.4.5. FRAP Experiments: Shuttling Between Nucleus and Cytoplasm

The nucleus–cytoplasm shuttling can be readily studied by imaging the recovery of fluorescence after bleaching ERK-GFP in the nucleus. The basic idea is that when a fraction of the nuclear protein is bleached, the following recovery of fluorescence can only occur in the presence of exchange between the nucleus and the cytoplasm (but see Note 7). This experiment requires three distinct phases. Initially the fluorescence of the cell is imaged at low laser power. This prebleach image is necessary for the following analysis. It is important that a good signal-to-noise ratio is reached, so it is useful to acquire several images to be averaged. The photobleaching itself is realized by illuminating at maximum laser power in an area inside the nucleus: this can be obtained by two different methodologies, depending on the features offered by the specific microscope. If the excitation lasers are controlled by an acusto-optic modulator (AOM) the intensity can be controlled very rapidly, and this allows to bleach an arbitrary region inside the nucleus. In the absence of an AOM, bleaching can be produced by operating the microscope in the line scan modality for a few seconds with the laser at full power (Fig. 4a): since the protein is able to move inside the nucleus, bleaching affects the entire nuclear pool, and, indeed, when normal imaging is resumed, the entire nucleus appears bleached (Fig. 4b). The third phase of the experiment consists of a time lapse sequence that records the recovery of fluorescence. The excitation intensity must be low enough to cause little or no bleach, even if the following analysis can correct for this loss. The period of the acquisition must be fast enough to record several points during the early, steeper part of the recovery.

The time course of the recovery follows very closely an increasing exponential (see Note 7). If the fluorescence is measured in the cytoplasm, the signal shows a corresponding decrease, with identical time constant (Fig. 4c). The recovery curve is described by three parameters (Fig. 4d): the initial value is obtained by measuring the fluorescence immediately after bleaching and it is determined by the experimental parameters (duration and brightness of the bleaching and size of the nucleus), the time constant, which depends on the speed of the protein trafficking thought the nuclear membrane, and the asymptotic value of the recovery, which depends on the presence of a fraction of bleached protein that cannot be exchanged between the two compartments (the Immobile Fraction (IF)). It is important to point out that this interpretation of the data requires that the rate of intracellular diffusion is much faster than the shuttling. In practice, this condition is satisfied in most situations: indeed the first postbleach image of Fig. 4b shows that, even if the bleached area was a short

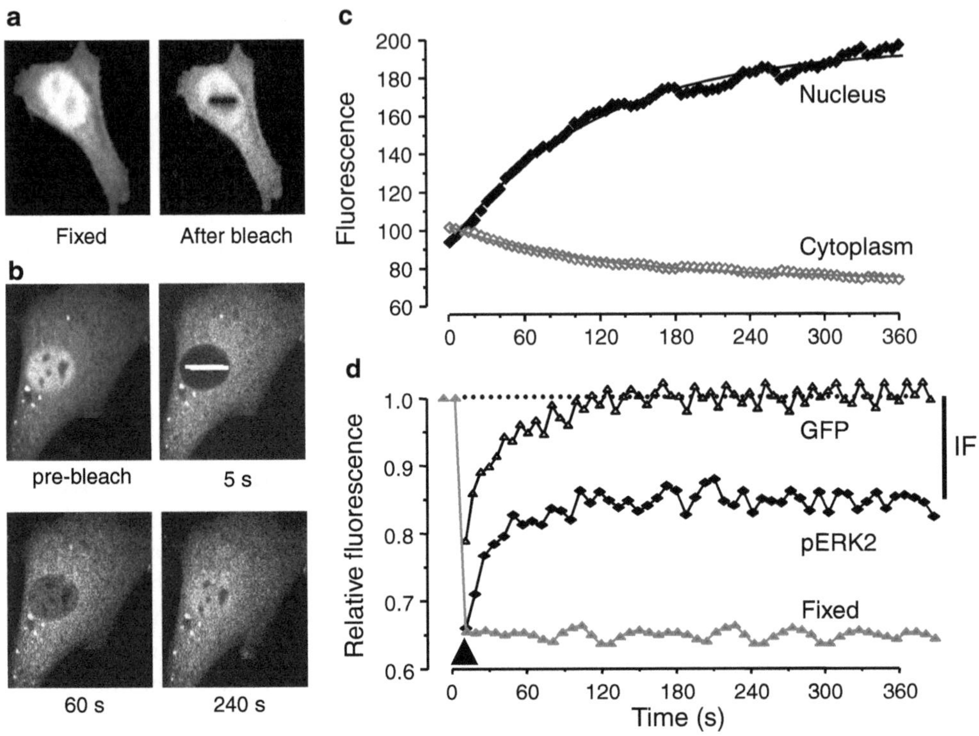

Fig. 4. (**a**) Line scan bleaching in the nucleus, shown on a fixed cell: the path travelled by the laser is clearly shown. (**b**) The same procedure applied on a living cell shows that the nucleus is uniformly bleached as in control cells: this indicates that the typical time for the mixing of ERK2-GFP in the nucleus is <5 s. The *white line* in the panel relative to 5 s indicates the bleach location. (**c**) The time course of the fluorescence changes is identical when measured in the nucleus and in the cytoplasm. Both sets of data points are fitted by an exponential of time constant 115 s. (**d**) This data shows how the recovery curves appears after the normalization described in the text (17.2). *The arrowhead* pointing at the first point of the recovery indicates the depth of bleaching. In the absence of immobile fraction (IF), the recovery tends to the asymptotic value of 1. pERK2 shows a substantial immobile fraction. The third set of data points (*gray*) show the absence of recovery in a fixed specimen.

line, there is no memory of this indicating that the compartment is completely mixed within a time window much shorter than the bleaching period (5 s). If diffusion in the cytoplasm is not rate limiting for the nucleus–cytoplasmic shuttling, this technique can be used to measure shuttling for any protein of interest. As an example, it has been used successfully to measure the trafficking between nucleus and cytoplasm of MeCP2 (23).

The recovery sequence must be at least two or three times the time constant of the recovery: a long sequence improves the accuracy of the estimate of the immobile fraction. If the software allows to do that, it is a good practice to reduce the period of acquisition after the initial rapid phase of recovery, since this will decrease the bleaching due to imaging. However it must be noted that longer sequences are more prone to be affected by cell movements or changes of morphology that affect the fluorescence distribution (see Note 8).

The imaging data are analyzed by measuring the fluorescence from the nucleus (F_{Nuc}) and from the entire cell, including the nucleus (F_{Tot}). The recovery is then described by the function:

$$F(t) = \frac{F_{Tot}^{PB}}{F_{Nuc}^{PB}} \times \frac{F_{Nuc(t)}}{F_{Tot}(t)}, \tag{2}$$

where F^{PB} indicates the fluorescence measured before bleaching in the nucleus (Nuc) or on the entire cell (Tot). All data must be corrected for background fluorescence. This normalization corrects for bleaching caused by imaging, since the two terms F_{Nuc} and F_{Tot} are affected by bleaching identically. In the absence of IF, the data computed by this normalization would converge to an asymptotic value of one. Figure 5a shows results from a typical experiment. The same cell has been imaged before and after

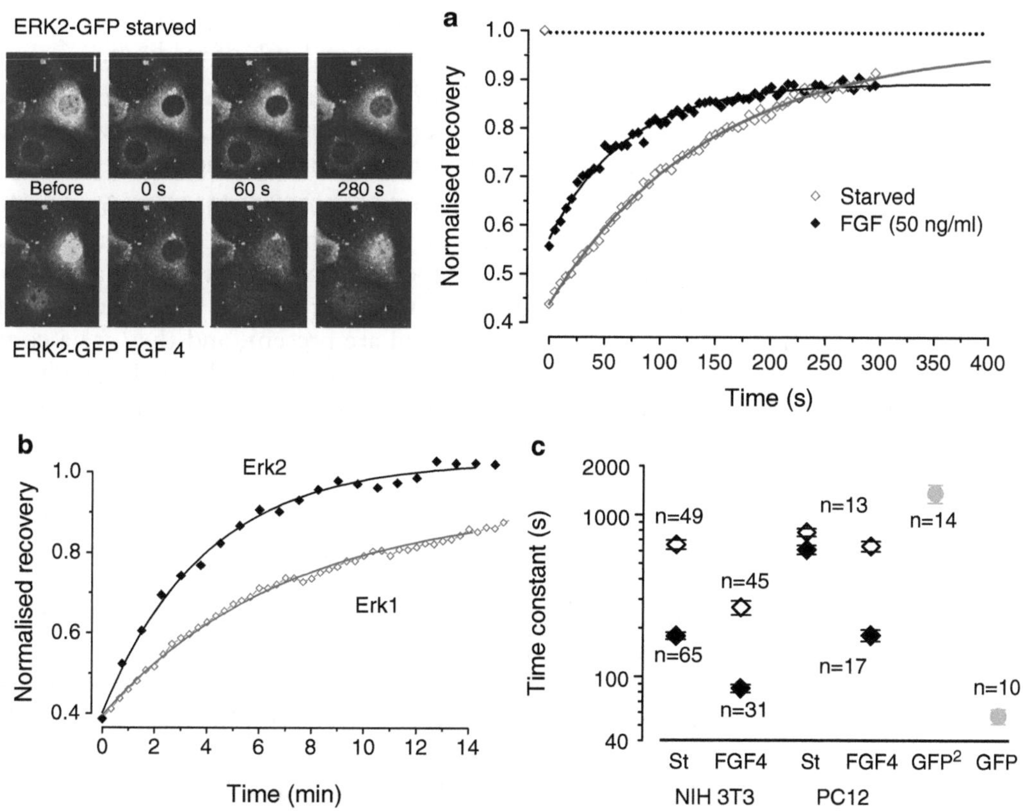

Fig. 5. (**a**) The same cell was bleached before and after stimulation with FGF4. This revealed the increased speed of trafficking of phospho-ERK and the increased immobilization in the nucleus. Data were normalized with (17.2) and fitted extremely well by a single exponential. (**b**) Trafficking studies also reveal an hitherto unsuspected difference between ERK1 and ERK2. (**c**) The cumulative data show the dependency of trafficking on ERK phosphorylation and the slower permeation of ERK1 (*empty symbols*) compared to ERK2 (*filled symbols*). Each symbol represents the mean of the group and the standard error; the number of measures is indicated in the graph. Data from fibroblasts and PC12 are shown. For comparison, the shuttling speed of a GFP dimer (GFP2) and of a GFP monomer is also reported. The dimer, which is smaller than ERK-GFP, is larger than the size limit for passive diffusion through the nuclear pore.

stimulation with FGF4. The analysis of the recovery curves shows that, upon phosphorylation, the speed of nucleus–cytoplasmic shuttling increases and that the IF increases. Likely, this indicates the formation of complex large enough to be unable to permeate the nuclear membrane. The method is sensitive enough that differences in the shuttling rates can be detected within the ERK family (Fig. 5b, c).

5. Notes

1. Cells are cultured in different ways according to experimental necessities. In these experiments, we have plated the cells either on 14 mm glass disks or on 35 mm dish with 12 mm coverslip on the bottom (Willco). The disks were cultured in multiwell plates (12-well plate, BD Falcon, 351143). In both cases the cells were kept in about 1 mL of medium.
2. $CaCl_2$ should be added only after buffering the solution to pH 7.3 to avoid possible precipitation of Ca. Afterwards, the solution can be brought to the final volume.
3. If a microscope stage incubator is available, imaging can be performed directly in the culture medium. However, it has to be noted that some compounds present in the standard media can produce significant autofluorescence. In DMEM, both Riboflavin and Phenol Red are present, and they are a common source of problems. Minimum media without Phenol Red can be used to good effect, but the autofluorescence of the media in the specific imaging conditions has to be carefully controlled.
4. Depending on the cell type, age, and plasmids used, these quantities may vary and should be adjusted according to the experimental conditions and to the desired level of expression. Transfection in the Willco dishes can be made to function using only 150 μL of transfecting solution carefully placed on the central well. Therefore a higher concentration of expensive reagents (the plasmids and Lipofectamin) can be used if necessary.
5. Since cells are extremely delicate (especially PC12 because of their thin neurites), every procedure of medium replacement should be performed very carefully. When removing the medium, a small film of fluid is left inside the well, to avoid complete desiccation of cells. When replacing the medium, the fresh one is added very gently to the side of the dish, to avoid cell damaging or detaching. The fresh medium should be placed in the incubator beforehand to equilibrate pH and temperature.

6. During long imaging sessions, it is very important to keep the preparation very stably. If the cells are cultured in petri dishes, it is possible to find or manufacture stable holders. If the cells are cultured on coverslips, it is quite important to block the disk very stably, but without causing any deformation of the glass that might disturb the cells. Finally, the microscope should be carefully checked for possible instabilities of the sample holder. Check that variable friction at the microscope focus knob is tight enough to prevent any slow vertical drift of the objective during the time lapse sequence.

7. Unless two-photon excitation is used, the fluorescent protein is bleached also on the out-of-focus planes. If the nucleus occupies most of the vertical volume of the cell, there is only a negligible loss of fluorescence from the cytoplasm. However, if the nucleus is small, the volume interested by the bleaching might include some cytoplasm. Therefore the recovery comprises two different processes: the trafficking between nucleus and cytoplasm, as seen above, and the diffusion in the cytoplasm. The latter process occurs much more rapidly than the former (time constant of a few seconds), and this causes a rapid component of the recovery curve seen right after the bleach. The presence of this component does not hinder the measure of shuttling, provided that the fast component is not taken into account. The obvious solution is excluding the first few points during the computation of the exponential fit. The single exponential describes the mixing up of two well-stirred compartments, where, well stirred means that there are no concentration gradients within each compartment: this condition is satisfied when (1) the intra-compartment diffusion is much faster than the shuttling between nucleus and cytoplasm and (2) the cytoplasmic compartment is compact enough to allow the perfect mixing. If the cells are very polarized, and neurons are a good example of this situation, it might take a long time to see perfect mixing between the protein in the perinuclear region and the protein in the processes farther away. In these conditions, one might observe a small and very slow tail of the recovery which is not due to exchange through the nuclear membrane but is rather due to the slow equilibration of the entire cell. Great care should be exercised when analyzing the FRAP experiments performed in polarized cells.

8. Almost invariably, the cells move during long imaging sessions. This is problematic since the areas to measure in each frame change with time. One solution is to hand draw the area in each frame, but of course that is a very time consuming procedure. A better solution is to realign by translating the frames of the sequence to compensate the cell movements. This can

be done in several ways: a practical solution is a plug-in available in ImageJ (Stack Reg, see http://bigwww.epfl.ch/thevenaz/stackreg/). In most cases this procedure returns very satisfactory results.

References

1. Brunet, A., D. Roux, P. Lenormand, S. Dowd, S. Keyse, and J. Pouyssegur, (1999) *Nuclear translocation of p42/p44 mitogen-activated protein kinase is required for growth factor-induced gene expression and cell cycle entry.* Embo J 18(3):664–74.
2. Kim, K., K. Nose, and M. Shibanuma, (2000) *Significance of nuclear relocalization of ERK1/2 in reactivation of c-fos transcription and DNA synthesis in senescent fibroblasts.* J Biol Chem 275(27):20685–92.
3. Cowley, S., H. Paterson, P. Kemp, and C.J. Marshall, (1994) *Activation of MAP kinase kinase is necessary and sufficient for PC12 differentiation and for transformation of NIH 3 T3 cells.* Cell 77(6):841–52.
4. Robinson, M.J., S.A. Stippec, E. Goldsmith, M.A. White, and M.H. Cobb (1998) *A constitutively active and nuclear form of the MAP kinase ERK2 is sufficient for neurite outgrowth and cell transformation.* Curr Biol 8(21):1141–50.
5. Fukuda, M., I. Gotoh, M. Adachi, Y. Gotoh, and E. Nishida, (1997) *A novel regulatory mechanism in the mitogen-activated protein (MAP) kinase cascade. Role of nuclear export signal of MAP kinase kinase.* J Biol Chem 272(51):32642–8.
6. Marchi, M., A. D'Antoni, I. Formentini, R. Parra, R. Brambilla, G.M. Ratto, and M. Costa, (2008) *The N-terminal domain of ERK1 accounts for the functional differences with ERK2.* PLoS One 3(12):e3873.
7. Costa, M.C., F.; Matilde, M.; Roy, A.; Maffei, L.; Ratto, G. M. (2003) *Dynamics of ERK nuclear translocation in living cells.* in *Society for Neuroscience.* New Orleans.
8. Ando, R., H. Mizuno, and A. Miyawaki, (2004) *Regulated fast nucleocytoplasmic shuttling observed by reversible protein highlighting.* Science 306(5700):1370–3.
9. Costa, M., M. Marchi, F. Cardarelli, A. Roy, F. Beltram, L. Maffei, and G.M. Ratto, (2006) *Dynamic regulation of ERK2 nuclear translocation and mobility in living cells.* J Cell Sci 119:4952–63.
10. Fujioka, A., K. Terai, R.E. Itoh, K. Aoki, T. Nakamura, S. Kuroda, E. Nishida, and M. Matsuda, (2006) *Dynamics of the Ras/ERK MAPK cascade as monitored by fluorescent probes.* J Biol Chem 281(13):8917–26.
11. Boulton, T.G., S.H. Nye, D.J. Robbins, N.Y. Ip, E. Radziejewska, S.D. Morgenbesser, R.A. DePinho, N. Panayotatos, M.H. Cobb, and G.D. Yancopoulos, (1991) *ERKs: a family of protein-serine/threonine kinases that are activated and tyrosine phosphorylated in response to insulin and NGF.* Cell 65(4):663–75.
12. Seger, R. and E.G. Krebs, (1995) *The MAPK signaling cascade.* Faseb J 9(9):726–35.
13. Lefloch, R., J. Pouyssegur, and P. Lenormand, (2009) *Total ERK1/2 activity regulates cell proliferation.* Cell Cycle 8(5):705–11.
14. Mazzucchelli, C., C. Vantaggiato, A. Ciamei, S. Fasano, P. Pakhotin, W. Krezel, H. Welzl, D.P. Wolfer, G. Pages, O. Valverde, A. Marowsky, A. Porrazzo, P.C. Orban, R. Maldonado, M.U. Ehrengruber, V. Cestari, H.P. Lipp, P.F. Chapman, J. Pouyssegur, and R. Brambilla, (2002) *Knockout of ERK1 MAP kinase enhances synaptic plasticity in the striatum and facilitates striatal-mediated learning and memory.* Neuron 34(5):807–20.
15. Pages, G., S. Guerin, D. Grall, F. Bonino, A. Smith, F. Anjuere, P. Auberger, and J. Pouyssegur, (1999) *Defective thymocyte maturation in p44 MAP kinase (Erk 1) knockout mice.* Science 286(5443):1374–7.
16. Vantaggiato, C., I. Formentini, A. Bondanza, C. Bonini, L. Naldini, and R. Brambilla, (2007) *ERK1 and ERK2 mitogen-activated protein kinases affect Ras-dependent cell signaling differentially.* J Biol 5(5):14.
17. Bessard, A., C. Fremin, F. Ezan, A. Fautrel, L. Gailhouste, and G. Baffet, (2008) *RNAi-mediated ERK2 knockdown inhibits growth of tumor cells in vitro and in vivo.* Oncogene 27(40):5315–25.
18. Fremin, C., F. Ezan, P. Boisselier, A. Bessard, G. Pages, J. Pouyssegur, and G. Baffet, (2007) *ERK2 but not ERK1 plays a key role in hepatocyte replication: an RNAi-mediated ERK2 knockdown approach in wild-type and ERK1 null hepatocytes.* Hepatology 45(4):1035–45.
19. Lenormand, P., C. Sardet, G. Pages, G. L'Allemain, A. Brunet, and J. Pouyssegur, (1993) *Growth factors induce nuclear translocation of MAP kinases (p42mapk and p44mapk) but not of their activator MAP kinase kinase*

(p45mapkk) in fibroblasts. J Cell Biol 122(5):1079–88.

20. Rubinfeld, H., T. Hanoch, and R. Seger, (1999) *Identification of a cytoplasmic-retention sequence in ERK2.* J Biol Chem 274(43):30349–52.
21. Burack, W.R. and A.S. Shaw, (2005) *Live Cell Imaging of ERK and MEK: simple binding equilibrium explains the regulated nucleocytoplasmic distribution of ERK.* J Biol Chem 280(5):3832–7.
22. Horgan, A.M. and P.J. Stork, (2003) *Examining the mechanism of Erk nuclear translocation using green fluorescent protein.* Exp Cell Res 285(2):208–20.
23. Marchi, M., A. Guarda, A. Bergo, N. Landsberger, C. Kilstrup-Nielsen, G.M. Ratto, and M. Costa, (2007) *Spatio-temporal dynamics and localization of MeCP2 and pathological mutants in living cells.* Epigenetics 2(3):187–97.

Part IV

Studies on the Regulation of MAP Kinase Cascades

Chapter 18

Studying the Regulation of MAP Kinase by MAP Kinase Phosphatases In Vitro and in Cell Systems

Céline Tárrega, Caroline Nunes-Xavier, Rocío Cejudo-Marín, Jorge Martín-Pérez, and Rafael Pulido

Abstract

Signaling through MAPK pathways involves a network of activating kinases and inactivating phosphatases. While single MAPK kinases account for specific activation of the distinct MAPKs, inactivation of MAPKs by phosphatases involves a wider spectrum of enzymes, with phosphatases from distinct families displaying specificity toward MAPKs. The dual-specificity family of MAPK phosphatases, MKPs, constitutes the major group of MAPK inactivating phosphatases. MKPs are widely expressed, in a tissue- and development-regulated manner, and the control of their expression and function is crucial for the regulation of MAPK signaling. Here, we present three methods to analyze the regulation of MAPKs by MKPs, using transient and stable-inducible MKP overexpression cell systems and in vitro phosphatase experiments.

Key words: MAPK phosphatases, MKPs, MAPK dephosphorylation, Phosphatase assay, Tetracycline-inducible cells

1. Introduction

Inactivation of MAP kinase (MAPK) cascades by protein phosphatases is crucial to control the strength, duration, and location of MAPK signaling events (1). Specific serine/threonine phosphatases (PPs), tyrosine phosphatases (PTPs), and dual-specificity phosphatases (DSPs) exist that inactivate MAPKs by direct dephosphorylation of their TXY motifs. PPs that dephosphorylate MAPKs include several PP2A and PP2C enzymes (2, 3). However, the activities of PP2A and PP2C toward components of the MAPK pathways are not only restricted to MAPK

Céline Tárrega and Caroline Nunes-Xavier contributed equally to this work.

Rony Seger (ed.), *MAP Kinase Signaling Protocols: Second Edition*, Methods in Molecular Biology, vol. 661,
DOI 10.1007/978-1-60761-795-2_18, © Springer Science+Business Media, LLC 2010

dephosphorylation, but also target other upstream kinases. In addition, both inhibitory and activating effects of these phosphatases on the MAPK pathways have been documented, depending on the PP2A holoenzyme, the PP2C isozyme, and the MAPK pathway (4, 5). PP2A forms functional complexes with ERK1/2 and the PTP HePTP (6), and the PP2C enzymes PP2Cα and Wip1 associate physically with p38 (7, 8), which could account for substrate specificity; however, the molecular basis of these associations are mostly unknown.

In general, tyrosine and dual-specificity MAPK phosphatases are regarded as specific negative regulators of MAPK pathways by direct dephosphorylation and inactivation of the MAPKs. MAPK phosphatases of the tyrosine-specific classical PTP family are encoded by three genes in mammals, whose products are the non-nuclear proteins PTP-SL/PTPBR7, HePTP, and STEP. HePTP is mostly expressed in lymphoid tissues, whereas PTP-SL/PTPBR7 and STEP (both of which display multiple protein variants) are mostly expressed in the brain (9, 10). The MKP family of dual-specificity MAPK phosphatases, which is distinct from the classical PTP family, constitutes the major group of direct MAPK inactivating enzymes. There are ten MKP genes in mammals encoding active phosphatases, both nuclear and non-nuclear (MKP1, MKP2, DUSP2, DUSP5, MKP3, MKP4, MKPX, DUSP8, MKP7, and MKP5) (11, 12). The expression of several MKPs (MKP1, MKP3, DUSP2, and DUSP5) is induced as a consequence of the activation of the MAPK pathways, and serves as a negative feedback regulatory loop for these pathways (13, 14). Finally, a group of low molecular weight atypical DSPs (LMW-aDSPs) also dephosphorylate and inactivate MAPKs (15, 16). VHR, which dephosphorylates ERK1/2 and JNK, is the LMW-aDSP most extensively studied (17). For a more comprehensive list of PTP- and dual-specificity MAPK phosphatases, and for the standardized protein and gene names, see refs. 18, 19.

MAPK-specific PTPs and MKPs possess a regulatory MAPK-binding domain that provides substrate specificity and promotes efficient dephosphorylation of MAPKs. Binding specificity is mainly achieved by the interaction between a docking motif in the MAPK-binding domain of the phosphatases (KIM motif) and a docking groove in the MAPKs (20–23). In addition, the catalytic activity of some MKPs (MKP1, MKP2, MKP3, MKP4, DUSP2, and DUSP7) is increased upon MAPK binding (24–28), and in the case of MKP3, the N-terminal regulatory domain has been shown to inhibit the activity of the catalytic domain by interdomain binding (29). In this regard, the protein stabilizer dimethyl sulfoxide (DMSO) enhances, in vitro, the catalytic activity of

MKP3 to a similar extent than its binding to ERK (30). LMW-aDSPs lack a MAPK-binding domain, and the molecular mechanisms that account for the control of their catalytic activity and substrate specificity toward MAPKs remain uncertain. In this regard, VHR forms complexes with ERK1/2 and the vaccinia-related kinase VRK3, and VHR phosphatase activity is enhanced upon binding to VRK3 (31).

As mentioned above, the major regulatory mechanisms of the activity of MAPK phosphatases include their transcriptional control by the MAPK pathways, intradomain interactions, and allosteric activation upon binding to MAPKs or other regulatory proteins. Phosphorylation and acetylation of MAPK phosphatases also regulate directly their function. Examples include the negative regulation of PTP-SL/PTPBR7, HePTP and STEP by PKA phosphorylation (32–34), and the positive regulation of MKP1 upon acetylation by p300 (35). Remarkably, these posttranslational modifications target the KIM motifs of PTP-SL/PTPBR7, HePTP, STEP, and MKP1, underscoring the importance of high affinity specific docking in the functional interplay between MAPKs and MAPK phosphatases. Also, phosphorylation of MKP1, MKP3, and MKP7 by ERK1/2 regulates the protein stability of the MKPs, providing an additional feedback regulatory checkpoint (36–38).

In a previous issue of these series, we have described methods to analyze the effects of PTPs on MAPK activities in intact cells and in vitro (39). Here, we present three additional methods to study the regulation of MAPKs by MKPs. Two methods are presented to analyze comparatively, both in vitro and in intact cells, the phosphatase activity of MKPs toward MAPKs. A third method is provided that describes the generation of stable, Tet-On inducible cell lines expressing MKPs, and its utility to monitor the role of these enzymes in the regulation of cell functions controlled by MAPKs. Specific examples are provided using several MKPs, and the protocols are suitable for any other MAPK phosphatase.

2. Materials

2.1. Analyzing the Activity and the Specificity of MKPs Toward MAPKs In Vitro

All solutions are prepared in double-distilled, deionized MilliQ filtered water. Cell culture and transfection procedures require sterile conditions.

1. Mammalian cell lines suitable for transfection and transient overexpression of recombinant proteins.
2. Tissue culture media, transfection reagents, and cell activating agents.

3. cDNAs of the tagged-MAPKs and MKPs of interest, cloned into mammalian expression vectors.
4. Phosphate-buffered saline (PBS): 1.85 mM NaH_2PO_4, 8.4 mM $NaHPO_4$, and 150 mM NaCl.
5. Lysis buffer: 50 mM Tris–HCl pH 7.5, 150 mM NaCl, 1% Igepal (Nonidet P-40), 1 mM PMSF, and 1 μg/mL aprotinin. Two buffers are used, with or without phosphatase inhibitors (Ser/Thr phosphatase inhibitors: 20 mM $Na_4P_2O_7$, 100 mM NaF; PTP inhibitor: 2 mM Na_3VO_4).
6. Bradford protein concentration assay reagents.
7. Anti-tag, anti-MKPs, anti-phosphoactive-MAPKs (optionally, anti-MAPKs antibodies), peroxidase-conjugated secondary antibodies, protein A-Sepharose, and prestained molecular weight standard marker. The antibodies used in our experiments are: anti-HA, 12CA5 monoclonal antibody; anti-MKP3 and anti-MKP4, our polyclonal antibodies; anti-DUSP5, polyclonal antibody provided by Keyse (40); anti-ERK1/2 and anti-p38α, polyclonal antibodies from Santa Cruz Biotechnology; anti-phosphoactive-ERK1/2 (monoclonal), and anti-phosphoactive-p38α (polyclonal) from Cell Signaling Technology.
8. Washing buffer for immunoprecipitation: 20 mM HEPES pH 7.5, 150 mM NaCl, 10% glycerol, 10 mM $Na_4P_2O_7$, and 0.1% Triton X-100.
9. Phosphatase reaction buffer: 25 mM HEPES pH 7.5, 5 mM EDTA, and 10 mM DTT.
10. Sodium dodecyl sulfate-polyacrylamide gel electrophoresis (SDS-PAGE) loading buffer (4×): 250 mM Tris–HCl pH 6.8, 40% glycerol, 8% SDS, 0.02 mg/mL bromophenol blue, and 4% 2-mercaptoethanol.
11. Polyvinylidene fluoride (PVDF) protein transfer membranes.
12. Transfer buffer: 48 mM Tris Base, 39 mM glycine, 0.037% SDS, and 20% methanol.
13. Incubation buffer for immunoblot analysis: 50 mM Tris–HCl pH 7.5, 150 mM NaCl, 5 mM EDTA, 0.05% Triton X-100, and 0.025% gelatin.
14. Stripping solution: NaOH 0.2 M and 1% SDS.
15. Western blotting chemiluminescence kit.

2.2. Analyzing the Activity and the Specificity of MKPs Toward MAPKs in Intact Cells

See materials in Subheading 2.1. Note that the lysis buffer used contains the phosphatase inhibitors, and that the phosphatase reaction buffer is not necessary.

2.3. Generation of Tet-On Inducible MKP-Cell Lines to Study Cell Functions Regulated by MAPK

1. Mammalian Tet-On cell line suitable for transfection (see Note 9).
2. cDNA of the MKP, or the tagged-MKP of interest, cloned into a mammalian expression pTRE vector.
3. Tissue culture media, trypsin–EDTA solution, culture dishes of different sizes (10 cm, 6 cm, 12-well, 24-well, and 96-well plates), transfection reagents, and 3 mm cloning discs for isolation of single colonies.
4. Doxycycline and the appropriate antibiotics, depending on the expression vector (see Notes 9 and 13).
5. Anti-MKP or anti-tag antibodies, lysis buffer, SDS-PAGE buffers, and Western blot buffers and reagents, as in Subheading 2.1.

3. Methods

The first method (Subheading 3.1) is based on the immunoprecipitation of the MKP with specific antibodies, followed by an in vitro phosphatase assay toward an active recombinant MAPK, which is obtained separately from stimulated, MAPK-transfected cells. The second method (Subheading 3.2) relies on the coexpression of both the recombinant MAPK and MKP in the same cells, and is intended to monitor the dephosphorylation of the MAPK in intact cells. Note that in this case the MKP is also subjected to putative regulation by the stimulus that activates the MAPK (see Note 4). The final readout of these experiments is the pTpY content of the MAPK at its TXY loop, measured by immunoblot with specific anti-phosphoactive-MAPK antibodies. Examples are given using MKP3 (DUSP6), MKP4 (DUSP9), and DUSP5. When performed in combination, Subheadings 3.1 and 3.2 are useful to compare the specificity of an MKP toward distinct MAPKs, and of different MKPs toward a particular MAPK. In addition, the methods allow the unveiling of indirect regulatory effects of MKPs on MAPKs, or of putative regulatory mechanisms of MKP activity (regulation of expression, posttranslational modifications) in cells. The third method (Subheading 3.3) describes the generation of stable cell lines that express MKPs under the control of the tetracycline-inducible (Tet-On) transcription factor (41), and their use to investigate the role of these phosphatases in the regulation of cellular functions controlled by MAPKs. In this method, clonal populations of cells, inducibly expressing the MKPs of interest, are obtained. The clones can be used to study processes such as differentiation or proliferation, or other adaptative responses to external stimuli, in whole cell populations. The induction of mRNA and protein expression with a tetracycline derivative, doxycycline, allows control of the timing and intensity of MKP expression, which facilitates the mimicking of physiologic conditions. In addition, inducibility circumvents putative cell toxicity of ectopic overexpression of the MKPs. Examples are provided using the MCF-7 human breast carcinoma cell line (42), and the MKPs, MKP3 and DUSP5.

3.1. Analyzing the Activity and the Specificity of MKPs Toward MAPKs In Vitro

1. Transfect cells with a plasmid containing the tagged-MAPK (see Note 1). After 24–72 h of culture, treat the cells with the adequate stimulus to activate the MAPK (see Note 2), rinse the cells with ice-cold PBS, lyse the cells in lysis buffer containing phosphatase inhibitors (see Note 3), transfer the lysate to a 1.5 mL Eppendorf tube, centrifuge at 14,000 × *g* for 5 min, 4°C, and save the supernatant. Keep an aliquot (50 μg) to check the expression of the tagged-MAPK. Immunoprecipitate the kinase from the rest of the lysate with an anti-tag antibody and protein A-Sepharose beads. Alternatively, endogenous MAPKs can be immunoprecipitated from non-transfected cells using specific anti-MAPK antibodies.
2. At the same time, transfect a separate set of cells with the MKPs of interest. After 24–72 h of culture, rinse the cells with ice-cold PBS, lyse the cells in lysis buffer lacking phosphatase inhibitors (see Note 3) and proceed as in step 1. Keep an aliquot (50 μg) to check the expression of the MKP. Immunoprecipitate the MKPs with specific antibodies and protein-A Sepharose beads. Before lysis, cells can be treated as required to test putative changes in MKP amount, substrate recognition, or phosphatase activity (see Note 4).
3. Wash the immunoprecipitated pellets four times with 1 mL of ice-cold washing buffer. Perform a last washing with 1 mL of ice-cold phosphatase reaction buffer (see Note 5).
4. Split the MAPKs in separate Eppendorf tubes (see Note 6): one tube as a control of the amount of immunoprecipitated active MAPK (see Fig. 1a, lanes 1) and one tube for each MKP tested. Add to each tube (except to the control one) the immunoprecipitated MKP, splitting the pellet (see Note 6) between the distinct MAPKs analyzed.
5. Incubate the MAPK–MKP mixture with constant shaking at 37°C for 30 min (see Note 7).
6. Add the appropriate volume of 4× SDS-PAGE loading buffer. Boil the Eppendorf tube for 2 min, spin, and load the mix in a 10% SDS-PAGE gel. Include a lane with prestained molecular weight markers.
7. Run the gel and transfer to a PVDF membrane. Cut a piece of the membrane in the range of 30–60 kDa.

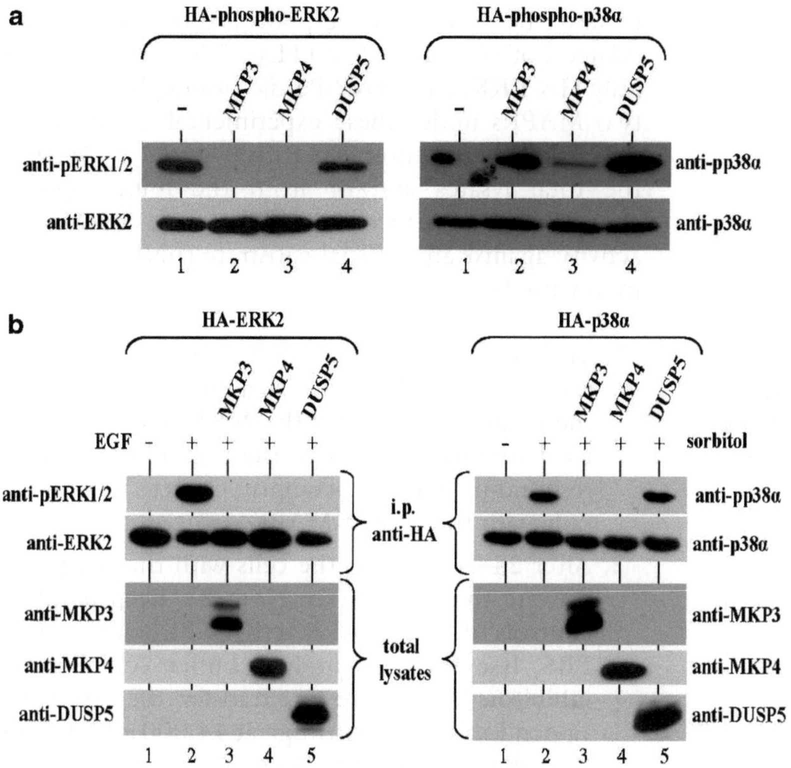

Fig. 1. (**a**) Dephosphorylation of MAPKs by MKPs in vitro. Phosphoactive HA-ERK2 and HA-p38α were incubated in phosphatase reaction buffer at 37°C for 30 min, in the presence of MKP3, MKP4, or DUSP5, as indicated. In *lane 1*, an equivalent sample of activated HA-MAPK, nontreated with phosphatase, was loaded as a control of the amount of immunoprecipitated active MAPK. The samples were resolved on 10% SDS-PAGE under reducing conditions and subjected to immunoblot with anti-phosphoactive-ERK1/2 (anti-pERK1/2), anti-ERK2, anti-phosphoactive-p38α (anti-pp38α), or anti-p38α antibodies. The active-HA-MAPKs and the MKPs were separately obtained from HEK293-overexpressing cell lysates by immunoprecipitation with anti-HA and anti-MKPs antibodies, respectively. (**b**) Effect of MKPs on MAPKs dephosphorylation in intact cells. HEK293 cells were co-transfected with HA-ERK2 or with HA-p38α and MKP3, MKP4, or DUSP5, as indicated. After 48 h, cells were left untreated (–) or were treated (+) with EGF (50 ng/mL, 5 min; to activate ERK2) or with sorbitol (0.5 M, 30 min; to activate p38α). Cells were lysed with ice-cold lysis buffer containing the complete cocktail of phosphatase inhibitors and HA-ERK2 and HA-p38α were immunoprecipitated (i.p.) from cell lysates with the anti-HA mAb. Immune complexes were resolved on 10% SDS-PAGE under reducing conditions and subjected to immunoblot with the antibodies indicated, as in (**a**) (*upper panels*). To monitor the overexpression of each MKP, total lysates were resolved on 10% SDS-PAGE under reducing conditions and subjected to immunoblot with specific anti-MKPs antibodies, as indicated. In *lanes 1 and 2*, HA-ERK2 and HA-p38α were co-transfected with empty vector.

8. Detect the phosphoactive MAPKs by standard Western blot and chemiluminescence techniques, using anti-phosphoactive-MAPK antibodies as the primary antibodies.
9. Strip the membranes with stripping solution and reprobe with anti-MAPK or anti-tag antibodies (see Note 8).

An example of comparative in vitro dephosphorylation of HA-ERK2 and HA-p38α by MKP3, MKP4, and DUSP5, is shown in Fig. 1a. As observed, the intrinsic in vitro phosphatase activity of the three MKPs toward HA-phosphoactive-ERK2 and toward HA-phosphoactive-p38α is different: MKP4 dephosphorylated both HA-ERK2 and HA-p38α, MKP3 dephosphorylated only HA-ERK2, and DUSP5 did not dephosphorylate any of the two MAPKs under these experimental conditions. The correct expression of the three MKPs was tested by Western blot from the total lysates and/or from the immunoprecipitates (not shown). It is also recommended to test the MKPs for in vitro activity against an artificial substrate (pNPP or OMFP are commonly used).

3.2. Analyzing the Activity and the Specificity of MKPs Toward MAPKs in Intact Cells

1. Transfect cells, in 6 cm dishes, with the plasmid containing the tagged-MAPK and with a mixture of plasmids containing the tagged-MAPK plus the MKP (see Note 1). To keep the total amount of DNA constant in all the wells, add the corresponding amount of empty vector to those wells transfected only with the tagged-MAPK.
2. After 24–72 h, treat the cells with the adequate stimulus to activate the MAPK (see Note 2). Keep points untreated as controls of basal MAPK activity. Rinse the cells with ice-cold PBS, lyse the cells in lysis buffer containing phosphatase inhibitors (see Note 3), transfer the lysate to a 1.5 mL Eppendorf tube, centrifuge at 14,000 × *g* for 5 min, 4°C, and save the supernatant. Keep 50 μg of lysates to check the expression of the tagged-MAPKs and/or the MKPs. Immunoprecipitate the tagged-MAPK with an anti-tag antibody and protein A-Sepharose beads. This allows to analyze selectively the recombinant MAPK, which is coexpressed with the MKP of interest (see Note 1).
3. Wash the immunoprecipitated pellets four times with 1 mL of ice-cold washing buffer. Perform a last washing with 1 mL of PBS.
4. Process the samples as in Subheading 3.1 (steps 6–8) to perform Western blot with anti-phosphoactive-MAPK and anti-MAPK antibodies.
5. Load SDS-PAGE gels in parallel to monitor by Western blot the expression of the MKPs from the total lysates, using anti-MKP antibodies.

An example of the analysis of the content of phosphoactive-HA-ERK2 and HA-p38α in the presence of MKP3, MKP4, or DUSP5, in intact cells, is shown in Fig. 1b. MKP3 and MKP4 were highly effective at dephosphorylating HA-ERK2 and HA-p38α under these conditions, whereas DUSP5 dephosphorylated HA-ERK2, but not HA-p38α. Note the differences observed between the results obtained in vitro (Fig. 1a) and in intact cells (Fig. 1b). Coexpression of HA-p38α with MKP3 resulted in HA-p38α dephosphorylation, whereas MKP3 did not dephosphorylate HA-p38α in vitro. Also, coexpression of HA-ERK2 with DUSP5 resulted in HA-ERK2 dephosphorylation, whereas DUSP5 did not dephosphorylate HA-ERK2 in vitro. This may suggest that, in intact cells, the effects of MKP3 on HA-p38α dephosphorylation, or of DUSP5 on HA-ERK2 dephosphorylation, are indirect. It may also happen that MKP3 and DUSP5 activity toward HA-p38α and HA-ERK2, respectively, are positively regulated in intact cells, and that such regulation is lost after cell disruption. Alternatively, it is possible that some of the effects observed in these experiments are biased by an excess of phosphatase, due to its overexpression. This putative promiscuity of MKPs can be analyzed by performing the experiments with titrated amounts of MKPs, obtained after transfection of the cells with decreasing amounts of the MKP plasmids. In any case, the combined analysis of in vitro dephosphorylation and coexpression in intact cells, allows a better definition of functional specificity of MKPs toward MAPKs.

3.3. Generation of Tet-On Inducible MKP-Cell Lines to Study Cell Functions Regulated by MAPK

1. Perform transient transfection assays on the cell line of interest, containing the Tet-On plasmid, with pTRE plasmids containing the MKPs (see Notes 1 and 9). Transfections can be performed in 6-well plates, using a range of 0.5–2 μg of plasmid per well (see Note 10), to optimize the transfection efficiency.
2. Culture for 24 h and induce the expression of ectopic proteins with 100 ng/mL doxycycline during additional 24 h (see Note 11). Rinse the cells with ice-cold PBS, lyse the cells in lysis buffer, transfer the lysate to a 1.5 mL Eppendorf tube, centrifuge at 14,000 × *g* for 5 min, 4°C, and save the supernatant. Check for MKP expression by Western blot, using 50–100 μg of total lysates and anti-tag or anti-MKP antibodies. Figure 2a shows an example of doxycycline-induced transient expression of wild type (wt) and inactive mutant (C293S mutation) forms of MKP3 in MCF-7 cells (see Note 12).
3. Perform the transfections for clone selection in 10 cm dishes, using the optimized conditions from the pilot transfection experiments (see Note 12).
4. Culture for 24 h and add 100 μg/mL hygromycin (if pTRE-2hyg is being used) (see Note 13). Wash twice with PBS to remove cell debris, and change media twice a week, as necessary. Culture cells until individual, well-separated colonies are visible and big enough to be isolated and expanded (see Note 14).
5. Wash cells with PBS and isolate the colonies using 3-mm cloning disks soaked in trypsin–EDTA solution: lay the soaked cloning disk on the individual colony and transfer the disk to 24-well plates.

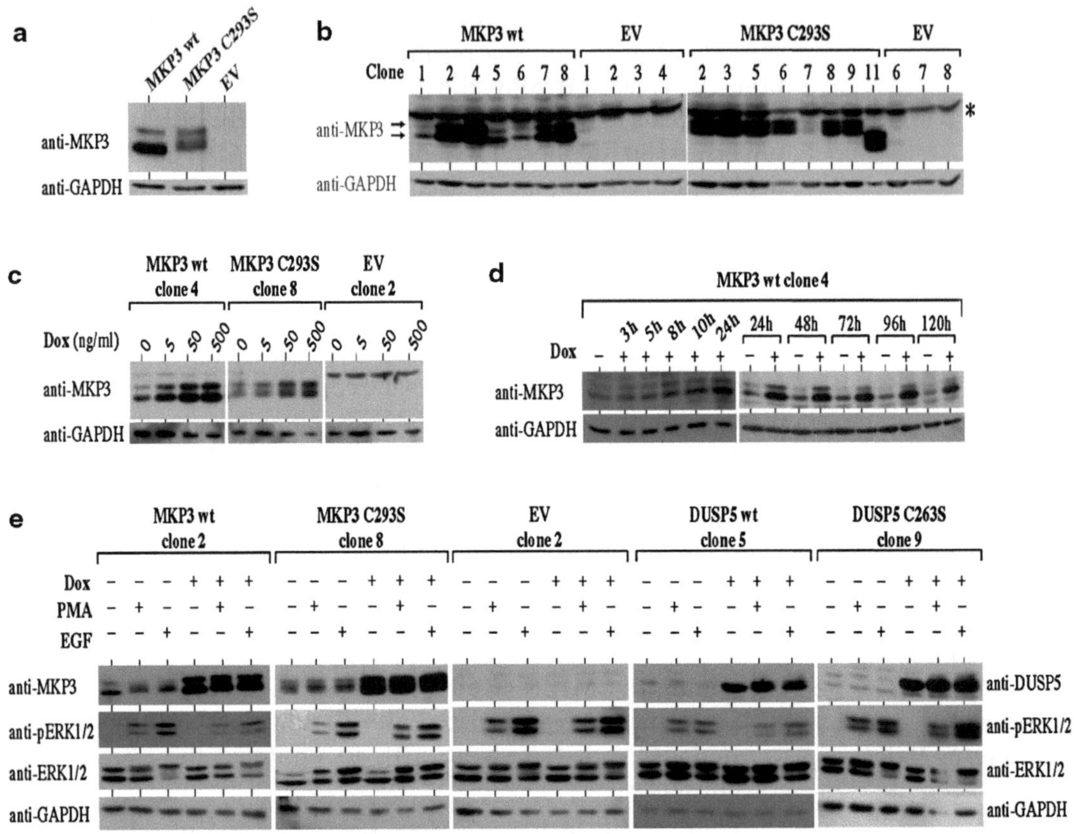

Fig. 2. (**a**) Transient expression of MKPs in MCF-7 cells upon doxycycline induction. pTRE-MKP3 wild type (wt), C293S mutation, and pTRE empty vector (EV), were transfected into MCF-7-Tet-On cell line, and transient expression of MKP3 was tested after 24 h of induction with doxycycline (Dox; 100 ng/mL). Cells were lysed with ice-cold lysis buffer, and total lysates were resolved on 10% SDS-PAGE under reducing conditions and subjected to Western blot with the antibodies indicated. Note the slower migration of the MKP3 C293S mutation, due to hyper-phosphorylation by ERK1/2 (not observed in MKP3 wt because it dephosphorylates and inactivates ERK1/2). (**b**) Screening of stable MCF-7-Tet-On clones for MKP expression. Individual MCF-7-Tet-On clones containing MKP3 wild type (wt), MKP3 C293S, or empty vector (EV), were tested for doxycycline-induced (100 ng/mL, 24 h) expression of MKP3. Cells were processed by Western blot as in (**a**). The two *arrows* indicate the migration of MKP3 (two starting Met are used, ref. 25). The *asterisk* indicates a background band. Note the variable levels of expression for each clone, as well as the presence of one MKP3 C293S clone (number 11) with wrong MKP3 migration. (**c**) Dose–response of doxycycline in the induction of MKP3 in MCF-7-Tet-On cells. MKP3 expression in the indicated clones (as shown in (**b**)) was tested after induction with increasing concentrations of doxycycline, as indicated, for 24 h. Cells were processed by Western blot as in (**a**). 100 ng/mL was chosen as the optimal doxycycline concentration for further experiments. (**d**) Time response of doxycycline in the induction of MKP3 in MCF-7-Tet-On cells. MKP3 expression in the clone number 4 (as shown in (**b**)) was tested after induction with 100 ng/mL of doxycycline during increasing periods of time, as indicated. Cells were processed by Western blot as in (**a**). MKP3 expression was maximal after 24 h, and was maintained up to 96 h. (**e**) Stable MCF-7-Tet-On MKP3 wt and DUSP5 wt cell lines show decreased ERK1/2 activity. The content of phosphoactive-ERK1/2 in MCF-7-Tet-On-MKP3, -DUSP5, or -empty vector (EV) clones was tested by Western blot using anti-phosphoactive-ERK1/2. MKP3 and DUSP5 expression was induced with 100 ng/mL of doxycycline for 24 h, prior to stimulation with PMA (50 ng/mL, 30 min) or EGF (50 ng/mL, 5 min) to activate ERK1/2, as indicated. Total ERK1/2 content was monitored with anti-ERK1/2 antibodies. Cells were processed by Western blot as in (**a**). In all experiments, the expression of GAPDH was monitored, as a control of protein loading.

6. Culture cells until semiconfluence is reached (see Note 15). Harvest the cells and transfer: (a) one-third to 24-well plate to check for MKP expression and (b) two-thirds to 12-well plates.

7. (a) Culture cells until confluence is reached, and add 100 ng/mL of doxycycline (see Note 11) to the 24-well plates to induce MKP expression. Harvest and process the cells as in step 2 to check for MKP expression. Figure 2b shows an example of the testing of several clones for expression of MKP3 wt and C293S mutation in MCF-7 cells. Note the differences of expression between the distinct clones, as well as the artefactual expression of proteins of wrong apparent molecular weight in one clone (MKP3 C293S clone 11, in Fig. 2b). (b) Expand positive cells for MKP expression to six-well plates, and then to 10 cm plates (see Note 16).

Before doing experiments with the clones expressing the proteins of interest, it is recommended to test the optimal expression conditions upon doxycycline induction. Figure 2c, d shows examples of dose- and time responses of doxycycline-induced expression of MKP3 wt and C293S mutation in MCF-7 cells. Note that in the absence of doxycycline, weak expression of the MKPs can be detected, indicating leaking of the inducible system (see Note 17). Examples of the activation status of ERK1/2 in MCF-7 clones that express wild type or catalytically inactive forms of MKP3 and DUSP5 are provided in Fig. 2e. As shown, the activation of ERK1/2 (determined by phosphoactive-ERK1/2 content, or by the shift on the migration of ERK1/2 bands) by two distinct stimuli (PMA and EGF) was diminished, in a doxycycline inducible manner, in the clones expressing wild type MKP3 or DUSP5, in comparison with the empty vector (EV) clones or the clones expressing MKP3 C293S or DUSP5 C263S.

4. Notes

1. There are different suitable protocols for transfection of adherent mammalian cells that work well with some given cell lines. We routinely use HEK293 (Subheadings 3.1 and 3.2) or COS-7 cells that can be transfected with very high efficiency by the calcium phosphate method or by DEAE-dextran, respectively, with vectors containing the SV40 origin of replication. In addition, a variety of commercial lipid reagents are available that provide good transfection efficiency in a wide range of cell lines. For transfection of MCF-7 cells (Subheading 3.3), FuGENE 6™ (Roche Diagnostics) was used. For MAPK transfections, we use hemagglutinin (HA) N-terminal-tagged MAPKs, cloned into the mammalian expression vector pCDNA3. Myc and Flag N-terminal-tagged MAPKs are also commonly utilized. For transient MKP transfections of HEK293 cells, we use MKPs cloned into the mammalian expression vector pRK5. Tagged-MKPs can also be

used, although we have observed low levels of expression of certain HA-N-terminal-tagged MKPs, such as MKP3. Note that the expression plasmids for MKPs and MAPKs are the same in both Subheadings 3.1 and 3.2. For the assays in intact cells, it is important that both the recombinant MAPK and the MKP are coexpressed in a representative number of cells. Perform control transfections using different ratios of MAPK:MKP plasmid DNAs, and check coexpression by two-color immunofluorescence. For both, in vitro and in intact cells assays, suitable controls include the use of catalytically inactive versions of the MKPs. Catalytic Cys to Ser mutations, or catalytic Asp to Ala mutations, are the inactive forms more commonly used (in the MKPs used in this study, catalytic Cys and Asp residues are as follows: MKP3, Cys293, and Asp262, accession NM_053883, rat sequence; MKP4, Cys342, and Asp311, accession BC004738, mouse sequence; DUSP5, Cys263, and Asp232, accession BC062545, human sequence). Note that these mutations may act as substrate-trapping molecules (43). Alternative catalytically defective mutations such as catalytic Arg to Met mutation, which do not have substrate-trapping properties, can also be used (catalytic Arg in the MKPs used in this study are as follows: MKP3, Arg299; MKP4, Arg348; DUSP5, Arg269).

2. Activation of MAPKs is cell type-, stimulus- and time-dependent. As a general rule, to obtain an optimal activation of ERK2, we stimulate COS-7 or HEK293 cells with 50 ng/mL of epidermal growth factor (EGF) at 37°C for 5 min; to obtain an optimal stimulation of p38α, we stimulate COS-7 or HEK293 cells with 0.5 M sorbitol at 37°C for 30 min. In the experiments with MCF-7 cells, we activated ERK1/2 with 50 ng/mL of EGF at 37°C for 5 min, or with 50 ng/mL of PMA at 37°C for 30 min. In all our experiments, no serum starvation was made previous to the stimulation, but in some cell lines serum starvation is recommended when basal MAPK activity is wanted to be minimized. Activation conditions should be optimized for each MAPK and cell line assayed.

3. Recommended volumes of lysis buffer are: 10 cm cell culture dish, 1 mL; 6 cm dish, 400 μL. All the protease and phosphatase inhibitors are added fresh to the lysis buffer before use, from concentrated stock solutions. The stocks we use are: 100 mM PMSF in ethanol, −20°C; 0.2 mg/mL aprotinin, −20°C; 0.5 M NaF, 20°C; 200 mM $Na_4P_2O_7$, 20°C; and 100 mM Na_3VO_4, −20°C. The phosphatase inhibitors (NaF and $Na_4P_2O_7$ as Ser/Thr phosphatase inhibitors; and Na_3VO_4 as a PTP inhibitor) are required to preserve the phosphorylation of the MAPKs after cell lysis (44). In the case of the cells

transfected with the MKPs, the cell lysis buffer lacks phosphatase inhibitors to avoid inhibition of the MKPs.

4. The separate transfection of MAPKs and MKPs allows testing variations in the steady-state levels (changes in stability) or in the substrate binding and/or catalytic activity of the MKPs upon distinct cell growth conditions, without interference between the stimulus used to activate the substrate MAPK and the conditions in which the activity of the MKPs needs to be tested. Changes in the MKP activity due to posttranslational modifications preserved after the cell lysis could be detected by this method. Note that putative phosphorylation of MKPs is likely not preserved in this assay after the cell lysis, due to the absence of phosphatase inhibitors in the lysis buffer (see Note 3). Alternatively, to preserve Ser/Thr phosphorylation, lysis of MKP-transfected cells can be made in the presence of Ser/Thr phosphatase inhibitors.
5. PTPs require reducing conditions to be active. To keep the MKPs reduced, add fresh DTT to the phosphatase reaction buffer before use.
6. The protein A-Sepharose beads are resuspended in an appropriate volume of phosphatase reaction buffer (at least twice of the bead volume). To split them, it is convenient to cut out the end of the micropipet tip. For the in vitro phosphatase assays, we split the tagged-MAPK from one transfected 10 cm dish in three to four alliquots, and the MKP from one transfected 10 cm dish in two alliquots.
7. The tube used as a control for the amount of activated MAPK immunoprecipitated is kept on ice during the phosphatase assay. In our experience, the stability of the immunoprecipitated HA-MAPK, incubated at 37°C in phosphatase buffer in the absence of any other immune complex pellet, is low. It is convenient to run in parallel a control in which the HA-MAPK is incubated at 37°C in phosphatase buffer in the presence of immune complex pellets from mock-transfected or catalytically inactive phosphatase-transfected cells.
8. This reprobing step is recommended to monitor that equal amounts of MAPKs have been subjected to the phosphatase assay.
9. Pilot experiments to generate the double-stable MCF-7-Tet-On cell lines were done according to the Tet-On® Advanced Inducible Gene Expression System User Manual (Clontech). A variety of cell lines which have integrated the Tet-On regulator plasmid (containing the *E. coli* "reverse" Tet repressor fused to the Herpes virus VP16 activation domain, rtTA) are commercially available. Alternatively, Tet-On plasmid can be transfected into the desired cell line to

obtain stable cell clones expressing the Tet repressor. Proceed as indicated for the pTRE-transfections and -clone selection, using the parental cell line of interest and the Tet-On plasmid. Generate several clones and test them for response efficiency performing transient transfections with a pTRE-Luciferase (or other suitable marker) plasmid. Note that, to allow for double-selection, the Tet-On and the pTRE plasmids have to contain different antibiotic resistance genes (pTet-On, neomycin resistance; and pTRE2hyg, hygromycin resistance, in the experiments in Subheading 3.3).

10. Transient transfections with the response plasmids pTRE (containing the cDNA of the protein of interest) are recommended to check that the transfection conditions and the pTRE plasmids are correct, and that protein species of the expected sizes are detected by Western blot, before starting the time-consuming isolation and testing of stable clones. In addition, these transient transfections may serve to optimize the transfection conditions of the cell line and the pTRE plasmids of interest.
11. This time of induction and doxycycline dose is a starting indication for the pilot expression experiments. The conditions to obtain maximal induction (if required) have to be adjusted for each cell line and ectopic protein to be expressed (see below).
12. Transfection of the Tet-On cell line with the empty pTRE vector is necessary to obtain negative control clones. Transfection with a catalytically inactive mutated form of the MKP is recommended as an additional control of specificity.
13. Note that neomycin and hygromycin have to be present in the cultures during all the cloning procedures and during the growing of the clones, to keep the selection for the presence of the Tet-On plasmid on the cells. The optimal concentration of neomycin and hygromycin (below toxicity) needs to be tested for each cell line (orientate range for neomycin: 100–500 μg/mL; for hygromycin: 50–200 μg/mL).
14. The time necessary for the isolation of the clones depends on the cell line and the antibiotics used (orientate range: 2–4 weeks). In our experiments with MCF-7-Tet-On cells, individual colonies were visible after 3 weeks, and clones were usually isolated and expanded after 2 additional weeks.
15. The time to reach semiconfluence is variable depending on the cell line. Also, intrinsic variability in this step is due to variations in the number of viable cells transferred to the wells from the original colonies.
16. Although the clones expressing higher levels of the protein of interest are usually chosen for further studies, clones can also be chosen at this step that express different levels of the protein of interest. To avoid artifact results due to clonality in the

experiments performed with the clones, it is recommended to expand, at least, two distinct clones displaying each desired level of protein expression. If colonies were not well separated when they were isolated from the plate, it is recommended to perform an additional cloning step by limiting dilution using 96-well plates.

17. Leaking (expression in the absence of the inducer) is minimized by growing the cells in medium containing tetracycline-free fetal calf serum.

Acknowledgments

This work was supported in part by grants SAF2006-08319 and SAF2009-10226 from Ministerio de Educación y Ciencia and Ministerio de Ciencia e Innovación, AP-117/08 and ACOMP/2009/363 and ACOMP/2010/222 from Generalitat Valenciana, and ISCIII-RETIC RD06/0020/0049 from Instituto de Salud Carlos III (Spain and Fondo Europeo de Desarrollo Regional), by grants from Fundación de Investigación Médica Mutua Madrileña (Spain), and by EU Research Training Network MRTN-CT-2006-035830. Caroline Nunes-Xavier has been funded by EU Marie Curie Research Training Program. We thank Jack Dixon for providing MKP3 cDNA, Steve Keyse for DUSP5 antibody and comments to the manuscript, and Robert P. Shiu for MCF-7 Tet-On cells. MKP4 and DUSP5 cDNAs were from Mammalian Gene Collection (Geneservice, UK). We thank Isabel Roglá for expert technical assistance.

References

1. Murphy, L. O., and Blenis, J. (2006) MAPK signal specificity: the right place at the right time. *Trends Biochem. Sci.* **31**, 268–275.
2. Janssens, V., and Goris, J. (2001) Protein phosphatase 2A: a highly regulated family of serine/threonine phosphatases implicated in cell growth and signalling. *Biochem. J.* **353**, 417–439.
3. Tamura, S., Li, M. G., Komaki, K., Sasaki, M., and Kobayashi, T. (2004) Roles of mammalian protein phosphatase 2C family members in the regulation of cellular functions. In *Topics in Current Genetics: Protein Phosphatases.* (Ariño, J., and Alexander, D. R., eds.) Springer-Verlag, Berlin Heidelberg, pp. 91–105.
4. Tamura, S., Toriumi, S., Saito, J., Awano, K., Kudo, T., and Kobayashi, T. (2006) PP2C family members play key roles in regulation of cell survival and apoptosis. *Cancer Sci.* **97**, 563–567.
5. Junttila, M. R., Li, S-P., and Westermarck, J. (2008) Phosphatase-mediated crosstalk between MAPK signaling pathways in the regulation of cell survival. *FASEB J.* **22**, 954–965.
6. Wang, P., Liu, P., Weng, J., Sontag, E., and Anderson, R. G. W. (2003) A cholesterol-regulated PP2A/HePTP complex with dual specificity ERK1/2 phosphatase activity. *EMBO J.* **22**, 2658–2667.
7. Takekawa, M., Maeda, T., and Saito, H. (1998) Protein phosphatase 2Cα inhibits the human stress-responsive p38 and JNK MAPK pathways. *EMBO J.* **17**, 4744–4752.
8. Takekawa, M., Adachi, M., Nakahata, A., Nakayama, I., Itoh, F., Tsukuda, H., Taya, Y.,

and Imai, K. (2000) p53-inducible Wip1 phosphatase mediates a negative feedback regulation of p38 MAPK-p53 signaling in response to UV radiation. *EMBO J.* **189**, 6517–6526.
9. Barr, A. J., and Knapp, S. (2006) MAPK-specific tyrosine phosphatases: new targets for drug discovery? *Trends Pharmacol. Sci.* **27**, 525–530.
10. Hendriks, W. J., Elson, A., Harroch, S., and Stoker A. (2007) Protein tyrosine phosphatases: functional inferences from mouse models and human diseases. *FEBS J.* **275**, 816–830.
11. Camps, M., Nichols, A., and Arkinstall, S. (2000) Dual specificity phosphatases: a gene family for control of MAP kinase function. *FASEB J.* **14**, 6–16.
12. Theodosiu, A., and Ashworth, A. (2002) MAP kinase phosphatases. *Genome Biol.* **3**, Reviews 3009.
13. Owens, D. M., and Keyse, S. M. (2007) Differential regulation of MAPK kinase signalling by dual-specificity protein phosphatases. *Oncogene.* **26**, 3203–3213.
14. Pulido, R., and Hooft van Huijsduijnen, R. (2007) Protein tyrosine phosphatases: dual-specificity phosphatases in health and disease. *FEBS J.* **275**, 848–866.
15. Farooq, A., and Zhou, M-M. (2004) Structure and regulation of MAPK phosphatases. *Cell. Signal.* **16**, 769–779.
16. Kondoh, K., and Nishida, E. (2007) Regulation of MAP kinases by MAP kinase phosphatases. *Biochim. Biophys. Acta.* **1773**, 1227–1237.
17. Cerignoli, F., Rahmouni, S., Ronai, Z., and Mustelin, T. (2006) Regulation of MAP kinases by the VHR dual-specific phosphatase. *Cell Cycle.* **5**, 2210–2215.
18. Alonso, A., Sasin, J., Bottini, N., Friedberg, I., Frieberg, I., Osterman, A., Godzik, A., Hunter, T., Dixon, J., and Mustelin, T. (2004) Protein tyrosine phosphatases in the human genome. *Cell.* **117**, 699–711.
19. Patterson, K. I., Brummer, T., O'Brien, P. M., and Daly, R. J. (2009) Dual-specificity phosphatases: critical regulators with diverse cellular targets. *Biochem. J.* **418**, 475–489.
20. Pulido, R. Zúñiga, Á., and Ullrich, A. (1998) PTP-SL and STEP protein tyrosine phosphatases regulate the activation of the extracellular signal-regulated kinases ERK1 and ERK2 by association through a kinase interaction motif. *EMBO J.* **17**, 7337–7350.
21. Zhou, B., Wu, L., Shen, K., Zhang, J., Lawrence, D.S., and Zhang, Z-Y. (2001) Multiple regions of MAP kinase phosphatase 3 are involved in its recognition and activation by ERK2. *J. Biol. Chem.* **276**, 6506–6515.
22. Tárrega, C., Blanco-Aparicio, C, Muñoz, J. J., and Pulido, R. (2002) Two clusters of residues at the docking groove of mitogen-activated protein kinases differentially mediate their functional interaction with the tyrosine phosphatases PTP-SL and STEP. *J. Biol. Chem.* **277**, 2629–2636.
23. Tanoue, T., and Nishida, E. (2003) Molecular recognitions in the MAP kinase cascades. *Cell. Signal.* **15**, 455–462.
24. Camps, M., Nichols, A., Gillieron, C., Antonsson, B., Muda, M., Chabert, C., Boschert, U, and Arkinstall, S. (1998) Catalytic activation of the phosphatase MKP-3 by ERK2 mitogen-activated protein kinase. *Science.* **280**, 1262–1265.
25. Dowd, S., Sneddon, A. A., and Keyse, S. M. (1998) Isolation of the human genes encoding the Pyst1 and Pyst2 phosphatases: characterisation of Pyst2 as a cytosolic dual-specificity MAP kinase phosphatase and its catalytic activation by both MAP and SAP kinases. *J. Cell Sci.* **111**, 3389–3399.
26. Hutter, D., Chen, P., Barnes, J., and Liu, Y. (2000) Catalytic activation of mitogen-activated protein (MAP) kinase phosphatase-1 by binding to p38 MAP kinase: critical role of the p38 C-terminal domain in its negative regulation. *Biochem. J.* **352**, 155–163.
27. Chen, P., Hutter, D., Yang, X., Gorospe, M., Davis, R. J., and Liu. Y. (2001) Discordance between the binding affinity of mitogen-activated protein kinase subfamily members for MAP kinase phosphatase-2 and their ability to activate the phosphatase catalytically. *J. Biol. Chem.* **276**, 29440–29449.
28. Farooq, A., Plotnikova, O., Chaturvedi, G., Yan, S., Zeng, L., Zhang, Q., and Zhou, M-M. (2003) Solution structure of the MAPK phosphatase PAC-1 catalytic domain: insights into substrate-induced enzymatic activation of MKP. *Structure.* **11**, 155–164.
29. Mark, J. K., Aubin, R. A., Smith, S., and Hefford, M. A. (2008) Inhibition of mitogen-activated protein kinase phosphatase 3 activity by interdomain binding. *J. Biol. Chem.* **283**, 28574–28583.
30. Fjeld, C. C., Rice, A. E., Kim, Y., Gee, K. R., and Denu, J. M. (2000) Mechanistic basis for catalytic activation of mitogen-activated protein kinase phosphatase 3 by extracellular signal-regulated kinase. *J. Biol. Chem.* **275**, 6749–6757.
31. Kang, T-H., and Kim, K-T. (2006) Negative regulation of ERK activity by VRK3-mediated

activation of VHR phosphatase. *Nat. Cell Biol.* **8**, 863–869.

32. Blanco-Aparicio, C., Torres, J., and Pulido, R. (1999) A novel regulatory mechanism of MAP kinases activation and nuclear translocation mediated by PKA and the PTP-SL tyrosine phosphatase. *J. Cell Biol.* **147**, 1129–1135.
33. Saxena, M., Williams, S., Taskén, K., and Mustelin, T. (1999) Crosstalk between cAMP-dependent kinase and MAP kinase through a protein tyrosine phosphatase. *Nat. Cell Biol.* **1**, 305–311.
34. Paul, S., Snyder, G. L., Yokakura, H., Picciotto, M. R., Nairn, A. C., and Lombroso, P. J. (2000) The dopamine/D1 receptor mediates the phosphorylation and inactivation of the protein tyrosine phosphatase STEP via a PKA-dependent pathway. *J. Neurosci.* **20**, 5630–5638.
35. Cao, W., Bao, C., Padalko, E., and Lowenstein, C. J. (2008) Acetylation of mitogen-activated protein kinase phosphatase-1 inhibits Toll-like receptor signalling. *J. Exp. Med.* **205**, 1491–1503.
36. Brondello, J. M., Pouysségur, J., and McKenzie, F. R. (1999) Reduced MAP kinase phosphatase-1 degradation after p42/p44MAPK-dependent phosphorylation. *Science.* **286**, 2514–2517.
37. Marchetti, S., Gimond, C., Chambard, J-C., Touboul, T., Roux, D., Pouysségur, J., and Pagès, G. (2005) Extracellular signal-regulated kinases phosphorylate mitogen-activated protein kinase phosphatase 3/DUSP6 at serines 159 and 197, two sites critical for its proteasomal degradation. *Mol. Cell. Biol.* **25**, 854–864.
38. Katagiri, C., Masuda, K., Urano, T., Yamashita, K., Araki, Y., Kikuchi, K., and Shima, H. (2005) Phosphorylation of Ser-446 determines stability of MKP-7. *J. Biol. Chem.* **280**, 14716–14722.
39. Torres, J., Blanco-Aparicio, C., and Pulido, R. (2004) Regulation of MAPK cascades by protein tyrosine phosphatases. In *Methods in Molecular Biology, vol. 250: MAP Kinase Signaling Protocols.* (Seger, R, ed.) Humana Press, Totowa, pp. 103–112.
40. Mandl, M., Slack, D. N., and Keyse, S. M. (2005) Specific inactivation and nuclear anchoring of extracellular signal-regulated kinase 2 by the inducible dual-specificity protein phosphatase DUSP5. *Mol. Cell. Biol.* **25**, 1830–1845.
41. Gossen, M., Freundlieb, S., Bender, G., Muller, G., Hillen, W., and Bujard, H. (1995) Transcriptional activation by tetracycline in mammalian cells. *Science.* **268**, 1766–1769.
42. Venditti, M., Iwasiow, B., Orr, F. W., and Shiu, R. P. (2002) C-myc gene expression alone is sufficient to confer resistance to antiestrogen in human breast cancer cells. *Int. J. Cancer* **99**, 35–42.
43. Flint, A. J., Tiganis, T., Bardford, D., and Tonks, N. K. (1997) Development of "substrate-trapping" mutants to identify physiological substrates of protein tyrosine phosphatases. Proc. Natl. Acad. Sci. U. S. A. **94**, 1680–1685.
44. Tárrega, C., and Pulido, R. (2009) A one-step method to identify MAP kinase residues involved in inactivation by tyrosine- and dual-specificity protein phosphatases. *Anal. Biochem.* **394**, 81–86.

Chapter 19

Proteomic Analysis of Scaffold Proteins in the ERK Cascade

Melissa M. McKay and Deborah K. Morrison

Abstract

ERK cascade scaffolds serve as docking platforms to coordinate the assembly of multiprotein complexes that contribute to the spatial and temporal control of ERK signaling. Given that protein–protein interactions are essential for scaffold function, determining the full repertoire of scaffold binding partners will likely provide new insight into the regulation and activities of the ERK cascade scaffolds. In this chapter, we describe methods to identify scaffold interacting proteins using a proteomics approach. This protocol is based on the affinity purification of scaffold complexes from tissue culture cells and utilizes mass spectrometry to identify the protein constituents of the complex.

Key words: ERK cascade scaffolds, Scaffold binding partners, Proteomics, Mass spectrometry, Affinity purification, Pyo/Glu-Glu tag

1. Introduction

The ERK cascade is activated in response to many extracellular and intracellular signaling cues and, in collaboration with other effector cascades, functions to transduce these signals into specific biological responses (1, 2). Like other MAPK cascades, the ERK cascade is a three-tiered kinase module, comprised of the Raf, MEK, and ERK protein kinases (3). In addition to these core components, a number of protein scaffolds have been identified that provide crucial spatial and temporal control of ERK cascade signaling (4, 5). In mammals, these proteins include MP-1, IQGAP, Paxillin, Sef, Kinase Suppressor of Ras 1 (KSR1), and KSR2 (6–11). By interacting with some or all of the core kinase components of the cascade, ERK scaffolds can increase the efficiency and specificity of ERK activation. These scaffolds can also recruit positive and negative regulators and coordinate feedback loops to modulate the intensity and duration of ERK cascade signaling.

Rony Seger (ed.), *MAP Kinase Signaling Protocols: Second Edition*, Methods in Molecular Biology, vol. 661,
DOI 10.1007/978-1-60761-795-2_19,

Moreover, an emerging theme for the ERK scaffolds is that through their distinct subcellular localizations, they can target the ERK module to specific cellular compartments and/or targets (12). Notably, each of these scaffold functions is mediated by the association of the scaffold with a particular set of proteins. Thus, determining the full repertoire of proteins that a scaffold interacts with will likely provide further insight into the regulation and activities of the ERK scaffolds.

In this chapter, we describe a protocol for the analysis of ERK scaffold complexes using a proteomics approach. Our laboratory has successfully used this methodology to identify new binding partners for the KSR1 and KSR2 scaffolds, and further analysis of these interactions has revealed important regulatory mechanisms and functional properties of the KSR scaffold family (11, 13–15). This protocol involves the affinity purification of scaffold complexes from tissue culture cells and utilizes mass spectrometry to identify the protein constituents of the complex. Our laboratory exclusively uses the Pyo-epitope tag (16, 17) for the affinity purification of protein complexes for mass spectrometry analysis. We find that the mouse monoclonal antibody recognizing the Pyo tag is highly specific and results in lower nonspecific protein binding, in comparison to other antibody/tag combinations such as FLAG, HA, or Myc-9E10. The methodology presented allows for the large-scale and unbiased identification of scaffold binding partners and should prove useful for the study of other ERK cascade scaffolds, as well as the JIP family of scaffolds that regulate the p38 and JNK MAPK cascades.

2. Materials

2.1. Generation of Pyo-Affinity Resin

Unless specifically noted, all solutions are prepared using deionized, distilled water and chemical reagents from Sigma-Aldrich (St. Louis, MO).

1. Culture supernatant from hybridoma cells producing the Pyo monoclonal antibody.
2. Protein G Sepharose 4 Fast Flow affinity resin (GE Healthcare, Piscataway, NJ). Stored at 4°C.
3. Phosphate-buffered saline (PBS): 10 mM sodium phosphate buffer (pH 7.4), 137 mM NaCl, 1.5 mM KCl. Stored at 4°C.
4. 0.2 M sodium borate (pH 9.0), made fresh on the day of use. To prepare, dissolve 3.81 g of sodium borate in 40 mL of water by warming to 60°C. Allow solution to cool to room temperature. If sodium borate precipitates out of solution,

add a small volume of water, rewarm, and cool completely. Adjust pH to 9.0 using concentrated HCl, and bring the final volume up to 50 mL with water.

5. Dimethyl pimelimidate (DMP).
6. 0.2 M ethanolamine (pH 8.0). To prepare, dissolve 1.95 g of ethanolamine in 75 mL of water, adjust the pH to 8.0 with 1 N NaOH, and bring the final volume up to 100 mL with water.
7. 0.01% merthiolate in PBS. Note that merthiolate is very toxic, and care should be taken to avoid inhalation, ingestion, or contact with skin.

2.2. Cell Culture, Protein Expression, and Cell Lysis

1. Cells: 293T (see Note 1).
2. Mammalian expression vector encoding Pyo-tagged protein of interest. Typically, we add two tandem copies of the Pyo-tag (EYMPME: 5′-GAG TAT ATG CCC ATG GAG-3′) to the N-terminus of a protein (following an ATG start codon); however, depending on the functional properties of the protein, it may be more desirable to place the tag at the C-terminus.
3. FuGENE 6 (Roche, Indianapolis, IN) for transfection of cells.
4. Dulbecco's Modified Eagle Medium (DMEM) (Invitrogen, Carlsbad, CA).
5. Growth Medium: DMEM supplemented with 10% heat-inactivated fetal bovine serum, 1% penicillin/streptomycin (Invitrogen), and 2 mM L-glutamine (Invitrogen).
6. PBS (see Subheading 2.1, item 3).
7. Low Salt Triton X-100 lysis buffer: 30 mM Tris–HCl (pH 8.0), 75 mM NaCl, 10% glycerol, 1% Triton X-100 containing protease inhibitors (0.15 units/mL aprotinin, 20 μM leupeptin, 1 mM phenylmethylsulfonyl fluoride (PMSF)) and phosphatase inhibitors (0.5 mM sodium vanadate, 0.1 μM calyculin A). Prepare fresh using high purity Triton X-100 (Surfact Amps X-100, supplied as a 10% solution, cat# 28314, Thermo Scientific, Waltham, MA) and prechilled HPLC-grade water. Protease and phosphatase inhibitors should be added from concentrated stock solutions immediately prior to use. In particular, PMSF has a very short half-life in aqueous solutions (~20 min).
   (a) 200 mM PMSF: dissolve 0.348 g of PMSF in 10 mL ethanol. Aliquot and store protected from light at −20°C. Note, PMSF will precipitate from solution at −20°C, so briefly warm at 37°C and vortex prior to use.
   (b) 10 mg/mL leupeptin: dissolve 50 mg leupeptin (Roche) in 5 mL of water. Aliquot and store at −20°C.

(c) 100 mM sodium vanadate: for 10 mL dissolve 183.9 mg sodium vanadate in 10 mL water and adjust to pH 10 (yellow in color). Boil until solution turns colorless then cool to room temperature and adjust the pH to 9. Repeat boil/cool/pH step until solution remains at pH 9. Make 1 mL aliquots and store at −20°C or, if in frequent use, at 4°C. Note, if precipitate is observed after thawing, warm at 37°C and vortex until the solution is clear.

(d) 20 μM calyculin A: resuspend 10 μg calyculin A (Cell Signaling Technology, Danvers, MA) in 500 μL of 50% ethanol/50% water. Store at 4°C.

2.3. Affinity Purification of Scaffold Complexes for Mass Spectrometry Analysis

The following materials and solutions should be designated for use exclusively in proteomic analysis and should always be handled with gloved hands to prevent keratin contamination (see Note 2).

1. Pyo affinity resin (*from* Subheading 2.1). Alternatively, Pyo/Glu-Glu Affinity Matrix (Cat # AFC-115P, Covance Inc., Princeton, NJ).
2. Low Salt Triton X-100 lysis buffer (see Subheading 2.2, item 7).
3. HPLC-grade water (CHROMASOLV for HPLC, Sigma).
4. 2× gel sample buffer: prepared by diluting 5× sample buffer (250 mM Tris–HCl (pH 6.8), 30% glycerol, 10% sodium dodecyl-sulfate (SDS), 500 mM dithiothreitol (DTT), 0.2% bromophenol blue) with water. 5× and 2× are stored at −20°C.
5. 4–20% Tris–glycine polyacrylamide minigel, 1.0 mm × 10 wells (Invitrogen) (see Note 3).
6. SDS–polyacrylamide gel electrophoresis (PAGE) apparatus. Our laboratory has dedicated a gel electrophoresis apparatus for use only in the preparation of samples for mass spectrometry analysis, and it is kept stored in a plastic container to reduce keratin contamination from dust.
7. 1× Novex Tris–glycine SDS running buffer: prepared fresh in a clean, well-rinsed graduated cylinder by diluting 10× Novex Tris–glycine SDS running buffer (Invitrogen) with HPLC-grade water.
8. Prestained SDS–PAGE protein standards (e.g., BioRad Broad Range, BioRad, Hercules, CA).
9. Nalge S700 plastic box (cat# 1917H42, Thomas Scientific, Swedesboro, NJ) for staining and destaining of polyacrylamide gels.
10. Fix/Stain solution: prepare 100 mL by pipetting 70 mL HPLC-grade water, 20 mL methanol, 5 mL acetic acid, and

5 mL Coomassie Brilliant Blue R250 (BioRad) directly into the plastic gel box using sterile pipettes (see Note 4).

11. Destain solution: for 100 mL combine 75 mL HPLC-grade water, 20 mL methanol, and 5 mL acetic acid. Use sterile pipettes to prepare directly in the plastic gel box.
12. 5% acetic acid solution in HPLC-grade water. Use sterile plastic pipettes to prepare directly in the plastic gel box.
13. 1.5 mL Eppendorf-style plain microfuge tubes. Do not use tubes that are colored, have O-rings, or have been chemically treated. We use Eppendorf Safe-Lock Tubes (Eppendorf part #022363204), which are available from many commercial sources.
14. Sterile #10 blade scalpels.

3. Methods

3.1. Generation of Pyo-Affinity Resin

To affinity purify Pyo-tagged protein complexes for proteomic analysis, we routinely use protein G sepharose beads in which the Pyo monoclonal antibody has been covalently coupled to the bead matrix. The following is a protocol for the generation of the Pyo-affinity resin; however a Pyo/Glu-Glu affinity resin is commercially available from Covance, Inc. that gives comparable results. This methodology can also be adapted for the coupling of other antibodies to the bead matrix.

1. Wash 500 μL of protein G sepharose beads with 14 mL PBS in a 15 mL conical tube.
2. Pellet beads by spinning at 228×*g* for 10 min and aspirate wash buffer.
3. Resuspend beads in 1 mL of PBS and transfer to a 50 mL conical tube containing 48 mL Pyo hybridoma tissue culture supernatant. Rinse 15 mL tube with 500 μL PBS and transfer residual beads to 50 mL tube containing the Pyo supernatant.
4. Incubate for 4 h at 4°C with rocking.
5. Pellet beads by spinning at 228×*g* for 10 min. Decant the supernatant into a clean tube, leaving 3 mL to resuspend the bead pellet. Transfer resuspended beads to a clean 15 mL conical tube. Rinse 50 mL tube with 3 mL of the decanted supernatant and transfer any residual beads to the 15 mL tube.
6. Pellet beads by spinning at 228×*g* for 5 min (use these conditions for all subsequent bead pelleting steps). Aspirate

supernatant and wash beads twice with 12 mL 0.2 M sodium borate [pH 9.0].

7. Pellet beads, aspirate supernatant, and resuspend beads in 14 mL 0.2 M sodium borate [pH 9.0]. Take a 200 μL aliquot for test sample #1.
8. To crosslink the antibody to beads, add 72.52 mg dimethyl pimelimidate (DMP) (final concentration 20 mM) and incubate for 30 min at room temperature with rocking.
9. Pellet beads, aspirate supernatant, and wash beads once with 0.2 M ethanolamine [pH 8.0].
10. Pellet beads, aspirate supernatant, and resuspend bead pellet in 14 mL 0.2 M ethanolamine [pH 8.0]. Incubate for 2 h at room temperature on a rocker. Take a 200 μL aliquot for test sample #2.
11. Pellet beads, aspirate supernatant, and wash beads once with PBS.
12. Pellet beads, aspirate supernatant, and resuspend bead pellet in 1 mL PBS containing 0.01% merthiolate. Store beads at 4°C.
13. Test antibody coupling efficiency.
    (a) Pellet beads in test samples and wash three times with PBS.
    (b) Resuspend bead pellet in 20 μL 2X SDS-gel sample buffer and analyze by standard SDS-PAGE and Coomassie Brilliant blue staining. If the antibody has been successfully coupled, the IgG heavy chain at 50 kDa will be visible in sample #1, but not in sample #2.

3.2. Cell Culture, Protein Expression, and Cell Lysis

1. Seed 16 10 cm tissue culture dishes with 293T cells at a concentration of 1.0×10^6 cells/dish in Complete Media (see Note 5). Incubate cells overnight at 37°C with 5% CO_2.
2. 16–18 h after plating, take eight dishes and transfect cells with the plasmid for expression of the Pyo-tagged protein using FuGENE 6 reagent. Transfect 6 μg of plasmid DNA per 10 cm dish of cells using a 2:1 DNA:FuGENE 6 ratio according to the manufacturer's instructions. After transfection, incubate cells at 37°C with 7% CO_2 for 48 h. The remaining eight dishes will serve as the untransfected cell control.
3. 48 h after transfection, decant media from each dish and carefully wash the cell monolayer with 5 mL ice-cold PBS. 293T cells may be very loosely adherent at this stage, and if many of the cells detach from the dish during the wash step, scrape the cells into the PBS using a rubber policeman and proceed as indicated in the next step.
4. Decant wash, add 4 mL PBS to each dish, and scrape cells into the PBS solution using a rubber policeman. Transfer the

cell suspension to a prechilled 50 mL conical tube, pooling all eight dishes per condition in one tube. Pellet cells by spinning at 3,000 × *g* for 10 min at 4°C.

5. Carefully aspirate supernatant and resuspend each cell pellet in 1.5 mL Low Salt Triton X-100 lysis buffer. Transfer each lysate to a clean prechilled microfuge tube (designated for proteomics work) and incubate on ice for 20 min, vortexing every 5 min to mix.
6. Centrifuge the samples at 14,000 × *g* for 10 min in a 4°C microfuge to pellet unbroken cells, nuclei, and insoluble debris.
7. Transfer lysate to a clean prechilled microfuge tube (designated for proteomics work) and centrifuge samples at 14,000 × *g* for 5 min in a 4°C microfuge.
8. Transfer lysate to a clean prechilled microfuge tube (designated for proteomics work), making sure not to transfer any residual debris.

3.3. Affinity Purification of Scaffold Complexes for Mass Spectrometry

The following is a protocol that we routinely use for the preparation of samples for mass spectrometry (MS) analysis. However, because MS facilities can differ in their requirements for sample preparation and submission, the facility that will be processing the samples should be contacted prior to sample preparation to discuss their specific protocols and requirements. It is also imperative that precautions are taken to avoid keratin contamination during all steps of the protocol.

1. Determine the protein concentration of the lysates from Subheading 3.2 using a standard Bradford assay. Typically, we find that the lysate from control untransfected cells has a higher protein concentration than does the lysate from transfected cells expressing the Pyo-tagged protein. If this is the case, normalize the lysates by transferring to a clean prechilled microfuge tube a volume of control untransfected cell lysate that is equivalent in total protein amount to that present in the ~1.5 mL transfected cell lysate. Equalize the volume of the two samples using Low Salt Triton-X100 lysis buffer.
2. To each lysate sample, add 25–40 μL Pyo-affinity resin (from Subheading 3.1), taking the aliquot directly from the resin pellet. Note, to prevent loss of the Pyo-affinity resin, we do not resuspend the resin by mixing. Spin samples at 3,000 × *g* for 2 min in microfuge and check pellet to ensure that an equivalent amount of resin was added to each sample. Incubate samples for 4 h at 4°C with rocking.
3. After incubation, spin samples at 3,000 × *g* for 2 min in a microfuge to pellet resin. Using a clean pipette tip for each

sample, carefully aspirate the supernatant making sure not to disturb the resin pellet.

4. Wash resin pellet, by adding 1 mL of Low Salt Triton X-100 lysis buffer and inverting each tube several times. Note, do not vortex samples as vortexing may cause the sepharose bead matrix to rupture, allowing nonspecific proteins to be trapped inside the bead.
5. Repeat resin wash step 4 more times.
6. On final wash, carefully aspirate the supernatant such that only the resin pellet remains. Immediately resuspend resin pellet in 25 μL boiling hot 2× gel sample buffer, and boil samples for 6 min.
7. Briefly spin samples in a microfuge to pellet beads. Load entire sample into one well of a 10 well 4–20% Tris–glycine polyacrylamide minigel. Load protein markers and run gel at 15–20 mAmps.
8. After electrophoresis, the following steps put the gel in direct contact with the lab environment and are particularly susceptible to keratin contamination. Extra care should be taken to make sure that hands are double-gloved and that all solutions are handled appropriately. Avoid directly touching the gel even when wearing gloves, and if the gel needs to be manipulated, do so using use a sterile pipette tip.
9. Carefully separate the gel plates, allowing the gel to remain on one plate. Cut and remove stacking gel using a clean razor blade. Carefully invert the plate containing the gel over the plastic gel box that contains 100 mL Fix/Stain solution and allow the gel to drop into the Fix/Stain solution. Incubate at room temperature with rocking until protein bands are visualized.
10. Decant Fix/Stain solution and destain gel in 100 mL Destain solution at room temperature with rocking. Change solution as needed to destain gel such that the background is clear but protein bands are still visible.
11. After destaining, incubate gel in 100 mL 5% acetic acid for at least 1 h (gel can be stored in this buffer). An example of a Coomassie-stained gel showing Pyo-affinity purified KSR2 scaffold complexes is showed in Fig. 1.
12. To cut protein bands for MS analysis, transfer the gel to a sterile 150 mm tissue culture dish containing 10 mL HPLC-grade water and place on a light box. Number Eppendorf tubes S1, S2, etc., for bands cut from the Pyo-tagged protein sample lane and C1, C2, etc., for bands cut from the untransfected control lane. Using a sterile scalpel, cut slices of approximately 5 mm in width moving from the top of the gel to the bottom and transfer the gel slice into the Eppendorf

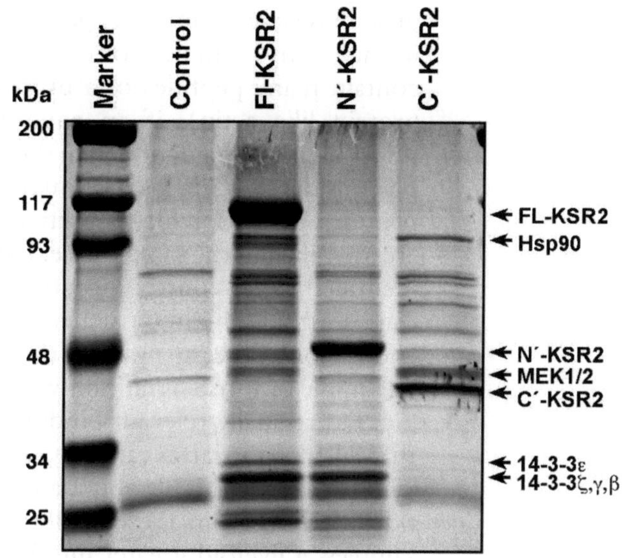

Fig. 1. Coomassie stained gel of KSR2 scaffold complexes. Pyo-tagged full-length (FL) KSR2 or Pyo-tagged *N*-terminal (*N*′) or *C*-terminal (C′) domains of KSR2 were expressed in 293T cells. Scaffold complexes were isolated from cell lysates using the Pyo-affinity resin and resolved by electrophoresis on a 4–20% Tris–glycine polyacrylamide gel. Arrows indicate the KSR2 proteins, as well as some of the KSR2 binding partners identified by MS analysis.

tube using the scalpel. Cut slices from both the sample and control lanes, making sure to match the size and position of each control slice with the corresponding sample slice. Typically, we take a picture of the gel before cutting and then label the picture to indicate the corresponding gel slice/tube number. By cutting out the entire lane, one does not limit the identification of potential scaffold binding partners based upon size or band staining. However, if this is cost-prohibitive, a smaller number of sample bands can be excised, making sure to cut a corresponding slice from the control lane.

13. Submit samples to designated facility for MS analysis or store frozen at −20°C.

3.4. Analysis of Mass Spectrometry Data

Proteins are identified by matching the MS ion spectra of the peptides to SwissProt and NCBI databases using search algorithms like SEQUEST. Therefore, it is important to provide the MS facility with the species of the cell line from which the protein complexes were derived so that proper databases can be searched. From the MS analysis, a list will be generated of the peptides derived from a particular protein that is present in each gel slice sample. Determining which of the identified proteins may be a functionally relevant scaffold binding partner requires a comparison to be made between the number of peptides for a particular protein recovered in the experimental sample and that recovered in the control sample. For example, although a sample slice may contain many peptides of a protein (such as a highly abundant protein like actin), if an equivalent number of peptides for this protein is also identified in the control slice, the protein would be considered a nonspecific interactor. In contrast, if only a few peptides of a protein (in particular, a low abundance protein) are identified in the sample slice, but no peptides of this protein are identified in the control slice, this protein may represent a significant scaffold binding partner. Typically, the proteins that we choose for further analysis are those found only in the experimental sample or those significantly enriched in the experimental sample. Functionally relevant binding partners for the KSR scaffolds that fall into the latter category are members of the 14-3-3 family. The 14-3-3s are highly abundant proteins, and though a few peptides derived from these molecules are often found in control samples, a fivefold greater number of 14-3-3 peptides are found in KSR samples. It is also important to note that the MS analysis will detect proteins associated with the translation and folding of the overexpressed Pyo-tagged scaffold, such as Hsp70, chaperonins, and ribosomal binding proteins, and these interacting proteins are not routinely investigated further.

3.5. Confirmation of Interactions

Binding interactions identified in the proteomics analysis will need to be further validated. Typically, we confirm these interactions by co-immunoprecipitation assays in which the scaffold complex is immunoprecipitated under conditions similar to those described in Subheading 3.3, and the presence of the putative binding partner is detected by immunoblot analysis. Co-immunoprecipitation assays can be conducted using transfect cells or cell lines stably expressing the Pyo-tagged protein, or cells with endogenous expression of the proteins of interest. Once the interactions have been confirmed, these assays can also be adapted to evaluate the interactions under various signaling conditions (e.g., serum starvation or growth factor treatment) or using mutant proteins to determine the residues/domains required for binding.

4. Notes

1. We typically use 293T cells for proteomic studies because they are easily transfectable, exhibit good protein expression, and have a broad endogenous protein expression pattern. In particular, 293T cells are preferred over HeLa cells, in that though HeLa cells are easily transfectable, the endogenous protein expression profile is more restricted and the expression of

certain genes has been silenced by methylation in this carcinoma line (e.g., LKB1). We have, however, successfully performed proteomic studies in other cell lines that are difficult to transfect by using adenoviral or retroviral expression systems to transiently or stably express the Pyo-tagged protein.

2. MS protein analysis is a highly sensitive technique, and the major cause of sample contamination is keratin found on skin and in dust. We have found that the following precautions can significantly reduce the risk of keratin contamination. First, purchase fresh reagents for use exclusively in mass spectrometry sample preparation and keep them stored away from general lab use. Second, wear double gloves at all times when working with samples and handling proteomics reagents. Third, rinse all surfaces that may contact the gel with HPLC-grade water. These surfaces include the outside of gloved hands, gel electrophoresis apparatus, staining boxes, razor blade, and scalpels.
3. Optimal MS analysis requires high-quality gel electrophoresis. It is important to have well-focused bands so that the protein is concentrated in a minimum amount of gel. In addition, a gel of 1.0 mm thickness is preferred given that the recovery of peptides after gel digestion is reduced in thicker gels and that there is greater loss of protein during the staining/destaining process in thinner gels. In our studies, we have had excellent success using commercially available 4–20% gradient polyacrylamide minigels. Moreover, the use of commercially available precast minigels can help reduce keratin contamination.
4. Excessive stain in a gel slice can interfere with the MS analysis. Diluting the Coomassie Brilliant Blue stain in Fix/Destain solution allows the gel to be fixed and stained in one step and also permits enough staining to visualize protein bands, while reducing the time required for destaining.
5. Each Pyo-tagged construct should be tested for expression levels in the cell line that will be used for affinity purification of the scaffold complexes, to determine the number of cells/dishes that will be required for MS sample preparation. In general, we have found that for most well-expressed proteins, lysate from 6 to 10 10-cm dishes of transfected 293T cells is sufficient for sample preparation.

References

1. Shaul, Y. D., and Seger, R. (2007) The MEK/ERK cascade: from signaling specificity to diverse functions. *Biochim Biophys Acta* **1773**, 1213–26.
2. Yoon, S., and Seger, R. (2006) The extracellular signal-regulated kinase: multiple substrates regulate diverse cellular functions. *Growth Factors* **24**, 21–44.
3. Qi, M., and Elion, E. A. (2005) MAP kinase pathways. *J Cell Sci* **118**, 3569–72.
4. Morrison, D. K., and Davis, R. J. (2003) Regulation of MAP kinase signaling modules

by scaffold proteins in mammals. *Annu Rev Cell Dev Biol* **19**, 91–118.

5. Kolch, W. (2005) Coordinating ERK/MAPK signalling through scaffolds and inhibitors. *Nat Rev Mol Cell Biol* **6**, 827–37.
6. Schaeffer, H. J., Catling, A. D., Eblen, S. T., Collier, L. S., Krauss, A., and Weber, M. J. (1998) MP1: a MEK binding partner that enhances enzymatic activation of the MAP kinase cascade. *Science* **281**, 1668–71.
7. Roy, M., Li, Z., and Sacks, D. B. (2005) IQGAP1 is a scaffold for mitogen-activated protein kinase signaling. *Mol Cell Biol* **25**, 7940–52.
8. Ishibe, S., Joly, D., Liu, Z. X., and Cantley, L. G. (2004) Paxillin serves as an ERK-regulated scaffold for coordinating FAK and Rac activation in epithelial morphogenesis. *Mol Cell* **16**, 257–67.
9. Torii, S., Kusakabe, M., Yamamoto, T., Maekawa, M., and Nishida, E. (2004) Sef is a spatial regulator for Ras/MAP kinase signaling. *Dev Cell* **7**, 33–44.
10. Therrien, M., Michaud, N. R., Rubin, G. M., and Morrison, D. K. (1996) KSR modulates signal propagation within the MAPK cascade. *Genes Dev* **10**, 2684–95.
11. Dougherty, M. K., Ritt, D. A., Zhou, M., Specht, S. I., Monson, D. M., Veenstra, T. D., and Morrison, D. K. (2009) KSR2 is a calcineurin substrate that promotes ERK cascade activation in response to calcium signals. *Mol Cell* **34**, 652–62.
12. Casar, B., Arozarena, I., Sanz-Moreno, V., Pinto, A., Agudo-Ibanez, L., Marais, R., Lewis, R. E., Berciano, M. T., and Crespo, P. (2009) Ras subcellular localization defines extracellular signal-regulated kinase 1 and 2 substrate specificity through distinct utilization of scaffold proteins. *Mol Cell Biol* **29**, 1338–53.
13. Muller, J., Ory, S., Copeland, T., Piwnica-Worms, H., and Morrison, D. K. (2001) C-TAK1 regulates Ras signaling by phosphorylating the MAPK scaffold, KSR1. *Mol Cell* **8**, 983–93.
14. Ory, S., Zhou, M., Conrads, T. P., Veenstra, T. D., and Morrison, D. K. (2003) Protein phosphatase 2A positively regulates Ras signaling by dephosphorylating KSR1 and Raf-1 on critical 14-3-3 binding sites. *Curr Biol* **13**, 1356–64.
15. Ritt, D. A., Zhou, M., Conrads, T. P., Veenstra, T. D., Copeland, T. D., and Morrison, D. K. (2007) CK2 Is a component of the KSR1 scaffold complex that contributes to Raf kinase activation. *Curr Biol* **17**, 179–84.
16. Schaffhausen, B., Benjamin, T. L., Pike, L., Casnellie, J., and Krebs, E. (1982) Antibody to the nonapeptide Glu-Glu-Glu-Glu-Tyr-Met-Pro-Met-Glu is specific for polyoma middle T antigen and inhibits in vitro kinase activity. *J Biol Chem* **257**, 12467–70.
17. Grussenmeyer, T., Scheidtmann, K. H., Hutchinson, M. A., Eckhart, W., and Walter, G. (1985) Complexes of polyoma virus medium T antigen and cellular proteins. *Proc Natl Acad Sci U S A* **82**, 7952–4.

Chapter 20

Analysis of ERKs' Dimerization by Electrophoresis

Adán Pinto and Piero Crespo

Abstract

Signals transmitted by ERK MAP Kinases regulate the functions of multiple substrates present in the nucleus and the cytoplasm. Once phosphorylated, ERKs dimerize. The functions of these dimers had remained elusive until recently when we demonstrated that ERK dimers are assembled using scaffolds proteins as platforms. Dimerization is critical for connecting the scaffolded ERK complex to cognate cytoplasmic substrates. Contrarily, nuclear substrates associate to ERK monomers. These results identify dimerization as a key determinant of the spatial specificity of ERK signals. Moreover, we showed that preventing ERK dimerization, without affecting ERK phosphorylation, is sufficient for attenuating cellular proliferation, transformation, and tumor development. Thus, analyzing ERK dimerization will be an important factor in the future for determining, for example, the real impact on the ERK pathway of some drugs that do not affect ERK phosphorylation. Herein, we describe user-friendly methods for such purpose.

Key words: MAP kinases, ERK, Dimerization, PAGE-electrophoresis, Native gels

1. Introduction

ERK 1 and 2 mitogen-activated protein kinases (MAPKs) are cytoplasmic serine/threonine kinases that participate in the transduction of signals from the surface to the interior of the cell. ERKs become activated in response to multiple stimuli, including those that regulate cellular proliferation, differentiation, and survival. Once activated, they phosphorylate a broad spectrum of substrates that are distributed throughout different subcellular compartments, including the nucleus and the cytoplasm.

Upon phosphorylation, ERKs dimerize. These dimers are primarily ERK1 and ERK2 homodimers, ERK1–ERK2 heterodimers being unstable (1). However, until recently, a clear notion of the biochemical and biological significance of ERK dimerization was missing.

Rony Seger (ed.), *MAP Kinase Signaling Protocols: Second Edition*, Methods in Molecular Biology, vol. 661, DOI 10.1007/978-1-60761-795-2_20, © Springer Science+Business Media, LLC 2010

ERK signals are modulated by several types of regulatory proteins. Scaffold proteins serve such a purpose by linking the different components of the signaling cascade into a multienzymatic complex by which ERK signals amplitude, duration, and spatial specificity are regulated (2), but the precise mechanisms whereby scaffolds can achieve spatial specificity for ERK signals have been elusive.

Recently, we have demonstrated that upon activation, scaffold proteins serve as platforms in which ERK dimers are assembled, forming complexes capable of interacting with and activating cognate cytoplasmic substrates. Conversely, scaffolds and ERK dimers are dispensable for the activation of ERK nuclear substrates. In addition, preventing the formation of ERK dimers, without ERKs total activation levels being altered, has a profound, inhibitory impact on cellular processes essential for tumor progression (3), pointing to ERKs dimerization as a potential target for antineoplastic therapeutic intervention.

Herein, we describe two easy methods for analyzing ERK dimerization "in vivo". The first one, based on "traditional" PAGE-SDS electrophoresis, is the one that we utilized in our aforementioned publication (3). It was originally described by Philipova and Whitaker, (4) and they deserve full credit for its design. The second one is based on native nondenaturing electrophoresis, and it has been set up in our laboratory. Both of them serve as reliable methods for studying ERK dimerization in cells in culture.

2. Materials

2.1. Cell Culture and Lysis

1. Dulbecco's Modified Eagle's Medium (DMEM) (LONZA, Verviers, Belgium) supplemented with 10% fetal bovine serum (Gibco/BRL, Bethesda, MD).
2. Phosphate buffered saline (PBS).
3. Solution of trypsin (0.25%) and ethylenediamine tetraacetic acid (EDTA) (1 mM) from Gibco/BRL.
4. Epidermal growth factor (EGF, Sigma, USA). Working solution is prepared to 50 μg/mL with BSA and acetic acid in water.
5. MAPK assay cell lysis buffer: 20 mM Hepes, pH 7.5, 10 mM ethylene glycol tetraacetic acid (EGTA), 40 mM glycerol 2-phosphate, 1% (w/v) NP-40 (Sigma), 2.5 mM $MgCl_2$, 2 mM orthovanadate (Sigma), 1 mM dithiothreitol. Store at 4°C and supplement with phosphatase and protease inhibitors (Roche) (see Note 1).
6. Teflon cell scrapers (Sarsted).

2.2. SDS-Polyacrylamide Gel Electrophoresis for ERK2 Monomer/Dimer Separation

1. Lower buffer: 1.5 M Tris–HCl, pH 8.8.
2. Upper buffer: 1 M Tris–HCl, pH 6.8.
3. Thirty percent acrylamide/bis solution (37.5:1) (acrylamide is a neurotoxin when unpolymerized, so care should be taken to avoid exposure) and *N,N,N,N'*-Tetramethyl-ethylenediamine (TEMED).
4. Ammonium persulfate (APS): prepare 10% solution and store at 4°C.
5. Running buffer: 25 mM Tris–HCl, 200 mM glycine, and 0.1% (w/v) SDS. Store at room temperature.
6. Laemli 5× SDS-PAGE sample buffer: 10 mM Tris–HCl 1 M pH 6.8, 2% SDS, 50% glycerol, 0.04% bromophenol blue in water.
7. Prestained molecular weight markers: Novex® Sharp Pre-Stained Protein Standards (INVITROGEN).

2.3. Native Gel Electrophopresis for ERK2 Monomer/Dimer Separation

1. Lower buffer: 1.5 M Tris–HCl, pH 8.8.
2. Upper buffer: 1 M Tris–HCl, pH 6.8.
3. Thirty percent acrylamide/bis solution (37.5:1) (acrylamide is a neurotoxin when unpolymerized, so care should be taken to avoid exposure) and *N,N,N,N'*-Tetramethyl-ethylenediamine (TEMED).
4. Ammonium persulfate (APS): prepare 10% solution and store at 4°C.
5. Running buffer: 25 mM Tris–HCl, 200 mM glycine. Store at room temperature.
6. Native 2× sample buffer: 20% glycerol, 0.02% (w/v) bromophenol blue and 0.126 M Tris–HCl pH 6.8.

2.4. Western Blotting for ERK2 Monomer/Dimer

1. Transfer buffer: 25 mM Tris–HCl, 200 mM glycine. Store in the transfer apparatus at room temperature (with cooling during use, see Note 2).
2. Nitrocellulose transfer membrane (PROTRAN).
3. Tris-buffered saline with Tween (TBS-T): 20 mM Tris–HCl, pH 7.4, 150 mM NaCl and 0.06% (v/v) Tween-20 (Sigma).
4. Blocking buffer: 4% (w/v) heat-shock fractionated bovine serum albumin (BSA) in TBS-T.
5. Primary antibody dilution buffer: TBS-T supplemented with 2% (w/v) BSA.
6. Anti-ERK2 mouse monoclonal antibody (D-2), # sc-1647 (Santa Cruz Biotechnology) (see Note 3).
7. Secondary antibody: Antimouse IgG conjugated to horseradish peroxidase (Biorad). Diluted at 1:10,000 in 2% (w/v) nonfat dry milk in TBS-T.

8. Enhanced chemiluminescent (ECL) reagents (Amersham/GE healthcare) and X-ray film (Konika/Minolta).

3. Methods

As in the assay of active ERK by Western blotting, to obtain reliable and reproducible results, it is important to terminate the samples rapidly and effectively at the end of the treatment protocol. Treatment with EGF (100 ng/mL, 5 min) is a positive control for dimerization of ERK in many cell types.

To identify the forms of ERK2, it is possible to use two methods of electrophoresis: SDS-Polyacrylamide Gel Electrophoresis (SDS-PAGE), where samples are not boiled, (4) and native gel electrophoresis (5–7). It is important that in native gel electrophoresis, the ERK2 dimer appears lower than the monomer in the blot. In SDS-PAGE method, ERK2 monomer appears with a molecular mass of 42 kDa, and ERK2 dimer appears with a molecular mass of 90 kDa.

3.1. Preparation of Samples for ERK2 Monomer/Dimer Separation by Western Blotting

1. Cells are passaged when approaching confluence (grown in the appropriate culture medium in a tissue culture incubator, 37°C and 5% CO_2) with trypsin/EDTA to provide new maintenance cultures on 100-mm dishes and experimental cultures on 35-mm dishes. One 35-mm dish is required for each experimental data point. A 1:12 split of HEK 293T or Hela cells will provide experimental cultures that are approaching confluence after 24 h. At this point, the cultures are rinsed once with sterile PBS and incubated for 18 h in DMEM supplemented with 0.1% fetal bovine serum (see Notes 4 and 5).
2. All the materials required for the lysis have to be made ready: cell lysis buffer at 4°C (supplemented with fresh phosphatase and protease inhibitors), a labeled microcentrifuge tube for each sample, a vacuum aspirator, ice-cold PBS, and a cell scraper.
3. The cultures are treated with agonists according to the protocol and the medium then removed by aspiration. Immediately, they are rinsed with ice-cold PBS 2×. 200 μL of cell lysis buffer is added, cells scraped, and transferred to precooled labeled Eppendorf tubes.
4. Samples are centrifuged at 17,000 *g* for 10 min at 4°C, and supernatants are collected in a new labeled Eppendorf tube.
5. Preparation of samples: 100 μg protein extract is mixed with the appropriate volume of 5× SDS-PAGE sample buffer or with an equivalent volume of 2× native sample buffer. Samples are ready for separation by electrophoresis.

3.2. SDS-PAGE for ERK2 Monomer/Dimer Separation

1. This protocol is performed in a Mini Protean Gel system (Bio-Rad). Glass plates are cleaned with rinsable detergent and 96% ethanol before use. The plates are assembled with 1.5 mm spacers.
2. Prepare 10% polyacrylamide (10 mL, for one gel) separating gel by mixing 4 mL of water, 3.3 mL of acrylamide stock solution, 2.5 mL of lower buffer, 100 μL of 10% SDS, 100 μL of APS, and 10 μL of TEMED. Mix the solution and cast the gel and overlay separating gel with water. Allow to polymerize (about 10 min).
3. Prepare stacking gel by mixing 3.4 mL of water, 830 μL of acrylamide solution, 630 μL of upper buffer, 50 μL of 10% SDS, 50 μL of APS, and 5 μL of TEMED. Mix the solution and cast the gel. Insert the comb and allow to polymerize (5–10 min).
4. Prepare the running buffer by diluting 100 mL of 10× Tris-Glycine and 10 mL of 10% SDS with 890 mL of water in a measuring cylinder. Cover with Parafilm and invert to mix.
5. Remove the comb and assemble the gel in the apparatus, add running buffer, and watch out for leaks.
6. Load samples (45 μL in a 10-well comb), one sample in each well. Include one for prestained molecular weight markers.
7. Run gel: Connect the wire leads to the power supply on and run the gel at 100 V (constant voltage) for about 120 min. End run when bromophenol blue dye reaches the bottom, turn off the power supply.

3.3. Native Gel for ERK2 Monomer/Dimer Separation

1. This protocol is performed in a Mini Protean Gel system (Bio-Rad). Glass plates are cleaned with rinsable detergent and 96% ethanol before use. Assemble the plates with 1.5 mm spacers.
2. Prepare 8% polyacrylamide (10 mL, for one gel) separating gel by mixing 4.6 mL of water, 2.7 mL of acrylamide stock solution, 2.5 mL of lower buffer, 100 μL of APS, and 10 μL of TEMED. Mix the solution and cast the gel, and overlay separating gel with water. Allow to polymerize (about 10 min).
3. Prepare the stacking gel by mixing 3.4 mL of water, 830 μL of acrylamide solution, 630 μL of upper buffer, 50 μL of APS, and 5 μL of TEMED. Mix the solution and cast the gel. Insert the comb and allow to polymerize (5–10 min).
4. Prepare the running buffer by diluting 100 mL of 10× Tris-Glycine with 900 mL of water in a measuring cylinder. Cover with Parafilm and invert to mix.

5. Remove the comb and assemble the gel in the apparatus, add running buffer, and watch out for leaks.
6. Load samples (45 μL in a 10-well comb), one sample in each well.
7. Run gel: Connect the wire leads to the power supply on and run the gel at 60 V (constant voltage) for about 150 min. End run when bromophenol blue dye reaches the bottom, and turn off the power supply.

3.4. Western Blotting for ERK2 Monomer/ Dimer Separation

1. The gel (containing the samples separated by electrophoresis) is transferred to nitrocellulose membranes electrophoretically in a Mini Protean (Bio-Rad) system. The transfer sandwich is opened and a wet foam piece placed on the black side of the sandwich, and a 3 mm paper is added on top of it.
2. The gel between the glass plates is taken out. The stacking gel is removed and discarded and one corner cut from the separating gel to allow its orientation to be tracked. The separating gel is then laid on top of the 3 mm paper.
3. The wet nitrocellulose membrane (previously submerged in methanol, 1 min) is placed on the gel, ensuring that no bubbles are trapped. Another sheet of 3 mm paper is wetted and carefully laid on top of the nitrocellulose membrane. The second wet foam sheet is laid on top and the cassette transfer is closed.
4. The cassette is placed into the transfer tank in such a way that the nitrocellulose membrane faces the anode side (black to black). It is vital to ensure this orientation, or else the proteins will be lost from the gel into the buffer instead of being transferred to the membrane.
5. A block of ice is placed in the tank.
6. The lid is put on the tank and the power supply activated. Transfers can be accomplished at either 125 mA overnight or 100 V for 1 h.
7. At the end of the transfer, the nitrocellulose membrane is removed from the transfer sandwich and incubated in 4% of BSA solution for 1 h at room temperature for blocking on a rocking platform.
8. Once quenched, the blot is incubated for 1 h at room temperature with the primary antibody (overnight incubation at 4°C is also OK) on a rocking platform.
9. The primary antibody is then removed and the membrane washed three times with TBS-T (see Note 6).
10. Secondary antibody solution is freshly prepared for each experiment and added to the membrane for 1 h at room temperature on a rocking platform.

11. The secondary antibody is discarded and the membrane washed three times for 10 min with TBS-T.
12. The ECL reagents are mixed together and then immediately added to the blot and incubated for 1 min.
13. The blot is removed from the ECL reagents and placed between the leaves of an acetate sheet protector that has been cut to the size of an X-ray film cassette (see Note 7).
14. The acetate containing the membrane is then placed in an X-ray film cassette with film for a suitable exposure time, typically a few minutes.

4. Notes

1. Unless stated otherwise, all solutions should be prepared in water that has a resistivity of 18.2 MΩ cm and a total organic content of less than five parts per billion. This standard is referred to as "water" in this text.
2. Adequate cooling to keep the buffer no warmer than room temperature by the use of a refrigerated bath (in this case with an ice block) is essential to prevent heat-induced damage to the experiment.
3. This is the most specific antibody we have found to detect only ERK2 isoform by Western blotting (most of the commercial antibodies recognize both isoforms of ERK).
4. This protocol can be used for separation of ERK2 monomer/dimer in many other cell culture systems.
5. During serum starvation, plates should be placed in a tissue culture incubator (37°C, 5% CO_2). It is important that the plates remain flat and the medium covers all the plates equally. The aim of this starvation is to make the cells enter quiescence, which, under these conditions, can be achieved within 14–24 h depending on the cell type. Starving for too long a period, or any change in temperature or pH, may be stressful to the cells and thereby, induce activation of one or more signaling pathways.
6. The primary antibody can be saved for subsequent experiments by storage at −20°C. Avoid using it if the solution is not clear.
7. Exact alignment of the subsequent film with the membrane can be difficult. We, therefore, apply an adhesive and luminescent piece of tape to the edge of the acetate sheet to provide an alignment mark for the film and membrane and thus allow identification of the signals with the lanes and determine the right molecular weight by comparison with the prestained marker lane.

References

1. Khokhlatchev, A. V., Canagarajah, B., Wilsbacher, J., Robinson, M., Atkinson, M., Goldsmith, E., and Cobb, M. H. (1998). Phosphorylation of the MAP kinase ERK2 promotes its homodimerization and nuclear translocation. *Cell* **93**, 605–615.
2. Kolch, W. (2005). Coordinating ERK/MAPK signalling through scaffolds and inhibitors. *Nat Rev Mol Cell Biol* **6**, 827–837.
3. Casar, B., Pinto, A., and Crespo, P. (2008). Essential role of ERK dimers in the activation of cytoplasmic but not nuclear substrates by ERK-scaffold complexes. *Mol Cell* **31**, 708–721.
4. Philipova, R., and Whitaker, M. (2005). Active ERK1 is dimerized in vivo: bisphosphodimers generate peak kinase activity and monophosphodimers maintain basal ERK1 activity. *J Cell Sci* **118**, 5767–5776.
5. Schagger, H., and von Jagow, G. (1991). Blue native electrophoresis for isolation of membrane protein complexes in enzimatically active form. *Anal Biochem* **199**, 223–231.
6. Stefan, R., and Thomas, W. (2009). Oligomeric structure and functional characterization of the urea transporter from Actinobacilus pleuropneumoniae. *J Mol Biol* **387**, 619–627.
7. SERVA Electrophoresis SERVAGel™ TG instruction manual.

Chapter 21

MAP Kinase: SUMO Pathway Interactions

Shen-Hsi Yang and Andrew D. Sharrocks

Abstract

The convergence and coordinated cross talk of different signalling pathways forms a regulatory network which determines the biological outcome to environmental cues. The MAPK pathways are one of the important routes by which extracellular signals are transduced into intracellular responses. Through protein phosphorylation mechanisms, they can play a pivotal role in regulating other posttranslational modifications such as protein acetylation and ubiquitination. In addition, protein sumoylation has emerged as an important pathway which also functions through post-translational modification. The SUMO pathway modulates a diverse range of cellular processes including signal transduction, chromosome integrity, and transcription. Interestingly, recent studies have provided links between the SUMO and MAPK signalling pathways which converge to modulate transcription factor activity. This was first demonstrated by the observation that the activation of the ERK pathway caused de-sumoylation of the transcription factor, Elk-1. Furthermore, a growing number of links are now being made between the MAPK pathway and protein sumoylation. Given the nature of protein sumoylation in diverse biological functions, it is not surprising that the effect of MAPK pathways on sumoylation varies between different proteins. Here, we describe protocols that can be used in studying the cross talk between the MAPK and SUMO pathways, particularly at the level of gene regulation.

Key words: MAP kinase, ERK, Transcription factor, Sumoylation, SUMO, Elk-1, Ni-NTA, ChIP assay

1. Introduction

SUMO conjugation has emerged as an important post-translational modification and is part of an enzymatic cascade, involving an E1 activating enzyme (SAE1/2) and an E2-conjugating enzyme (Ubc9) Fig. 1a. A number of distinct classes of E3 ligases (including RanBP2, the PIAS family, and Pc2) have been identified that enhance the efficiency and specificity of sumoylation of target substrates in vivo (1, 2). Importantly, sumoylation is a reversible process. Several hydrolases [SUMO-specific proteases (SENPs)] that function to produce the mature SUMO paralogues, to remove the modifications

Rony Seger (ed.), *MAP Kinase Signaling Protocols: Second Edition*, Methods in Molecular Biology, vol. 661, DOI 10.1007/978-1-60761-795-2_21, © Springer Science+Business Media, LLC 2010

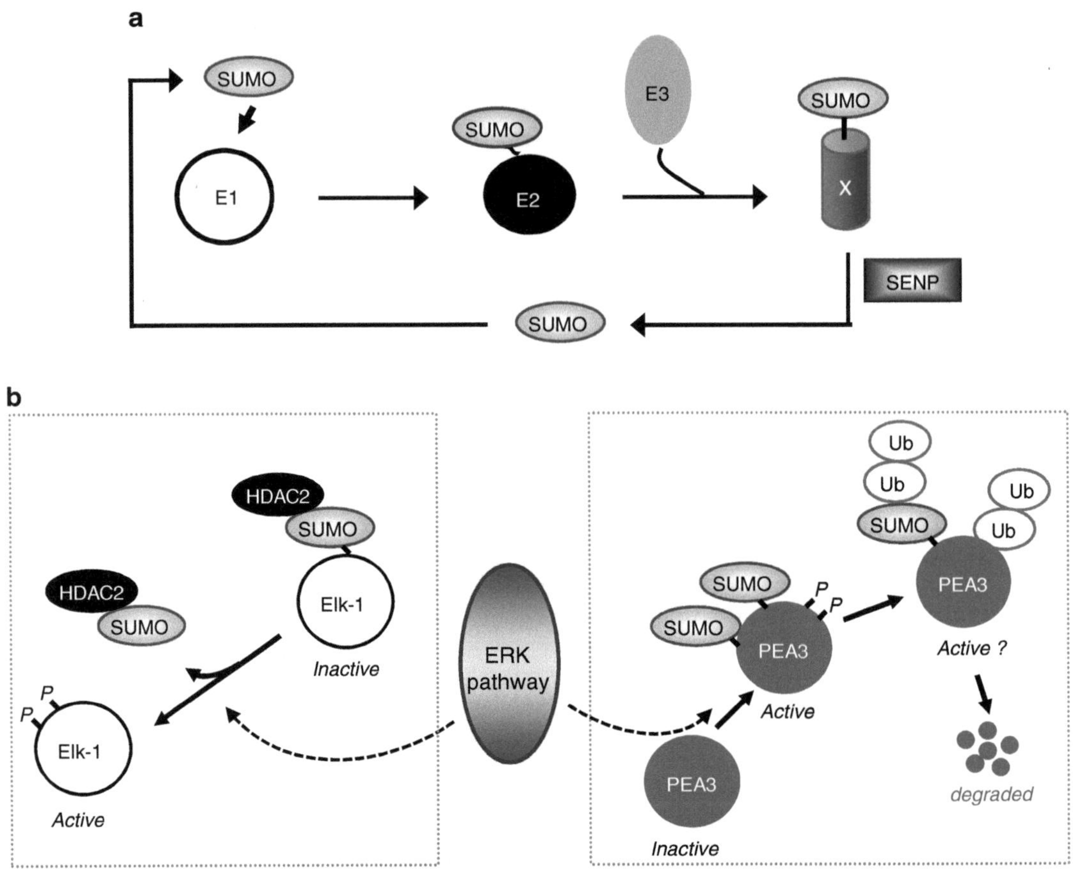

Fig. 1. (**a**) The SUMO Cycle. The mature SUMO is activated by the SAE1/SAE2 enzyme (E1). The SUMO ~ Ubc9 is formed after transesterification of SUMO onto the SUMO conjugating enzyme, Ubc9, (E2). Subsequently, the target protein (X) is selected in an E3 ligase-dependent or independent manner, and the SUMO is ligated to the substrate through isopeptide-bound linkage (SUMO-X). SUMO modification can be reversed from target proteins by the action of SUMO-specific proteases and potentially be reused in the SUMO cycle. (**b**) A diagram depicting the example routes which encompass the interplay between the SUMO and MAPK pathways in the regulation of gene transcription. The sequential action of these pathways on Elk-1 (left panel) and PEA3 (right panel) are shown. *Left panel*: The SUMO and ERK pathways combine to regulate the transcriptional activity of Elk-1. The transcriptionally inert state of Elk-1 is controlled by its sumoylation via recruiting a HDAC2-containing corepressor complex. The ERK MAPK triggers a series of molecular events, which result in the loss of Elk-1 sumoylation, and hence, release of the co-repressor complex. At the same time, Elk-1 is phosphorylated (-P) and this converts Elk-1 into a transient, highly active state. *Right panel*: MAP kinase signaling initiates a cascade that initially promotes PEA3 sumoylation (PEA3-SUMO) and enhances its transactivation properties. This is important for polyubiquitination (Ub) and eventual turnover of PEA3. The polyubiquitination might potentially occurs on the SUMO moiety and/or PEA3 itself. The question mark represents the unknown transcriptional activity of these forms of PEA3.

completely or edit chains of SUMO moieties, have been identified (3). Sumoylation usually occurs on an acceptor lysine residue of a substrate at the core SUMO consensus motif, ΨKxE (4, 5). In addition to this core motif, a number of extended SUMO consensus motifs have been identified including the KEPE motif, SC motif, PDSM, and NDSM, which serve to further increase the specificity of substrate modification (6–9). However, sumoylation of several substrates has also been demonstrated to take place on sites, that do not conform to these motifs such as found in E2-25K (10).

The available structural data suggest that the SUMO modification is likely to alter the overall structure of its substrates and hence regulate their properties in controlling DNA–protein interactions, protein–protein interactions, and sub-cellular localisation. The modification of proteins by SUMO conjugation can modulate a diverse range of cellular processes including important roles in controlling genome stability, DNA repair and replication, and gene transcription (1). The sumoylation of transcriptional regulatory proteins often imparts repressive properties by altering protein stability, sub-nuclear localisation, DNA binding activity, and co-repressor recruitment. However, in some cases, sumoylation can cause an increase in transcriptional activation. Importantly, the changes in transcription factor activity controlled by the SUMO pathway have been shown in several cases to be regulated by signalling pathways including the MAPK cascades (11–13).

MAPK pathways provide a common route to transmit extra-cellular signals into intra-cellular responses. Different pathways can transduce stress (JNK and p38 pathways) or mitogenic (ERK pathway) signals. They are one of the principal mechanisms for the regulation of cell-cycle progression and proliferation (14). For example, the ETS-domain transcription factor Elk-1 is regulated by phosphorylation in response to activation of the MAPK pathways. This phosphorylation triggers a series of molecular events which convert Elk-1 from a transcriptionally inert state into a transient, highly active state for rapid gene activation before returning to a basal repressive state (15–19).

SUMO modification of Elk-1 is required for its transcriptional repressive activity via SUMO-dependent recruitment of histone deacetylases (20). Importantly, activation of the ERK MAPK pathway leads to Elk-1 de-sumoylation and HDAC loss. At the same time, this pathway converts Elk-1 into its fully activated status through its phosphorylation (17). This dynamic interplay between the repressive SUMO pathway and activating ERK pathway provides an excellent solution for fine tuning the final outputs of gene activity and hence orchestrating responses to environmental cues. In addition, PIASxα, an E3 ligase, has recently been shown to function as a key integrator that determines the differential response of the transcription factor Elk-1 to the ERK and the stress-activated p38 MAP kinase pathways in a manner independent from its E3 ligase activity (21, 22). Furthermore, a growing number of links are now being made between the MAPK pathways and protein sumoylation including the transcription factors STAT1, ATF7, c-Fos, progesterone receptor (PR), and the ETS-domain transcription factor PEA3 (23–27). In contrast to the Elk-1, sumoylation of several proteins can be enhanced by ERK MAPK pathway activation. A good example of this is PEA3, where

ERK-mediated increases in sumoylation are important for enhancing its transcriptional activation properties and also for its ubiquitination and subsequent degradation (27). The diagram shown in Fig. 1b uses Elk-1 and PEA3 to illustrate the different routes that the interplay between SUMO and MAPK pathway can take to regulate transcription factor activity. Given the widespread nature of protein sumoylation and phosphorylation that is involved in diverse biological functions, it is expected that the effect of MAPK pathways on sumoylation will be a more common event but that the functional outcomes will differ between different proteins. Here, we describe protocols to facilitate further studies on the coordinated cross talk between MAPK and SUMO pathways in response to extracellular signals.

2. Materials

2.1. Detection of SUMO-Modified Conjugates by SUMO Assay

2.1.1. Cell Culture, Cell Lysis, and Immunoprecipitation

1. Dulbecco's Modified Eagles's Medium (DMEM) (GIBCO/Invitrogen) supplemented with 10% foetal bovine serum (FBS, GIBCO/Invitrogen) and 1% penicillin–streptomycin (GIBCO/Invitrogen).
2. Trypsin-ethylenediamine tetraacetic acid (EDTA) solution (0.05%, GIBCO/Invitrogen).
3. DPBS without $CaCl_2$ and $MgCl_2$ (GIBCO/Invitrogen).
4. Direct SUMO lysis buffer: 1.72% sodium dodecylsulfate (SDS), 50 mM Tris–HCl [pH 6.7], 10% glycerol, 0.33% NP-40, 0.33% sodium deoxycholate, 10 mM *N*-ethylmaleimide (NEM), 10 μg/ml of E64, and Complete Mini protease inhibitor cocktail (Roche) (see Note 1).
5. 5× Laemmli sample buffer: 250 mM Tris–HCl [pH 6.8], 10% SDS, 50% glycerol, 0.5% bromophenol blue, and 0.5 M 1,4-Dithiothreitol (DTT).
6. IP dilution buffer: 50 m M [*N*-(2-hydroxyethyl)piperazine *N*′-(2-ethane sulfonic acid)] (HEPES) [pH 7.9], 5% glycerol, 0.1% NP-40, 150 mM NaCl, 0.1 mM EDTA, 10 mM NEM, 10 μg of E64/ml, and Complete Mini protease inhibitor cocktail (Roche) (see Note 1).
7. Appropriate primary antibodies.
8. Dynal Magnetic Particle Concentrator (MCP-S). Dynabeads protein G or A (see Note 2).
9. Cell scrapers (Costar).
10. Ultrasonic cell disruptor (Microson).
11. Refrigerated Microcentrifuge (MSE).

2.1.2. Cell Transfection, Cell Lysis and Nickel Affinity His-Tagged Protein Purification Under Denaturing Conditions

1. Dulbecco's Modified Eagles's Medium (DMEM) (GIBCO/Invitrogen) supplemented with 10% foetal bovine serum (FBS, GIBCO/Invitrogen) and 1% penicillin–streptomycin (GIBCO/Invitrogen).
2. Dulbecco's Modified Eagles's Medium (DMEM) (GIBCO/Invitrogen) and 1% penicillin–streptomycin (GIBCO/Invitrogen).
3. Trypsin-EDTA solution (0.05%, GIBCO/Invitrogen).
4. Polyfect transfection reagent (QIAGEN) (see Note 3).
5. OPTI-MEM (GIBCO/Invitrogen).
6. DPBS without $CaCl_2$ and $MgCl_2$ (GIBCO/Invitrogen).
7. Ni-NTA agarose (QIAGEN) (see Note 4).
8. Guanidine lysis solution: 6 M Guanidine-hydrochoride, 50 mM $NaH_2PO_4{\cdot}H_2O$, 75 mM Tris–HCl [pH 8.0], 100 mM NaCl, 0.05% NP40, 20 mM imidazole, and 10 mM β-mercaptoethanol (see Note 5).
9. Urea Wash Buffer: 8 M Urea, 100 mM $NaH_2PO_4{\cdot}H_2O$, 10 mM Tris.Cl [pH 6.3], 100 mM NaCl, 20 mM imidazole, and 0.05% NP40.
10. Elution Buffer: 250 mM imidazole, 200 mM Tris–HCl [pH 6.8], 16.5% (v/v) glycerol, 3.3% (w/v) SDS, 0.00165% (w/v) bromophenol blue, and 0.42 mM β-mercaptoethanol (see Note 6).
11. Cell scrapers (Costar).
12. Bioruptor (Diagenode).
13. Refrigerated Microcentrifuge (MSE).
14. Transfection tube (Greiner bio-one).

2.1.3. The Manipulation of MAPK Pathway Activation for Examining Changes in Protein Sumoylation

1. Phorbol 12-myristate 13-acetate (PMA, Sigma) is dissolved at 1 mM in dimethylsulphoxide (DMSO) and stored in single-use aliquots at –80°C. Typical working concentration is 10 nM.
2. Epidermal Growth Factor (EGF, Sigma) is dissolved at 0.1 mM in H_2O and stored at –20°C. Typical working concentration is 50 nM.
3. 1,4-Diamino-2,3-dicyano-1,4-*bis*(2-aminophenylthio)butadiene (U0126, Calbiochem) is dissolved at 10 mM in DMSO and stored at –80°C. Typical working concentration is 10 μM.

2.2. SDS-Polyacrylamide Gel Electrophoresis (SDS-PAGE)

1. Separating buffer (3.55×): 1.5 M Tris–HCl [pH 8.8], 0.4% (w/v) SDS. Store at 4°C.
2. Stacking buffer (4×): 0.5 M Tris–HCl [pH 6.8], 0.4% (w/v) SDS. Store at 4°C.

3. 30% acrylamide/bisacrylamide solution (37.5:1, Amresco) (see Note 7).
4. *N,N,N,N′*-Tetramethyl-ethylenediamine (TEMED, Sigma). Store at 4°C.
5. Ammonium persulphate: prepare 10% solution in water and store at 4°C.
6. Propan-2-ol (Fisher Scientific).
7. SDS running buffer (10×): 250 mM Tris, 1.9 M Glycine, and 1% (w/v) SDS. Store at room temperature.
8. 5× Laemmli sample buffer: 250 mM Tris–HCl [pH 6.8], 10% SDS, 50% glycerol, 0.5% bromophenol blue, and 0.5 M 1,4-Dithiothreitol (DTT).
9. Page ruler pre-stained protein ladder (Fermentas).

2.3. Western Blotting

1. Towbin transfer buffer: Prepare 10× stock with 250 mM Tris, 1.9 M Glycine, and autoclaved. For each 1 l of 1× working solution is prepared by diluting 100 ml of 10× stock solution with 700 ml of water and 200 ml of methanol (20% (v/v)) (see Note 8).
2. Nitrocellulose membrane (Millipore) and 3MM chromatography paper (Whatman).
3. Tris-buffered saline with Tween (TBS-T): Prepare 10× stock with 0.5 M Tris–HCl [pH 8.0], 1.5 M NaCl, and autoclaved. Dilute 50 ml stock solution with 450 ml water and add Tween 20–0.05% (v/v) before use.
4. Blocking buffer: 5% (w/v) low-fat dried skimmed milk (Marvel) in TBS-T.
5. Primary antibodies.
6. Secondary antibodies: Horseradish peroxidase (HRP)-labelled Goat anti-mouse and Goat anti-rabbit Ig (BD Pharmingen) (see Note 9).
7. SuperSignal West Dura Extended Duration Substrate (Thermo Scientific).
8. Ponceau solution (Thermo Scientific).
9. X-ray film (Kodak).

2.4. Functional Analysis

2.4.1. Reporter Assays

1. DPBS without $CaCl_2$ and $MgCl_2$ (GIBCO/Invitrogen).
2. Dual-light Chemiluminescent Reporter Gene Assay System (TROPIX) containing Lysis solution: 100 mM potassium phosphate [pH 7.8] and 0.2% Triton X-100 (Store at 4°C); Buffer A: Lyophilised power [reconstitute in 5 ml of water (store at −20°C)] and Galacton-Plus substrate (100× concentrate): dilute to 1× in Buffer A before each assay; Buffer B: Lyophilised

luciferin [reconstitute in 22 ml of water (store at –20°C)] and light emission accelerator – II (store at 4°C).

3. Cell scrapers (Costar).
4. Refrigerated Microcentrifuge (MSE).
5. Luminometer (TD-20/20, Turner designs).

2.4.2. Chromatin Immunoprecipitation

1. Dilute 37% formaldehyde solution (Sigma–Aldrich) to 1% with DMEM.
2. Freshly prepare 10 mM dimethyl adipimidate (DMA, Pierce) in ice-cold DPBS, containing 0.25% DMSO (see Note 10).
3. 125 mM glycine in DPBS.
4. The ice-cold DPBS containing 10 mM NEM, 10 μg/ml of E64, and Complete Mini protease inhibitor cocktail (Roche) (see Note 1).
5. Buffer I: 10 mM HEPES [pH 6.5], 0.5 mM (ethylene glycol-bis[β-amino ethylether]-*N*,*N*,*N′*,*N′*-tetraacetic acid) (EGTA), 10 mM EDTA, 0.25% (v/v) Triton X-100, 10 mM NEM, 10 μg/ml of E64, and Complete Mini protease inhibitor cocktail (Roche) (see Note 1).
6. Buffer II: 10 mM HEPES [pH 6.5], 0.5 mM EGTA, 1 mM EDTA, 200 mM NaCl, 10 mM NEM, 10 μg/ml of E64, and Complete Mini protease inhibitor cocktail (Roche) (see Note 1).
7. Lysis Buffer: 50 mM Tris–HCl [pH 8.1], 10 mM EDTA, 1% (w/v) SDS, 10 mM NEM, 10 μg/ml of E64, and Complete Mini protease inhibitor cocktail (Roche) (see Note 1).
8. Dynal Magnetic Particle Concentrator (MCP-S). Dynabeads protein G or Dynabeads protein A (see Note 2).
9. Bovine Serum Albumin (BSA, 20 mg/ml, Roche) and Salmon testes DNA (9.9 mg/ml, Sigma).
10. Primary antibodies and normal mouse IgG (1 mg/ml, Upstate) or normal rabbit IgG (1 mg/ml, Upstate).
11. ChIP dilution buffer: 16.7 mM Tris–HCl [pH 8.1], 167 mM NaCl, 1.2 mM EDTA, 0.01% (w/v) SDS, 1.1% (v/v) Triton-X-100, 10 mM NEM, 10 μg/ml of E64, and Complete Mini protease inhibitor cocktail (Roche) (see Note 1).
12. Low salt buffer: 20 mM Tris–HCl [pH 8.1], 150 mM NaCl, 2 mM EDTA, 0.1% (w/v) SDS, and 1% (v/v) Triton-X-100.
13. High salt buffer: 20 mM Tris–HCl [pH 8.1], 500 mM NaCl, 2 mM EDTA, 0.1% (w/v) SDS, and 1% (v/v) Triton-X-100.
14. LiCl solution: 10 mM Tris–HCl [pH 8.1], 250 mM LiCl, 1 mM EDTA, 1% (v/v) NP-40, and 1% (v/v) sodium deoxycholate.

15. TE buffer: 10 mM Tris–HCl [pH 8.0], 1 mM EDTA, and 0.1% (v/v) NP-40.
16. Elution buffer: 100 mM $NaHCO_3$ and 1% (w/v) SDS (see Note 11).
17. 5 M NaCl.
18. 500 mM EDTA, 1 M Tris–HCl [pH 6.5], and proteinase K (10 mg/ml, recombinant PCR grade, Roche).
19. PCR clean-up kit (QIAGEN).
20. Quantitect SYBR Green PCR Kit (QIAGEN).
21. Cell scrapers (Costar).
22. Ultrasonic cell disruptor (Microson).
23. Refrigerated Microcentrifuge (MSE).

3. Methods

Protein sumoylation has emerged as an important modification that controls a number of cellular processes including the regulation of gene expression. The identification of transcriptional regulators as substrates for SUMO modification and their functional roles is a critical step for the understanding of the biological roles modulated by the SUMO pathway. Moreover, several examples of cross talk between MAPK and SUMO pathway have been uncovered (23–27). This provides an additional layer of the regulation and hence allows coordinated control for fine tuning of the biological output in response to the environmental cues. The flow chart presented in Fig. 2 illustrates the stepwise approaches which can be used to facilitate studies into protein sumoylation with particular emphasis on probing transcription regulators and their subsequent functions in transcriptional control.

3.1. Detection of SUMO-Modified Conjugates by SUMO Assays

Sumoylation is a reversible post-translational modification process. In general, the stoichiometry of SUMO conjugates is low at any given time within cells. Moreover, sumoylation is very labile in cell lysates due to the presence of multiple SUMO-specific proteases. It is therefore important to carefully design an experimental strategy in order to capture the steady-state of SUMO-conjugated proteins. A number of issues that determine the success of experiments are summarised in Fig. 2. These are dependent on the circumstances of the individual proteins of interest, including the relative stoichiometry of SUMO conjugates, antibody availability, and protein expression levels. In addition, the sumoylation status of a particular protein can also be regulated by MAPK activation, hence the appropriate manipulations of signalling pathway induction should

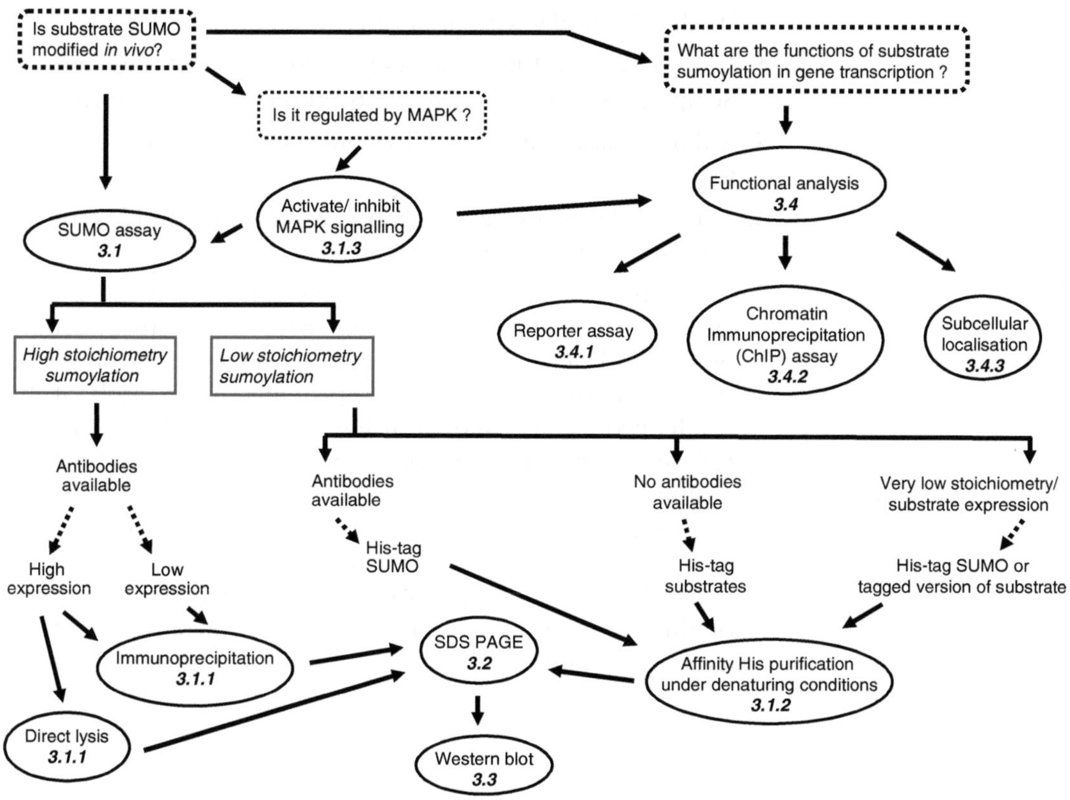

Fig. 2. A flow chart of the approaches that can be used to study protein sumoylation and its potential downstream functions. The dashed boxes represent examples of questions posed for studying the interplay between the MAPK and SUMO pathway in regulating protein sumoylation and function. The circles outline the experimental approaches to be followed and refer to the relevant section in the text. The considerations of available reagents and protein expression and the subsequent routes taken in identifying the correct experimental approach are indicated.

also be considered in the experimental procedures. Here, we describe a number of common approaches used in determining the status of protein sumoylation.

3.1.1. Analysis of Endogenous Protein Sumoylation

With the above-mentioned caveats, the detection of endogenous protein sumoylation is often a challenging task. The first task is to determine the level of substrate sumoylation and then select the appropriate strategy (Fig. 2). The first assumption is that high level sumoylation will be present. Two approaches by either direct lysis or immunoprecipitation from cell lysates can be taken.

Sample preparation by direct lysis: This approach is most suitable for highly expressed proteins.

1. Cells are seeded at the appropriate density (5×10^5 293T cells per well of 6-well-plate in 3 ml of DMEM/10% serum) the day before harvesting.
2. The dishes should be >80% confluent on the day of harvest.

3. Wash cells briefly with 3 ml of DPBS on ice.
4. Lyse cells with 200 μl of ice-cold direct SUMO lysis buffer. Scrape cells and transfer to a 1.5-ml microcentrifuge tube.
5. Mildly sonicate using an Ultrasonic cell disruptor at power 7 for 15 s.
6. Remove debris by centrifugation for 10 min at 13 krpm at 4°C and save the supernatant.
7. Add 5 μl of 5× sample buffer immediately to 20 μl of each sample. The samples are boiled for 5 min before loading onto SDS PAGE (see Subheading 3.2) and Western blotting (see Subheading 3.3) for target proteins. An example of the results produced is shown in Fig. 3. Sumoylation is characterised by the appearance of an additional band whose mobility change upon conjugation of a single SUMO moiety is over 10 kDa from that observed for the unmodifed protein. Often bands of even slower mobility are seen due to either the anomalous migration of the branched polypeptides or due to multiple conjugation events (on either separate sites or in chains).

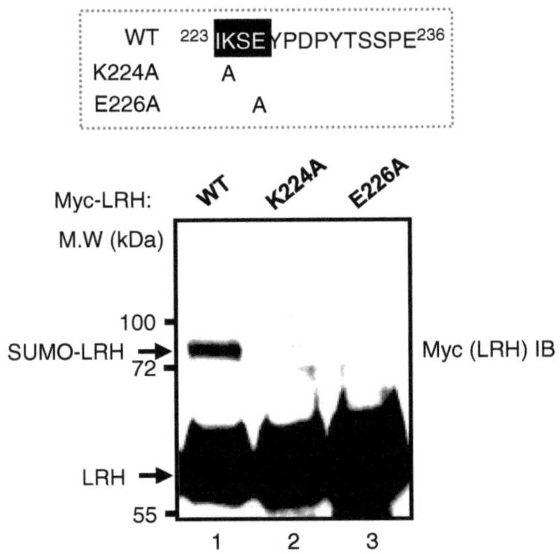

Fig. 3. Sumoylation of LRH-1 in vivo. The amino acids sequence of LRH-1 containing the core SUMO consensus site ψKXE is indicated by a *black box*. The residues altered in each mutant protein are indicated. The indicated wild-type and mutant myc-tagged LRH-1 proteins were coexpressed in 293 cells with HA-tagged SUMO-2. The sample preparation for SUMO assay was performed by the direct lysis method. The LRH-1 and SUMO-conjugated species were detected by immunoblotting (IB) with an anti-myc antibody. Sumoylation of wild-type LRH is clearly detectable as a high molecular weight species (lane 1) but mutation of the acceptor lysine or the downstream glutamic acid residue reduces the sumoylation of LRH (lanes 2 & 3). The mobility shift is more than the 10 kDa expected from the addition of one SUMO molecule probably due to, at least in part, the migration of a branched polypeptide in the gel. The degree of anomalous migration is dependent on the position of SUMO linkage to the protein.

Immunoprecipitation: The appearance of high molecular weight conjugates is consistent with sumoylation but could arise due to alternative modifications. Thus, the direct lysis method described above (see Subheading 3.1.1) will indicate, but not prove, the modification of the protein of interest with SUMO. In addition, this method might not be sufficient to detect SUMO conjugates when only low levels of endogenous protein are expressed. To determine the status of protein sumoylation under these conditions, immunoprecipitation experiments can be performed.

1. Sample preparation is as described for procedure 1–6 above (see Subheading 3.1.1) except half of the volume of Direct SUMO lysis buffer is added (see Note 12).
2. Prior to immunoprecipitation, the extracts are diluted 17-fold with IP dilution buffer (see Note 12).
3. Diluted lysates are incubated with 1 μg of appropriate primary antibody at 4°C overnight on a rotating platform (see Note 13).
4. A 7.5 μl of Dynabeads protein A or G (see Note 2) is added to each sample to collect the immune complexes by incubating at 4°C for 1 h on a rotating platform.
5. Pellet the beads in a magnetic rack and then wash the beads four times with 1 ml of IP dilution buffer. Beads are rotated at room temperature for 5 min and left in the magnet for 1 min in between washes to precipitate the beads.
6. Elute immune complexes by adding 15 μl of 2× sample buffer. The samples are boiled for 5 min before loading onto SDS PAGE (see Subheading 3.2) followed by Western blotting (see Subheading 3.3) with anti-SUMO antibodies against the target protein.

3.1.2. Analysing Protein Sumoylation by Using Over-Expression Systems

The detection of a SUMO-modified protein might be hampered by a number of issues. For example, very low stoichiometry of protein sumoylation might be obtained, corresponding antibodies might not be available or low levels of endogenous protein expression might be observed. Here, we describe an alternative approach using mammalian over-expression systems followed by Nickel affinity purification of SUMO conjugates under denaturing conditions.

Cell transfection: The following suggestions are aimed at increasing the success of protein sumoylation experiments (see Fig. 2). Mammalian expression vectors carrying hexahistidine-tagged SUMO paralogues can be constructed and expressed in cells if low stoichiometry of SUMO conjugates becomes apparent. Alternatively, in the event that the availability of an antibody is an issue or a target protein is expressed at low endogenous levels, a plasmid containing the protein of interest fused to a hexahistidine tag can be constructed and expressed. In the case of very low protein expression and low stoichiometry of SUMO-modified species, co-expression of the hexahistidine-tagged SUMO paralogues and the protein of interest fused to a short-peptide tag (i.e. Flag, HA, or Myc) or reciprocally a combination of tagged SUMO and a hexahistidine-tagged target might be required.

1. The day before transfection, seed 5×10^5 293T cells per well of a 6-well-plate in 3 ml of DMEM/10% serum and incubate the cells at 37°C and 5% CO_2 (see Note 14).
2. The dishes should be >80% confluent on the day of transfection.
3. Dilute 4 μg of plasmid DNA [i.e. $(His)_6$-SUMO paralogues, $(His)_6$-tagged protein of interest or a combination of $(His)_6$-SUMO paralogue and a tagged version of protein of interest (see above)] and 15 μl of polyfect transfection reagent in 100 μl of OPTI-MEM (see Note 3) in a transfection tube. The mixtures are vortexed for 10 s and incubated for an additional 12 min at room temperature to allow complex formation.
4. Add 600 μl of cell growth medium (DMEM/10% serum) to the transfection tube containing the transfection complexes. Remove medium and immediately transfer the complexes to the cells.
5. Incubate the cells with the complexes for 1.5 h at 37°C and 5% CO_2 in an incubator.
6. Remove the medium by gentle aspiration. Wash cells once with serum-free medium.
7. Add fresh DMEM medium and grow the cells overnight before harvesting (see Note 15).

Nickel affinity His-tagged protein purification under denaturing conditions: The labile nature of SUMO conjugates in cell lysates often obstructs the detection of protein sumoylation. The inhibition of multiple SUMO-specific proteases can be achieved by performing nickel affinity His-tagged protein purification under denaturing conditions. In addition, the disruption of protein complexes in denaturing conditions will further demonstrate modification of particular protein of interest with SUMO rather than the co-purification of other sumoylated proteins with similar sizes.

1. Wash cells briefly with 3 ml of DPBS.
2. Lyse cells with 300 μl of guanidine lysis buffer. Scrape cells and transfer to 1.5 ml microcentrifuge tubes and leave at 4°C for 10 min.
3. Sonicate lysate at 4°C for 30 s on/off for 15 min using a Bioruptor.
4. Remove debris by centrifugation for 30 min at 13 krpm at 4°C and save the supernatant.

5. Add 25 µl of equilibrated 50% slurry of Ni-NTA agarose to 250 µl of cleared lysate and incubate at room temperature for 2 h using a thermal mixer.
6. Let beads settle and carefully remove the supernatant. Wash beads once with 0.8 ml of guanidine lysis buffer and three times with 0.8 ml of urea wash buffer (see Note 16).
7. After the final wash, the samples can then be spun down at 2 krpm for 1 min. Remove as much liquid as possible.
8. Elute samples by adding 15 µl of elution buffer and incubate at 100°C for 10 min before loading onto SDS-PAGE (see Subheading 3.2) followed by protein detection by Western blotting (see Subheading 3.3). An example of the results produced is shown in Fig. 4 (see Subheading 3.1.1 for an explanation of the expected mobilities of the sumoylated species).

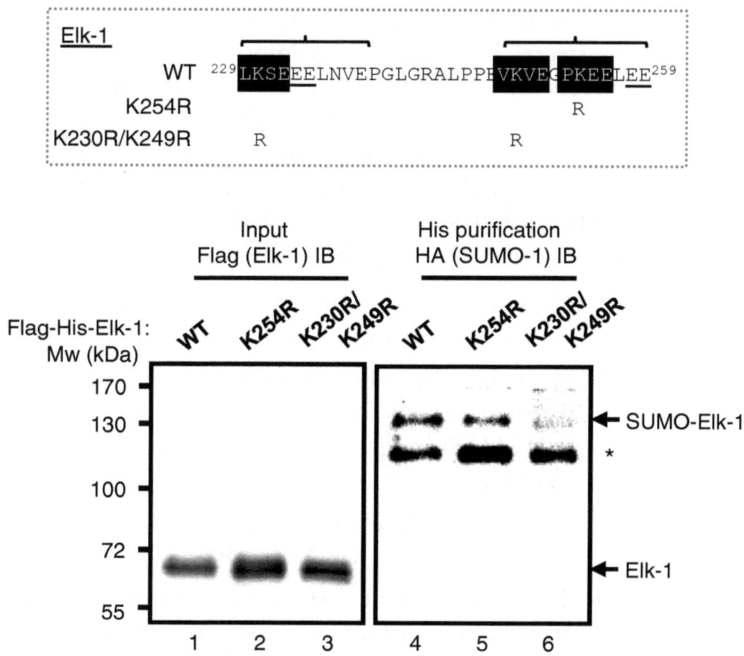

Fig. 4. Sumoylation of Elk-1 in vivo. The sequence of the Elk-1 region containing the core SUMO consensus site is indicated by a *black box* and the NDSM is bracketed. The residues altered in each mutant protein are indicated. The indicated wild-type (WT) or mutant versions of His-Flag-tagged Elk-1(1–428) were coexpressed with HA-tagged SUMO-1. His-tagged Elk-1 was precipitated from cell lysates using Nickel affinity His-tagged protein purification under denaturing conditions. Sumoylated-Elk-1 was detected by immunoblotting (IB) with an anti-HA antibody (lanes 4–6). Total lysates were blotted with an anti-Flag antibody for Elk-1 as the input control (lanes 1–3). The asterisk represents a non-specific band. The appearance of a high molecular weight form of wild-type Elk-1 (lane 1) in the precipitated samples is indicative of sumoylation. These sumoylated species are reduced upon mutation of the key lysine acceptor sites in Elk-1 (lane 3). In the case of Elk-1, the mobility shift is more than the 10 kDa expected from the addition of one SUMO molecule, probably due at least in part, to the migration of a branched polypeptide in the gel.

3.1.3. The Manipulation of MAPK Pathway Activation for Examining Changes in Protein Sumoylation

A number of studies have provided links between the SUMO and MAPK signalling pathways. It is therefore important to implement experimental approaches that involve combinations of SUMO assays and MAPK signalling pathway induction. These approaches will further our understanding of potential areas of cross talk between these pathways. Here, we describe three common approaches involved in manipulating MAPK activation in this context.

Over-expression of constitutively activated forms of upstream MAPKK: A number of constitutively activated upstream MAPKKs have been used to activate the MAPK signalling pathways. There are mammalian expression constructs encoding MEK–DN (for ERK MAPK), MKK6(D/E) (for p38 MAPK), and MKK7–JNK fusion (for JNK MAPK) proteins (see Note 17). Over-expression of these constructs will lead to the potent stimulation of a particular MAPK pathway and hence will allow the analysis of the potential cross-talk between the SUMO and MAPK pathways. The details of the transfection protocol are as described in Subheading 3.1.2.

Activation of MAPK pathways by growth factors, cytokines, or stress: Ultimately, it is important to determine whether the effects on protein sumoylation levels can be achieved by the activation of the endogenous MAPK pathways. This can be done by challenging cells with growth factors (e.g. EGF) or phorbol esters (e.g. PMA) for ERK MAPK activation (see Note 18) or with cytokines (e.g. TNFα, IL-1) or stress stimuli (e.g. sorbitol, H_2O_2) for JNK and p38 MAPK activation.

1. Cells are seeded at the appropriate density (4×10^5 293T cells per well of a 6-well-plate in 3 ml of DMEM/10% serum).
2. Next day, the cells are washed three times with 3 ml of pre-warmed serum-free DMEM and left in 3 ml serum-free DMEM. Subsequently, cells are cultured for a further 24 h in a 37°C, 5% CO_2.
3. Stimulate cells with growth factors, cytokines, or stress for typically 30 min (see Note 19).
4. The subsequent steps in sample preparations (see Subheading 3.1.1), immunoprecipitations (see Subheading 3.1.1) and nickel-affinity purification (see Subheading 3.1.2) are as described previously.

Identification of the specific MAPK pathway involved by using pharmacological inhibitors: In order to identify a specific MAPK pathway involved in linking cellular stimuli to changes in protein sumoylation, it is important to use a combinatorial treatment with both inducers and pathway inhibitors. A number of inhibitors exist commercially, including U0126 (for ERK MAPK pathway inhibition), SB203580 (for p38 MAPK pathway inhibition), and SP600125 (for JNK MAPK pathway inhibition (28) (see Note 20). The experimental procedures are essentially the same as described in Subheading 3.1.3 with the exception that an hour pretreatment with inhibitors should be made before adding the inducers (see Note 21).

3.2. SDS-PAGE

SDS-PAGE is a widely used biochemical approach to separate proteins according to their charge–mass ratios that results in different migration under electrophoresis. Due to this property, the proteins that are subject to post-translational modifications that add significant mass can be readily separated from their non-modified forms.

1. These instructions assume the use of a BIO-Rad mini-protean II electrophoresis gel system. It is essential that the glass plates are kept clean by rinsing extensively with ethanol.
2. Prepare a pair of 0.75-mm thick, 10% gel by mixing 2.25 ml of 3.55× separating buffer, with 2.88 ml of 30% Acrylamide/Bis-acrylamide solution, 2.86 ml water, 50 μl of 10% ammonium persulphate solution, and 20 μl of TEMED. Pour the gel, leaving space for a 1 cm stacking gel, and overlay with propan-2-ol (see Note 22).
3. Pour off the propan-2-ol and rinse the top of gel with plenty of water. Finally withdraw as much water as possible by blotting with tissue paper.
4. Prepare the stacking gel by mixing 1.88 ml of 4× stacking buffer, with 1 ml of 30% Acrylamide/Bis-acrylamide soultion, 4.65 ml water, 50 μl of 10% ammonium persulphate solution, and 15 μl of TEMED. Pour the stacking gel and insert the comb.
5. Prepare 1× running buffer from 10× stock solution.
6. Once the stacking gel has set, carefully remove the comb and use a 10-ml syringe fitted with a 22-gauge needle to wash the wells with running buffer.
7. Add the running buffer to the upper and lower chambers of gel unit.
8. 20 μl of each sample plus 5 μl of 5× Laemmli sample buffer is boiled at 100 °C for 5–10 min before loading onto SDS-PAGE. Load 3 μl of prestained protein marker in one of the wells.
9. Complete the assembly of the gel unit and connect to a Bio-Rad power pack. The gel can be run at 200 V for 1–2 h (see Note 23).

3.3. Western Blotting

The Western blot is an analytical tool for the detection of specific proteins in a given cell lysate that are separated by SDS-PAGE and transferred onto a membrane. This technique uses antibodies specific to the target protein as probes. Typically, a two-step method is used in the detection process. This involves the incubation of a primary antibody and then a labelled (e.g. chemiluminescent-linked) secondary antibody.

1. After electrophoresis, the samples are subsequently transferred onto a nitrocellulose membrane. The following protocol assumes the use of a Bio-Rad Mini Trans-Blot electrophoretic transfer cell system.
2. Prepare 1× transfer buffer. For each of the gels to transfer, cut four sheets of 3MM paper and a sheet of the nitrocellulose membrane to the dimensions of gel.
3. Remove the stacking gel before transfer. Equilibrate the separating gel in transfer buffer and soak the membrane, 3MM paper, and fibre pads in the same buffer for 5 min (see Note 24).
4. Prepare the gel sandwich. Place the cassette, with the black side down, on a clean surface. Lay on one of the pre-wetted fibre pads on the cassette, and then place two sheets of 3MM paper on top. Place the equilibrated gel on top of 3MM paper and then the pre-wetted membrane on the gel. Complete the sandwich by placing two pieces of 3MM paper on the membrane and add the fibre pad (see Note 25). Place the cassette in the module.
5. Add a Frozen Bio-Ice cooling unit. Fill the tank with transfer buffer (see Note 26).
6. The transfer should be carried out at 250 mA for 1 h. Upon completion of the run, disassemble the blotting sandwich. The gel is left in place on top of the nitrocellulose and the shape of the gel can then be cut into the membrane using a razor blade. The membrane is now ready for the further processing (see Note 27).
7. The nitrocellulose membrane is first rinsed with water and then incubated in 15 ml of blocking buffer for 20 min at room temperature on a rocking platform. Repeat this procedure once more.
8. The blocking buffer is now replaced with a solution containing the appropriate dilution of primary antibody (see Note 28) in blocking buffer and incubated for 1 h at room temperature on a rocking platform.
9. The primary antibody is then removed and the membrane washed twice for 15 min with 20 ml of TBS-T buffer.

10. The secondary antibody is freshly prepared as 1:10,000-fold dilution in blocking buffer and added to membrane for 1 h at room temperature on a rocking platform (see Note 9).
11. The secondary antibody is decanted off and discarded and the membrane washed four times for 15 min each with TBS-T buffer.
12. Prepare a working solution by mixing equal parts of the stable peroxide solution and the Luminol/Enhancer solution (see Note 29).
13. Incubate blot with working solution for 1 min and remove blot from working solution. Subsequently, sandwich it in acetate sheets or cling film. Carefully press out any bubbles from between the blot and sheets.
14. The blot is now ready for developing using either X-ray film on a detection system or Fluor-S MultiImager (Bio-Rad). Data from the MultiImager are analysed by Quantity One software (Bio-Rad).

3.4. Functional Analysis

Protein sumoylation modulates a diverse range of cellular functions. Importantly, it often functions in the regulation of transcriptional properties. In addition, it is common that the SUMO pathway participates in the dynamic localisation of proteins (1, 29). Here, we describe examples of functional analyses that can be used as an indirect readout for changes in protein sumoylation. There readouts include reporter gene assays, chromatin immunoprecipitation, and localisation analysis.

3.4.1. Reporter Assays

In addition to analysing endogenous gene transcription by RT-PCR methods, reporter gene assays are widely used for studying gene regulation and function. Due to their high sensitivity, genes encoding luciferase and β-galactosidase are the most popular source for reporter gene assays. To elucidate whether the sumoylation of a particular transcription factor might play a role in controlling its ability to regulate gene transcription, a number of strategies based on reporter gene assays can be considered. In addition, in some cases, sumoylation might be difficult to detect and additional evidence for SUMO function might be required. For example, mutations in the potential SUMO conjugation sites in the protein of interest can be created using site-directed mutagenesis approaches (see Note 30). These mutants can be monitored for their transcriptional readout in reporter gene assays in comparison to the wild-type protein. As an alternative way of implicating the SUMO pathway, the SUMO pathway can be blocked by overexpressing either the viral protein, Gam-1 (to interfere with the E1 enzyme) (30) or a dominant-negative mutant of the E2 SUMO conjugation enzyme (Ubc9). These reagents provide invaluable tools for the functional analysis of the SUMO pathway in controlling gene transcription. Further studies where pathway components are depleted such as the E1, E2 enzymes, or the SUMO paralogues by small interference RNA (siRNA) technology (see Note 31) can often lead to more substantiated and reliable conclusions.

1. These instructions assume the use of a Dual-Light chemiluminescent reporter Gene Assay System (TROPIX, PE Biosystems).
2. The procedures for cell transfection are described as in Subheading 3.1.2 (see Note 32).
3. 24 h after transfection, cells are briefly washed with 3 ml of DPBS and subsequently lysed in 75 μl of Lysis Solution.
4. Detach cells from plate with a cell scraper and transfer the cell lysate to a 1.5 ml microfuge tube.
5. Leave lysate on ice for 10 min and centrifuge at 4°C, 13 krpm for 10 min.
6. Transfer supernatants to a fresh tube and use immediately.
7. Add 12.5 μl of Buffer A to the extract samples in assay tubes and immediately add 50 μl of Buffer B. Typically 0.5–5 μl of cell extracts are used for chemiluminescent detection. After a 5 s delay, read the luciferase signal for 10 s (see Note 33).
8. Leave for 30–60 min at room temperature (see Note 34).
9. Add 50 μl of Accelerator-II. Read the β-galactosidase signal for 10 s after incubating for 5 s.

3.4.2. Chromatin Immunoprecipitation (ChIP)

ChIP assays can provide an excellent route for determining functional changes in transcription factor activity caused by the SUMO pathway. For example, this allows the direct evaluation of the involvement of sumoylation in transcription factor binding to promoters or the more indirect effects of sumoylation on the recruitment of transcriptional coregulatory factors. To establish such links, ChIP assays can be performed in combination with the reagents and approaches described above (see Subheading 3.4.1) to disrupt the SUMO pathway.

1. Stimulate or treat 5×10^6 cells on a 10-cm dish.
2. Remove media, then cross-link SUMO, histones or other transcriptional regulatory factors by adding 1% formaldehyde solution in culture media to the cells and incubate for 2 min at 37°C. Aspirate formaldehyde solution, wash briefly with DPBS; freshly prepared 10 mM DMA solution is added and incubated for a further 2 min at 37°C (see Note 10).
3. Aspirate the DMA solution and then incubate cells with 125 mM glycine in DPBS at room temperature for 5 min to terminate the cross-linking reaction.

4. Wash cells three times with 20 ml of DPBS.
5. Scrape cells into 1 ml of ice-cold DPBS containing a Complete Mini protease inhibitor cocktail, 10 μg/ml of E64 and 10 mM NEM (see Note 1). Pellet cells for 5 min at 700 × *g* at 4°C.
6. Wash pellet with 1 ml of Buffer I by gentle pipetting, pellet as in step 5.
7. Wash pellet with 1 ml of Buffer II by gentle pipetting, pellet as in step 5.
8. Re-suspend cell pellet in 550 μl of lysis buffer (approx. 10^6 cells/100 μl) and leave at 4°C for 10 min.
9. Sonicate lysate 3 × 15 s (power 8) to reduce DNA length to between 200 and 600 base pairs (see Note 35). Remove debris by centrifugation for 15 min at 13 krpm at 4°C.
10. Samples are aliquoted to 5 × 100 μl, 1 × 10 μl for input control and 1 × 10 μl for checking the DNA length (see Note 36).
11. Dilute lysate (100 μl) tenfold in ChIP dilution buffer (900 μl) and incubate with 2 μg of appropriate primary antibody (see Note 37) at 4°C overnight on a rotating platform.
12. Pre-incubate 7.5 μl of Dynabeads protein A or G (see Note 2) with 50 μg ssDNA, 50 μg BSA in 100 μl of ChIP dilution buffer at 4°C for 1 h.
13. The immune complexes are collected by adding the pre-blocked beads to the lysate prepared in step 11 and incubated for a further 1 h at 4°C.
14. Wash beads (by rotating at room temperature for 5 min and left in the magnet for 1 min in between washes to precipitate the beads) sequentially with 1 ml of low salt, high salt, LiCl, and 2 × TE buffer. Beads are rotated at room temperature for 5 min in each solution and left in a magnet for 1 min in between washing.
15. Elute immune complexes by adding 150 μl freshly made elution buffer and incubate at room temperature for 15 min with rotation. Repeat the elution and combine the eluates.
16. Add 12 μl 5 M NaCl to the combined the eluates and also treat the input control and gel sample (10 μl of sample and 290 μl of elution buffer) and reverse the crosslinks by incubating at 65°C overnight.
17. Add 6 μl of 0.5 M EDTA, 12 μl of 1 M Tris.Cl [pH 6.5] and 1.2 μl of 10 mg/ml proteinase K to the eluates and incubate at 45°C for 1 h.
18. Recover DNA using QIAGEN PCR clean-up kit and finally elute precipitated and input DNA in 50 μl of PCR grade water.

19. To detect bound promoter DNA, qPCR is performed with a Rotor-gene RG-300 thermal cycler (Corbett research) using the Quantitect SYBR Green PCR Kit (QIAGEN) (see Note 38). Briefly, the reaction mixes are prepared by adding 4.84 μl of template DNA from step 18, 5 μl of PCR mix, 0.06 μl of 50 μM of primer pair, and 0.1 μl of 50 mM $MgCl_2$. The PCR programme is conducted as follows: 95°C for 15 min, 40 cycles of: 95°C for 10 s, 53°C for 10 s, and 72°C for 20 s. A melting curve is performed from 72°C to 99°C in a 1°C increment.

3.4.3. Sub-cellular Localisation

The sub-cellular localisation of transcriptional regulatory factors often plays an important role in controlling gene expression. The SUMO pathway has been shown to participate in the regulation of dynamic localisation of proteins and hence might have a role in modulating gene transcription. A number of strategies can be used for the analysis of protein localisation. These include standard immunofluorescence methods, cell imaging experiments using fluorescent-tagged fusion proteins, and biochemical fractionation assays. However, the detailed protocols are out of the scope of this chapter.

4. Notes

1. Add Complete Mini protease inhibitor cocktail, E64, and NEM before use. E64 and NEM are cysteine protease inhibitors that prevent the hydrolysis reaction of SUMO conjugates during cell lysis.
2. Dynabeads protein A is used to collect the immunocomplexes from primary antibodies against rabbit and protein G is used for mouse-generated antibodies. In addition, other types of beads such as protein A/G agarose or sepharose beads can also be used.
3. The choice of transfection reagent will be dependent on the cell type used in the experiment. For examples, polyfect reagent is most suitable for 293T cell transfection and lipofectamine 2000 for the Hela cell line. Other transfection reagents can also be used. For example, oligofectamine is designed for efficient siRNA oligo delivery.
4. Ni-NTA agarose should be washed and equilibrated as a 50% slurry in the guanidine lysis solution before use.
5. β-mercaptoethanol should be added immediately before use.

6. For each 500 μl of elution buffer, the solution is freshly prepared by adding 210 μl of water, 125 μl of 1 M imidazole, and 165 μl of 5 × Laemmli sample buffer.
7. This is a neurotoxin when unpolymerised.
8. Transfer buffer can be used for a number of transfers as long as the voltage is maintained constant for each successive run. The buffer should be kept cold through the inclusion of an ice-cooling block during the transfer.
9. The choice of HRP-conjugated secondary antibody will be dependent on the primary antibody used. In addition, some primary antibodies that covalently conjugate to HRP can also be used. In this case, the secondary antibody incubation can be eliminated.
10. DMA should be prepared in non-amine-containing buffers, such as phosphate and HEPES to prevent crosslinking to the buffer.
11. Prepare freshly before use.
12. The high percentage of SDS during cell lysis is to preserve the SUMO conjugates by inhibiting the activity of the SENPs. However, the further dilution step is required for the subsequent immunoprecipitation experiments. This dilution step is to bring down the SDS to 0.1% (w/v) and avoid antibody denaturation.
13. In some cases, lower incubation times can be used depending on the nature of antibody used in the experiments.
14. Due to their property of high efficiency for DNA transfection, 293T cells are chosen as a convenient cell type for the experiments. Other cell types might be needed for various reasons.
15. At this stage, the cells can be maintained in DMEM/10% serum. However, for subsequent induction of the MAPK pathway, it is essential to carry out an overnight incubation in serum-free medium to down-regulate basal signalling levels and allow the cells to adopt a quiescent state (G_0) before activating the pathway.
16. Avoid centrifugation of the samples. Instead, let the beads settle to the bottom of the tube; this normally takes about 7–10 min. However, if it is necessary, give samples a brief centrifugal pulse at a speed less than 2 krpm.
17. These constructs are mainly used to activate their corresponding MAPKs. However, it is worth noting that some degree of cross talk between MAPK pathways at the level of the terminal kinases might occur by this approach. Subsequent verification (see Subheading 3.1.3) will be required.

18. PMA can also activate other kinases such as the protein kinase A (PKA) and protein kinase C (PKC).
19. The kinetics of pathway activation and the downstream effects on protein sumoylation can vary between cell types and according to the stimulus used, so treatment times might be varied (ranging from 10 min to 8 h).
20. SP600125 is not a very specific inhibitor for JNK MAPK.
21. Prior to MAPK induction, it is important to pre-treat with the corresponding MAPK inhibitors for at least 30–60 min. Prolonged treatment is not recommended.
22. The percentage of gel prepared will be dependent on the size of protein of interest and can be lowered for high molecular weight proteins.
23. The gel running time will need to be adjusted according to the size of protein of interest and length of separation required.
24. In some cases, the SUMO conjugates might run with a high apparent molecular weight. Sometimes these conjugates might even be in the stacking gel, so removal of the stacking gel will hamper the detection of these conjugates.
25. Removing any air bubbles which may have formed is very important for good results. Use a glass tube to gently roll air bubbles out.
26. The buffer temperature and ion distribution in the tank can be maintained by continuously mixing with a magnetic stirring bar throughout the transfer.
27. At this stage, the membrane can be stained by Ponceau solution to ensure the completion of transfer and equal loading of lysates.
28. Typical antibody dilutions range from 1:500 to 1:5,000.
29. Use 0.1 ml working solution per cm^2 of membrane. Exposure to the sun or any other intense light source can damage the working solution. For the best results, the working solution should be kept in a light-proof bottle.
30. Sumoylation often takes place on an acceptor lysine residue of a substrate at the core SUMO consensus motif, ΨKxE (where Ψ is hydrophobic amino acids). Whereas mutation on an acceptor lysine residue (K) can disrupt substrate sumoylation and its functions, the effects cannot necessarily be fully attributed to the sumoylation. This is because lysine residues can also be modified by many other post-translational modifications such as methylation, acetylation, and ubiquitination. It is therefore important to also generate mutations of the glutamic acid residue (E) to disrupt the consensus motif for

sumoylation while leaving the lysine residue intact. Typically, mutagenesis will alter the K or E to alanine (A), as this will result in the minimal disturbance in the protein structure. Alternatively, K to arginine (R) can be used to preserve the charged nature of this residue. In addition, the effects of these mutants on sumoylation can also be tested as described in Subheading 3.1.2.

31. Either plasmid- or oligo-based siRNA can be used in this type of experiment. However, to ensure the optimal knockdown of protein of interest, a two-step transfection procedure is often used. Briefly, the siRNA is firstly transfected for 24 h prior to the transfection of plasmids containing the reporter gene set (the additional siRNA can then be included in this plasmid mix). The oligo-based siRNA can be delivered by oligofect reagent (QIAGEN) in the first step of transfection.
32. The plasmids used for the reporter gene set often include either multiple-GAL4 DNA binding sites fused to the luciferase reporter gene and the GAL4 DNA binding domain fused to the protein of interest or a natural promoter which responds to the specific transcription factor under study and an expression vector encoding the transcription factor (in some cases, the endogenous expression of the transcription factor is sufficient). In addition, an expression construct encoding the β-galactosidase gene is co-transfected to monitor the transfection efficiency.
33. The enhanced luciferase reaction produces a light signal which decays with a half-life of approximately 1 min. It is important to maintain consistent timing of addition of Buffer B. In addition, the remaining lysates should be stored at −20°C for further analysis. For example, the expression levels of protein of interest can be examined by Western blotting.
34. Light signal from β-galactosidase reaction is initially negligible due to the lower initial pH (7.8). After a 30–60 min incubation (to ensure the complete loss of luciferase light signal), the β-galactosidase signal is initiated by addition of light emission Accelerator-II which raises the pH to generate optimal reaction conditions and provides luminescence enhancer to increase light intensity. It is important to maintain consistent timing of measurement of β-galactosidase signal before and after adding Accelerator-II.
35. The samples can be cooled on dry ice between pulses, but do not freeze the samples.
36. Check the size of sonicated DNA by agarose gel electrophoresis. This should be performed after reversal of crosslinks (top up with 300 μl elution buffer followed by step 16–18).

20 μl of cleaved DNA is loaded onto the 1% agarose gel for gel electrophoresis.

37. Depending on the source of primary antibody used, either normal mouse or rabbit IgG is selected as a negative control for the background immunoprecipitation.
38. Other methods for performing qPCR are also available.

Acknowledgments

We thank members of our laboratory and Ling-I Su for comments on the manuscript and stimulating discussions and Ron Hay for reagents. The authors are supported by grants from the Wellcome Trust and a Royal Society-Wolfson award to ADS.

References

1. Johnson, E.S. (2004) Protein modification by SUMO. *Annu. Rev. Biochem.* **73**, 355–382.
2. Hay, R.T. (2005) SUMO: a history of modification. *Mol. Cell.* **18**, 1–12.
3. Kim, J.H., Baek, S.H. (2009) Emerging roles of desumoylating enzymes. *Biochim Biophys Acta* **1792**, 155–162.
4. Rodriguez, M.S., Dargemont, C. & Hay, R.T. (2001) SUMO-1 conjugation in vivo requires both a consensus modification motif and nuclear targeting. *J. Biol. Chem.* **276**, 12654–12659.
5. Melchior, F. (2000) SUMO-nonclassical ubiquitin. *Annu. Rev. Cell Dev. Biol.* **16**, 591–626.
6. Diella, F., Chabanis, S., Luck, K., Chica, C., Ramu, C., Nerlov, C., & Gibson, T. J. (2009) KEPE – a motif frequently superimposed on sumoylation sites in metazoan chromatin proteins and transcription factors. *Bioinformatics.* **25**, 1–5.
7. Subramanian, L., Benson, M.D., & Iñiguez-Lluhí, J.A. (2003) A synergy control motif within the attenuator domain of CCAAT/enhancer-binding protein alpha inhibits transcriptional synergy through its PIASy-enhanced modification by SUMO-1 or SUMO-3. *J. Biol. Chem.* **14**, 9134–9141.
8. Hietakangas, V., Anckar, J., Blomster, H.A., Fujimoto, M., Palvimo, J.J., Nakai, A., Sistonen, L. (2006) PDSM, a motif for phosphorylation-dependent SUMO modification. *Proc. Natl. Acad. Sci. U S A.* **103**, 45–50.
9. Yang, S.H., Galanis, A., Witty, J., & Sharrocks, A.D. (2006) An extended consensus motif enhances the specificity of substrate modification by SUMO. *EMBO J.* **25**, 5083–5093.
10. Pichler, A., Knipscheer, P., Oberhofer, E., van Dijk, W.J., Körner, R., Olsen, J.V., Jentsch, S., Melchior, F., Sixma, T.K. (2005) SUMO modification of the ubiquitin-conjugating enzyme E2-25K. *Nat Struct Mol Biol* **12,** 264–269.
11. Yang, S.H., Jaffray, E., Senthinathan, B., Hay, R.T., & Sharrocks, A.D. (2003) SUMO and transcriptional repression: dynamic interactions between the MAP kinase and SUMO pathways. *Cell Cycle.* **2**, 528–530.
12. Yang, S.H., & Sharrocks, A.D. (2006) Interplay of the SUMO and MAP kinase pathways. *Ernst Schering Res Found Workshop.* **57**, 193–209.
13. Guo, B., Yang, S.H., Witty, J., & Sharrocks, A.D. (2007) Signalling pathways and the regulation of SUMO modification. *Biochem Soc Trans.* **35**, 1414–1418.
14. Yang, S.H., Sharrocks, A.D., Whitmarsh, A.J. (2003) Transcriptional regulation by the MAP kinase signaling cascades. *Gene* **27,** 3–21.
15. Sharrocks, A.D. (2002) Complexities in ETS-domain transcription factor function and regulation; lessons from the TCF subfamily. *Biochem. Soc. Trans.* **30**, 1–9.
16. Yang SH, Bumpass DC, Perkins ND, Sharrocks AD. The ETS domain transcription factor Elk-1 contains a novel class of repression domain (2002) *Mol Cell Biol.* **22**, 5036–5046.

17. Yang SH, Jaffray E, Hay RT, Sharrocks AD. (2003) Dynamic interplay of the SUMO and ERK pathways in regulating Elk-1 transcriptional activity. Mol Cell. **12**, 63–74.
18. Li, Q.J., Yang, S.H., Maeda, Y., Sladek, F.M., Sharrocks, A.D., Martins-Green, M. (2003) MAP kinase phosphorylation-dependent activation of Elk-1 leads to activation of the co-activator p300. *EMBO J.* **22,** 281–291.
19. O'Donnell, A., Yang, S.H., Sharrocks, A.D. (2008) MAP kinase-mediated c-fos regulation relies on a histone acetylation relay switch. *Mol Cell.* **29,** 780–785.
20. Yang, S.H., Sharrocks, A.D. (2004) SUMO promotes HDAC-mediated transcriptional repression. *Mol Cell.* **13,** 611–617.
21. Yang, S.H., Sharrocks, A.D. (2005) PIASx acts as an Elk-1 coactivator by facilitating derepression. *EMBO J.* **24,** 2161–2171
22. Yang, S.H., Sharrocks, A.D. (2006) PIASxalpha differentially regulates the amplitudes of transcriptional responses following activation of the ERK and p38 MAPK pathways. *Mol Cell.* **22,** 477–487.
23. Vanhatupa, S., Ungureanu, D., Paakkunainen, M., Silvennoinen, O. (2008) MAPK-induced Ser727 phosphorylation promotes SUM Oylation of STAT1. *Biochem J.* 409, 179–185.
24. Camuzeaux, B., Diring, J., Hamard, P.J., Oulad-Abdelghani, M., Donzeau, M., Vigneron, M., Kedinger, C., Chatton, B. (2008) p38beta2-mediated phosphorylation and sumoylation of ATF7 are mutually exclusive. *J Mol Biol.* **384,** 980–991.
25. Bossis G, Malnou CE, Farras R, Andermarcher E, Hipskind R, Rodriguez M, Schmidt D, Muller S, Jariel-Encontre I, Piechaczyk M. (2005) Down-regulation of c-Fos/c-Jun AP-1 dimer activity by sumoylation. *Mol Cell Biol.* **25,** 6964–6979.
26. Daniel, A.R., Faivre, E.J., Lange, C.A. (2007) Phosphorylation-dependent antagonism of sumoylation derepresses progesterone receptor action in breast cancer cells. *Mol Endocrinol.* **21**, 2890–2906.
27. Guo, B., Sharrocks, A.D. (2009) Extracellular signal-regulated kinase mitogen-activated protein kinase signaling initiates a dynamic interplay between sumoylation and ubiquitination to regulate the activity of the transcriptional activator PEA3. *Mol Cell Biol.* **29,** 3204–3218.
28. Bain, J., Plater, L., Elliott, M., Shpiro, N., Hastie, C.J., McLauchlan, H., Klevernic, I., Arthur, J.S., Alessi, D.R., Cohen, P. (2007) The selectivity of protein kinase inhibitors: a further update. *Biochem J.* **408,** 297–315.
29. Watts, F.Z. (2004) SUMO modification of proteins other than transcription factors. *Semin Cell Dev Biol.* **15,** 211–220.
30. Pozzebon, M., Segré, C.V., Chiocca, S. (2009) Inhibition of the SUMO pathway by Gam1. *Methods Mol Biol.* **497,** 285–301.

Chapter 22

Computational Modelling of Kinase Signalling Cascades

David Gilbert, Monika Heiner, Rainer Breitling, and Richard Orton

Abstract

In this chapter, we describe general methods used to create dynamic computational models of kinase signalling cascades, and tools to support this activity. We focus on the ordinary differential equation models, and show how these fit into a general framework of qualitative and quantitative (stochastic and continuous) models. The modelling we describe is part of the activity of BioModel engineering which provides a systematic approach for designing, constructing, and analyzing computational models of biological systems.

Key words: Computational modelling, Ordinary differential equations, Continuous Petri nets

1. Introduction

Computational modelling of intracellular biochemical networks has become a growth topic in recent years, due to advances both in the power and availability of software systems for the simulation and analysis of such networks, as well as an increase in the quality and amount of experimentally determined parameter data available for modelling.

Modelling biochemical systems is the core part of the process of BioModel Engineering (1) which is at the interface of computing science, mathematics, engineering, and biology, and provides a systematic approach for designing, constructing, and analyzing computational models of biological systems. BioModel Engineering does not aim at engineering biological systems per se (in contrast to synthetic biology), but rather aims at describing their structure and behaviour, in particular at the level of intracellular molecular processes, using computational tools and techniques.

Rony Seger (ed.), *MAP Kinase Signaling Protocols: Second Edition*, Methods in Molecular Biology, vol. 661,
DOI 10.1007/978-1-60761-795-2_22, © Springer Science+Business Media, LLC 2010

The most useful kinds of models for signalling pathways are dynamic models that describe the time course behaviour of molecular concentrations or even individual molecules. This contrasts with static models which merely describe the topology of the system, i.e. the molecular species involved and their relationships or wiring diagram. In addition to simulation, dynamic models permit a range of analytical techniques that give insight about system-level features that emerge from the elementary interactions of the components. Emergent properties such as bifurcations, robustness to interference, or oscillations are not obvious from the network topology and their discovery requires computational methodologies. Dynamic models provide a powerful framework for hypothesis generation and testing and the identification of inconsistencies between a model and experimental data. They are often used by life scientists as a means to explore their ideas about the organisation and control of a biological system.

The "correctness" of a model can be established in several ways. *Biological model validation* establishes whether a model contradicts our knowledge of a biological system and hence requires experimental data about the behaviour of the system. A special technique contributing to model validation is *model checking*, which establishes whether a set of formal properties hold for a model, and is often automated using computer programs. A biologically valid model can be *incomplete* and hence may not describe all the observations we can potentially make of a system, but should not incorrectly describe those behaviours of the system for which it allows predictions.

As the EGFR-activated ERK (EGFR/ERK) pathway is such an important signalling pathway, the deregulation of which has long been implicated in various forms of cancer, it has become a popular target for computational modelling strategies (2–4). Currently, there are a wide variety of models of the EGFR/ERK pathway available which have led to novel insights and interesting predictions as to how this system functions (5). A large number of the models of this pathway are based on the ordinary differential equation (ODE) approach (6). Examples of popular models of the EGFR/ERK pathway include the models described by Brightman et al. (7), Schoeberl et al. (8), and Brown et al. (9), all of which use an ODE-based approach. In general, models have been developed to illustrate particular aspects of pathway behaviour, and may not be consistent between each other. For instance, the Brightman and Brown models predict that the negative feedback loop from ERK to SOS is essential for a transient activation of ERK to be achieved, whereas the Schoeberl model predicts that the negative feedback loop is not required for the transient activation of ERK. Orton et al. (10) have developed a model which overcomes these inconsistencies and suggest some corrections to the Schoeberl model on which their work is based.

Aspects of behaviour of the MAPK pathway which have been investigated using computational models include:

- Ultrasensitivity of the ERK cascade as a result of a two-step distributive activation mechanisms (11–13).
- Oscillatory behaviour of the ERK cascade due to embedded negative feedback loops (14).
- The effects of receptor location, trafficking, and degradation on downstream ERK signalling (8, 15).
- The dynamic differences between the transient and sustained activation of ERK by different growth factors (7, 16).
- The influence of Raf kinase inhibitor protein (RKIP) on the ERK pathway (17).

1.1. Modelling Techniques for Biochemical Networks

In general, biochemical networks can always be modelled using qualitative information, describing the molecular species and their interconnections; if information about the reaction kinetics is known, quantitative models can also be employed. Modelling approaches based on ordinary differential equations (ODEs) have been described in detail before (e.g. (45)). Here we focus on approaches inspired by computing science concepts, which have recently gained popularity in systems biology (18, 19).

In the qualitative representation, a biochemical network is described by its topology; usually as a bipartite graph with two types of nodes representing either biochemical entities or reactions, and arcs being optionally annotated by stoichiometric information. This depiction corresponds to a qualitative Petri net where biochemical entities are places (Fig. 1; places are depicted as circles) and reactions are transitions (depicted as squares). These descriptions are "time-free", i.e. they do not describe the real-time dynamics of the system. The qualitative description can be further enhanced by the abstract representation of discrete quantities of species, achieved in Petri nets by the use of tokens at places; see for example Fig. 1. These can represent the number of molecules, or the level of concentration, of a species, and a particular arrangement of tokens over a network describes a system state, called a marking. Figure 1 illustrates this for a simple enzymatic reaction;

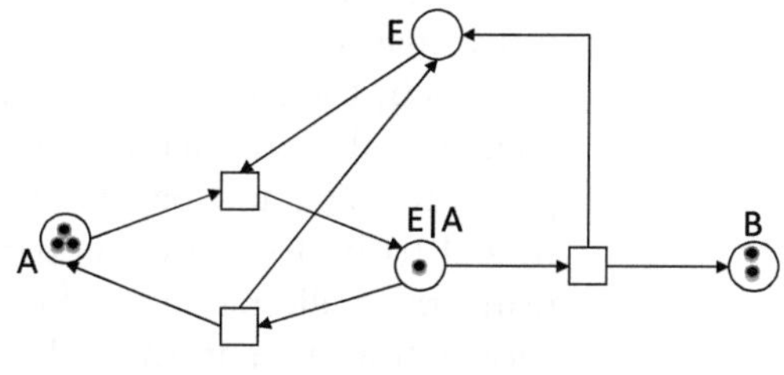

Fig. 1. Qualitative Petri net for a simple enzymatic reaction.

note that the default stoichiometric weights have been omitted. The current state (marking) of the system, indicated by the black tokens in each place (circle), is as follows: there are no free molecules of enzyme E, three molecules of substrate A, one molecule of the enzyme–substrate complex $E|A$, and two molecules of the product B. Equally, we could say that there are three units of concentration of A, one unit of $E|A$, and two of B. Clearly, due to the lack of free enzyme, the enzyme–substrate complex will have to be broken up by decomplexation to release one molecule of E and B (forward reaction) or one molecule of E and A (reverse reaction) in order for any further complexation of A with E to take place. From a computing science perspective, the behaviour of such a net forms a discrete state space, which can be analysed in the bounded case, for example, by a branching time temporal logic, one instance of which is computational tree logic (CTL) (20).

Timed information can be added to the qualitative description in two ways – stochastic and continuous. The stochastic description preserves the discrete description of the values of biochemical entities, but in addition associates an exponentially distributed rate with each reaction. Thus stochastic approaches represent the individual behaviour of molecules and hence variability in the overall (averaged) behaviour of a system. Special behavioural properties can be expressed using, e.g. continuous stochastic logic (CSL), see (21), a probabilistic counterpart of CTL.

In the continuous approach, discrete values of species are replaced with continuous values, and hence only overall (averaged) behaviour can be described by concentrations. A particular deterministic rate information is associated with each reaction, and thus the concentration of a particular species in such a model will have the same value at each point of time for repeated simulations. This approach permits the continuous model to be represented as a set of ODEs. The state space of such models can be analysed by, for example, linear temporal logic with constraints (LTLc) in the manner of (22).

Priami et al. showed how the stochastic π-calculus could be used to model biomolecular processes (23) and Regev et al. showed in more detail how the π-calculus can be used to model and simulate the MAPK pathway (24) using a continuous approach; subsequently, Phillips and Cardelli (25) used the stochastic π-calculus to model the MAPK pathway by simulating the behaviour of individual molecules using the Gillespie algorithm (26). A related approach, the stochastic process algebra PEPA (27), was used by Calder et al. to model the influence of RKIP on the ERK signalling pathway (28), and permits different alternative formulations of a model to be formally compared. PEPA models can be simulated using the Gillespie algorithm or ODE solvers: Calder et al. have shown how to automatically derive ODEs from process algebra models (29).

The stochastic and continuous models are mutually related by approximation, and the qualitative model can be regarded as an abstraction of both quantitative descriptions. For more details of a formal framework that explains these relationships within the context of Petri nets, see (30). The advantages of using Petri nets as a kind of umbrella formalism include (31):

- An intuitive and executable modelling style
- A true concurrency (partial order) semantics, which may be weakened to inter-leaving semantics to simplify analyses
- Mathematically founded analysis techniques based on formal semantics
- Coverage of structural and behavioural properties as well as their relations
- Integration of qualitative and quantitative analysis techniques
- Reliable tool support

2. Tools

Several tools are available which permit the construction of quantitative biochemical pathway models using kinetic descriptions and their simulation and analysis; these often read and write in SBML (32) format which is one de-facto standard for the description of quantitative models of biochemical pathways. Such tools include BioNessie (33), and Copasi (34). There are also databases of biochemical models, often with descriptions in SBML format, for example Biomodels (http://www.biomodels.org) (35), and more specialized databases such as the MAPK database at Brunel (mapk.brunel.ac.uk), which was developed as part of the SIMAP project (http://www.eurtd.org/simap).

MATLAB (36) is a high-level language and interactive environment that contains a large number of ODE solvers which can be used to numerically solve and analyse ODEs. The SimBiology toolbox extends MATLAB with tools for modelling, simulating, and analyzing biochemical pathways, and has graphical interface as well as a facility to read and write SBML. The systems biology workbench (SBW) (37), is a software framework that includes Jarnac, a fast simulator of reaction networks, permitting time course simulation (ODE or stochastic), steady-state analysis, basic structural properties of networks, dynamic properties like the Jacobian, elasticities, sensitivities, and eigenvalues, and JDesigner, a friendly GUI front end to an SBW compatible simulator. Bifurcation analysis can be performed conveniently using Xppaut (38). CellDesigner (39) has a graphical interface using SBGN (Systems Biology Graphical Notation), is SBML compliant, and SBW-enabled so that it can integrate with other SBW-enabled simulation/analysis software packages. CellDesigner also supports simulation and parameter search, using the SBML ODE solver.

Petri net models can be designed using several software tools. Snoopy (40) supports qualitative as well as quantitative Petri nets, among them both continuous and stochastic Petri nets, and has an SBML interface. There are several analytical tools for Petri nets which can be accessed by the export feature of Snoopy, including Charlie (41), as well as a variety of model checkers, e.g. the IDD-MC (42) and DSSZ-MC (43) tools.

For more discussion of types of software to support modelling (see Note 1).

3. Modelling the MAPK Pathway

In the following, we describe a method for building a model of an MAPK pathway, given some knowledge about the topology of the pathway, and basic assumptions about the kinetics involved. This is part of a larger process of BioModel Engineering – see (1) and Note 2.

3.1. Modelling Biochemical Equations Using ODEs

ODE descriptions are so far the most widely used approach when modelling signalling pathways formally, and we now focus on this methodology. First we briefly recall the use of ODEs to model basic biochemical reactions; for more details refer to a standard text on biochemistry, e.g. (44).

Given a reaction $A \rightarrow B$ which occurs at a rate k, we can compute the time course of the concentrations of A and B by the following equation, where $[A]$ stands for the concentration of A etc:

$$\frac{\mathrm{d}[A]}{\mathrm{d}t} = -\frac{\mathrm{d}[B]}{\mathrm{d}t} = -k[A]$$

Enzymatic reactions are the basis of MAPK cascades. These reactions are represented in biochemical notation by

$$A \xrightarrow{F} B$$

where E is the enzyme which catalyses the conversion of substrate A into product B.

The arrow in this representation implicitly stands for a whole set of elementary chemical reactions. In the most general case, these can be described using *mass-action* kinetics, e.g. as follows:

$$E + A \underset{k_{-1}}{\overset{k_1}{\rightleftarrows}} E \mid A \xrightarrow{k_2} E + B$$

Note that here we use $E|A$ to denote the enzyme–substrate intermediate complex, with a reversible reaction to describe the association of the enzyme and the substrate with rate k_1 for the association and k_{-1} for dissociation, and k_2 as the production rate of B from the complex. There are more complex representations in which the association of the enzyme and the product are reversible, and even the conversion of the substrate to the product whilst complexed with the enzyme are reversible. For more details, see (45).

We can decompose the basic enzymatic reaction into three constituent reactions:

$$E + A \xrightarrow{k_1} E \mid A$$
$$E \mid A \xrightarrow{k_{-1}} E + A$$
$$E \mid A \xrightarrow{k_2} E + B$$

We can then derive the following four differential equations for each of the species, i.e. the substrate A, the product B, the enzyme E, and the substrate–enzyme complex $A|E$:

$$\frac{\mathrm{d}[A]}{\mathrm{d}t} = -k_1 \times [A] \times [E] + k_{-1} \times [A \mid E]$$
$$\frac{\mathrm{d}[A \mid E]}{\mathrm{d}t} = k_1 \times [A] \times [E] - k_{-1} \times [A \mid E] - k_2 \times [A \mid E]$$
$$\frac{\mathrm{d}[B]}{\mathrm{d}t} = k_2 \times [A \mid E]$$
$$\frac{\mathrm{d}[E]}{\mathrm{d}t} = -k_1 \times [A] \times [E] + k_{-1} \times [A \mid E] + k_2 \times [A \mid E]$$

This can also be achieved using a Petri net approach (see Fig. 1) where each of the three reactions is represented as a component of the Petri net, and then the components are composed by merging their places. The ODEs can then be read directly from the merged network, see (45, 46) (see Note 3).

Another common description of enzyme kinetics is the *Michaelis–Menten* equation, which can be derived from the mass-action description based on a number of simple assumptions:

$$V = V_{\mathrm{max}} \times \frac{[A]}{(K_{\mathrm{M}} + [A])}$$

Here, V is the reaction velocity, V_{max} is the maximum reaction velocity, and K_{M}, the Michaelis constant, i.e., the concentration of the substrate at which the reaction rate is half its maximum value. With the total (free and substrate-bound) enzyme concentration $[E_{\mathrm{T}}]$ and the equation

$$k_{\mathrm{cat}} = \frac{V_{\mathrm{max}}}{[E_{\mathrm{T}}]}.$$

We can write the differential equations describing the consumption of the substrate and the production of the product as follows:

$$\frac{\mathrm{d}[A]}{\mathrm{d}t} = -\frac{\mathrm{d}[B]}{\mathrm{d}t} = -k_{\mathrm{cat}} \times [E_{\mathrm{T}}] \times \frac{[A]}{(K_{\mathrm{M}} + [A])}$$

This form has the advantage that $[E_{\mathrm{T}}]$ might be dynamic, for example for an activated form of a kinase from the prevous stage in a cascade, while k_{cat} and K_{M} are constants. However, the assumptions underlying the Michaelis–Menten equation have been derived for in vitro conditions and strictly apply at the initial stage of an enzyme assay, but are often inappropriate for the in vivo situation of a kinase signalling cascade. In particular, the assumptions that the concentration of the product is close to zero, no product reverts to the substrate, and the concentration of the enzyme is much less than that of the substrate, are violated in the in vivo context (47) (see Note 3).

The mass-action equation for an enzymatic reaction is related to the Michaelis–Menten description by the following:

$$V = k_{\mathrm{cot}} \times [s] \times \frac{[MK]}{(K_{M1} + [MK])} - k^{1}_{\mathrm{cot}} \times \frac{[MK_F]}{(K_{M2} + [MK_F])}$$

$$\frac{d[MK]}{dt} = -\frac{d[MK_F]}{dt} = -V$$

For a more detailed discussion of the use of mass-action vs. Michaelis–Menten kinetics in modelling signalling pathways see (45).

3.2. Modelling Signal Transduction Cascades

In the course of signal transduction, extracellular signalling molecules bind to specific trans-membrane proteins (receptors) such as G-protein-coupled receptors (GPCRs) and receptor tyrosine kinases (RTKs), changing their conformation. This conformation change leads to a change in enzymatic activity of the receptor, which in turn affects the concentration of downstream compounds which are the substrates and products of the reaction catalysed by the receptor. In most cell types, MAPK cascades are activated through RTK and/or GPCR activation, which appear to function as central integration modules in signal processing. MAPK modules are evolutionarily conserved in cells from yeast to mammals. They typically consist of three kinases, which are activated sequentially by phosphorylating each other in response to stimuli, forming a three-tiered cascade, (48). The kinase in the first tier of the cascade is typically activated at the plasma membrane, whereas the third kinase is typically translocated from the cytoplasm to the nucleus upon activation, where it can regulate gene transcription through affecting chromatin structure and modifying the activity of transcription factors. The downstream compounds may themselves be enzymes that in a cascade of enzymatic reactions ultimately lead to a change in gene expression or some other major adjustment of cellular physiology.

From the modelling point of view, the main characteristic feature of the core of signal transduction pathways is that the product of one reaction becomes the enzyme for the next, and in general it is the dynamic behaviour, which is of interest in a signalling pathway, as opposed to the steady-state in a metabolic network.

A further feature of such cascades is that a mechanism has evolved to ensure that the non-phosphorylated form is regenerated from the phosphorylated form, enabling the signal to be deactivated where necessary. This is achieved by employing a phosphatase, an enzyme promoting the de-phosphorylation of the corresponding phosphorylated protein. Hence one step in a classical signal transduction cascade comprises a pair of forward and reverse enzymatic reactions in a so-called phosphorylation–de-phosphorylation loop. We can represent this in the following biochemical equation, where *MK* stands for a MAP kinase, *S* for the incoming signal (itself a kinase), and *P* is the phosphatase:

$$MK + S \underset{k_{-1}}{\overset{k_1}{\rightleftarrows}} MK \mid S \xrightarrow{k_2} MK_{\mathrm{p}} + S$$

$$MK + P \xleftarrow{kr_2} MK_{\mathrm{p}} \mid P \underset{kr_1}{\overset{kr_{-1}}{\rightleftarrows}} MK_{\mathrm{p}} + P$$

We can then encode this step as six differential equations using mass-action kinetics based on our approach above, i.e. either deriving the ODEs from a continuous Petri net (46), or by hand from our knowledge of standard kinetics.

$$\frac{\mathrm{d}[MK]}{\mathrm{d}t} = -k_1 \times [MK] \times [S] + k_{-1} \times [MK \mid S] + kr_2 \times [MK_{\mathrm{p}} \mid P]$$

$$\frac{\mathrm{d}[MK \mid S]}{\mathrm{d}t} = k_1 \times [MK] \times [S] - k_{-1} \times [MK \mid S] - k_2 \times [MK \mid S]$$

$$\frac{\mathrm{d}[MK_{\mathrm{p}}]}{\mathrm{d}t} = k_2 \times [MK \mid S] - kr_1 \times [MK_{\mathrm{p}}] \times [P] + kr_{-1} \times [MK_{\mathrm{p}} \mid P]$$

$$\frac{\mathrm{d}[S]}{\mathrm{d}t} = -k_1 \times [MK] \times [S] + k_{-1} \times [MK \mid S] + k_{-2} \times [MK \mid S]$$

$$\frac{\mathrm{d}[MK_{\mathrm{p}} \mid P]}{\mathrm{d}t} = kr_1 \times [MK_{\mathrm{p}}] \times [P] - kr_{-1} \times [MK_{\mathrm{p}} \mid P] - kr_2 \times [MK_{\mathrm{p}} \mid P]$$

$$\frac{\mathrm{d}[P]}{\mathrm{d}t} = -kr_1 \times [MK_{\mathrm{p}}] \times [P] + kr_{-1} \times [MK_{\mathrm{p}} \mid P] + kr_2 \times [MK_{\mathrm{p}} \mid P]$$

Alternatively, the step can be modelled more succinctly using Michaelis–Menten kinetics where K_{M1} and K_{M2} are the Michaelis constants for the forward and reverse reactions respectively. This results effectively in only one differential equation; traditionally the reference to the phosphatase is omitted because its total concentration can be considered constant as it does not need to be activated by the pathway (indeed, it may even be unknown); its existence is assumed in the term k_{cat}:

$$V = k_{cat} \times [S] \times \frac{[MK]}{(K_{M1} + [MK])} - k'\text{cat} \times \frac{[MK_p]}{(K_{M2} + [MK_p])}$$

$$\frac{d[MK]}{dt} = -\frac{d[MK_p]}{dt} = -V$$

In practice, many models employ mass-action kinetics for the phosphorylation step, and Michaelis–Menten kinetics for the de-phosphorylation step.

The central component of an *MAPK* cascade is a set of these phosphorylation–de-phosphorylation steps chained together by the phosphorylated product of one acting as the kinase enzyme for the next. In mammals these cascades have at least three such steps, where the proteins involved are Mitogen-activated Kinase-Kinase-Kinase denoted variously *MKKK* or MAP3K, Mitogen-activated Kinase-Kinase (MAP2K or *MKK*), and Mitogen-activated Kinase (*MAPK*) (see Fig. 2). Examples of *MKKK*, *MKK*, and *MAPK* are the proteins Raf, MEK, and ERK, respectively.

The mass-action description of the three-stage cascade is as follows, based on our initial description of one stage:

$$MKKK + S \underset{k1_2}{\overset{k1_1}{\rightleftharpoons}} MKKK|S \xrightarrow{k1_3} MKKK_P + S$$

$$MKKK + P_1 \xleftarrow{k1r_3} MKKK_P|P_1 \underset{k1r_2}{\overset{k1r_1}{\leftrightharpoons}} MKKK_P + P_1$$

$$MKK + MKKK_P \underset{k2_2}{\overset{k2_1}{\rightleftharpoons}} MKK|MKKK_P \xrightarrow{k2_3} MKK_P + MKKK_P$$

$$MKK + P_2 \xleftarrow{k2r_3} MKK_P|P_2 \underset{k2r_2}{\overset{k2r_1}{\leftrightharpoons}} MKK_P + P_2$$

$$MAPK + MKK_P \underset{k3_2}{\overset{k3_1}{\rightleftharpoons}} MAPK|MKK_P \xrightarrow{k3_3} MAPK_P + MKK_P$$

$$MAPK + P_3 \xleftarrow{k3r_3} MAPK_P|P_3 \underset{k3r_2}{\overset{k3r_1}{\leftrightharpoons}} MAPK_P + P_3$$

and the Michaelis–Menten form is:

$$V_1 = k1_{cat} \times [S] \times \frac{[MKKK]}{(K1_{M1} + [MKKK])} - k1\text{cat} \times \frac{[MKKKp]}{(K_{M2} + [MKKKp])}$$

$$V_2 = k2_{cat} \times [MKKKp] \times \frac{[MKK]}{(K2_{M1} + [MKK])} - k2\text{cat} \times \frac{[MKKp]}{(K2_{M2} + [MKKp])}$$

$$V_3 = k3_{cat} \times [MKKp] \times \frac{[MAPK]}{(K3_{M1} + [MAPK])} - k3\text{cat} \times \frac{[MAPKp]}{(K3_{M2} + [MAPKp])}$$

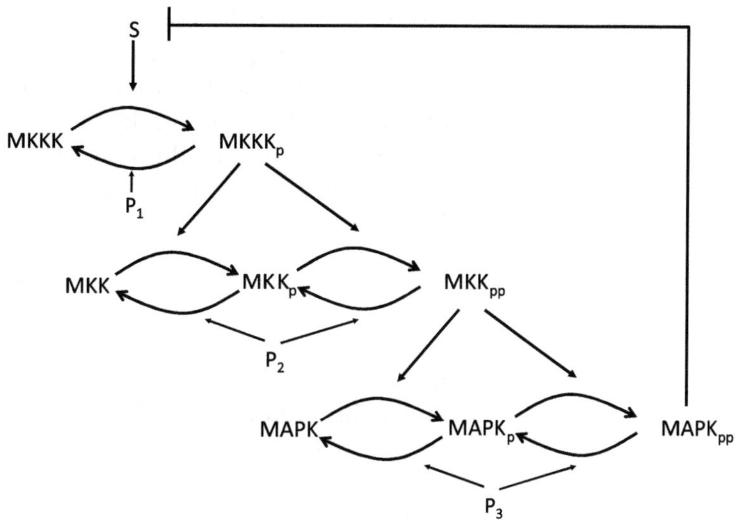

Fig. 2. Three-stage MAPK cascade with negative feedback.

The ODE systems for the mass-action and Michaelis–Menten forms can be easily derived from the biochemical equations above; the connections between the elements of the cascades are achieved by employing the phosphorylated product of one level of the cascade as the enzyme for the next stage.

3.3. Horizontal Composition: Double Phosphorylation

In some models, a stage in the cascade has the double phosphorylated form of the kinase as its product, i.e. as the enzyme for the next stage. This is usually represented in computational models by the horizontal chaining of two phosphorylation–de-phosphorylation units, as illustrated by the following equations for the *MKK* stage (see Fig. 2), where for the sake of simplicity we assume that both units share the same kinase and phosphatase, as well as the same reaction rates:

$$MKK + MKKK_P \underset{k2_2}{\overset{k2_1}{\rightleftarrows}} MKK|MKKK_P \xrightarrow{k2_2} MKK_P + MKKK_P$$

$$MKK + P_2 \xleftarrow{k2r_3} MKK_P|P_2 \underset{k2r_2}{\overset{k2r_1}{\leftrightarrows}} MKK_P + P_2$$

$$MKK_P + MKKK_P \underset{k2_2}{\overset{k2_1}{\rightleftarrows}} MKK_P|MKKK_P \xrightarrow{k2_3} MKK_{PP} + MKKK_P$$

$$MKK_P + P_2 \xleftarrow{k2r_3} MKK_{PP}|P_2 \underset{k2r_2}{\overset{k2r_1}{\leftrightarrows}} MKK_{PP} + P_2$$

3.4. Feedback in a Signalling Network

Feedback loops occur in signalling networks and are implemented in several different ways in cellular systems, using a variety of inhibition and activation mechanisms. Negative feedback can be modelled by sequestration of the input signal by the product (phosphorylated protein) of a subsequent stage, under the assumption that the complex formed will not catalyse the phosphorylation (49, 50). This will normally be a reversible reaction, with the rates ki_1 and ki_2 for the complexation and decomplexation respectively determining the strength of the inhibition. Thus, for example we can add the following equation to the mass-action description of the three-stage cascade above, and illustrated in Fig. 2.

$$S + MAPK_p \underset{ki_{-1}}{\overset{ki_1}{\rightleftarrows}} S \mid MAPK_p$$

Similarly, positive feedback can also be achieved by the sequestration of the input signal by the product of a subsequent stage, under the additional condition that the resulting complex catalyses phosphorylation more actively than does the input signal alone with rates kp_1, kp_2, and kp_3 . In this case, we can further add the following equation to the cascade in addition to the sequestration equation:

$$MKKK + S \mid MAPK_p \underset{kp_{-1}}{\overset{kp_1}{\rightleftarrows}} MKKK \mid S \mid MAPK_p \xrightarrow{kp_2} MKKK_p + S \mid MAPK_p$$

Many other molecular mechanisms achieving feedback can be envisaged. Each of these requires small modifications of the basic formalism and can be used to model the diversity of feedback mechanisms observed in *MAPK* kinases, such as deactivation of the active upstream component by the active downstream one, or desensitization by an additional change of state of the protein (such as phosphorylation at a second site); see e.g. (51).

3.5. Receptors and Other Complexes

There are many other components that need to be described in a model that attempts to capture more than the main signaling cascade in a *MAPK* pathway. For example, there are receptors and the signaling molecules that bind to them, as well as dimerisation of the receptors. These can simply be modeled using mass-action kinetics, in the manner of (8) and (52), where (EGFR|EGF)2 stands for a dimer.

$$\text{EGFR} + \text{EGF} \underset{k_1}{\overset{k_1}{\rightleftarrows}} \text{EGFR} \mid \text{EGF}$$

$$\text{EGFR} \mid \text{EGF} + \text{EGFR} \mid \text{EGF\_} \underset{k_2}{\overset{k_2}{\rightleftarrows}} (\text{EGFR} \mid \text{EGF})2$$

Similarly, the complexation and decomplexation of other species such as Shc, Grb2, Sos, GAP, Ras-GTP, and various scaffolding proteins can be modelled using mass-action kinetics. Incorporating additional details, such as protein trafficking and subcellular localization requires specialized methods not discussed in this review (3).

3.6. Tuning the Model

A model of a signalling pathway is of no use unless it both describes accurately the current data available and correctly predicts the behaviour when the system is perturbed. Thus, the model should be fitted to the available data. This can be sometimes achieved by modifying kinetic parameters by hand and possibly aided by automated scanning through parameter ranges. However, this can be a time consuming process, and scanning multiple parameters can be both computationally expensive as well as producing a very large amount of data for interpretation by the modeler. There is now more use of automated techniques for parameter fitting, employing various optimization strategies, which attempt to find some best values of parameters which cause the model to acceptably fit to the data, for example (53). Of particular interest are those techniques which fit models to overall descriptions of the time-series behaviour of species, rather than to exact values which may vary between laboratories and experiments (54, 55).These techniques not only provide a useful "sanity check" of model behaviour, but can also produce quantitative results.

4. Notes

1. There are several different kinds of software which can be used to construct models of biochemical systems. One type has a graphical interface which permits users to construct models starting with the network wiring diagram (topology), and then to add the equations and kinetic parameters to the model under construction; such software can directly generate a set of ODEs which can be solved using an ODE solver. Alternatively, the graphic-based software can be based on the Petri net formalism, so that users create a Petri net representation of the model, initially qualitative (i.e. the wiring diagram, stoichiometry, and initial marking), and then to add kinetic information in terms of the rates, so that an ODE-based or stochastic model can be generated and simulated. Another class of approach permits users to describe the model using biochemical equations (e.g. mass-action, Michaelis–Menten), and the ODEs are generated automatically from these, for subsequent simulation. Finally, some modellers prefer to directly describe models in terms of ODEs, and then to solve these using an ODE solver.
2. A major activity in BioModel Engineering is that of identifying a (quantitative) model by means of (1) finding the structure, (2) obtaining an initial state, and (3) parameter fitting. An important part of this will be detailed studies of the literature and public databases, as well as intense interactions between computational modellers and experimentalists.

Attempts at automating the first step are underway and consequently, the field of text mining of biomedical information is a very active one (56). Automated systems for extracting useful information from journals and patents will become more widely used as online full-text versions of experimental papers become more widely available.

3. While Michaelis–Menten kinetics are suitable for modelling biochemical systems which predominantly exhibit steady-state kinetics, such as metabolic pathways, there are problems in applying such an approach uncritically to systems which are characterised by dynamic behaviour, such as signalling pathways (47). Thus many modellers use mass-action kinetics to describe such systems. Michaelis–Menten kinetics should not simply be used because of the smaller number of parameters in the model; instead it can be useful to implement a suitably relaxed version of the assumptions of Michaelis–Menten kinetics in the parameters of the mass-action equations, thus avoiding unrealistic model behaviour.

References

1. Gilbert, D.R., Breitling, R., Heiner, M., and Donaldson, R. (2009) An introduction to BioModel Engineering, illustrated for signal transduction pathways, in membrane computing. Springer LNCS, Springer, New York 5391 13–28.
2. Kolch, W., Calder, M., and Gilbert, D.R. (2005) When kinases meet mathematics: the systems biology of MAPK signalling. FEBS Lett 579(8), 1891–1895.
3. Breitling, R., and Hoeller, D. (2005) Current challenges in quantitative modeling of epidermal growth factor signaling. FEBS Lett 579(28), 6289–6294. Wiley.
4. Wiley, H.S., Shvartsman, S.Y., and Lauffenburger, D.A. (2003) Computational modeling of the EGF-receptor system: a paradigm for systems biology. Trends Cell Biol 13(1), 43–50.
5. Orton, R.J., Sturm, O., Gormand, A.M., Kolch, W., and Gilbert, D.R. (2005) Computational modelling of the receptor-tyrosine-kinase-activated MAPK pathway. Biochem J 392(Pt 2), 249–261.
6. Butcher, J.C. (2003) Numerical methods for ordinary differential equations. Wiley, Philadelphia
7. Brightman, F.A., and Fell, D.A. (2000) Differential feedback regulation of the MAPK cascade underlies the quantitative differences in EGF and NGF signalling in PC12 cells. FEBS Lett 482(3), 169–174.
8. Schoeberl, B., Eichler-Jonsson, C., Gilles, E.D. et al. (2002) Computational modeling of the dynamics of the MAP kinase cascade activated by surface and internalized EGF receptors. Nat Biotechnol 20(4), 370–375.
9. Brown, K.S., Hill, C.C., Calero, G.A., Myers, C.R., Lee, K.H., Sethna, J.P., and Cerione, R.A. (2004) The statistical mechanics of complex signaling networks: nerve growth factor signaling. Phys Biol 1(3–4), 184–195.
10. Orton, R.J., Sturm, O., Gormand, A., Kolch, W., and Gilbert, D.R. (2008) Computational modelling reveals feedback redundancy within the EGFR/ERK signalling pathway. IET Syst Biol 2(4), 173–183.
11. Huang, C.Y., and Ferrell, J.E. Jr. (1996) Ultrasensitivity in the mitogen-activated protein kinase cascade. Proc Natl Acad Sci U S A 93, 10078–10083.
12. Burack, W.R., and Sturgill, T.W. (1997) The activating dual phosphorylation of MAPK by MEK is nonprocessive. Biochemistry 36, 5929–5933.
13. Ferrell, J.E. Jr., and Bhatt, R.R. (1997) Mechanistic studies of the dual phosphorylation of mitogen-activated protein kinase. J Biol Chem 272, 19008–19016.
14. Kholodenko, B.N. (2000) Negative feedback and ultrasensitivity can bring about oscillations in the mitogen-activated protein kinase cascades. Eur J Biochem 267, 1583–1588.

15. Resat, H., Ewald, J.A., Dixon, D.A., and Wiley, H.S. (2003) An integrated model of epidermal growth factor receptor trafficking and signal transduction. Biophys J 85, 730–743.
16. Sasagawa, S., Ozaki, Y., Fujita, K., and Kuroda, S. (2005) Prediction and validation of the distinct dynamics of transient and sustained ERK activation. Nat Cell Biol 7, 365–373.
17. Cho, K.H., Shin, S.Y., Kim, H.W., Wolkenhauer, O., McFerran, B., and Kolch, W. (2003) Mathematical modeling of the influence of RKIP on the ERK signaling pathway. LNCS 2602, 127–141.
18. Heiner, M., Uhrmacher, and Adelinde M. (Eds.) (2008).In: Proc. computational methods in systems biology. LNCS, Springer, New York 5307.
19. Breitling, R., Gilbert, D., Heiner, M., and Priami, C. (2009) Formal methods in molecular biology, Dagstuhl Seminar 09091 Proceedings.
20. Clarke, E.M., Grumberg, O., and Peled, D.A. (1999) Model checking. MIT Press, Cambridge.
21. Parker, D., Norman, G., and Kwiatkowska, M. (2006) PRISM 3.0.beta1 users' guide, Cambridge University Press, Cambridge.
22. Calzone, L., Chabrier-Rivier, N., Fages, F., and Soliman, S. (2006) Machine learning biochemical networks from temporal logic properties. Trans on Computat Syst Biol VI, LNBI 4220, 68–94.
23. Priami, C., Regev, A., Shapiro, E., and Silverman, W. (2001) Application of a stochastic name-passing calculus to representation and simulation of molecular processes. Inform Process Lett 80(1), 25–31.
24. Regev, A., Silverman, W., and Shapiro, E. (2001) Representation and simulation of biochemical processes using the π-calculus process algebra. Pac Symp Biocomput 6, 459–470.
25. Phillips A., and Cardelli L.A. (2005) Graphical representation for the stochastic pi-calculus. Trans on Computat Syst Biol VII, 4230, 123–152.
26. Gillespie, D.T. (1977) Exact stochastic simulation of coupled chemical reactions. J Phys Chem, 81(25), 2340–2361.
27. Gilmore S, Hillston J. (1994) The PEPA workbench: a tool to support a process algebra-based approach to performance modelling. LNCS 794, 353–368.
28. Calder, M., Gilmore, S., and Hillston, J. (2006) Modelling the influence of RKIP on the ERK signalling pathway using the stochastic process algebra PEPA. Transactions on computational systems biology. Springer, New York 4230, 1–23.
29. Calder, M., Gilmore, S., and Hillston, J. (2005) Automatically deriving ODEs from process algebra models of signalling pathways. In: Proc. computational methods in systems biology 2005. 204–215.
30. Gilbert, D.R., Heiner, M., and Lehrack, S. (2007) A unifying framework for modelling and analysing biochemical pathways using Petri nets. In: Proc. computational methods in systems biology 2007. Springer LNCS/LNBI, Springer, New York 4695, 200–216.
31. Heiner, M., Gilbert, D.R., and Donaldson, R. (2008) Petri nets for systems and synthetic biology in formal methods for systems biology. Springer LNCS, Springer, New York 5016, 215–264.
32. Hucka, M., Finney, A., Sauro, H.M. et. al. (2003) The system biology markup language (SBML): a medium for representation and exchange of biochemical network models'. Bioinformatics 19(4), 524–531.
33. BioNessie. A biochemical pathway simulation and analysis tool. University of Glasgow, http://www.bionessie.org
34. Hoops, S., Sahle, S., Gauges, R., Lee, C., Pahle, J., Simus, N., Singhal, M., Xu, L., Mendes, P., and Kummer, U. (2006) COPASI – a complex pathway simulator. Bioinformatics, 22, 3067–3074.
35. Le Noviere, N. et al. (2006) Biomodels database: a free, centralized database of curated, published, quantitative kinetic models of biochemical and cellular systems. Nucleic Acids Res 34(Database Issue), D689–D691.
36. Shampine, L.F., and Reichelt, M.W. (1997) The matlab ode suite. SIAM J Sci Comput, 18, 1–22.
37. Bergmann, F.T., and Sauro, H.M. (2006) SBW – a modular framework for systems biology. In Perrone, L. F., Lawson, B., Liu, J., and Wieland, F. P. (eds.).In: Proc. Winter Simulation Conference 2006, 1637–1645.
38. Ermentrout, B. Simulating, analyzing, and animating dynamical systems: a guide to XPPAUT for researchers and students, (Software, Environments, and Tools 14). SIAM publishers, 2002 ISBN-13: 978-0-898715-06-4.
39. Funakashi, A., Matsuoka, Y., Jouraku, A., Kitano, H., and Kikuchi, N. (2006) Cell designer: a modelling too for biochemical networks, Proc Winter Simulation Conference 2006. 1707–1712.
40. Snoopy. A tool to design and animate hierarchical graphs. BTU Cottbus, CS Dep., http://www-dssz.informatik.tu-cottbus.de.

41. Charlie Website. A tool for the analysis of place/transition nets. BTU Cottbus, http://www-dssz.informatik.tu-cottbus.de/software/charlie/charlie.html.
42. Schwarick, M., and Heiner, M. (2009) CSL model checking of biochemical networks with interval decision diagrams. In: Proc. computational methods in systems biology 2009. Springer LNCS/LNBI, Springer, New York 5688, 296–312.
43. Heiner, M., Schwarick, M., and Tovchigrechko, A. (2009) DSSZ-MC – a tool for symbolic analysis of extended Petri nets. In: Proc. Petri nets 2009. Springer LNCS, Springer, New York 5606, 323–332.
44. Berg, J.M., Tymoczko, J.L., and Stryer, L. (2009) Biochemistry. W H Freeman, New York.
45. Breitling, R., Gilbert, D.R., Heiner, M., and Orton, R. (2008) A structured approach for the engineering of biochemical network models, illustrated for signalling pathways. Brief Bioinform 9(5), 404–442.
46. Gilbert, D.R., and Heiner, M. (2006) From Petri nets to differential equations – an integrative approach for biochemical network analysis. Springer LNCS, Springer, New York 4024, 181–200.
47. Klipp, E., and Liebermeister, W. (2006) Mathematical modeling of intracellular signaling pathways. BMC Neurosci 7(Suppl 1), S10.
48. Orton, R.J., Sturm, O.E., Vyshemirsky, V., Calder, M., Gilbert, D.R., and Kolch, W. (2005) Computational modelling of the receptor-tyrosine-kinase-activated MAPK pathway. Biochem J 392, 249–261.
49. Blüthgen, N., Bruggeman, F.J., Legewie, S., Herzel, H., Westerhoff, H.V., and Kholodenko, B.N. (2006) Effects of sequestration on signal transduction cascades. FEBS J 273(5), 895–906.
50. Legewie, S., Schoeberl, B., Blüthgen, N., and Herzel, H. (2007) Competing docking interactions can bring about bistability in the MAPK cascade. Biophys J 93(7), 2279–2288.
51. Orton, R., Sturm, O., Gormand, A., Kolch, W., and Gilbert, D.R. (2008) Computational modelling reveals feedback redundancy within the EGFR/ERK signalling pathway. IET Syst Biol 1(4), 173–183.
52. Hornberg, J.J., Binder, B., Bruggeman, F.J., Schoeberl, B., Heinrich, R., and Hans, V.W. (2005) Control of MAPK signalling: from complexity to what really matters. Oncogene 24, 5533–5542.
53. Moles, C.G., M., Mendes, P., and Bang, J.R. (2003) Parameter estimation in biochemical pathways: a comparison of global optimization methods. Genome Res 13, 2467–2474.
54. Donaldson, R., and Gilbert, D.R. (2008) A model checking approach to the parameter estimation of biochemical pathways. In Proc. computational methods in systems biology. Springer LNCS, Springer, New York 5307, 269–287.
55. Rizk, A., Batt, G., Fages, F., and Soliman, S. (2008) On a continuous degree of satisfaction of temporal logic formulae with applications to systems biology. In Proc. computational methods in systems biology. Springer LNCS, Springer, New York 5307, 251–268.
56. Cohen, A.M., and Hersh, R.W. (2005) A survey of current work in biomedical text mining. Brief Bioinform 6(1), 57–71.

Part V

Use of Lower Organisms, Animal Models, and Human Genetics in the Study of MAP Kinases

Chapter 23

Analysis of Mitogen-Activated Protein Kinase Activity in Yeast

Elaine A. Elion and Rupam Sahoo

Abstract

Mitogen-activated protein (MAP) kinases play central roles in transmitting extracellular and intracellular information in a wide variety of situations in eukaryotic cells. Their activities are perturbed in a large number of diseases, and their activating kinases are currently therapeutic targets in cancer. MAPKs are highly conserved among all eukaryotes. MAPKs were first cloned from the yeast *Saccharomyces cerevisiae*. Yeast has five MAPKs and one MAPK-like kinase. The mating MAPK Fus3 is the best characterized yeast MAPK. Members of all subfamilies of human MAPKs can functionally substitute *S. cerevisiae* MAPKs, providing systems to use genetic approaches to study the functions of either yeast or human MAPKs and to identify functionally relevant amino acid residues that enhance or reduce the effects of therapeutically relevant inhibitors and regulatory proteins. Here, we describe an assay to measure Fus3 activity in immune complexes prepared from *S. cerevisiae* extracts. The assay conditions are applicable to other MAPKs, as well.

Key words: Mitogen-activated protein kinase, Fus3, Kss1, Mpk1/Slt2, Hog1, ERK, JNK, p38, Scaffold, Substrate, Protein kinase assay

1. Introduction

This chapter focuses on the use of immune complex assays for the analysis of Mitogen-activated protein kinase (MAPK) in yeast. The methods described for MAPK Fus3 are similar to those of other kinases and are generally applicable. We first describe the utility of immune complex assays and then provide some general background on MAPK signaling in yeast. The utility of an immune complex kinase assay is that it allows one to capture kinase molecules from their native environment after exposure to a particular stimulus or perturbant or alteration in the genetic background of the cells being studied. This allows one to monitor their activity as a result of changes in physiological conditions. By keeping the

Rony Seger (ed.), *MAP Kinase Signaling Protocols: Second Edition*, Methods in Molecular Biology, vol. 661,
DOI 10.1007/978-1-60761-795-2_23, © Springer Science+Business Media, LLC 2010

conditions native, one can maintain the activity of the kinase as influenced by associated factors, most typically proteins. These proteins may still be able to co-immunoprecipitate with the kinase depending on the strength of association. The method provides a way to define sets of proteins, nearest neighbors, that might have direct or indirect functions with the kinase. Factors of interest might include substrates, activating or inhibiting protein kinases and phosphoprotein phosphatases, regulatory subunits such as scaffolds, and needed chaperonins. The method allows one to monitor the phosphorylation status of the MAPK of interest, the phosphorylation status of its substrates, the relative amount of associated regulatory proteins, and the substrates under various conditions. Thus, it is much more direct than monitoring levels of downstream target genes/reporter genes and biological assays. These powerful advantages set immune complex assays apart from in vitro analysis of purified kinases.

MAP kinases play central roles in transmitting information in a wide variety of situations in eukaryotic cells and function as sensors of many extracellular and intracellular stimuli. Perturbation of MAPK signaling is associated with a number of diseases including cancer. They are highly conserved among all eukaryotes and share a feature of being activated through three- to four-tier protein kinase cascades that involve a MAPK kinase (MAPKK, MAP2K, or MKK), MAPK kinase kinase (MAPKKK, MAP3K, or MKKK), and sometimes a MAPK kinase kinase kinase (MAPKKKK, MAP4K, or MKKKK). MAPKs were first cloned from the yeast *S. cerevisiae* (1, 2) but were originally described biochemically in mammalian extracts (3). A key feature of MAPKs is that they must be phosphorylated by a highly specific dual-specificity MAPK kinase on conserved threonine and tyrosine residues within the TXY motif in the phosphorylation loop to be fully active. A second key feature of MAPKs is that they are frequently regulated by scaffold proteins which can serve to promote the activation of MAPK by MAPKK in response to specific stimuli, to localize MAPK to targets, and serve as a vehicle for feedback control of associated proteins that frequently regulate the localization and activity of MAPK cascades (4, 5).

The Ras-dependent ERK1/ERK2 MAPK pathway has long been targeted for the development of drugs in cancer therapy. Human MAPKK MEK1 and MEK2 are mutated in a developmental disease and are preferentially required for the survival of cancer cells as they are downstream and upstream of transforming oncogenes (6–8). MEK1 and MEK2 are therapeutic targets for several inhibitors used in cancer therapy (9, 10).

MAPKs have often been referred to as workhorses of eukaryotic cells because they recognize such a wide range of substrates. Although MAPKs are often described as having their major role to be the phosphorylation of transcription factors and regulators of chromatin, they were originally identified as regulators of cytoskeletal proteins (such as microtubule-associated protein-2) and protein kinases (such as RSK1 and RSK2 S6 kinase)(3, 5, 7, 8).

Yeast has five MAPKs (mating MAPK Fus3/Dac2, mating and invasive growth/pseudohyphal development MAPK Kss1, high osmolarity glycerol (HOG) response MAPK Hog1, protein kinase C-mediated cell integrity pathway MAPK Slt2/Mpk1/Bcy2/Slk2 (most often referred to as Mpk1) and Smk1, a MAPK required for spore-wall development. Lastly, there is a PKC pathway associated MAPK-like kinase called Mlp1 that is highly homologous to Mpk1 but has a Lys-Gly-Tyr in place of the characteristic Thr-X-Tyr found in MAPKs and two other invariant residue exchanged.

Fus3 and Kss1 are both activated by the MEK-type MAPKK Ste7. Mpk1 is activated by the MEK-type MAPKKs, MKK1, and MKK2, whereas Hog1 is activated by the scaffold-type MAPKK Pbs2. It is not known what MAPKK activates Smk1 (5, 11–13).

Fus3, Kss1, and Mpk1/Erk1 all belong to the ERK family of MAP kinases which is epitomized by the presence of a TEY motif within the phosphorylation loop, whereas Hog1 belongs to the p38 family of MAPKs and has a TGY motif. Smk1 is in a distinct family and has a TNY motif. Mlp1 is most homologous to Mpk1/Slt2. Details on the various MAPK pathways can be found in some reviews (5, 11–13). A key feature of many MAPKs is that they can stably associate with their regulators and targets in both inactive and active forms, with the inactive form serving a regulatory function (14, 15). This is known to be the case for Fus3, Kss1, and Mpk1.

MAPKs from a variety of organisms can substitute for the *S. cerevisiae* MAPKs. For example, the mammalian SAPKs p38a and JNK1 and the *C. elegans* JNK-1 may also functionally replace *HOG1* (16, 17). Mammalian ERK1 partially replaces Kss1 for invasive growth (18), whereas rat ERK2, and human ERK5 partially replace Mpk1 (17, 18). Mammalian MEKK1 can substitute for Bck1, the MAPKKK for Mpk1 (19). Co-expression of K-Ras and Raf-1 can substitute for the Ste11 MAPKKK, leading to Fus3 and/or Kss1 activation (20). In addition, the only purported oncogenic allele of protein kinase B, Akt1(E17K), leads to enhanced activation of Mpk1/Slt2 when overexpresed in yeast (21). Yeast can also be used to identify key amino acid residues that influence catalytic activity. For example, intrinsically hyperactive Mpk1 mutations isolated genetically in *S. cerevisiae* will hyperactivate human ERK2 when engineered into human ERK2 (22). Human MAPKs appear to have greater intrinsic activity when expressed in yeast than yeast MAPKs (22), perhaps due to relaxed specificity for MAPKKs or reduced phosphatase recognition.

Fus3 is the best-characterized *S. cerevisiae* MAPK. Fus3 is activated by a mechanism that requires participation of the Ste5 scaffold, which also binds Ste7 MAPKK and Ste11 MAPKKK and upstream G protein that is activated in response to peptide pheromone binding to a seven transmembrane receptor. Fus3 is the major MAPK of the mating pathway and provides essential functions for transcriptional activation, G1 arrest, projection formation, and cell fusion (5, 23). The related MAPK Kss1 has overlapping functions for many of these outputs and is also activated during mating; however, Fus3 is the more critical kinase. Kss1 functions in three additional pathways; the invasive growth/pseudohyphal development pathways that are activated during nutrient limitation (13, 15) and the SVG (sterile vegetative growth) pathway that regulates vegetative growth and cell wall integrity in response to glycosylation defects and cell-wall stress and is functionally equivalent to the invasive growth pathway (23). Both Fus3 and Kss1 are activated by unusual stimuli such as 1% butanol (24). The Kss1 pathways are inhibited by Fus3 to prevent cross-activation during pheromone signaling (23).

Fus3 is largely inactive during vegetative growth, with its activity repressed by the HOG pathway (25, 26). Fus3 kinase activity is rapidly induced within minutes in the presence of exogenous α-factor pheromone, with optimal activation occurring within 5 min followed by downregulation. Although the activation of Fus3 is dependent upon phosphorylation of conserved threonine 180 and tyrosine 182 residues within its T-loop, Fus3 can autophosphorylate tyrosine 182, and the amount of autophosphorylation affects its basal activity (27). Ste5 scaffold stimulates Fus3 through a VWA domain that serves as a coactivator with Ste7 (28). Fus3 is continuously inactivated by phosphatases and is rapidly inactivated upon a factor withdrawal. Fus3 is inactivated by at least three phosphatases that have specificity towards phosphotyrosine, Msg5, Ptp2, and Ptp3 (29) and possibly phosphothreonine phosphatases.

The activation of Fus3, in addition to Kss1, Mpk1, and Hog1 MAPKs, can be monitored using commercially available antibodies to mammalian phospho-p44/42 MAPKs and phospho-p38 MAPK (30). The kinase assay described here measures Fus3 activation in immune complexes prepared from *S. cerevisiae* extracts. The immune complex kinase assay described here detects in vivo stimuli and overexpressed proteins that play direct and indirect roles in stimulating the MAPK pathway at or upstream of Fus3 (31–34). Fus3 has a number of substrates, perhaps the best known being Ste5, Far1 (a cyclin-dependent kinase inhibitor), Ste12, and associated repressors Dig1/Rst1 and Dig2/Rst2 (31, 32, 35). This method thus provides a convenient approach to identify substrates and associated regulators such as scaffold proteins, assess overall level of MAPK activation and relative specific activity with respect to in vivo stimuli and mutations, and assess phosphorylation of exogenous substrate. The assay conditions are applicable to other yeast MAPKs (36) and should also be generally applicable to nonyeast MAPKs as the buffer conditions are quite similar to those used for all members of human MAPK subfamilies (37, 38).

2. Materials

2.1. Whole Cell Extracts

1. Yeast cells.
2. SC selective medium.
3. 5 mM α factor peptide in methanol (store at −20°C in aliquots).
4. Acid-washed glass beads 400–600 μm (Sigma G8772).
5. Ice bath.
6. Dry ice/ethanol bath.
7. Ice-cold water that has been autoclaved.
8. Ice-cold lysis buffer (Modified H buffer with glycerol).
9. 50 ml conical plastic disposable tubes.
10. 15 ml conical plastic disposable tubes.
11. Vortexer.
12. Timer.
13. Eppendorf tubes.
14. 1-ml and 0.2-ml pipetmen and pipet tips.
15. Lysis buffer (Modified H buffer with glycerol, chilled in ice bath): 25 mM Tris–HCl pH 7.5, 250 mM NaCl, 15 mM EGTA, 15 mM $MgCl_2$, 0.1% Triton X-100, 10% glycerol, 1 mM NaN_3, added just before use 0.5 mM sodium vanadate (100× stock), 1× Protease Inhibitor Mix (1,000× stock), 1 mM phenylmethylsulfonyl fluoride (250× stock), 1 mM DTT (1 M stock, stored at −20°C).
16. 1,000× Protease Inhibitor Mix: 1. 5 mg/ml chymostatin, 2.5 mg/ml pepstatin A, 3.5 mg/ml leupeptin, 4.5 mg/ml antipain, 5.5 mg/ml aprotinin (Dissolve in DMSO and store in aliquots at −20°C).
17. 250× PMSF: 250 mM PMSF in 95% ethanol, made fresh.
18. 50 mM sodium vanadate: 25 mM meta-vanadate, 25 mM *ortho*-vanadate, pH 7.0 with NaOH.

2.2. Materials to Assay Kinase Activity

1. Whole-cell extracts.
2. 12CA5 monoclonal antibody.
3. Protein A Sepharose (Pharmacia).
4. Modified H buffer made without glycerol and without NaCl.
5. 5 M NaCl.
6. 20 ml syringe and 18 guage needle.
7. Eppendorf tubes.
8. Water bath warmed to 30°C.
9. Timer.
10. 4× loading buffer.
11. ATP: 1 mM ATP cold stock in Tris–HCl pH 7.4 final, γ-^{32}P-ATP 10 μCi/μl at ~5,000 Ci/mmole (ICN). Make 20 μM and 50 μM hot ATP stocks by diluting the cold ATP stock directly into the manufacturer's hot ATP bottle and gently pipet up and down to mix.
12. 5× Kinase Buffer: 250 mM Tris pH7.4, 100 mM $MgCl_2$, 25 mM EGTA, add just before use 1 mM DTT, 1× protease inhibitor cocktail, 0.5 mM sodium vanadate.
13. Protein A-Sepharose: Recipe can either be scaled up or down. Swell 1.5 g Protein A-Sepharose beads in 30 ml of 50 mM Tris–HCl pH 7.5 for 1–2 h on ice. Pellet beads by gravity or very gentle centrifugation (1 min at 230 × *g*) and then wash them four times with modified H buffer without glycerol or protease inhibitors. Resuspend the beads in 15 ml of this buffer to yield a final slurry concentration of approximately 100 mg/ml. Store at 4°C. Slurry is stable for months.
14. Casein (Sigma): 5 mg/ml in H_2O stored at –20°C, warm to 37°C and vortex to dissolve.

3. Methods

3.1. Preparing Whole-Cell Extracts

The most important step in a successful kinase assay is generating whole-cell extracts in which the yield and activity of the kinase is optimal, with buffer conditions that permit association of the kinase with needed regulatory proteins and avoidance of loss of activity by inhibitors such as phosphatases. A general small-scale glass bead breakage protocol for preparing a basic whole-cell extract is described below as a starting point, with salt and detergent conditions that have been found to be optimal for Fus3. Gloves should be worn during all manipulations because of the inclusion of hazardous chemicals like PMSF and sodium azide.

Buffer conditions are a modified form of H buffer described in refs. 37, 38. Additional phosphatase inhibitors can be added during extract preparation, but they should be washed out from immune complexes prior to performing the kinase assay.

3.1.1. Step 1. Grow and Harvest Cells

The mating MAPK cascade is activated by peptide pheromones that bind to cell surface receptors on haploid *a* and α type cells. *a* cells secrete a-factor which binds to the a-factor receptor on α cells, while α cells secrete α-factor which binds to the α-factor receptor on *a* cells. Because of the ease of synthesizing the α-factor peptide pheromone, Fus3 is typically analyzed in *a* type cells. The analysis is done in *a* cells that harbor a deletion mutation in the *BAR1/SST1* gene that encodes a secreted protease that degrades α factor and reduces by ~100-fold the amount of α factor needed to stimulate cells. The epitope-tagged Fus3 is expressed from its native *FUS3* promoter in yeast on a low-copy centromeric plasmid to ensure that its expression is close to native levels (31). However, it is also possible to express Fus3 from a constitutive promoter to avoid the influence of pheromone on its own expression levels. Fus3 has been successfully tagged internally after residue 269 (Fus3-HA#5, hereafter referred to as Fus3-HA) and at the C-terminus; the internally tagged derivative is more readily immunoprecipitated from native whole-cell extracts and is of the same abundance as wild-type Fus3 (31). Tagging Fus3 with either the HA (31) or MYC (35) epitope does not block its kinase activity in vitro. However, the C-terminally tagged Fus3-Myc derivative may be more difficult to detect in immunoprecipitates, necessitating the use of 5–10 more whole-cell extracts in an immune complex kinase assay. Both myelin basic protein (MBP) and casein serve as good in vitro substrates in the kinase assay; however, of the two, casein is the better substrate (31). One can also use *Escherichia coli* expressed N-terminal fragments of Far1, Ste12, or Ste5 as physiological substrates; however, this requires advanced preparation and may be a limiting reagent. An important control for the assays is the inclusion of a catalytically impaired Fus3-HA. Fus3R42-HA and Fus3Y182F-HA have been shown to have a greatly reduced kinase activity in the assay and serve as appropriate negative controls (31). Another important control is a strain that lacks an epitope-tagged Fus3.

1. Cells should be grown under conditions optimal for the study. In the absence of specific information, it is advisable to harvest logarithmically growing cells as follows: A *MATa bar1Δ* strain harboring *FUS3-HA* on a *CEN* plasmid is grown in SC-selective media to an OD_{600} of 0.6–0.8 at 30°C with vigorous shaking. If multiple samples are being grown, equalize their densities, then split the cultures and induce one half

with 10–50 nM α factor (from 5 mM stock in methanol), the other half with methanol alone for 2–5 min for optimal activity, 1 h at 30°C for a longer term time point that will incorporate the effects of α factor on gene expression and protein levels (including Fus3). Cells are kept shaking. Aim to have 100 OD units for each sample.

2. Pellet cells by 5 min centrifugation at 4°C using a prechilled rotor.
3. Wash cells once in 30 ml ice water and pellet in 50 ml sterile plastic conical tube with screw cap using an ice bucket.
4. Drain pellet rapidly and thoroughly and immediately freeze pellet by immersing tube into a dry ice/ethanol bath.
5. Store frozen pellets at −80°C or process immediately.

3.1.2. Step 2. Prepare Extracts

1. Speed and maintenance of ice-cold conditions and use of a strong vortexer are key components of good extract preparation. Extract preparation can be done at room temperature using ice buckets.
2. Thaw pellets in an ice bath. Begin extract preparation while pellets are still partially frozen.
3. Add 0.8 ml lysis buffer and 0.8 ml of glass beads to each sample of 100 OD units (Use 1.2 ml lysis buffer and 1.2 ml beads for 200 OD units of cells). The beads should come to just below the meniscus.
4. Vortex cells vigorously for a total of five 30-s pulses, chilling samples on ice between pulses for at least 30 s.
5. Add 0.2 ml more lysis buffer and vortex for 30 s (Use 0.3 ml lysis buffer for 200 OD units). Check cells under a microscope for complete or nearly complete lysis. Vortex again, if necessary.
6. Centrifuge samples 10 min at 2,300 × *g* in a centrifuge and rotor that have been prechilled to 4°C.
7. Transfer supernatant to an Eppendorf tube. Don't worry if a few beads come along for the ride. Supernatant should be very milky with lipids floating on top.
8. Centrifuge supernatant in a microfuge chilled to 4°C at 12,000 × *g* for 10 min.
9. Transfer supernatant to new Eppendorf tube. Cap tube and invert gently to mix contents, which may still be slightly milky. Reserve 10 μl on ice to quantitate total protein in a Bio-Rad assay and distribute remainder into four Eppendorf tubes and store at −80°C and/or assay an aliquot immediately. Fus3 kinase activity will survive one thawing of the whole-cell extract.

10. Assay total protein using the Bio-Rad protein assay and calculate protein concentration using a BSA standard curve. Yield is generally 3–5 mg/ml total protein for 100 OD units of cells. Two 200 μg aliquots are needed per kinase assay.

3.2. Method to Immunoprecipitate Fus3-HA

1. Prepare duplicate samples in Eppendorf tubes on ice; one sample will be assayed for kinase activity and the other will be used to assess the amount of immunoprecipitated Fus3-HA in a separate immunoblot. Mix 200 μg of the whole-cell extract with 1 μg of antibody in a final volume of 0.5 ml of modified H buffer that has been adjusted to 100 mM NaCl with 5 M NaCl stock solution. As with extract preparation, all buffers and tubes are kept cold on ice.
2. Invert tube gently several times and incubate on ice for 90 min.
3. Centrifuge in a chilled microfuge at 12,000 × *g* for 10 min to pellet aggregates.
4. Transfer supernatant to a new Eppendorf tube and add 30 μl of Protein A-Sepharose slurry. Be sure to evenly suspend the sepharose slurry before distributing it to tube.
5. Rotate tube gently at 4°C for 1 h. (Rocking is much less efficient).
6. Gently pellet protein A-sepharose by spinning 30 s at 230 × *g* in a chilled centrifuge.
7. Wash pellet three times with 1 ml modified H buffer without glycerol or NaCl. For each wash, gently invert tube three times before pelleting.
8. After the last wash, aspirate away as much liquid as possible without touching the beads. To one of the duplicate samples add 25 μl of 2× loading buffer for subsequent immunoblot analysis to determine the amount of immunoprecipitated Fus3-HA. Assay the second sample for kinase activity as described below.

3.3. Kinase Assay Method

All manipulations are done with a Geiger counter, gloves, and appropriate shielding.

1. Gently wash beads twice with 0.5 ml 1× kinase buffer. Remove as much supernatant as possible after the second wash and incubate beads on ice (see Notes 1 and 2).
2. Resuspend beads in 18 μl of 1× kinase buffer without exogenous substrate or ATP: Warm sample to 30°C. Start reaction by adding 1 μl of 5 μg/μl casein and 1.3 μl of 20 μM ATP stock containing 10 μCi γ-^{32}P-ATP. Gently mix the sample and incubate in a 30°C water bath for 8 min (see Note 3).

3. Stop reaction by adding 10 μl of 4× loading buffer. Samples can be frozen at this point or run directly on a polyacrylamide gel as described below.
4. Incubate sample in a boiling water bath for 10 min then vortex and centrifuge briefly prior to loading supernatant on a 8% SDS–polyacrylamide gel with an acrylamide: bis-acrylamide ratio of 30:0.8 and a thickness of 0.75 mM. Run gel until the bromophenol blue dye front just runs off the gel.
5. Soak gel for 1 h or more in a mixture containing 10% acetic acid, 10% methanol, and 0.5% phosphoric acid added to a 10 mM potassium phosphate pH 6.8 buffer (potassium phoshate buffer made from mixing appropriate amounts of 0.1 M K_2HPO_4 and 0.1 M KH_2PO_4 stock solutions) and then for 30 min in 10% acetic acid and 10% methanol with several changes. Dry gel onto Whatman paper and expose to X-ray film, without an autoradiography enhancer screen, at 80°C for best resolution of bands. Typically, a 2–4 h exposure with a screen is long enough to reveal a profile of associated substrates ranging from 40 to 170 kDa and casein (see Notes 4 and 5).

3.4. Analyzing the Results – Important Controls

Given that this method involves co-immunoprecipitation of associated proteins, it is important to include controls to demonstrate that the kinase activity being monitored is from the kinase being immunoprecipitated. The most straightforward approach is to include a catalytically inactive derivative of the kinase in a parallel assay. In the absence of being able to do this, it is possible to immunoprecipitate the kinase under more stringent conditions to strip off associated proteins and only use exogenous substrate as a monitor. To demonstrate that the substrates being phoshporylated are directly due to the kinase in question it is possible to engineer the kinase to incorporate a novel ATP analog to circumvent the problem of associated kinases. In the case of Fus3 this would be Fus3Q93G, which will accept phenylethyl-ATP (39). Fus3Q93G (but not wild type Fus3) is also inhibited by 1-napthyl PP1 which has been modified with a similar bulky adduct (40, 41) see Note 5.

3.5. Assaying Kss1 Activity

To assay Kss1, we follow the method outlined for Fus3, except that we also add 0.3 mg/ml of the peptidase inhibitor benzamidine and 4 mM of the metalloprotease inhibitor *o*-phenanthroline during extract preparation and immunoprecipitation. Some of the Kss1-associated substrates have the same mobility as Fus3-associated substrates, suggesting they are the same. Kss1 may need to be expressed at higher levels using a stronger promoter, but overexpression is not ideal as it inhibits pathway activation. The hyperactive Ste11-4 kinase will preferentially activate Kss1 and provides a means to readily monitor Kss1 activation state while expressed at native levels.

4. Notes

1. Additional phosphatase inhibitors can be added during extract preparation but should be washed out from immune complexes prior to performing the kinase assay.
2. Fus3 is actively dephosphorylated, so it is important to limit the exposure of extracts to warm temperature and to keep them frozen at −80°C or in liquid nitrogen prior to assaying.
3. The in vivo concentration of ATP is an important consideration in thinking about protein kinase activation. For example, inhibitors of protein kinases will be more potent when the concentration of ATP is at or below the K_m for ATP for a given kinase (37, 38). In the assay described here, the range of ATP is 20–50 μM (similar to generally described conditions); however, the amount of radioactive ^{32}P-ATP is quite a bit higher to permit detection of low abundance proteins present in the immune complexes. Modifications to the described conditions can be optimized for individual kinases, and it is recommended that the researcher start first by varying ATP concentration and reaction time. It is also possible to use MBP as a substrate (10 mg/ml MBP in H_2O is a 10× stock).
4. If your ^{32}P signal is very low, consider the following reasons:
   (a) The WCE has been frozen and thawed more than once.
   (b) The level of Fus3-HA in your whole-cell extract is low.
5. See ref. 38 for details on preparation of yeast extracts and immune complexes, genetic and biochemical controls along with tips on troubleshooting.

References

1. Courchesne, W.E., Kunisawa, R., Thorner, J. (1989) A putative protein kinase overcomes pheromone-induced arrest of cell cycling in *S. cerevisiae*. *Cell.* **58**, 1107–1119.
2. Elion, E.A., Grisafi, P.L., Fink, G.R. (1990) FUS3 encodes a cdc2+/CDC28-related kinase required for the transition from mitosis into conjugation. *Cell.* **60**, 649–664.
3. Avruch, J. (2007) MAP kinase pathways: the first twenty years. *Biochim Biophys Acta.* **773**, 1150–1160.
4. Morrison, D.K. and Davis, R.J. (2003). Regulation of MAP kinase signaling modules by scaffold proteins in mammals. *Annu Rev Cell Dev Biol.* **19**, 91–118.
5. Qi, M., Elion, E.A. (2005) MAPK Pathways and their regulation. *J Cell Sci.* **118(Pt 16)**, 3569–3572.
6. Seger, R., Krebs, E.G. (1995) The MAPK signaling cascade. *FASEB J.* **9**, 726–735.
7. Robinson, M.J. and Cobb, M.H. (1997) Mitogen-activated protein kinase pathways. *Curr Opin Cell Biol.* **9**, 180–186.

8. Chang, L. and Karin, M. (2001) Mammalian MAP kinase signalling cascades. *Nature.* **410**, 37–40.
9. English, J.M. and Cobb, M.H. (2002) Pharmacological inhibitors of MAPK pathways. *Trends Pharmacol Sci.* **23**, 40–45.
10. Sebolt-Leopold, J. (2008) Advances in the development of cancer therapeutics directed against the RAS-Mitogen-Activated Protein Kinase Pathway. *Clin Cancer Res.* **14**, 3651–3656.
11. Schwartz, M. and Madhani, H. (2004) Principles of MAP kinase signaling specificity in *Saccharomyces cerevisiae. Annu Rev Genet.* **38**, 725–748.
12. Sheikh-Hamad, D. and Gustin, M.C. (2004) MAP kinases and the adaptive response to hypertonicity: functional preservation from yeast to mammals. *Am J Physiol Renal Physiol.* **287**, F1102–F1110 (review).
13. Chen, R.E. and Thorner, J. (2007) Function and regulation in MAPK signaling pathways: lessons learned from the yeast *Saccharomyces cerevisiae. Biochim Biophys Acta.* **1773(8)**, 1311–1340.
14. Dérijard, B., Hibi, M., Wu, I.H., Barrett, T., Su, B., Deng, T., Karin, M., Davis, R.J. (1994) JNK1: a protein kinase stimulated by UV light and Ha-Ras that binds and phosphorylates the c-Jun activation domain. *Cell.* **76**, 1025–1037.
15. Madhani, H.D. and Fink, G.R. (1998) The riddle of MAP kinase signaling specificity. *Trends Genet.* **14**, 151–155.
16. Alonso-Monge, R., Urena, T., Nombela, C., Pla, J. (2007) Functional characterization of human and fungal MAP kinases in *Saccharomyces cerevisiae. Yeast.* **24**, 715–722.
17. Levin-Salomon, V., Maayan, I., Avrahami-Moyal, L., Marbach, I., Livnah, O., Engelberg, D. (2009) When expressed in yeast, mammalian mitogen-activated protein kinases lose proper regulation and become spontaneously phosphorylated. *Biochem J.* **417**, 331–340.
18. Atienza, J.M., Suh, M., Xenarios, I., Landgraf, R., Colicelli, J. (2000) Human ERK1induces filamentous growth and cell wall remodeling pathways in *Saccharomyces cerevisiae. J Biol Chem.* **275**, 20638–20646.
19. Blumer, K.J., Johnson, G.L., Lange-Carter, C.A. (1994) Mammalian mitogen-activated protein kinase kinase kinase (MEKK) can function in a yeast mitogen-activated protein kinase pathway downstream of protein kinase C. *Proc Natl Acad Sci U S A.* **91**, 4925–4929.
20. Soga S, Kozawa T, Narumi H, Akinaga S, Irie K, Matsumoto K, Sharma SV, Nakano H, Mizukami T, Hara M. 1998 Tokyo Research Laboratories, Kyowa Hakko Kogyo Co. Ltd., Asahi-machi 3-6-6, Machida-shi, Tokyo 194, Japan.
21. Rodríguez-Escudero, I., Andrés-Pons, A., Pulido, R., Molina, M., Cid, V.J. (2009) Phosphatidylinositol 3-kinase-dependent activation of mammalian protein kinase B/Akt in *Saccharomyces cerevisiae*, an in vivo model for the functional study of Akt mutations. *J Biol Chem.* **284**, 13373–13383.
22. Levin-Salomon, V., Kogan, K., Ahn, N.G., Livnah, O., Engelberg, D. (2008) Isolation of intrinsically active (MEK-independent) variants of the ERK family of mitogen-activated protein (MAP) kinases. *J Biol Chem.* **283(50)**, 34500–34510.
23. Elion, E.A., Qi, M., Chen, W. (2005) Signal transduction. Signaling specificity in yeast. *Science.* **307**, 687–688.
24. Andersson, J., Simpson, D.M., Qi, M., Wang, Y., Elion, E.A. (2004) Differential input by Ste5 scaffold and Msg5 phosphatase route a MAPK cascade to multiple outcomes. *EMBO J.* **23**, 2564–2576.
25. Hall, J.P., Cherkasova, V., Elion, E., Gustin, M.C., Winter, E. (1996) The osmoregulatory pathway represses mating pathway activity in *Saccharomyces cerevisiae*: isolation of a FUS3 mutant that is insensitive to the repression mechanism. *Mol Cell Biol.* **16**, 6715–6723.
26. O'Rourke, S.M., Herskowitz, I. (2002) A third osmosensing branch in *Saccharomyces cerevisiae* requires the Msb2 protein and functions in parallel with the Sho1 branch. *Mol Cell Biol.* **22**, 4739–4749.
27. Brill, J.A., Elion, E.A., Fink, G.R. (1994) A role for autophosphorylation revealed by activated alleles of FUS3, the yeast MAP kinase homolog. *Mol Biol Cell.* **5**, 297–312.
28. Good, M., Tang, G., Singleton, J., Reményi, A., Lim, W.A. (2009) The Ste5 scaffold directs mating signaling by catalytically unlocking the Fus3 MAP kinase for activation. *Cell.* **136**, 1085–1097.
29. Zhan, X.L., Deschenes, R.J., Guan, K.L. (2009) Differential regulation of FUS3 MAP kinase by tyrosine-specific phosphatases PTP2/PTP3 and dual-specificity phosphatase MSG5 in *Saccharomyces cerevisiae. Genes Dev.* **11**, 1690–1702.
30. Bardwell, L., Shah, K. (2006) Analysis of mitogen-activated protein kinase activation and interactions with regulators and substrates. *Methods.* **40**, 213–223 (review).
31. Elion, E.A., Satterberg, B. and Kranz, J.E. (1993) FUS3 phosphorylates multiple components of mating signal transduction cascade: evidence for Ste12 and Far1. *Mol Biol Cell.* **4**, 495–510.

32. Kranz, J.E., Satterberg, B., Elion, EA. (1994) The MAP kinase Fus3 associates with and phosphorylates the upstream signaling component Ste5. *Genes Dev.* **8**(3), 313–327.
33. Choi, K.Y., Satterberg, B., Lyons, D.M., Elion, E.A. (1994) Ste5 tethers multiple protein kinases in the MAP kinase cascade required for mating in *S. cerevisiae*. *Cell.* **78**, 499–512.
34. Lyons, D.M., Mahanty, S.K., Choi, K.Y., Manandhar, M., Elion, E.A. (1996) The SH3-domain protein Bem1 coordinates mitogen-activated protein kinase cascade activation with cell cycle control in *Saccharomyces cerevisiae*. *Mol Cell Biol.* **16**, 4095–4106.
35. Tedford, K., Kim, S., Sa, D., Stevens, K., Tyers, M. (1997) Regulation of the mating pheromone and invasive growth responses in yeast by two MAP kinase substrates. *Curr Biol.* 7, 228–238.
36. Cherkasova, V.A. (2006) Measuring MAP kinase activity in immune complex assays. *Methods.* **40**, 234–242. (review).
37. Bain, J., Plater, L., Elliott, M., Shpiro, N., Hastie, C.J., McLauchlan, H., Klevernic, I., Arthur, J.S., Alessi, D.R., Cohen, P. (2007) The selectivity of protein kinase inhibitors: a further update. *Biochem J.* **408**, 297–315.
38. Booher, R.N., Alfa, C.E., Hyams, J.S., Beach, D.H. (1989) The fission yeast cdc2/cdc13/suc1 protein kinase: regulation of catalytic activity and nuclear localization. *Cell.* **58**, 485–497.
39. Elion, E.A. (2006) Co-precipitation of proteins. Curr Protoc Mol Biol. Chapter 20: unit20.5. Ausubel,F, Brent R Kingston R, Moore D, Smith JA, Seidman J, Struhl K, editors. Wiley Interscience publishers.
40. Bishop, A.C., Ubersax, J.A., Petsch, D.T., Matheos, D.P., Gray N.S., Blethrow J., Shimizu, E., Tsien, J.Z., Schultz, P.G., Rose, M.D., et al. (2000). A chemical switch for inhibitor-sensitive alleles of any protein kinase. *Nature.* **407**, 395–401.
41. Matheos, D., Metodiev, M., Muller, E., Stone, D., Rose, M.D. (2004) Pheromone-induced polarization is dependent on the Fus3p MAPK acting through the formin Bni1p. *J Cell Biol.* **165**, 99–109.

Chapter 24

Detection of RTK Pathway Activation in *Drosophila* Using Anti-dpERK Immunofluorescence Staining

Aharon Helman and Ze'ev Paroush

Abstract

In *Drosophila*, like in other metazoans, receptor tyrosine kinase (RTK) signaling pathways control diverse cellular processes such as migration, growth, fate determination, and differentiation (Shilo, Development 132:4017–4027, 2005). Activation of RTKs by their extracellular ligands triggers a signal transduction cascade, mediated by the Ras/Raf/MEK cassette, which ultimately leads to dual phosphorylation and activation of the mitogen-activated protein kinase/extracellularly regulated kinase (MAPK/Erk). Once active, MAPK/Erk phosphorylates its cytoplasmic and nuclear substrates, consequently modulating (i.e., stimulating or inhibiting) their biological function (Murphy and Blenis, Trends in Biochemical Sciences 31:268–275, 2006). The currently available antibody specific for the doubly phosphorylated form of MAPK/Erk (dpERK) (Yung et al., FEBS Letters 408:292–296, 1997) provides a valuable readout for RTK signaling: it enables the spatiotemporal detection of RTK pathway activity in the developing organism, in situ (Gabay et al., Development 124:3535–3541, 1997; Gabay et al., Science 277:1103–1106, 1997). Here, we present a detailed protocol for anti-dpERK immunofluorescent staining that can be applied to the analysis of MAPK/Erk signaling in *Drosophila* embryogenesis.

Key words: dpERK, *Drosophila*, Immunofluorescent antibody staining, MAPK/Erk, Phosphorylation, RTK signaling

1. Introduction

The monoclonal antibody recognizing dpERK was originally raised using a synthetic 11-amino acid long phospho-peptide conforming to the vertebrate MAPK/Erk activation loop (3). Due to the high evolutionary conservation of the activation loop, this antibody crossreacts with active (but not inactive) MAPK/Erk in different species, including *Drosophila* (2, 4). In flies, a single gene called *rolled* encodes the only MAPK/Erk family member, which serves as a downstream effector for all known RTK pathways (5).

Rony Seger (ed.), *MAP Kinase Signaling Protocols: Second Edition*, Methods in Molecular Biology, vol. 661,
DOI 10.1007/978-1-60761-795-2_24,

Accordingly, anti-dpERK staining is currently the method of choice for tracking and investigating the kinetics and dynamics of RTK signaling during *Drosophila* development (6) (Fig. 1).

One feature renders the anti-dpERK antibody superior to related antibodies available for activated MAPK/JNK and MAPK/p38 (7, 8): it stains intact tissues at various stages of development (oogenesis, embryogenesis, larval, and adult) (9–13). Hence, this antibody has served as an effective tool for exploring RTK signal transduction pathways in vivo (1). For example, quantification of the dynamic dpERK gradients formed in response to the Torso RTK pathway at the termini of syncytial *Drosophila* embryos has pointed to nuclear trapping as a novel mechanism for restricting diffusion of activated MAPK (14). In this context, the nuclei function to limit the spatial extent of receptor activation and to sharpen dpERK gradients. Anti-dpERK staining has also proved useful in genetic epistasis experiments, e.g., to establish whether gene functions that impinge on RTK pathways do so upstream, in parallel or downstream of MAPK/Erk (15).

In general, combining fluorescent antibody staining together with confocal microscopy enables the simultaneous analysis, at high resolution, of two or more distinct signals in the same sample. Given that RTK signaling pathways regulate multiple

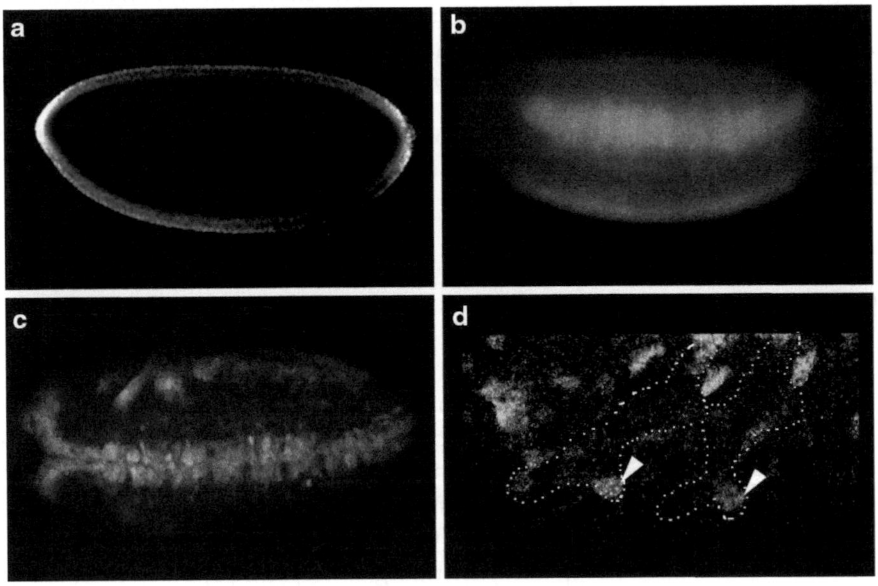

Fig. 1. Anti-dpERK staining is an effective readout for multiple RTK signaling pathways in *Drosophila* embryogenesis. (**a**–**d**) Confocal images of *Drosophila* embryos at different stages of development, stained for dpERK. In all panels, anterior is to the *left*. (**a**) Staining is evident at both termini of stage 4 syncytial blastoderm embryo, in regions where the Torso RTK pathway is active. (**b**, **c**) MAPK/Erk is phosphorylated and activated in response to two waves of EGFR-mediated signaling. Lateral view of stage 5 blastoderm (**b**) and ventral view of stage 10 (**c**) embryos; note the staining in the ventral ectoderm, in domains straddling the ventral midline. (**d**) At stage 12, staining is detected in tracheal branches (demarcated by *dashed lines*), specifically in posterior lateral migrating tip cells in which the FGFR pathway is active (*arrowheads*).

processes at different developmental stages in *Drosophila*, it is often useful to double stain for dpERK and other specific markers. This approach can help determine, for example, the level at which the phosphorylation by MAPK/Erk modifies the biological activity of a given substrate (16), or define the spatial domain of MAPK/Erk activation relative to other markers with known cell or tissue distribution (12, 17). Similarly, costaining with nuclear (e.g., DAPI) or other cellular markers (Lamin, FasIII, etc.) can establish the subcellular localization of activated MAPK/Erk under variable conditions, or in different developmental settings (18, 19). Staining of eye imaginal discs, for instance, revealed that in certain cells dpERK is actually retained in the cytoplasm for several hours before it translocates to the nucleus, suggesting that nuclear entry of MAPK sometimes requires a second stimulus besides phosphorylation (20, 21). Collectively, these and other studies illustrate the power of using the anti-dpERK antibody to uncover new facets of RTK signaling and MAPK/Erk biology.

Below, we describe a protocol for carrying out fluorescent anti-dpERK antibody staining on *Drosophila* embryos. We outline the different steps involved in the procedure, including the initial embryo collection and fixation, antibody staining, signal amplification and detection, and mounting, viewing, and storing of stained samples.

2. Materials

2.1. Embryo Collection and Fixation

All materials should be of high molecular and analytic grade.

1. 20-mL glass scintillation vials.
2. 1.5-mL micro-centrifuge tubes.
3. Chlorine bleach solution, diluted 1:1 in double-distilled water.
4. 37% formaldehyde solution (Sigma-Aldrich).
5. PBS solution (×10): 1.37 M NaCl, 27 mM KCl, 78.1 mM Na_2HPO_4, 14.7 mM KH_2PO_4.
6. Heptane (Merck).
7. Fix solution (freshly prepared), 10 mL for each scintillation vial: 5 mL of heptane, 1.35 mL of 37% formaldehyde, 0.5 mL of PBS (×10), 3.15 mL of double-distilled water.
8. Methanol.

2.2. Anti-dpERK Staining of Drosophila Embryos

1. 70, 50, and 30% methanol in PBS, 0.1% Tween-20 (Sigma-Aldrich).
2. TNT solution: 100 mM Tris–HCl pH 7.5, 150 mM NaCl, 0.05% Tween-20.

3. TNB solution: 100 mM Tris–HCl pH 7.5, 150 mM NaCl, 0.005 g/mL blocking reagent (Molecular Probes).
4. 30% hydrogen peroxide (H_2O_2).
5. Monoclonal mouse anti-dpERK antibody (Sigma) (see Note 1).
6. Goat anti-mouse IgG, biotin–SP conjugate (Millipore; Chemicon), diluted 1:1 in glycerol (see Note 1).
7. Streptavidin–horseradish peroxidase (HRP) conjugate (Molecular Probes).
8. Alexa Fluor 488 tyramide conjugate (Molecular Probes).
9. Amplification buffer (Molecular Probes).

2.3. Mounting, Viewing, and Storing Stained Embryos

1. Fluorescent mounting medium (Dako).
2. Standard microscope slides.
3. Coverslips (22 × 22 mm).
4. Fluorescence or confocal microscope.

3. Methods

3.1. Embryo Collection and Fixation

1. Collect embryos using a mesh sieve (or, as an alternative, using Falcon's Cell Strainer) and rinse them with double-distilled water (see Note 2).
2. Submerge the mesh sieve in a small volume of chlorine bleach solution for 2 min to dechorionate the embryos. Rinse the embryos immediately and thoroughly with double-distilled water to remove all residual bleach.
3. Transfer the embryos to a 20-mL glass scintillation vial, containing freshly prepared biphasic fix solution, by immersing the mesh sieve upside down in the upper (heptane) phase of the fix solution. Shake for 20 min at room temperature, using a roller mixer (see Note 3).
4. Discard the lower aqueous phase, taking care not to draw up the embryos found at the interface.
5. Add 5 mL of 100% methanol and shake vigorously for about 1 min to devitellinize the embryos, which should sink to the bottom of the scintillation vial.
6. Carefully collect the embryos from the bottom of the scintillation vial, and transfer them to a fresh 1.5-mL tube. Add another 2 mL of 100% methanol to the original scintillation vial and shake again vigorously for about 1 min to devitellinize more of the floating embryos. Collect all the settled embryos and transfer them to the 1.5-mL tube.
7. Rinse the embryos with 100% methanol (×3).

3.2. dpERK Staining of Drosophila Embryos

Unless otherwise stated, all of the washings specified below are performed at room temperature in 1 mL volumes. All washes are done using a roller mixer.

1. Rehydrate embryos, sequentially, in 70, 50, and 30% methanol. Wait for 2 min between washes, to allow the embryos to sink.
2. Incubate embryos in 300 μL of 3% H_2O_2 in amplification buffer, in a tube covered with aluminum foil, for 10 min.
3. Wash embryos in TNT, for 5 min.
4. Block embryos by incubating in TNB, for 30 min.
5. Incubate embryos in 300 μL of the anti-dpERK antibody, diluted 1:100 in TNB, overnight at 4°C (see Note 4).
6. Wash embryos in TNT, for 20 min (×3).
7. Incubate embryos in 300 μL of goat anti-mouse biotin-conjugated antibody, diluted 1:2,000 in TNB, for 60 min (see Notes 4 and 5).
8. Wash embryos in TNT, for 20 min (×3).
9. Incubate embryos in 300 μL of Streptavidin–HRP solution, diluted 1:100 in TNB, for 30 min.
10. Wash embryos in TNT, for 20 min (×3).
11. Add 150 μL of Alexa Fluor 488 tyramide conjugate, diluted 1:50 in amplification buffer. Incubate in a tube covered with aluminum foil, for 10 min (see Note 6).
12. Wash embryos in TNT, for 20 min (×3).

3.3. Mounting, Viewing, and Storing Stained Embryos

1. Use wide-aperture pipetor tips to pick up the embryos from the bottom of the tube, and spread them on a clean microscope slide. Use KimWipes to gently extract as much of the residual TNT as possible.
2. Submerge the embryos in a drop (approximately 60 μL) of mounting medium and cover with a 22 × 22 mm cover slip.
3. Seal the edges with transparent nail polish, to prevent movement of the cover slip. Slides can be stored this way for a few weeks, at 4°C in the dark.
4. Use conventional fluorescence or confocal microscopy to detect the pattern of dpERK staining in the mounted embryos.

4. Notes

1. Another possibility is to use rabbit anti-dpERK antibody (Phospho-p44/42 MAPK (Erk1/2) (Thr202/Tyr204) Rabbit

mAb; Cell Signaling). In this case, make sure to use goat anti-rabbit IgG biotin-SP conjugate as secondary antibodies.

2. Successful anti-dpERK antibody staining necessitates freshly fixed embryos; hence, embryos should be stained immediately following their fixation. For this reason, it is advisable to set up fly cages in a way that permits the collection of a sufficient number of embryos for staining.
3. When staining mutant embryos, it is sometimes impossible to collect many freshly fixed embryos. In such cases, use wide forceps to transfer the few embryos to a 1.5-mL microfuge tube, containing 0.5 mL of heptane (taken from the upper phase of a newly made fix solution). The embryos should sink to the bottom of the vial. Then, add 0.5 mL of the lower phase of the fix solution and shake for 20 min at room temperature using a roller mixer.
4. When double staining for dpERK and some other marker, add both primary antibodies together (step 5) and the appropriate fluorescent secondary antibody along with the goat anti-mouse biotin-conjugated antibody (step 7). In this case, the tube needs to be covered with aluminum foil from step 7 onwards.
5. When using a standard fluorescence antibody staining protocol, anti-dpERK staining usually results in a very weak signal. To augment the sensitivity of the signal, it is recommended to use tyramide signal amplification (TSA), an enzyme-mediated detection method that utilizes the catalytic activity of HRP to generate high-density labeling of a target protein in situ (22) http://probes.invitrogen.com/media/pis/mp20911.pdf. Note that when staining for dpERK at late stages of embryogenesis, often there is no need for TSA amplification. In this case, use the standard protocol for fluorescent staining, with the exception that fixation should be done with 8% formaldehyde (14).
6. When even a stronger enhancement of the signal is required, amplify using tyramide conjugated to biotin (Molecular Probes) diluted 1:50 in amplification buffer, for 10 min at room temperature; then, add streptavidin-Cy2 (Jackson ImmunoResearch Laboratories; code no. 016-220-084) diluted 1:500 in TNB, for 30 min at room temperature.

Acknowledgments

We are indebted to Einat Cinnamon for providing numerous invaluable tips that helped optimize the immunodetection of dpERK. We are grateful to Einat Cinnamon and Gerardo Jiménez for their comments on the manuscript. Work in our laboratory is supported by grants from the Israel Science Foundation (FIRST

Program, grant No. 1,215/07; and Center of Excellence, grant No. 180/09), German-Israeli Foundation (905-14.13/2,006), and the Król Charitable Foundation. A.H. is the recipient of a Ph.D. Fellowship awarded by the Rector of the Hebrew University.

References

1. Shilo, B. Z. (2005). Regulating the dynamics of EGF receptor signaling in space and time. *Development* **132**, 4017–4027.
2. Murphy, L. O. & Blenis, J. (2006). MAPK signal specificity: the right place at the right time. *Trends in Biochemical Sciences* **31**, 268–275.
3. Yung, Y., Dolginov, Y., Yao, Z., Rubinfeld, H., Michael, D., Hanoch, T., Roubini, E., Lando, Z., Zharhary, D., & Seger, R. (1997). Detection of ERK activation by a novel monoclonal antibody. *FEBS Letters* **408**, 292–296.
4. Gabay, L., Seger, R., & Shilo, B. Z. (1997). In situ activation pattern of *Drosophila* EGF receptor pathway during devèlopment. *Science* **277**, 1103–1106.
5. Biggs, W. H., III, Zavitz, K. H., Dickson, B., van der Straten, A., Brunner, D., Hafen, E., & Zipursky, S. L. (1994). The *Drosophila rolled* locus encodes a MAP kinase required in the *sevenless* signal transduction pathway. *The EMBO Journal* **13**, 1628–1635.
6. Gabay, L., Seger, R., & Shilo, B. Z. (1997). MAP kinase in situ activation atlas during *Drosophila* embryogenesis. *Development* **124**, 3535–3541.
7. Ishimaru, S., Ueda, R., Hinohara, Y., Ohtani, M., & Hanafusa, H. (2004). PVR plays a critical role via JNK activation in thorax closure during *Drosophila* metamorphosis. *The EMBO Journal* **23**, 3984–3994.
8. Inoue, H., Tateno, M., Fujimura-Kamada, K., Takaesu, G., Adachi-Yamada, T., Ninomiya-Tsuji, J., Irie, K., Nishida, Y., & Matsumoto, K. (2001). A *Drosophila* MAPKKK, D-MEKK1, mediates stress responses through activation of p38 MAPK. *The EMBO Journal* **20**, 5421–5430.
9. Williams, J. A., Su, H. S., Bernards, A., Field, J., & Sehgal, A. (2001). A circadian output in *Drosophila* mediated by *neurofibromatosis-1* and Ras/MAPK. *Science* **293**, 2251–2256.
10. Duchek, P. & Rorth, P. (2001). Guidance of cell migration by EGF receptor signaling during *Drosophila* oogenesis. *Science* **291**, 131–133.
11. Reich, A., Sapir, A., & Shilo, B. (1999). Sprouty is a general inhibitor of receptor tyrosine kinase signaling. *Development* **126**, 4139–4147.
12. Halfon, M. S., Carmena, A., Gisselbrecht, S., Sackerson, C. M., Jimenez, F., Baylies, M. K., & Michelson, A. M. (2000). Ras pathway specificity is determined by the integration of multiple signal-activated and tissue-restricted transcription factors. *Cell* **103**, 63–74.
13. Spencer, S. A. & Cagan, R. L. (2003). Echinoid is essential for regulation of EGFR signaling and R8 formation during *Drosophila* eye development. *Development* **130**, 3725–3733.
14. Coppey, M., Boettiger, A. N., Berezhkovskii, A. M., & Shvartsman, S. Y. (2008). Nuclear trapping shapes the terminal gradient in the *Drosophila* embryo. *Current Biology* **18**, 915–919.
15. Cinnamon, E., Gur-Wahnon, D., Helman, A., St Johnston, D., Jiménez, G., & Paroush, Z. (2004). Capicua integrates input from two maternal systems in *Drosophila* terminal patterning. *The EMBO Journal* **23**, 4571–4582.
16. Astigarraga, S., Grossman, R., Diaz-Delfin, J., Caelles, C., Paroush, Z., & Jiménez, G. (2007). A MAPK docking site is critical for downregulation of Capicua by Torso and EGFR RTK signaling. *The EMBO Journal* **26**, 668–677.
17. Yan, H., Chin, M. L., Horvath, E. A., Kane, E. A., & Pfleger, C. M. (2009). Impairment of ubiquitylation by mutation in *Drosophila* E1 promotes both cell-autonomous and non-cell-autonomous Ras-ERK activation in vivo. *Journal of Cell Science* **122**, 1461–1470.
18. Marenda, D. R., Vrailas, A. D., Rodrigues, A. B., Cook, S., Powers, M. A., Lorenzen, J. A., Perkins, L. A., & Moses, K. (2006). MAP kinase subcellular localization controls both pattern and proliferation in the developing *Drosophila* wing. *Development* **133**, 43–51.
19. Wang, S., Tsarouhas, V., Xylourgidis, N., Sabri, N., Tiklova, K., Nautiyal, N., Gallio, M., & Samakovlis, C. (2009). The tyrosine kinase Stitcher activates Grainy head and epidermal wound healing in *Drosophila*. *Nature Cell Biology* **11**, 890–895.
20. Kumar, J. P., Hsiung, F., Powers, M. A., & Moses, K. (2003). Nuclear translocation of activated MAP kinase is developmentally

regulated in the developing *Drosophila* eye. *Development* **130**, 3703–3714.

21. Kumar, J. P., Tio, M., Hsiung, F., Akopyan, S., Gabay, L., Seger, R., Shilo, B. Z., & Moses, K. (1998). Dissecting the roles of the *Drosophila* EGF receptor in eye development and MAP kinase activation. *Development* **125**, 3875–3885.
22. Adams, J. C. (1992). Biotin amplification of biotin and horseradish peroxidase signals in histochemical stains. *Journal of Histochemistry Cytochemistry* **40**, 1457–1463.

Chapter 25

Studying MAP Kinase Pathways During Early Development of *Xenopus laevis*

Aviad Keren and Eyal Bengal

Abstract

The following chapter describes several methods involved in the detection of MAPK activities and phosphorylated proteins during early development of *Xenopus laevis*. The *Xenopus* embryo provides a powerful platform for biochemical studies. We describe here basic methods of embryo manipulations such as egg fertilization, embryo growth and maintenance, microinjection of capped RNA and antisense morpholino oligonucleotides (AMOs), and isolation of explants. In addition, we describe methods to detect phosphorylated proteins, to analyze kinase activity, and to interfere with signaling pathways. Immunohistochemical staining performed on whole embryos or on tissue sections is an additional method for the detection of phosphorylated proteins in the developing embryo. Approaches to activate or inhibit MAPK activities including the ectopic expression of mutated isoforms of MAPK kinase, or the incubation of embryo explants with pharmacological inhibitors are described. Finally, we describe an in vitro kinase assay specifically designed for the *Xenopus* embryo.

Key words: *Xenopus* embryos, Embryo protein extracts, Detection of phosphoproteins in whole embryos, Embryo injections of MAPKK mutants, Inhibition of MAPK in explants

1. Introduction

One of the major challenges in developmental biology is to understand how cells integrate multiple signaling pathways to achieve different cell fates. Germ layer specification and patterning of the *Xenopus* embryo are regulated by secreted signaling proteins. For example, fibroblast growth factor (FGF) induces mesoderm formation at blastula, while dorsal-ventral gradients of Wnt and bone morphogenetic protein (BMP) determine the different cellular fates along the axis at gastrula. One interesting research question is how these signals are transduced and integrated within cells to

Rony Seger (ed.), *MAP Kinase Signaling Protocols: Second Edition*, Methods in Molecular Biology, vol. 661, DOI 10.1007/978-1-60761-795-2_25, © Springer Science+Business Media, LLC 2010

specify the embryo axis. Mitogen-activated protein kinases (MAPKs) function at the earliest cell specification and patterning events. To regulate these processes, MAPK together with other signaling molecules converge at the level of transcription factors to fine tune their activities. ERK MAPK is involved in mesoderm specification downstream to FGF signaling (1, 2). The p53 protein is a mediator of MAPK signaling in this process. FGF/MAPK causes the phosphorylation of p53 at serine residues 6 and 9, promoting a robust biochemical interaction of p53 with TGFβ-activated Smads. The p53-Smad complex is active in mesoderm induction (3). In the process of axis formation, FGF/MAPK and BMP/Smad1 pathways are integrated. At the ventral side of the embryo, BMPR phosphorylates C-terminal residues of Smad1 to induce its translocation into the nucleus, while at the dorsal side FGF/MAPK phosphorylates residues within the linker region of Smad1 to terminate its transcriptional activity (4, 5). Phosphorylation of the linker region of Smad1 by ERK MAPK serves as a priming event for phosphorylation at adjacent serine residues by GSK3. This additional phosphorylation targets the protein to the ubiquitin degradation system (reviewed in ref. 6). Therefore, dorsal activation of ERK MAPK signaling is an additional strategy of the organizer to terminate BMP activity via Smad1. Similarly, FGF/MAPK signaling is necessary for neuronal induction via the termination of Smad1 activity (4, 5). On the other hand, antagonism of BMP signaling at the dorsal end induces the ERK MAPK pathway (7, 8). Therefore, a loop between BMP and ERK MAPK is established at the dorsal side of the embryo. We have recently shown that p38 MAPK is also highly active in the dorsal organizer and functions via the phosphorylation of the CREB transcription factor in regulating the expression of *chordin* (9). Therefore, both p38 and ERK MAPK function by different strategies to prevent BMP signaling at the Spemann organizer.

The methods for studying MAPK in the *Xenopus* embryo were adopted from other systems and were modified for *Xenopus*. The *Xenopus* embryo has certain advantages and disadvantages with regard to the study of biochemical mechanisms. Among the advantages, one could mention the relatively big sized embryo, the availability of high numbers of embryos needed for biochemical analysis, the ability to isolate and grow explants, and the ectopic expression of mutated and epitope-tagged proteins. A major disadvantage is the relative shortage in commercial antibodies to *Xenopus* proteins. In this chapter, we describe several methods including detection of phosphorylated proteins by Western blotting and immunohistochemical staining of whole embryos, explants, and tissue sections. We present methods to activate MAPK pathways, by injection of transcripts encoding active alleles of MAPK kinase, and to inhibit MAPK by the expression of dominant negative alleles, injection of antisense morpholino oligonucleotides (AMOs), or treatment of embryo explants with pharmacological inhibitors. Finally, we describe a method to analyze the in vitro kinase activity of p38 MAPK.

2. Materials

2.1. Frogs

The South-African clawed frog *Xenopus laevis* were purchased from Nasco (Fort Atkinson, WI: http://www.nascofa.com). For review on the materials and conditions needed to maintain a healthy frog colony, see (10, 11).

2.2. Reagents and Kits

1. Cap analog (m7G(5′)ppp(5′)G) (New England Biolabs) stock prepared at 125 mM in water and stored at –20°C.
2. Human chorionic gonadotropin hormone is used to induce egg-laying (Pregnyl, Organon, Oss, Holland).
3. Ficoll (GE Healthcare) is used as 3% diluted in 1/3× MR.
4. AEC kit for immunohistochemistry on sections (ZYMED 95-9943, see Note 1).
5. Sp6, T7, or T3 for Capped RNA preparation was purchased from Promega Corp.
6. SB203580 powder was from Calbiochem. Powder is dissolved in DMSO to 10 mM.
7. A mixture of protease inhibitors (Roche, complete mini).
8. Hematoxylin (Pioneer Research Chemicals, see Note 2).

2.3. Buffers and Solutions

1. Modified Ringers' solution 1× MR: 2 mM KCl, 1 mM $MgCl_2$, 2 mM $CaCl_2$, 100 mM NaCl, and 5 mM Hepes pH 7.1. We use a stock of 1× MR which is diluted to 1/3× MR to grow the embryos.
2. 1× LCMR: low calcium and magnesium Ringer's solution: 66 mM NaCl, 1.33 mM KCl, 0.17 mM $MgCl_2$, 0.33 mM $CaCl_2$, and 5 mM Hepes pH 7.1.
3. 3% Ficoll: dilute Ficoll in 1/3× MR, add oxytetracyclin to a final concentration of 25 μg/ml, 0.2-mm filter and store at 15°C.
4. L-Cysteine: dissolve 2 g L-cysteine in 100 ml 1/3× MR and add 3.2 ml 5 N NaOH.
5. Reaction mixture for capped RNA preparation: to a final volume of 100 μl mix add: 42 μl DDW, 10 μg linearized DNA, 10 μl 10× transcription buffer (supplied with the polymerase), 10 μl of four NTPs mixture (2.5 mM), 25 μl of m7G(5′)

ppp(5′)G (10 mM) 1 μl of DTT (1 M), 32P-UTP 800 Ci/mmole 104 cpm, 1 μl of RNAsin (40 U), and 6 μl of Sp6, T7, or T3 polymerase (1,000 U).

6. Termination buffer for in vitro RNA transcription: 100 mM NaCl, 10 mM EDTA, 20 mM Tris-HCl pH 7.5, and 1% SDS.
7. Fixation buffer (MEMFA): 0.1 M MOPS pH 7.4, 2 mM EGTA, 1 mM $MgSO_4$, and 3.7% formaldehyde.
8. 10× PBS (1 l): 80-g NaCl, 2-g KCl, 2-g KH_2PO_4, and 7.65-g Na2HPO4.$2H_2O$.
9. PBST: 1× PBS and 0.1% Tween 20.
10. Whole Cell Extract buffer: 25 mM Hepes pH 7.5, 0.3 M NaCl, 1.5 mM $MgCl_2$, 0.2 mM EDTA, and 0.1% Triton X-100. Add protease inhibitor mixture.
11. Phosphatase inhibitors (per 1 ml Whole Cell Extract): 0.5 mM DTT, 20 mM PNPP, 1 mM $NaVO_4$, 1 mM β-glycerophosphate, and 2.5 mM sodium pyrophosphate.
12. 1× TBS: 10 mM Tris–HCl pH 8.0 and 150 mM NaCl.
13. Blocking buffer (Western): 5% non-fat dry milk, 0.1% Tween 20, and 1× TBS.
14. IP wash buffer: 25 mM Hepes pH 7.5, 0.2% Triton X-100, and 1 mM EDTA.
15. 2× Kinase buffer: 25 mM Hepes pH 7.5, 20 mM $MgCl_2$, 20 mM β-glycerol phosphate, 20 mM PNPP, 0.5 mM $NaVO_4$, and 2 mM DTT.

3. Methods

3.1. Ovulation, Fertilization, and Microsurgery

Female frogs are injected with human chorionic gonadotropin (hCG, 1,000 U per frog) 24 h prior to fertilization. *Xenopus* eggs are fertilized in vitro and maintained in salt solution (1× MR). Fertilized eggs are dejellied with 2% L-cysteine, washed in 1/3× MR, and injected in 3% Ficoll with different concentrations of capped in vitro transcribed RNA or AMOs into the marginal zone of one-cell stage embryos. Embryos are staged according to Nieuwkoop and Faber (12). At gastrula stage, embryos are dissected using an "eyebrow knife". Explants are grown in 1× LCMR and gentamicin (50 μg/ml) in Petri dishes prepared with 2% agarose (dissolved in 1× LCMR).

3.2. Preparation of In Vitro Transcribed RNA

Plasmids, containing cDNA for transcription, are linearized with restriction enzymes for 3 h at 37°C. Linearized plasmid DNA is extracted with phenol:chloroform (1:1) and precipitated, for

0.5 h, with 0.3 M sodium acetate (10% volume) and two volumes of ethanol. Following 30-min centrifugation at 4°C (12,000 × *g*), the precipitate is washed in 70% ethanol and dissolved in 30 μl DDW. For transcription, 10 μg of linearized DNA is incubated for 3 h at 37°C with reaction mixture for Capped RNA preparation. To terminate transcription, 1 μl of RNase free DNase is added, and the mixture is incubated for additional 15 min at 37°C. Then, 100 μl of termination buffer is added. Nucleic acids are extracted with phenol:chloroform (1:1). Following precipitation in 1.25 M ammonium acetate and 0.35 volume of isopropanol for 30 min at room temperature, RNA is centrifuged for 30 min at 12,000 × *g* and dissolved in DDW to a concentration of 1 μg/μl.

3.3. Immunobloting

1. To prepare cell extract, wash explants three times with 1× PBS.
2. To prepare protein lysate, add Whole Cell Extract buffer to explants (18 explants in 40 μl buffer). Keep on ice for 30 min with occasional vortexing of samples.
3. Clear extract by three consecutive centrifugations at 12,000 × *g* for 7 min each at 4°C.
4. Load equal amounts of extracted proteins (40 μg) on SDS–polyacrylamide gel electrophoresis (SDS–PAGE).
5. Transfer proteins to a nitrocellulose membrane. Block the membrane for 1 h at room temperature with blocking buffer.
6. Dilute antibodies in blocking buffer and incubate with membrane overnight at 4°C.
7. Wash membrane three times, 10 min each with 1× TBS/Tween 0.1%. Bound antibodies are detected with HRP-conjugated secondary antibodies (goat anti-rabbit; Pierce) diluted 1:2,500 in blocking solution and visualized using the enhanced chemiluminescence kit (Pierce).

3.4. Paraffin and Tissue Sections

1. Fix embryos and explants for 2 h in MEMFA on a nutator.
2. Dehydrate overnight at 4°C with 70% ethanol.
3. Dehydrate by successive 5-min washes: 70, 90, and 100% ethanol.
4. Wash for 10 min with chloroform (clearing step) and transfer the specimen to 100% paraffin and incubate at 60°C for 1.5 h (see Note 3).
5. Place specimen carefully in a mold under the binocular and let chill at 4°C for 1 h (see Fig. 1).
6. Section molds in a microtome (5 μm thin), place paraffin sections in water bath at 40°C, and mount on SuperFrost slides.
7. Incubate slides in a 37°C oven overnight.

3.5. Immunohistochemistry on Sections (According to ZYMED AEC Kit)

1. Incubate slides in Xylene for 20 min and wash slides twice with fresh isopropanol.
2. For antigen retrieval, put slides in a plastic coupling jar filled with 0.1 M sodium citrate buffer (pH 6.0) and microwave for 10 min at 90°C (heating times may vary with different antibodies). After the heating process, let the slides cool at room temperature for 30 min.
3. Wash slides three times with DDW and incubate with methanol and H_2O_2 (90:10) for 10 min (see Note 4). Wash slides three times with 1× PBS.
4. Place slides in a humidity chamber. Wipe slides to remove excess PBS (see Note 5). Drip 10 μl of blocking buffer (supplied with the kit) on specimen and incubate for 30 min.
5. Wipe blocking buffer, drip 10 μl of first antibody diluted in blocking buffer, and incubate overnight at 4°C.
6. Wash slides three times with 1× PBS and then incubate with goat biotinylated second antibody for 13 min. Wash slides three times with 1× PBS and incubate with streptavidin–HRP for 13 min.
7. Wash slides three times with 1× PBS and incubate with AEC until color is well developed.
8. Counterstain with hematoxylin and mount slides with Immumount (supplied with the kit).
9. Visualize staining under a standard light microscope.

3.6. Whole-Mount Immunohistochemistry

1. Fix embryos in MEMFA with rotation for 2 h at room temperature.
2. Dehydrate with methanol overnight at −20°C.
3. Rehydrate by successive 5-min washes: 100% methanol, 75% methanol–25% water, 50% methanol–50% water, 25% methanol–75% PBST and PBST.

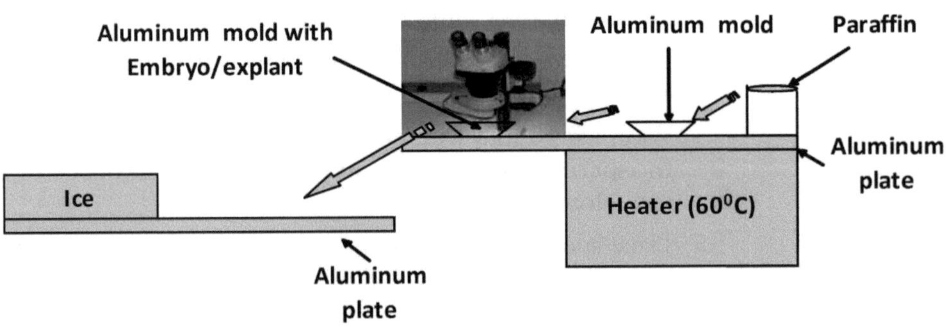

Fig. 1. Scheme presenting preparation of specimen in paraffin. Embryo is incubated in 100% paraffin at 60°C for 1.5 h on a heated aluminum plate. Then, embryo is placed carefully in a mold under the binocular at the desired orientation. The mold with the specimen is carefully placed on a cold aluminum plate to allow paraffin to solidify.

4. Inactivate endogenous peroxidase by rotation for 10 min with methanol:H_2O_2 (90:10), wash in PBST for 10 min, and block embryos for 1 h in 500 ml PBST supplemented with 10% goat serum.
5. Incubate overnight with 500 μl of PBST and 10% goat serum with the appropriate dilution of the primary antibody.
6. Wash three times with PBST, 2 h each and incubate overnight with 500 μl of PBST and 10% goat serum containing 1:100 dilution of HRP-conjugated secondary antibody (goat anti-rabbit IgM; Jackson).
7. Wash embryos three times, 2 h each, in PBST at room temperature.
8. Visualizing with color-developing reagent (DAB substrate tablets: Amersco) vigorously shake embryos for 10 min at 4°C in DAB solution (500 μl PBST and 500 μl 1 mg/ml DAB). Add 1 μl of 30% H_2O_2 diluted 1:1 in PBST and start shaking the tubes immediately. Once color is well developed, stop the reaction by adding PBST.

3.7. Activation of the p38 MAPK Pathway in *Xenopus laevis* Embryos

A useful approach to activate the p38 MAPK is by expressing its upstream kinase, MKK6. Constitutively activated epitope-tagged MKK6 (HA-MKK6E) is cloned into pCS2+ expression vector and is transcribed as mentioned above. The transcripts encoding activated MKK6 (100 pg) are injected into one-cell stage embryos, and the effect on development is followed at different stages. The expression of the protein can be verified by Western blot analysis with an antibody against HA tag. Protein lysates are prepared from explants and activation of the pathway is analyzed by Western blotting with antibodies to phospho-p38 (Cell signaling #9215) or antibodies to phospho-CREB (S133) (Cell signaling # 9198S), an indirect target of p38 MAPK (see Note 6).

3.8. Inhibition of the p38 MAPK Pathway

Three approaches can be used to inhibit the p38 MAPK pathway.

3.8.1. Incubation in SB203580: A Chemical Inhibitor of p38 MAPK

SB203580 is a relatively specific inhibitor of some p38 MAPK isoforms. SB203580 belongs to the pyridinyl imidazole compounds that function as specific inhibitors of p38α and p38β, but not p38γ and 38δ through competition with ATP for the same site on the p38 kinase (reviewed in (13)). Explants are dissected at early gastrula stage (as mentioned above) and are treated with 30 μM of SB302580 (dissolved in DMSO and diluted in LCMR medium) (see Note 7). Explants are kept in a dark place. When explants reach the proper developmental stage (several hours later), explants are washed twice in 1× PBS and immediately subjected to lysis. Proteins are separated in SDS–PAGE and analyzed by

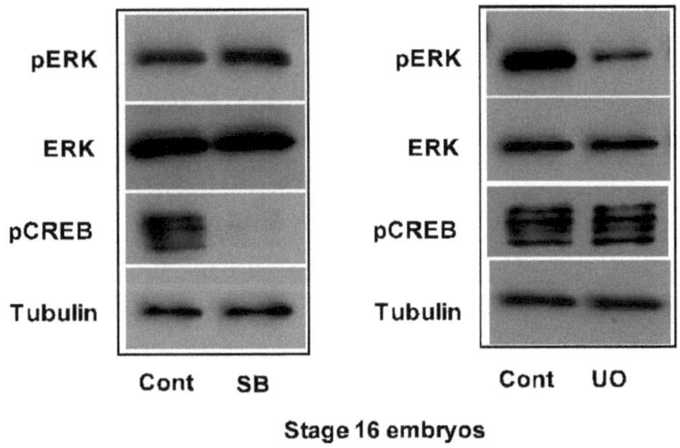

Fig. 2. Incubation of explants with MAPK pharmacological inhibitors. DMZ explants ($n=36$) were dissected at stage 10.5. At stage 11, isolated explants were incubated in SB203580 (30 μM) (left panel) or in U0126 (30 μM) (*right panel*). Explants were allowed to develop to stage 16. Proteins were extracted from explants to be further analyzed by Western blotting with the indicated antibodies.

Western blotting with anti-phospho-CREB antibody (Cell signaling #9198S) (Fig. 2). In *Xenopus* embryos, the p38 MAPK pathway induces phosphorylation of CREB serine 133 (9, 14).

3.8.2. Injection of Antisense Morpholino Oligonucleotides to Knockdown p38 MAPK In Vivo

The use of AMOs as a powerful tool to knockdown gene expression has been recently widely used. AMOs are short chains of about 25 morpholino subunits. Each subunit is comprised of a nucleic acid base, a morpholine ring, and a nonionic phosphorodiamidate intersubunit linkage. Morpholinos do not degrade their RNA targets, but instead act via RNAse H-independent steric blocking mechanism. They block translation initiation in the cytosol (by targeting the 5′ UTR through the first 25 bases of coding sequence) (reviewed in (15)). p38α AMO is stored in aliquots at 4°C and just before injection, AMO is slightly vortexed and warmed at 70°C for 5 min (see Note 8). After a short spin at room temperature, 10 nl of the p38α AMO is injected at concentration of 1 ng/1 nl. Embryos are carefully handled and are grown until sibling embryos reach later stages (14–16). Alternatively, explants are dissected at early gastrula stage (10.25) and cultured until sibling embryos reach stages 14–16. To analyze whether AMO inhibits p38 translation, proteins are extracted from explants or embryos and analyzed by Western blotting with anti-total p38 antibody (Santa Cruz Biotechnology #SC-535). Usually, a control AMO containing five mismatches relative to the experimental AMO is injected into embryos without any expected phenotypes. Another way to investigate the specificity of p38α knockdown is to inject a transcript encoding p38α that is not targeted by the AMO and it should rescue the phenotype. In our experiments, mouse p38α MAPK rescued morphological and molecular phenotypes of p38α-knockdown embryos.

3.8.3. Injection of a Transcript Encoding a Kinase Inactive MKK6 (MKK6A) to Inhibit the p38 MAPK Pathway

Another approach to inhibit the p38 MAPK pathway is to inject a transcript encoding for a kinase-dead MKK6 (HA-MKK6A). Preparation of the transcript is described in Subheading 3.2. Transcripts encoding HA-MKK6A (0.5 ng) are injected into one-cell stage embryos which are then dissected at early gastrula stage. Explants are grown until sibling embryos reach stage 14. The expression of the exogenous protein is verified by Western blot analysis with an antibody to HA epitope, and its activity by analyzing the phosphorylation of p38 MAPK and CREB (S133).

3.9. Analysis of p38 MAPK Pathway Activity in the Embryo by Detecting Phosphorylated CREB Protein

Although we were able to detect phospho-p38 by Western blot analysis (Cell signaling #9215), none of the tested commercial antibodies detected phospho-p38 in immunohistochemical staining. Therefore, we analyzed the phosphorylation of an indirect target of p38, the CREB protein. Our investigation revealed that at these early stages of *Xenopus* development, CREB phosphorylation is a consequence of p38 MAPK signaling and excludes the involvement of ERK MAPK and PKA (9, 14). An antibody to phospho-CREB that proved useful for immunohistochemistry in whole embryo staining (as is described in Subheading 3.6) and in tissue sections (as is described in Subheading 3.5) was purchased from Cell signaling (#9198S). In Fig. 3, a transverse section of *Xenopus* embryo (stage 30) immunostained with anti-phospho-CREB antibody and counterstained with hematoxylin (see Subheading 3) is presented. Phospho-CREB staining is mostly detected in the somite and the neural tube.

3.10. In Vitro Kinase Assay for p38 MAPK Activity

In vitro kinase assay for p38 with embryo extracts can be performed by immunoprecipitation (IP) of endogenous phosphorylated p38 or alternatively by IP of exogenous epitope-tagged p38. We find the latter much more sensitive, although it requires an additional step of injecting embryos with transcripts encoding epitope-tagged p38 MAPK.

3.10.1. RNA Microinjection to Embryos

Preparation of RNA for injections is described in Subheading 3.2. Embryo microinjection is performed as follows:

Fill mesh-bottomed dish with 3% Ficoll. The mesh-bottomed dishes keep the embryos from moving during injection. Using a flexible pipette tip, gently rotate embryos into isolated wells of the mesh until they are all ready to be injected. Carefully draw out the 3% Ficoll without disturbing the embryos. Inject in an organized manner–row by row. This will allow you to keep track of the injected embryos. At the end of each row, check that the needle is not clogged.

3.10.2. Immunoprecipitation

Isolation and growth of explants is described in Subheading 3.1. Protein extract preparation is described in Subheading 3.3. Usually, explants (DMZ and/or VMZ) are isolated at early gastrula (stage 10.25) and grown to stages 14–15, at which time proteins

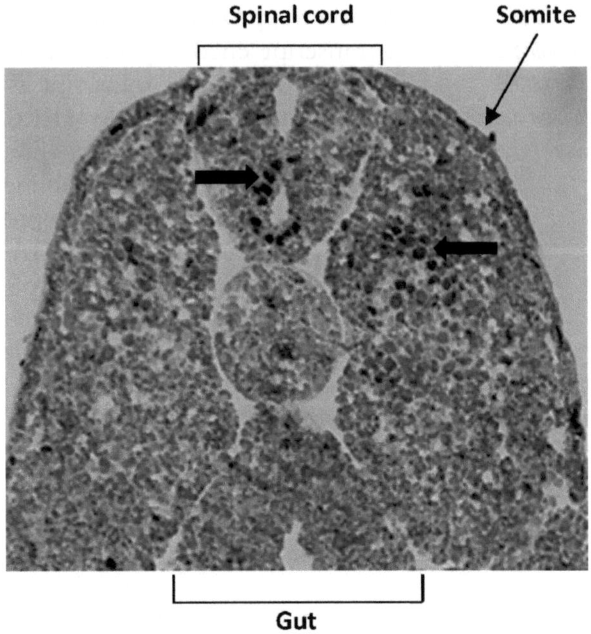

Fig. 3. Phospho-CREB is abundant in the somites. Embryos were allowed to develop to stage 30. Embryos were fixed and sectioned (transversely) as described under Subheading 3. Immunohistochemical staining of one transverse section using an antibody to phospho-CREB (S133) is presented. Sections were counterstained with hematoxylin. Nuclear staining of phospho-CREB is represented by *dark punctate staining*. *Arrow* is pointing at stained nuclei.

are extracted. Around 500 μg of protein extract is used per one IP reaction. The following steps are taken:

1. Protein extract is incubated with 20 μl of protein A beads (1:1) and rotated for 1 h at 4°C (preclearing). Beads are precipitated by slow centrifugation (1,000 × *g* for 1 min), and protein extract is kept on ice.
2. Preparation of protein A beads bound to antibodies: 40 μl of protein A beads (1:1) are rotated with 1–2 μg of purified antibody (per one IP reaction) for 1 h at RT. Beads are washed three times in 1 ml 1× PBS. Beads are washed once with Whole Cell Extract buffer.
3. Protein extract from step 1 is mixed with the prepared beads from step 2. Whole Cell Extract buffer is added to increase volume to 200–400 μl. Mixture is gently rotated for 2 h at 4°C.
4. Wash three times in 1-ml cold Whole Cell Extract buffer and precipitate beads between washes.
5. Wash once with IP wash buffer and spin (1,000 × *g*) for 1 min. Remove residual solution with a yellow tip, without disturbing the beads. Keep on ice.

3.10.3. Kinase Assay

Add to beads the following (per reaction): 21.5 μl of 2× kinase buffer, 3-μl cold ATP (0.2 mM), 0.5 μl [γ-^{32}P] ATP (3,000 Ci/mmol), and 5 μl substrate (GST-ATF2, 1 μg/μl). Reaction mixture is incubated for 30 min at 30°C. The kinase reaction is analyzed by SDS–PAGE and autoradiography.

4. Notes

1. We use the rabbit/mouse primary (AEC) and not the broad spectrum, which was found less specific.
2. The hematoxylin staining is a crucial stage in the staining process. Excess incubation time may mask your antibody staining. We usually dilute hematoxylin 1:5 with DDW and incubate with the specimen for 1 min.
3. Do not exceed the 1.5-h incubation time, otherwise protein may be degraded.
4. This step is necessary to inactivate the endogenous peroxidase.
5. When drying the specimen be careful not to dry it to completion. We usually divide the slides into two parts, handling each separately.
6. It is not always an easy task to find commercial antibodies that cross-react with *Xenopus* proteins. Therefore, in some cases we use antibodies to indirect substrates of a signaling pathway. The CREB protein is not directly phosphorylated by p38 MAPK, but by MSK, a substrate of p38 MAPK.
7. We found that higher concentrations of SB302580 than that typically used in tissue culture cells (10 μM) are necessary to inhibit p38 in explants. This is probably due to the low penetration of the chemical through the many cellular layers of the three-dimensional explant. Still, in our experiments the higher concentration of SB302580 did not result with nonspecific inhibition of ERK MAPK (see Fig. 25.2).
8. AMO is warmed at 70°C to prevent it from clogging the pipette during injection into embryos.

References

1. Gotoh, Y., Masuyama, N., Suzuki, A., Ueno, N. and Nishida, E. (1995) Involvement of the MAP kinase cascade in *Xenopus* mesoderm induction. *EMBO J*, **14**, 2491–2498.
2. Umbhauer, M., Marshall, C.J., Mason, C.S., Old, R.W. and Smith, J.C. (1995) Mesoderm induction in *Xenopus* caused by activation of MAP kinase. *Nature*, **376**, 58–62.
3. Cordenonsi, M., Montagner, M., Adorno, M., Zacchigna, L., Martello, G., Mamidi, A., Soligo, S., Dupont, S. and Piccolo, S. (2007) Integration of TGF-beta and Ras/MAPK signaling through p53 phosphorylation. *Science*, **315**, 840–843.
4. Kuroda, H., Fuentealba, L., Ikeda, A., Reversade, B. and De Robertis, E.M. (2005) Default neural induction: neuralization of

dissociated *Xenopus* cells is mediated by Ras/MAPK activation. *Genes Dev*, **19**, 1022–1027.

5. Pera, E.M., Ikeda, A., Eivers, E. and De Robertis, E.M. (2003) Integration of IGF, FGF, and anti-BMP signals via Smad1 phosphorylation in neural induction. *Genes Dev*, **17**, 3023–3028.
6. Eivers, E., Fuentealba, L.C. and De Robertis, E.M. (2008) Integrating positional information at the level of Smad1/5/8. *Curr Opin Genet Dev*, **18**, 304–310.
7. Goswami, M., Uzgare, A.R. and Sater, A.K. (2001) Regulation of MAP kinase by the BMP-4/TAK1 pathway in *Xenopus* ectoderm. *Dev Biol*, **236**, 259–270.
8. Zetser, A., Frank, D. and Bengal, E. (2001) MAP kinase converts MyoD into an instructive muscle differentiation factor in *Xenopus*. *Dev Biol*, **240**, 168–181.
9. Keren, A., Keren-Politansky, A. and Bengal, E. (2008) A p38 MAPK-CREB pathway functions to pattern mesoderm in *Xenopus*. *Dev Biol*, **322**, 86–94.
10. Lacal, J.C. (1998) Oocytes microinjection assay to study the MAP-kinase cascade. *Methods Mol Biol*, **84**, 139–152.
11. Wu, M. and Gerhart, J. (1991) Raising *Xenopus* in the laboratory. *Methods Cell Biol*, **36**, 3–18.
12. Nieuwkoop, P., and Faber, J. (1967) *Normal Table of Xenopus laevis (Daudin)*. North-Holland, Amsterdam.
13. Lee, J.C., Kassis, S., Kumar, S., Badger, A. and Adams, J.L. (1999) p38 mitogen-activated protein kinase inhibitors – mechanisms and therapeutic potentials. *Pharmacol Ther*, **82**, 389–397.
14. Keren, A., Bengal, E. and Frank, D. (2005) p38 MAP kinase regulates the expression of XMyf5 and affects distinct myogenic programs during *Xenopus* development. *Dev Biol*, **288**, 73–86.
15. Eisen, J.S. and Smith, J.C. (2008) Controlling morpholino experiments: don't stop making antisense. *Development*, **135**, 1735–1743.

Chapter 26

Deciphering Signaling Pathways In Vivo: The Ras/Raf/Mek/Erk Cascade

Gergana Galabova-Kovacs and Manuela Baccarini

Abstract

The Ras/Raf/MEK/ERK cascade is a highly conserved signal transduction module, whose activation results in a number of different physiological outcomes. Depending on the cell type or the stimulus used, the pathway has been implicated in proliferation, differentiation, apoptosis, and migration. Because of this wide range of activities, these kinases are considered attractive (anticancer) therapeutic targets. However, their essential functions in the context of the whole organism are still incompletely known. Here, we describe immunohistochemical and immunofluorescence methods that can be used to define the essential function(s) and the relevant downstream targets of Raf-1, B-Raf, and Mek-1 in in vivo models of organ development, remodeling, and neoplasia.

Key words: Erk signaling, Kinase cascades, In Vivo signaling

1. Introduction

The ERK-MAPK pathway was the first signal transduction cascade to be discovered and described from the cell membrane to the nucleus. The pathway has been intensely studied for about 20 years and is the paradigm for MAP kinase modules in general. Downstream of membrane receptors, the small G protein Ras recruits Raf from the cytosol to the cell membrane, where activation occurs (1). Activated Raf binds and phosphorylates MEK, which, in turn, activates ERK. An impressive roster of ERK substrates has been described, comprising membrane and cytoskeletal proteins, cytosolic enzymes, and transcription factors (2). The pathway is activated in 70% of all human cancers (3) and in a number of congenital progressing conditions (4). Clearly, the enzymes of the Raf/MEK/ERK cascade are prime candidates for molecule-targeted therapies.

Rony Seger (ed.), *MAP Kinase Signaling Protocols: Second Edition*, Methods in Molecular Biology, vol. 661, DOI 10.1007/978-1-60761-795-2_26,

Yet, little is known about their role in the whole organism. We use conventional and conditional ablation to identify the essential biological functions of Raf and Mek. To understand how the ERK pathway is wired in vivo, we combine phenotype and signaling analysis in the context of the whole organism (in vivo) with biochemical experiments in ablated cells. Using this combination, we could show that B-Raf is the nonredundant activator of the Erk pathway in vivo in the developing mouse placenta and in a mouse tumor model of insulinoma, in which cases it promotes angiogenesis (5, 6), and in the central nervous system, where it regulates the differentiation of dorsal root ganglia (7) and oligodendrocytes (8). C-Raf is crucial for Erk activation only in erythroid differentiation (9), but its essential roles in survival (10–12), migration, and tumorigenesis (13, 14) are independent of its enzymatic activity and are due to its ability to bind to, and inhibit, other serine/threonine kinases. Finally, we have discovered an unexpected essential role of MEK1 in downregulating Mek-2/Erk signaling in the context of a Mek1:Mek2 heterodimer, establishing Mek1 as the critical modulator of Mek/Erk signaling (15). The methods used to define the signaling pathways affected by the ablation of Raf or Mek in vivo rely essentially on the availability of (phospho) specific antibodies, reflecting the activation state of individual pathways. Used in immunohistochemistry or immunofluorescence, and combined with antibodies defining cells of a specific lineage, these antibodies are a powerful tool for pathway analysis in vivo.

2. Materials

2.1. Immunohistochemistry and Immunofluorescence

2.1.1. Fixation

1. PBS, phosphate buffer solution: 0.2 g KCl, 8 g NaCl, 1.4 g Na_2HPO_4, 0.2 g $KHPO_4$ in 1 L ddH_2O, adjusted to pH 7.4 with HCl, store at 4°C.
2. 4% (w/v) Paraformaldehyde (PFA) in PBS, adjust to pH 7.4, store at 4°C.
3. 70% Ethanol.

2.1.2. Dehydration and Paraffin Embedding According to Equipment Specifications

2.1.3. Slides

1. Menzel Slides: "Super Frost" (Thermo Scientific).
2. Microscope cover glasses (Marienfeld).

2.1.4. Solutions

1. TBS, tris buffered solution: 2.4 g Tris base and 8 g NaCl in 1 L ddH_2O, adjusted to pH 7.6 with HCl, store at 4°C.
2. TBST: 0.2% (v/v) Tween-20 in TBS, store at 4°C.

3. PBS/Triton X-100 0.1%: 0.1% (v/v) Triton X-100, store at 4°C.
4. Methanol.
5. Endogenous peroxidase activity blocking solution: 30% hydrogen peroxide (H_2O_2) diluted in Methanol/ddH_2O (50% v/v) to final H_2O_2 concentration of 1% (v/v), prepare fresh at all times.
6. 10 mM sodium citrate buffer: 2.94 g tri-sodium citrate dehydrate in 1 L ddH_2O, adjust to pH 6.0 and store at 4°C.
7. 1 mM EDTA Buffer/0.05% Tween-20 (pH 8.0): 0.37 g EDTA in 1 L ddH_2O, adjust to pH 8.0 and add 500 μl Tween-20 to final concentration of Tween-20 0.05% (v/v).
8. Blocking buffer: goat serum or horse serum are diluted to 1.5% (v/v) in TBST.
9. Primary antibody diluted in blocking buffer according to the manufacturer's specifications.
10. *Immunhistochemistry*: Secondary antibody, anti-rabbit EnVision+® System, ready-to-use.
11. *Immunhistochemistry*: 0.2 M Tris–HCl buffer solution, 24.2 g Tris base in 1 L ddH_2O, adjust to pH 7.6 with HCl.
12. *Immunhistochemistry*: DAB chromogen solution (3,3′-Diaminobenzidine tetrahydrochloride, one tablet in 50 mL 0.2 M Tris–HCl, pH 7.6. Add 0.03% H_2O_2, mix, and filter through Whatman® filter.
13. Harris Hematoxylin 1:5 in ddH_2O.
14. Ethanol.
15. Xylene.
16. Entelan.
17. *Immunofluorescence:* secondary antibody, diluted in TBS.
18. *Immunofluorescence:* 20 ng/mL DAPI (4′,6-diamidine-2-phenyl indole) in ddH_2O, store at 4°C.
19. *Immunofluorescence:* Dako fluorescent mounting medium.

2.2. Whole Mount Immunostaining for Anti-phosphoERK1/2 and Anti-phosphoMek 1/2

2.2.1. Embryo Isolation

Forceps, scissors, Falcon tissue culture dishes 30 × 10 mm, Nunc 24 well multidishes, stereo microscope.

2.2.2. Fixation and Staining

1. Ice-cold (4°C) 8% (w/v) PFA in PBS, adjust to pH 7.4. Fixation overnight at 4°C with mixing/rocking.
2. PBS, store at 4°C.

3. PBS/NP40, 0.5% (v/v) Nonidet-P40 in PBS.
4. PBS/Triton X-100 0.1%, 0.1% (v/v) Triton X-100 in PBS, store at 4°C.
5. PBSST, blocking buffer, 5% (v/v) goat serum, and 0.1% (v/v) Triton X-100 in PBS.
6. Methanol.
7. Endogenous peroxidase activity blocking solution: 30% hydrogen peroxide (H_2O_2) diluted 1:5 in Methanol.
8. Primary antibody diluted in PBSST: Affinity-purified rabbit phosphoERK1/2 (Cell Signaling, #9101); monoclonal rabbit phosphoMek1/2 antibody (Cell Signaling #2338); rabbit polyclonal anti-B-Raf (H-145; Santa Cruz, #sc-9002); mouse monoclonal βIV-tubulin antibody (Sigma, Cat. # T7941).
9. Secondary antibody for immunohistochemistry, EnVision+® System, ready-to-use; for immunofluorescence, Alexa 488 or Alexa 549-conjugated antibodies from Molecular Probes.
10. 0.2 M Tris–HCl buffer solution.
11. DAB chromogen solution.
12. Glycerol.

3. Methods

3.1. Fixation and Sectioning

1. Tissue samples are fixed in 4% (w/v) ice-cold PFA in PBS (pH 7.4) overnight at 4°C.
2. 3 μm sections are placed on Menzel Gläser “Super Frost”, left to adhere for 20–30 min, and placed overnight in the incubator at 55°C to dry .
3. Store the sections at room temperature (RT).

3.2. Determination of Activities by Staining with Anti-Phospho Antibodies

3.2.1. Analysis of MEK and ERK Activation by Immunohistochemistry

We use indirect immunohistochemistry, which consists in using an unlabeled primary antibody that will detect the antigen, and a labeled secondary antibody that will react with the primary antibody and make it visible. The secondary antibody recognizes the IgG of the animal species in which the primary antibody has been raised. This method amplifies the signal since several secondary antibodies react with different antigenic sites on the primary antibody. We use secondary antibodies labeled with horseradish peroxidase (HRP, immunoenzyme method). We have used this method to visualize the expression of the Erk activator B-Raf (Fig. 1) as well as Erk phosphorylation (Fig. 2) in dorsal root ganglia.

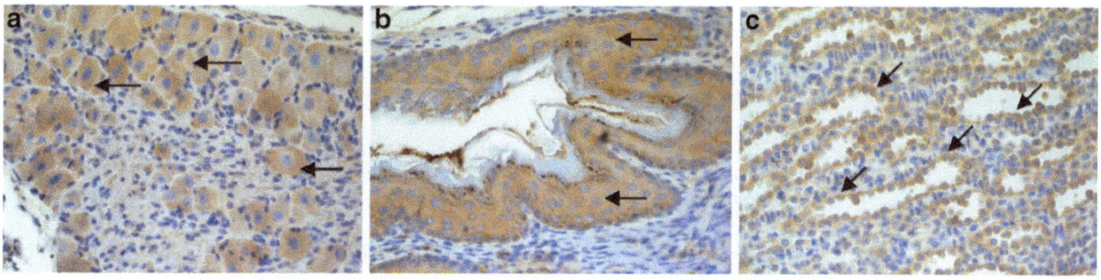

Fig. 1. B-Raf immunohistochemistry. 3 μm paraffin sections of dorsal root ganglia (**a**) stomach (**b**), and kidney (**c**) of an 18-day old mouse. The *arrows* show B-Raf-positive neurons (**a**), B-Raf-positive epithelium (**b**) and B-Raf-positive tubular epithelium. The sections were counterstained with hematoxylin.

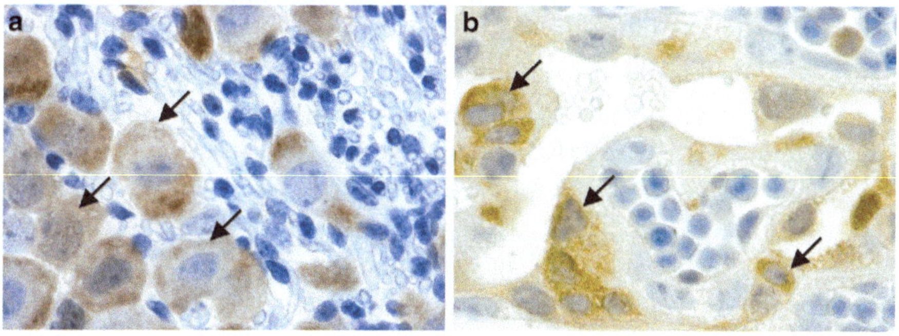

Fig. 2. PhosphoERK immunohistochemistry. 3 μm paraffin section of dorsal root ganglia from an 18-day old mouse (**a**) and of mouse placenta at embryonic day E10.5. The *arrows* highlight positive neurons (**a**) or throphoblast cells (**b**). Sections were counterstained with hematoxylin.

1. Deparaffinize sections.
   - Xylene 2 × 15 min
   - 100% Ethanol 2 × 10 min
   - 90% Ethanol 2 × 10 min
2. Wash with ddH_2O for 5 min.
3. Wash with PBS for 5 min.
4. Block endogenous peroxidase. Many cells and tissues contain endogenous peroxidase activity, which will turn over the DAB substrate, causing false positive staining. To eliminate endogenous peroxidase, the tissue section is treated with hydrogen peroxide (10 min at RT) prior to incubation with the primary antibody.
5. Wash briefly with tap water ten times.
6. Wash with ddH_2O for 5 min.
7. Antigen retrieval. Pretreatment with the antigen retrieval reagent will break the protein cross-links formed during fixation and expose hidden antigenic sites. Heat the section in 10 mM citrate buffer (pH 6.0) in microwave to unmask the antigens.

- Heat up at maximum power
- 1 × 5 min 560 W
- 10 min 400 W
- Cool down the slides for 30 min

8. Wash with TTBS 5 min.
9. Incubate for 1 h at RT in blocking buffer to saturate unspecific binding sites.
10. Incubate overnight at RT with primary antibody: affinity-purified rabbit phosphoERK1/2 (Cell Signaling, #9101) or monoclonal rabbit phosphoMek1/2 antibody (Cell Signaling #2338), working dilution 1:50 in the protein blocking solution.
11. Wash 3 × 5 min with TTBS.
12. Incubate for 30 min at RT with secondary antibody (anti-rabbit IgG, HRP-conjugated).
13. Wash 3 × 5 min with TTBS.
14. Incubate with DAB chromogen solution for 10 min at RT. DAB is oxidized at the site of peroxidase activity, forming an insoluble precipitate at the site of the secondary antibody. The brown product contrasts well when counterstained with Hematoxylin.
15. Wash with ddH_2O for 5 min at RT.
16. Counterstain with Harris Hematoxylin for 1 min.
17. Wash with ddH_2O for 5 min.
18. Wash for 10 min under running tap water.
19. Dehydration.
    - 50% Ethanol for 5 min
    - 70% Ethanol for 5 min
    - 80% Ethanol for 5 min
    - 90% Ethanol 2 × 10 min
    - 100% Ethanol 2 × 10 min
    - Xylene 2 × 10 min
20. Mount with Entelan and coverslips.

3.2.2. Analysis of B-Raf Expression by Immunohistochemistry

1. Deparaffinize and block endogenous peroxidase activity as above.
2. Antigen retrieval: incubation in 1 mM EDTA–Tween 20 buffer, pH 8.0.
    - Preheat water bath with the jar containing 1 mM EDTA–Tween 20 buffer until temperature reaches 95–100°C.

- Immerse slides in the jar. Close the lid and incubate for 20–40 min (optimal incubation time should be determined depending on the tissue sample).
- Cool down slides for 20 min.

3. Wash with 1× PBS for 5 min.
4. Incubate for 1 h at RT in blocking buffer.
5. Incubate overnight at 4°C with primary antibody: rabbit polyclonal anti-B-Raf (H-145; Santa Cruz, #sc-9002), working dilution 1:50 in PBS/Triton X-100 0.1%.
6. Wash 3 × 5 min with 1× PBS.
7. Incubation with secondary antibody, washes, incubation with chromogen, counterstaining, dehydration, and mounting as above.

3.2.3. Analyisis of ERK Activation in Oligodendrocytes In Vivo

We combine labeling with an antibody that recognizes a specific cell type (anti βIV-tubulin, expressed in differentiating oligodendrocytes), with that of an antibody recognizing activated Erk (Fig. 3). This double staining allows us to determine the activation status of the ERK pathway in a specific population of brain cells. As described above, we use the indirect method to amplify the signal. In the example below, however, the secondary antibodies are labeled with different fluorescent dyes, and this is called indirect immunofluorescence method.

1. Brain sections are treated as described in steps 1–11 of Subheading 3.2.1.
2. Incubate 45 min at RT in the dark with secondary antibody: anti-rabbit Alexa 488 (Molecular Probes, #A11008), working dilution 1:1,500 in TBS.
3. Wash 3 × 5 min with TTBS.

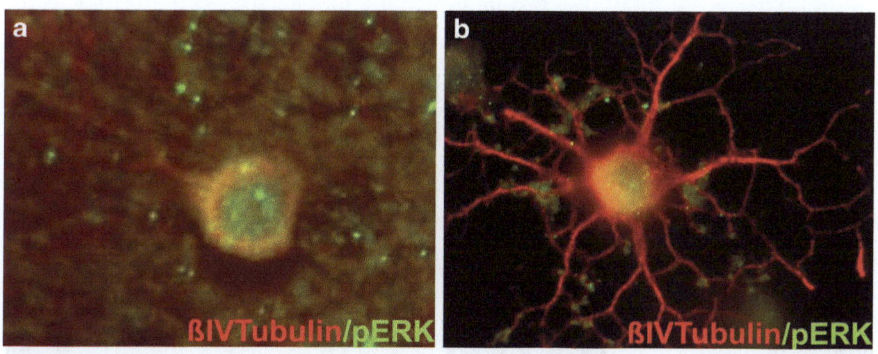

Fig. 3. ERK activation in oligodendrocytes in situ and in culture. (**a**), 3 μm paraffin section from the cortex of an 18-day-old mouse, stained with anti-βIV tubulin (oligodendrocyte-specific marker, *red*) and anti-phosphoERK (*green*). (**b**), Oligodendrocytes isolated from newborn mice double-stained with anti-βIV tubulin (*red*) and anti-phosphoERK (*green*).

4. Incubate 3 h at RT or overnight at 4°C in the dark with primary antibody: mouse monoclonal βIV-tubulin antibody (Sigma, Cat. # T7941), working dilution 1:100, diluted in the blocking buffer.
5. Wash 3 × 5 min with TTBS.
6. Incubate for 45 min at RT in the dark with secondary antibody, anti-mouse Alexa 594 (Molecular Probes, #A21201), working dilution 1:1,500 in TBS.
7. Wash 3 × 5 min with TTBS.
8. Counterstain with DAPI for 5 min at RT.
9. Wash with PBS 3 × 5 min.
10. Mount with Dako fluorescent mounting medium and coverslips.

3.3. Whole Mount Immunostaining for phosphoERK1/2 and phosphoMek 1/2

Whole mount preparations such as the one shown in Fig. 4 provide 3D information about the antigens without the need for reconstruction from sections. The major limitation of this technique is the penetration of the antibody. If this is incomplete, uneven staining or false negative staining may result. Our method, derived from a protocol by the Rossant lab (16), uses Triton X-100 to enhance penetration of the antibody. To increase sensitivity and minimize the background, we use the Dako Envision+ System, a technology based on a dextran backbone that permits the binding of a large number of HRP molecules to the secondary antibody. Important: the embryos must be fully immersed in the working solutions at all times.

1. Dissect the embryos in cold PBS and place them in Nunc 24-well multidishes. Work on ice to reduce phosphatase activity.
2. Fix the embryos in 8% PFA/PBS pH 7.4, shaking/rocking at 4°C overnight.
3. Wash 2 × 10 min with PBS/NP40.

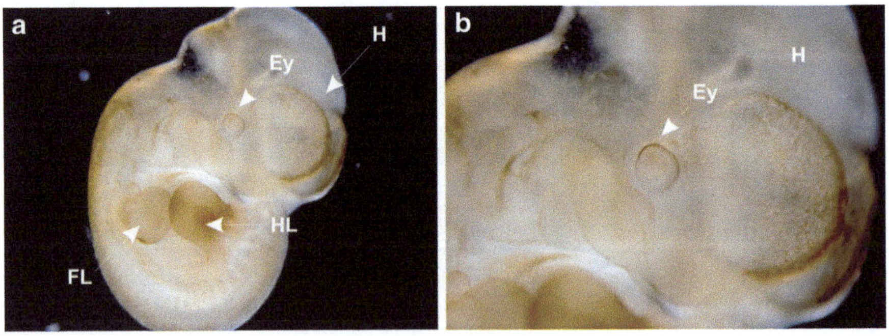

Fig. 4. (**a**) Whole mount phosphoERK staining of a mouse embryo at embryonic day E11.5 *H* head, *Ey* eye, *Fl* forelimb, *Hl* hind limb. (**b**) higher magnification.

4. Dehydrate on ice immersing the embryos sequentially in 25% methanol/75% PBS, 50% methanol/50% PBS, 75% methanol/25% PBS, and 100% methanol (each step 15 min).
5. Block endogenous peroxidase activity for 1–2 h at 4°C, with rocking/shaking.
6. Wash in methanol for 10 min on ice.
7. Rehydrate on ice with 75% methanol/25% PBS, 50% methanol/50% PBS, 25% methanol/75% PBS, and finally with PBS/0.1% Triton X-100 (each step 15 min).
8. Wash with PBSST 2 × 1 h rocking/shaking at 4°C.
9. Dilute anti-phosphoERK1/2 (Cell Signaling #9101) or anti-phosphoMek 1/2 (Cell Signaling, #2338) 1:350 in PBSST and incubate overnight rocking/shaking at 4°C.
10. Wash with PBSST 2 × 15 min at RT, followed by 5 × 1 h rocking/shaking at 4°C.
11. Incubate with secondary antibody: HRP-conjugated anti-rabbit Dako EnVision+® System (Dako Cytomation, #K4003) overnight rocking/shaking at 4°C.
12. Wash with PBSST 2 × 15 min at RT, followed by 5 × 1 h rocking/shaking at 4°C.
13. Incubate with DAB chromogen solution for 10–15 min.
14. Wash 3 × 5 min with PBS.
15. Postfix embryos in 8% paraformaldehyde for 1 h at RT.
16. Clear embryos in 50% glycerol/50% PBS rocking/shaking for 1–2 h at RT.
17. Clear embryos in 70% glycerol/30% PBS rocking/shaking for 1–2 h at RT.
18. Store in 70% glycerol/30% PBS at 4°C.
19. Take pictures with stereo microscope.

3.4. Immunofluorescence on Cells

3.4.1. Fixation

1. Fix cells in ice-cold 4% PFA for 10 min at RT.
2. Wash 2 × 5 min with PBS.

3.4.2. Double Immunofluorescent Staining for Anti-phosphoERK 1/2 and βIV-Tubulin

1. Permeabilize the cells by incubating with PBS/0.1% Triton X-100 for 5 min.
2. Wash 3 × 5 min with PBS.
3. Incubate for 1 h at RT in blocking buffer, 5% goat serum diluted in PBS.
4. Incubate overnight at 4°C with primary antibody: affinity-purified rabbit phosphoERK1/2 (Cell Signaling, #9101), 1:250 in PBS/0.1% Triton X-100.

5. Wash 3 × 5 min with PBS.
6. Incubate for 45 min at RT in the dark with secondary antibody, anti-rabbit Alexa 488 (Molecular Probes, #A11008) 1:1,500 in PBS.
7. Wash 3 × 5 min with PBS.
8. Incubate for 3 h at RT or overnight at 4°C with primary antibody: anti- βIV-tubulin (# T7941, Sigma) 1:600 PBS/0.1% Triton X-100.
9. Wash 3 × 5 min with PBS.
10. Incubate for 45 min at RT in the dark with secondary antibody anti-mouse Alexa 594 (Molecular Probes #A21201), 1:1,500 in PBS.
11. Wash 3 × 5 min with PBS.
12. Counterstain with DAPI for 5 min at RT. DAPI is the classic nuclear counterstain for immunofluorescence microscopy. DAPI is cell-permeable and emits bright fluorescence when bound to the minor groove of double-stranded DNA, staining the nuclei with minimal cytoplasmic background.
13. Wash 3 × 5 min with PBS.
14. Mount with Dako fluorescent mounting medium and coverslips.

4. Notes

1. Immunohistochemistry: antigen retrieval. When heating the sections in 10 mM citrate buffer (pH 6.0) in the microwave to unmask the antigens, minimize evaporation (would change the salt concentration in the buffer) – Never let the slides dry; this will compromise tissue morphology. For antigen retrieval in 1 mM EDTA–Tween 20 buffer, pH 8.0, the optimal incubation time should be determined depending on the tissue sample (usually between 20 and 40 min).
2. Immunohistochemistry/Immunofluorescence: choice of antibodies. This is the key to a successful experiment and the reason why the antibodies are scrupulously listed in this chapter. Many antibodies that work well in immunoprecipitation and/or immunoblotting are not specific enough for immunohistochemistry/immunofluorescence – for instance, several allegedly specific antibodies will give you a signal in knockout cells and tissues. The suitability of a given antibody for immunohistochemistry/immunofluorescence should always be tested in pilot experiments using fixed control cells and knockout cells, or if these are not available, cells in which the antigen has been knocked down by RNA interference.

References

1. Leicht DT, Balan V, Kaplun A, et al. (2007) Raf kinases: function, regulation and role in human cancer. Biochim Biophys Acta Mol Cell Res 1773(8):1196–1212.
2. Yoon S, Seger R. (2006) The extracellular signal-regulated kinase: multiple substrates regulate diverse cellular functions. Growth Factors 24(1):21–44.
3. Downward J. (2006) Cancer biology: signatures guide drug choice. Nature 439(7074):274–275.
4. Tidyman WE, Rauen KA. (2009) The RASopathies: developmental syndromes of Ras/MAPK pathway dysregulation. Curr Opin Genet Dev 19(3):230–236.
5. Galabova-Kovacs G, Matzen D, Piazzolla D, et al. (2006) Essential role of B-Raf in ERK activation during extraembryonic development. Proc Natl Acad Sci U S A 103(5):1325–1330.
6. Sobczak I, Galabova-Kovacs G, Sadzak I, Kren A, Christofori G, Baccarini M. (2008) B-Raf is required for ERK activation and tumor progression in a mouse model of pancreatic beta-cell carcinogenesis. Oncogene 27(35):4779–4787.
7. Zhong J, Li X, McNamee C, Chen AP, Baccarini M, Snider WD. (2007) Raf kinase signaling functions in sensory neuron differentiation and axon growth in vivo. Nat Neurosci 10(5):598–607.
8. Galabova-Kovacs G, Catalanotti F, Matzen D, et al. (2008) Essential role of B-Raf in oligodendrocyte maturation and myelination during postnatal central nervous system development. J Cell Biol 180(5):947–955.
9. Rubiolo C, Piazzolla D, Meissl K, et al. (2006) A balance between Raf-1 and Fas expression sets the pace of erythroid differentiation. Blood 108(1):152–159.
10. Piazzolla D, Meissl K, Kucerova L, Rubiolo C, Baccarini M. (2005) Raf-1 sets the threshold of Fas sensitivity by modulating Rok-{alpha} signaling. J Cell Biol 171(6):1013–1022.
11. O'Neill E, Rushworth L, Baccarini M, Kolch W. (2004) Role of the kinase MST2 in suppression of apoptosis by the proto-oncogene product Raf-1. Science 306(5705):2267–2270.
12. Yamaguchi O, Watanabe T, Nishida K, et al. (2004) Cardiac-specific disruption of the c-raf-1 gene induces cardiac dysfunction and apoptosis. J Clin Invest 114(7):937–943.
13. Ehrenreiter K, Piazzolla D, Velamoor V, et al. (2005) Raf-1 regulates Rho signaling and cell migration. J Cell Biol 168(6):955–964.
14. Ehrenreiter K, Kern F, Velamoor V, Galabova-Kovacs G, Sibilia M, Baccarini M. (2009) Raf-1 addiction in Ras-induced skin carcinogenesis. Cancer Cell 16(2):149–160
15. Catalanotti F, Reyes G, Jesenberger V, et al. (2009) A Mek1-Mek2 heterodimer determines the strength and duration of the Erk signal. Nat Struct Mol Biol 16(3):294–303.
16. Corson LB, Yamanaka Y, Lai KM, Rossant J. (2003) Spatial and temporal patterns of ERK signaling during mouse embryogenesis. Development 130(19):4527–4537.

Chapter 27

Mutational and Functional Analysis in Human Ras/MAP Kinase Genetic Syndromes

William E. Tidyman and Katherine A. Rauen

Abstract

The Ras/mitogen-activated protein kinase (MAPK) pathway is essential in regulation of the cell cycle, cell differentiation, growth, and cell senescence, each of which are critical to normal development. A class of developmental disorders, the "RASopathies," is caused by germline mutations in genes that encode protein components of the Ras/MAPK pathway which result in dysregulation of the pathway and profound deleterious effects on development. One of these syndromes, cardiofacio-cutaneous (CFC) syndrome, is caused by germline mutations in *BRAF*, *MAP2K1* (*MEK1*) and *MAP2K2* (*MEK2*), and possibly *KRAS* genes. Here, we describe the laboratory protocols and methods that we used to identify mutations in *BRAF* and *MEK1/2* genes as causative for CFC syndrome. In addition, we present the techniques used to determine the effect these mutations have on activity of the Ras/MAPK pathway through Western blot analysis of the phosphorylation of endogenous ERK1/2, as well as through the use of an in vitro kinase assay that measures the phosphorylation of Elk-1.

Key words: RASopathies, Cardio-facio-cutaneous syndrome (CFC), BRAF, MAP2K1 (MEK1), MAP2K2 (MEK2), Ras/MAPK, Signal transduction pathway, ERK1, ERK2, Elk-1, Kinase assay

1. Introduction

Genetic syndromes caused by germline mutations in genes that encode components of the Ras/mitogen-activated protein kinase (MAPK) pathway are collectively referred to as the RASopathies (1). Thus far, eight distinct syndromes have been described. These syndromes include Noonan syndrome, caused by mutations in the genes *PTPN11* (2); *SOS1* (3, 4), *RAF1* (5, 6)

Rony Seger (ed.), *MAP Kinase Signaling Protocols: Second Edition*, Methods in Molecular Biology, vol. 661,
DOI 10.1007/978-1-60761-795-2_27,

and *KRAS* (7) neurofibromatosis 1 (NF1), caused by mutations in the gene, neurofibromin (8–10); LEOPARD syndrome, caused by mutations in *PTPN11*, separate from those mutations that cause Noonan syndrome (11) and *RAF1* (6); gingival fibromatosis 1, caused by *SOS1* mutations (12); capillary malformation-AV malformation syndrome, caused by haploinsufficiency of *RASA1* (13); Costello syndrome, caused by mutations in *HRAS* (14); cardio-facio-cutaneous syndrome (CFC), caused by mutations in *BRAF* (15, 16), *MEK1* (*MAP2K1*) and *MEK2* (*MAP2K2*), (16) and possibly *KRAS* (15); and Legius (NF1-like) syndrome, caused by mutations in *SPRED1* resulting in haploinsufficiency (17). The majority of these mutations result in increased signal transduction down the Ras/MAPK pathway, but usually to a lesser extent than somatic mutations associated with oncogenesis. Some mutations have been shown to not only impair pathway signal transduction but also disrupt normal regulation of the pathway resulting in constitutive activation (18). Each syndrome exhibits unique phenotypic features; however, since they all cause dysregulation of the Ras/MAPK pathway, there are numerous overlapping phenotypic features between many of the syndromes. CFC syndrome is a rare developmental disorder with characteristic phenotypic features, including craniofacial features, cardiac defects, ectodermal abnormalities, growth deficiency, hypotonia, and neurocognitive delay. Approximately 80% of individuals with CFC syndrome have mutations in *BRAF*, whereas *MEK1* and *MEK2* mutations account for the remaining approximate 20% of affected mutation-positive individuals (16). In contrast, *KRAS* mutations associated with CFC syndrome appear to be very rare (15).

The laboratory techniques and methods we used to identify and characterize mutations in *BRAF*, *MEK1*, and *MEK2* genes as causative for CFC syndrome are presented here. These protocols entail the isolation of genomic DNA from the patient to DNA sequencing and analysis to indentify the presence of a mutation in these genes. In order to determine whether or not the specific mutation affects protein function and consequently Ras/MAPK pathway activity, the mutation is introduced into a cDNA encoding the protein in question and subsequently cloned into a eukaryotic expression plasmid. The cDNA plasmid is then used to transfect mammalian cells in culture, and MAPK pathway activity is assessed through the measurement of phosphorylation of the downstream pathway effectors ERK1/2 by Western blot analysis and an in vitro kinase assay. The protocols and analytical approach described here for *BRAF* and *MEK1/2* genes are also applicable for examining other genes associated with disorders of the Ras/MAPK pathway.

2. Materials

2.1. Genomic DNA Isolation and Purification

1. Oragene-DNA self-collection kit DNA Genotek Inc. (Ottawa, Ontario, Canada).
2. TE (10 mM Tris–HCl, 1 mM EDTA).

2.2. DNA Sequencing and Analysis

1. Big Dye v3.1 Cycle Sequencing Kit (Applied Biosystems, Foster City, CA).
2. Sequencer Analysis Software version 3.7 (Applied Biosystems, Foster City, CA).
3. PolyPhred Software v5.02 (University of Washington, Seattle, WA).
4. SeqScape® Software (Applied Biosystems, Foster City, CA).
5. Mutation Surveyor 3.00 (SoftGenetics LLC, State College, PA).

2.3. Insertion of Point Mutation into cDNA

1. pENTR vector (Invitrogen, Carlsbad, CA).
2. pcDNA3-FLAG vector (Invitrogen, Carlsbad, CA).
3. pcDNA3-MYC vector (Invitrogen, Carlsbad, CA).
4. Human MEK1 and MEK2 cDNAs (Origene, Rockville, MD).
5. LB agar plates.
6. QIAprep miniprep kit (Qiagen, Valencia, CA).
7. PfuUltra PCR master mix (Stratagene, La Jolla, CA).
8. *Dpn*I restriction enzyme.
9. DH10B competent cells.
10. SOC media (Invitrogen, Carlsbad, CA).

2.4. Transient Transfection and Expression of Mutated Gene

1. HEK-293T cells.
2. Lipofectamine 2000 (Invitrogen, Carlsbad, CA).
3. Dulbecco's modified Eagle's medium (DMEM).
4. OptiMEM media (Invitrogen, Carlsbad, CA).
5. Fetal bovine serum.
6. 0.05% Trypsin.
7. Phosphate buffered saline (PBS) 137 mM NaCl, 2.7 mM KCl, 4.3 mM Na_2HPO_4, 1.47 mM KH_2PO_4 (pH 7.4).
8. Six-well tissue culture plates (BD Biosciences, San Jose, CA).

2.5. Western Blot

1. Cell lysis buffer: 20 mM Tris–HCl (pH 7.5), 150 mM NaCl, 1 mM EDTA, 1 mM EGTA, 1% Triton X100, 2.5 mM sodium pyrophosphate, 1 mM β-glycerolphosphate, 1 mM Na_3VO_4, 1 μg/ml leupeptin, 1 mM PMSF.

2. Bio-Rad protein assay reagent (Bio-Rad, Hercules, CA).
3. XCell SureLock Mini Cell electrophoresis unit (Invitrogen, Carlsbad, CA).
4. 3× Sample buffer: 187.5 mM Tris–HCl (pH 6.8), 6% w/v SDS, 30% glycerol, 0.03% bromophenol blue.
5. 17 well 1 mm thick NuPAGE 4–12% Bis-Tris gel (Invitrogen, Carlsbad, CA).
6. NuPAGE MOPS SDS running buffer (Invitrogen, Carlsbad, CA).
7. NuPAGE transfer buffer (Invitrogen, Carlsbad, CA).
8. Immobilon-P PVDF membrane (Upstate/Millipore, Bellerica, MA).
9. TBST: 50 mM Tris–HCl, 150 mM NaCl, 0.05% Tween-20, pH 8.0.
10. 5% Blotto (5% nonfat dry milk in TBST).
11. Antibody diluent (1% nonfat dry milk in TBST).
12. Flag-M2 antibody (Sigma-Aldrich, Saint Louis, MO).
13. Anti-ERK antibody (Santa Cruz Biotechnology, Santa Cruz, CA).
14. Anti-p-ERK antibody (Santa Cruz Biotechnology, Santa Cruz, CA).
15. Anti-mouse-HRP antibody (Santa Cruz Biotechnology, Santa Cruz, CA).
16. Anti-rabbit-HRP antibody (Santa Cruz Biotechnology, Santa Cruz, CA).
17. Western blotting detection reagents (GE Healthcare/ Amersham Biosci.).
18. Visualizer Western Blot Detection Kit (UpstateMillipore, Billerica, MA).

2.6. Kinase Assay

1. p44/42 MAPK assay kit (Cell Signaling Technology, Danvers, MA).
2. Kinase buffer: 25 mM Tris (pH 7.5), 5 mM β-glycerolphosphate, 2 mM DTT, 0.1 mM Na_3VO_4, 10 mM $MgCl_2$.
3. Kinase buffer with substrates: Kinase buffer from above plus, 200 μM ATP and 1 μg/reaction.
4. Elk-1 fusion protein (Cell Signaling Tech., Danvers, MA).
5. p-Elk-1 antibody (Cell Signaling Tech., Danvers, MA).
6. Total Elk-1 antibody (Cell Signaling Tech., Danvers, MA).
7. LumiGLO ECL reagent (Cell Signaling Tech., Danvers, MA).

3. Methods

3.1. Genomic DNA Isolation and Purification

Genomic DNA is isolated from a saliva sample obtained from the patient and from immediate family members for comparison.

1. A patient's saliva sample is collected using the Oragene DNA self-collection kit following the manufacturer's directions.
2. Mix and incubate at 50°C for 2 h.
3. Transfer the sample to a 15-ml conical centrifuge tube.
4. Add 1/25th volume of OragenePurifer (OG-L2P) solution to the sample and mix.
5. Incubate on ice for 10 min.
6. Centrifuge at room temperature for 30 min at 3,500 × *g*.
7. Transfer supernatant to a new centrifuge tube, add an equal volume of 100% ethanol and let stand for 10 min at room temperature.
8. Centrifuge for 10 min at 3,500 × *g* and discard supernatant. Wash pellet by gently adding 1 ml of 70% ethanol and gently pour off without disturbing the pellet. Air-dry the pellet.
9. Resuspend DNA by adding 50 μl TE and gently pipet up and down. Do not vortex as this will shear the DNA. To insure the DNA is completely in solution, the sample may be incubated up to 2 days at room temperature.

3.2. DNA Sequencing and Analysis

BRAF, *MAP2K1* (*MEK1*), and *MAP2K2* (*MEK2*) coding regions are sequenced for the detection of mutations using direct, bidirectional sequencing. Briefly, exons and intronic flanking regions are amplified using standard PCR techniques. The PCR primers used for sequencing are designed to amplify all the coding exons and intronic flanking regions of *BRAF* (NM_004333.2), *MEK1* (NM_002755.2), and *MEK2* (NM_030662.2). They are given in Tables 1 and 2. For sequencing, the PCR primers are modified on the 5′ end to include M13 forward (GTAAAACGACGGCCAGT) and reverse (CAGGAAACAGCTATGACC) sequences. Bidirectional DNA sequencing is then performed using a Big Dye v3.1 Cycle Sequencing kit according to the manufacturer's directions and run on an ABI3730xl or ABI3700 capillary automatic sequencing instrument.

DNA sequence data is examined with Sequencer Analysis Software version 3.7. An electropherogram showing the most common CFC-causing mutation in *BRAF*, a heterozygous nt770A→G transition in exon 6, is compared to an electropherogram from an unaffected individual (Fig. 1). The DNA

Table 1
***BRAF* sequencing primers**

Exon	Forward	Reverse	bp
1	GCTCTCCGCCTCCCTTCC	GGCCATTGTGTGTGTTTACG	405
2	GAAGGCTGCTCACCAAACC	TCTTCCCAAATCTATTCCTAATCC	551
3	TGAAACTTAAAACCCTATCAACTGG	GATGCCTCTATTTGCATGACC	500
4	TGTAGAAATGGTGTTGTATCTGACC	CCAAATAAATCACACTCTGAATGG	515
5	TGTTACTAGCCCCTCGATAACC	GCTTACAGGTACAAGCACACG	577
6	TCTTCCTTTCACCTCTGTTTCC	AACAATCGTATGGAAGAAAAACC	589
7	TTTTAACAGTTGTTTCTGAGAATGG	CCAGGAGATCCAAAAGAAAGC	528
8	AGCAGCTTTGGCAGTATTGG	TCATCAGAGAGAAACCAGAAGC	518
9	CATTGGCAAGTGCTTCAAAA	TTGGGTTTCTCTACACATTTTTCTC	355
10	AATGAGGCCCCTTTTTGG	ATTCTGGACCAGCCTTTTCG	593
11	AGTAAGGGGATCTCTTCCTGTATCC	TGCTGTGAACAGTTTTTATTTCC	416
12	CATGGAACAAACAAGGTTGG	CATTGCATACTACTTAAAAGAATGTGG	511
13	TTTTTCTGACAACATTTTACCG	CCAGCCATTAGTTAGCATCC	417
14	GGCTTGACTGGAGTGAAAGG	AAAAGCAGGCTGTGGTATCC	506
15	GGAAAGCATCTCACCTCATCC	TGGTTTCAAAATATTCGTTTTAAGG	566
16	GAATCAGGAATGGGAAGTGG	TCTATCCTTCACGCTTACC	576
17	GAGAACCTTTGCCACAGTCC	AATTTCTAGGTGTGCCACTGC	541
18	TGCTTTCTTGTAAAGTGTGATGG	CCCGGAACAGAAAGTAAAGC	582

sequence data is subsequently analyzed using two sequence analysis programs, PolyPhred Software v5.02 and SeqScape® Software. In addition, the DNA sequence is manually reviewed using Mutation Surveyor 3.00. Further evaluation of detected nucleotide mutations consists of *S*orting *I*ntolerant *F*rom *T*olerant (SIFT; http://www.blocks.fhcrc.org/sift/SIFT.html) and screening against known databases such as: NCBI, UniProtKB/Swiss-Prot, Cosmic, JSNP (http://www.ncbi.nlm.nih.gov/SNP, http://www.ca.expasy.org/sprot, http://www.sanger.ac.uk/genetics/CGP/cosmic/, http://www.snp.ims.u-tokyo.ac.jp).

3.3. Insertion of Point Mutation into cDNA

Human *BRAF* cDNA that was used in our study was a kind gift from Martin McMahon. Human MEK1 and MEK2 cDNAs were purchased from Origene. All three cDNAs were cloned into the pENTR vector. The protocol described below outlines the insertion of point mutations identified through sequence analysis in the cDNAs by PCR. Primers containing the specific point mutation should be between 25 and 45 bases in length, with an

Table 2
***MEK1* and *MEK2* sequencing primers**

Gene	Exon	Forward	Reverse	bp
MEK1	1	GGTCCACTGAGACCGCTACC	CTGTACAGTGGCCAGGAACC	527
	2	CCTCTCTAGCCTCCCACTTTG	CAAACAGCACAAAAAGGTATTGA	452
	3	CCTGTTTCTCCTCCCTCTACC	ACACCCACCAGGAATACTGC	494
	4	AGTGCAGTGGTGTGATCTCG	CTTTCCCCTCATTGACTTCC	324
	5	GGAGGAAGGCAAATTTGTGA	GGTAGGAGTTGCTGCCTCAG	341
	6	TAACGGACTCCTTCCTGTGG	TCCTCCCTCACTTCTTGTCC	545
	7	AGGAGGCCAAATTCAAGAGG	ACAACACCCACCTGGAGACC	510
	8	TGGCTGTTTAATGTTTATTGTCC	TGGTGCTTAGTATAAAGCTGTGC	598
	9	GGATGGGGAGAGGAGATGG	ATCAGACGGGAGGGTAAAGG	252
	10	CTGTGGGCATGATACTGTGC	AACTGATGGGAGAGCAAAGG	370
	11	TTCCAAGTGCAGCACAAGC	ACACACAAATAGCCCCAAGC	461
MEK2	1	CCTGCCTCTCGGACTCG	GTGCACTCCTCGCGAACC	443
	2	CTTGAGGTCCTGAGGTCTGC	GCCTGGAGCTAATCAGAATGC	582
	3	TTGGTCTTGACCACTGTTGG	AGAAGGATCCCCTGGAAGC	380
	4	AGGCAGAACTGTCAGAAGACG	CTTGGCCACTCTCTTTCTGC	381
	5, 6	CAGCACTGTCTCGTCTCTGG	GAGAACTGGGAGGGACAGC	580
	7	GGTCATTAGCCATGGAGAGG	CACTGCTTCCAGCTCTGTCC	507
	8	CAATTTAGGCTGGCATGTGG	TGCAGCACAGTAGAAGATGG	563
	9	ATCCAGATGTCCCTCTGTGG	CTCTGGGAAAAGGAATCTGG	549
	10	CTCTCTGGTCGGAGAGATCC	TCTCATGAGGGCAAAGAAGC	490
	11	TTTGCTTTCTGTCCCGTACC	AGCTCAGGGATGTCCTCTCC	577

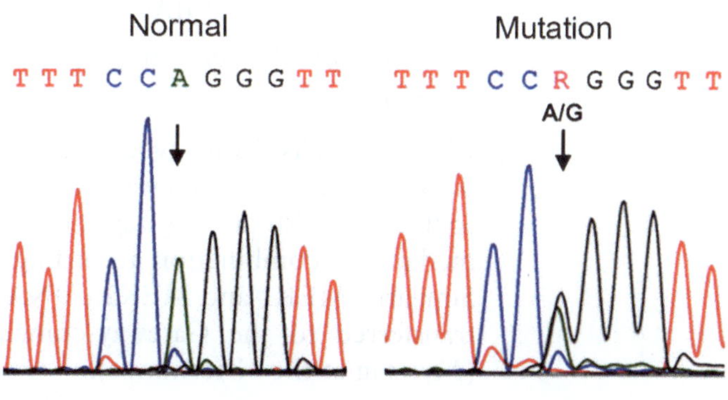

Fig. 1. Electropherograms of normal and mutated *BRAF* (**a**) Electropherogram showing normal *BRAF* nucleotide sequence. (**b**) Electropherogram of the same region in *BRAF* showing a heterozygous nt770A→G transition in exon 6, the most common CFC-causing mutation. This mutation results in a missense substitution of Q257R in the cysteine-rich domain.

approximate 40% G/C content and a melting temperature of at least 78°C.

1. 25 μl PfuUltra PCR master mix
2. 1 μl Forward primer containing point mutation (10 pmol/μl)
3. 1 μl Reverse primer containing point mutation (10 pmol/μl)
4. 1 μl cDNA plasmid template (50 ng/μl)
5. 22 μl Water, for a total reaction volume of 50 μl
   Cycling parameters:
   3 min at 94°C (one cycle)
   (15 cycles) 30 s at 94°C
   30 s at 55°C
   2 min/Kb template DNA at 72°C

Following PCR, add 1 μl *Dpn*I enzyme to the PCR product and incubate at 37°C for 2 h.

DH10B competent cells are transformed using the following protocol:

1. Add of 2 μl *Dpn*I digested PCR product to 40 μl DH10B competent cells.
2. Incubate 20 min on ice.
3. Heat shock by incubating 45 s in a 42°C water bath, followed by a 2-min incubation on ice.
4. Add 400 μl SOC media with no antibiotics and incubate 45 min at 37°C.
5. Transfer the entire volume onto LB agar plates containing the appropriate antibiotic.
6. Incubate overnight at 37°C.
7. Pick at least ten colonies and purify plasmid DNA by standard miniprep techniques or by using a suitable purification method such as QIAprep.
8. Confirm the presence of the point mutation by DNA sequencing.

BRAF cDNAs (wild type and mutants) are transferred from pENTR into the Gateway-compatible pcDNA3-FLAG vector (pcDNA3 with a Flag-tag at the N-terminus) by Gateway-mediated recombination according to manufacturer's directions. Similarly, MEK1 and MEK2 cDNAs (wild type and mutants) are transferred to the Gateway-compatible pcDNA3-MYC vector (Myc-tag at the N-terminus).

3.4. Transient Transfection and Expression of the Mutated Gene

In order to examine the functional consequences of the mutations in BRAF or MEK1 and MEK2 proteins on Ras/MAPK pathway activity, mutation harboring cDNA expression vectors generated above are transiently transfected into mammalian cells in culture. Empty expression vectors as well as wild-type cDNA expression vectors for BRAF and MEK1 and MEK2 proteins are used as control in these experiments. In addition, BRAF, MEK1, and MEK2 cDNAs harboring known activating and inactivating mutations are also used for comparison. Human embryonic kidney (HEK) 293T cells (a variant expressing SV40 Large T antigen) are used to analyze activation of the MAPK pathway by exogenously expressed genes. HEK-293T cells are particularly suitable for use because of their low basal level of activity of the endogenous MAPK pathway and their high transfection efficiency. The following transfection protocol utilizes Lipofectamine 2000 reagent and is optimized for HEK-293T cells and the expression plasmid pcDNA3-MYC.

1. HEK-293T cells are maintained in high-glucose DMEM supplemented with 10% fetal bovine serum (FBS) without antibiotics.
2. Trypsinize the maintenance culture using 0.05% trypsin. Determine the cell concentration using a hemocytometer and establish cultures for transfection in DMEM supplemented with 5% FBS at a density of 1×10^6 HEK-293T cells per well in six-well plates.
3. Incubate cells overnight at 37°C in 5% CO_2. The following day, aspirate media, rinse each well with PBS, and replace media with OptiMEM.
4. Transfect with Lipofectamine 2000 reagent by adding 2 μg of plasmid DNA per well to 100 μl of OptiMEM media in a 1.5 ml microfuge tube.
5. In a separate tube, add 5 μl of Lipofectamine 2,000–100 μl OptiMEM media, vortex and allow the tube to stand for 5 min. Combine, vortex, and incubate at room temperature for 20 min. Slowly add OptiMEM–Lipofectamine–DNA mixture to each well.
6. Twenty four hours following transfection, aspirate the media and wash each well with 2 ml PBS. Replace the media with DMEM supplemented with 5% FBS and incubate for an additional 24 h.
7. Cells are harvested 48 h following transfection. In order to use the same cell lysates in both Western blot analysis and in vitro kinase assay, the cells are harvested under nondenaturing conditions. Add 350 μl cell lysis buffer to each well, incubate for 5 min on ice, scrape cells, and transfer to a microcentrifuge tube on ice.
8. Vortex, or briefly sonicate on ice, to insure complete lysis of the nuclei.

9. Centrifuge at 14,000 × g for 10 min at 4°C and transfer supernatant to a new tube on ice.
10. The cell lysates may be used for Western blot analysis and kinase assays immediately, or snap-frozen in a dry ice/ethanol bath and stored at −80°C for later analysis.

3.5. Western Blot

Western blot analysis is carried out to determine the expression level of the transfected proteins and the endogenous activity of the Ras/MAPK by assessing the cellular levels of phospho-ERK. Exogenous protein expression is assessed through the detection of the appropriate epitope tag from the expression vector used. Exogenous BRAF expression is assessed by detection of the Flag epitope, whereas the expression levels of the MEK1/2 proteins are determined through detection of the Myc epitope. MAPK pathway activity resulting from overexpression of the BRAF- and MEK1/2-mutated proteins is assessed by examining the phosphorylation status of endogenous ERK proteins by Western blot, using total ERK and ERK-phospho-specific antibodies (Fig. 2).

1. Determine the protein concentration in the cell lysate using Bio-Rad protein assay, or another suitable method.
2. Aliquot 10 μg protein of cell lysate, 3× sample buffer, and water for a total volume of 10 μl.

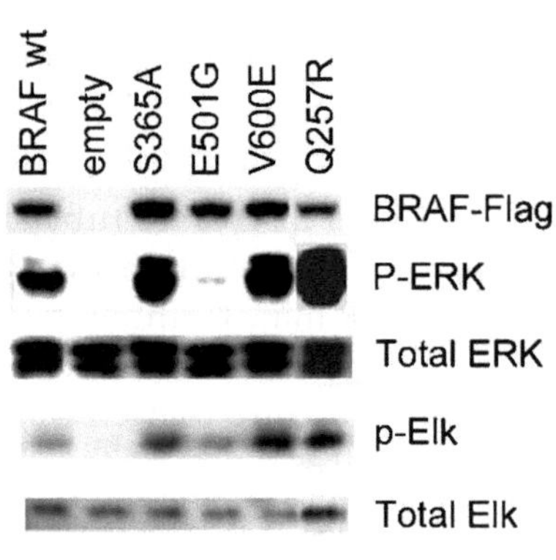

Fig. 2. Western blot and kinase assay western blot from human embryonic kidney (HEK) 293T cells over-expressing mutated BRAF proteins. HEK-293T cells were transiently transfected with, wild-type BRAF, empty vector, BRAF p.S365A, a positive control mutant which has known high activity [19], BRAF p.E501G, a CFC-causing mutation that is kinase impaired, BRAF p.V600E the most common activating mutation found in cancer and BRAF p.Q257R the most common activating CFC-causing BRAF mutation [16]. BRAF expression was assessed using an antibody to the Flag epitope. ERK1/2 phosphorylation was assayed by Western blotting using a phospho-specific antibody. The cell expressing BRAF mutants p.S365A, pV600E, and p.Q257R all exhibited increased ERK phosphorylation over that produced by BRAF wt. and the kinase impaired p.E501G mutant. Good correlation is observed between endogenous ERK phosphorylation and the in vitro kinase assay phosphorylation of Elk-1. BRAF mutants p.S365A, pV600E, and p.Q257R all produced more p-Elk-1 than BRAF wt and BRAF p.E501G.

3. Denature the protein sample by incubating at 70°C for 10 min.
4. Load 10 μl sample on a 17-well 1 mm thick NuPAGE 4–12% Bis-Tris gel.
5. Run 5 min at 50 V to stack samples, followed by 60–75 min at 125 V on an XCell SureLock Mini Cell electrophoresis unit with NuPAGE MOPS SDS running buffer.
6. Transfer proteins to Immobilon-P PVDF membrane using an XCell II Blot Module Western Blot transfer unit at 30 V for 60 min in NuPAGE transfer buffer containing 20% methanol.
7. Block the blot for 30 min in 5% Blotto.
8. Dilute primary antibodies 1:1,000 in TBST containing 1% nonfat dry milk.
9. Incubate the membrane with 10 ml of primary antibody overnight at 4°C with shaking.
10. Wash the membrane with shaking three times, for 5 min each, in TBST.
11. Dilute the appropriate HRP conjugated secondary antibody in 20 ml of TBST containing 2% nonfat dry milk. (anti-mouse-HRP 1:10,000, anti-rabbit-HRP 1:2,000).
12. Incubate at room temperature for 30–60 min.
13. Wash three times with shaking, for 5 min each, in TBST.
14. Drain the blot onto a paper towel for 10 s and place blot onto plastic wrap.
15. Detection of BRAF-Flag and total ERK is done by ECL using the Western Blotting Detection reagents from GE Healthcare/Amersham Biosciences according to the manufacturer's directions. Detection of pERK usually requires greater sensitivity; therefore, Visualizer Western Blot Detection reagent according to the manufacturer's directions.

3.6. Kinase Assay

In addition to the standard Western blot analysis, MAPK pathway activity is determined by an in vitro kinase assay that assesses the phosphorylation of one of the targets of activated ERK1/2, the transcription factor Elk-1. The kinase assay is based on the p44/42 MAPK assay kit that utilizes antibodies specific to p-Elk-1 and total Elk-1. Because of variable levels of expression of some of the transfected mutant BRAF and MEK proteins, the relative levels of BRAF expression are assessed in the preceding Western blot. The corresponding amounts of cell lysates used in the following kinase assay may be adjusted to normalize for the BRAF expression levels. However, transfections that resulted in very low levels of

BRAF expression should be repeated. The kinase assay is based on the immunoprecipitation of p-ERK1/2 followed by an in vitro kinase reaction using an Elk-1 fusion protein as substrate. Both p-Elk-1 and total Elk-1 are then assessed by Western blot analysis of the kinase reaction (Fig. 2).

1. Add 15 μl anti-p44/42 (ERK1/2) antibody/beads to 200 μl cell lysate (volume of lysate used may be adjusted based on the expression level of BRAF determined by Western blot described above). Make sure the anti-ERK1/2 antibody beads complex are in suspension before addition.
2. Incubate overnight at 4°C with rocking.
3. Centrifuge the antibody/bead complex at 14,000 × *g* for 30 s.
4. Wash pellet two times with 0.5 ml of cell lysis buffer. Flick and invert the tube vigorously to insure the bead complex is fully resuspended.
5. Wash pellet two times with 0.5 ml of kinase buffer. Flick and invert tube to insure that the antibody/bead/ERK complex is fully resuspended.
6. For each reaction, resuspend the antibody/bead/ERK complex in 50 μl of kinase buffer containing 200 μM ATP and 1 μg Elk-1 fusion protein.
7. Incubate the kinase reaction at 30°C, transfected for 30 min for MEK proteins, or for 45 min for BRAF proteins.
8. Terminate the reaction by adding 25 μl 3× sample buffer. Mix by vortexing. Make sure solution is at the bottom of the tube by microcentrifugation for 30 s.

The in vitro phosphorylation of Elk is then assessed by measuring the amount of total Elk and p-Elk in the kinase reaction by Western blot analysis.

1. Heat the kinase reaction at 100°C for 2–5 min.
2. Load 15 μl sample per lane in duplicate in a 17-well 1 mm thick NuPAGE 4–12% Bis-Tris gel.
3. Electrophoresis and protein transfer to PVDF membrane is carried out as described in the Western blotting protocol.
4. Block the membranes with TBST containing 5% nonfat dry milk for 1–2 h at room temperature.
5. Wash the membranes three times with shaking, for 5 min each, in TBST.
6. Dilute separately the anti-p-Elk-1 and the anti-total Elk-1 antibodies (1:1,000) in TBST containing 5% BSA.

7. Separately incubate the membranes in 10 ml diluted either anti-p-Elk-1 or anti-total Elk-1 antibody overnight at 4°C with gentle rocking.
8. Wash the membranes three times with shaking in TBST for 5 min each.
9. Incubate the membrane with anti-mouse HRP-conjugated secondary antibody diluted 1:2,000 in TBST.
10. Wash the membrane three times with shaking, for 5 min each, in TBST.
11. Drain the blot onto a paper towel and place onto plastic wrap. Add 10 ml LumiGLO ECL reagent to the membrane and incubate for 1 min at room temperature.
12. Drain the blot, cover with plastic wrap, and expose to ECL or X-ray film.

4. Notes

1. Heat treatment of the Oragene/saliva sample is necessary to release the DNA from the buccal epithelial cells and inactivation of nucleases.
2. Ethanol precipitation of the DNA sample is done at room temperature. Precipitation at −20°C is not recommended because impurities may coprecipitate.
3. In air drying of the DNA pellet, the DNA is dry when its appearance changes from white to clear. Excessive drying results in the DNA being very difficult to resuspend.
4. Resuspension of the DNA pellet is accomplished by gently pipetting the sample up and down. Vortexing will result in DNA shearing. If the DNA is difficult to resuspend, more TE may be added, or the sample may be heated for 1 h at 50°C.
5. The majority of DNA mutations associated with syndromes of the Ras/MAPK pathway are heterozygous that result in missense amino acid substitutions. A heterozygous substitution appears as two overlapping nucleotide peaks sharing the same nucleotide position on the electropherogram (Fig. 27.1). Note that the heterozygous peaks are usually half the height of the surrounding peaks.
6. It is important to establish that the detected heterozygous substitution is not a previously described single nucleotide polymorphism (SNP) by comparing to human SNP databases.

7. The sources of cDNAs used in these experiments, as well as the plasmids employed, will vary; consequently, the PCR cycle and subcloning protocols used may be adjusted to meet specific requirements.
8. Transfection experiments are set up using the mutation-harboring cDNAs being examined along with both positive and negative control cDNA plasmids carried out in parallel cell culture dishes.
9. Experiments should be carried out in triplicate using separate DNA samples and transfections for statistical analysis.
10. Western blot analysis is used to assess the level of expression of the transfected BRAF or MEK1/2 proteins. The level of expression of the exogenous protein may be variable due to differences in transfection efficiency and potential differences in the gene expression of the mutated cDNA. It is essential to assess the level of the transfected protein before one can establish the effect the exogenous protein has on Ras/MAPK pathway. Transfections resulting in low levels of exogenous protein expression should be repeated.
11. Western blot analysis is also used to assess how the activity of the Ras/MAPK pathway is affected by expression of the mutated exogenous BRAF or MEK1/2 proteins by comparing the levels of total ERK and p-ERK. ERK1/2 proteins are the terminal effectors of the Ras/MAPK pathway and are phosphorylated by MEK1/2 when the pathway is active.
12. Several downstream targets of active ERK1/2 have been identified, including the transcription factor Elk-1; therefore, Western blotting is a convenient means to assess the amount of active ERK1/2 in the cell lysate.
13. The in vitro kinase assay serves as a second means, distinct from the Western blot analysis of p-ERK, to assess the activity of the Ras/MAPK pathway. In addition, this assay, which is based on immunoprecipitation of p-ERK, provides greater sensitivity than standard Western blot analysis of p-ERK levels.
14. The results of the Western blot analysis on the levels of endogenous p-ERK should correlate with the in vitro phosphorylation of Elk. From these analyses, one can establish whether the mutant protein in question produces an increase or decrease in activity of the Ras/MAPK pathway compared to that produced by the wild-type protein. The example we have given in Fig. 27.2 shows that the BRAF p.Q257R, the most common CFC-causing mutation, produces greater pathway activity than wild-type BRAF but less activity than that produced by the most common cancer-causing mutation in BRAF (BRAF p.V600E).

Acknowledgments

We thank Anne Estep and Clarice Bunag for their expert technical assistance. Special thanks are due to Pablo Rodriguez-Viciana for his expert assistance in the development of some of these protocols.

References

1. Tidyman, W.E. and K.A. Rauen, (2009) The RASopathies: developmental syndromes of Ras/MAPK pathway dysregulation. *Curr Opin Genet Dev.* **19**, 230–236.
2. Tartaglia, M., et al., (2001) Mutations in PTPN11, encoding the protein tyrosine phosphatase SHP-2, cause Noonan syndrome. *Nat Genet.* **29**, 465–468.
3. Roberts, A.E., et al., (2007) Germline gain-of-function mutations in SOS1 cause Noonan syndrome. *Nat Genet.* **39**, 70–74.
4. Tartaglia, M., et al., (2007) Gain-of-function SOS1 mutations cause a distinctive form of Noonan syndrome. *Nat Genet.* **39**, 75–79.
5. Razzaque, M.A., et al., (2007) Germline gain-of-function mutations in RAF1 cause Noonan syndrome. *Nat Genet.* **39**, 1013–1017.
6. Pandit, B., et al., (2007) Gain-of-function RAF1 mutations cause Noonan and LEOPARD syndromes with hypertrophic cardiomyopathy. *Nat Genet.* **39**, 1007–1012.
7. Schubbert, S., et al., (2006) Germline KRAS mutations cause Noonan syndrome. *Nat Genet.* **38**, 331–336.
8. Cawthon, R.M., et al., (1990) Identification and characterization of transcripts from the neurofibromatosis 1 region: the sequence and genomic structure of EVI2 and mapping of other transcripts. *Genomics.* 7, 555–565.
9. Viskochil, D., et al., (1990) Deletions and a translocation interrupt a cloned gene at the neurofibromatosis type 1 locus. *Cell.* **62**, 187–192.
10. Wallace, M.R., et al., (1990) Type 1 neurofibromatosis gene: identification of a large transcript disrupted in three NF1 patients. *Science.* **249**, 181–186.
11. Digilio, M.C., et al., (2002) Grouping of multiple-lentigines/LEOPARD and Noonan syndromes on the PTPN11 gene. *Am J Hum Genet.* **71**, 389–394.
12. Hart, T.C., et al., (2002) A mutation in the SOS1 gene causes hereditary gingival fibromatosis type 1. *Am J Hum Genet.* **70**, 943–954.
13. Eerola, I., et al., (2003) Capillary malformation-arteriovenous malformation, a new clinical and genetic disorder caused by RASA1 mutations. *Am J Hum Genet.* **73**, 1240–1249.
14. Aoki, Y., et al., (2005) Germline mutations in HRAS proto-oncogene cause Costello syndrome. *Nat Genet.* **37**, 1038–1040.
15. Niihori, T., et al., (2006) Germline KRAS and BRAF mutations in cardio-facio-cutaneous syndrome. *Nat Genet.* **38**, 294–296.
16. Rodriguez-Viciana, P., et al., (2006) Germline mutations in genes within the MAPK pathway cause cardio-facio-cutaneous syndrome. *Science.* **311**, 1287–1290.
17. Brems, H., et al., (2007) Germline loss-of-function mutations in SPRED1 cause a neurofibromatosis 1-like phenotype. *Nat Genet.* **39**, 1120–1126.
18. Oishi, K., et al., (2009) Phosphatase-defective LEOPARD syndrome mutations in PTPN11 gene have gain-of-function effects during *Drosophila* development. *Hum Mol Genet.* **18**, 193–201.
19. Hmitou, I., et al., (2007) Differential regulation of B-raf isoforms by phosphorylation and autoinhibitory mechanisms. *Mol Cell Biol.* **27**, 31–43.

Part VI

Study of MAP Kinases in Specific Physiological Systems

Chapter 28

Implication of the ERK Pathway on the Post-transcriptional Regulation of VEGF mRNA Stability

Khadija Essafi-Benkhadir, Jacques Pouysségur, and Gilles Pagès

Abstract

Vascular Endothelial Growth Factor-A (VEGF-A) is one of the most important regulators of physiological and pathological angiogenesis. Constitutive activation of the ERK pathway and over-expression of VEGF-A are common denominators of tumours of different origins. Understanding VEGF-A regulation is of primary importance to better comprehend pathological angiogenesis. VEGF-A expression is regulated at all steps of its synthesis including transcription, mRNA stability, an under estimated way of VEGF regulation and translation. In this chapter, we present the link between VEGF mRNA stability through AU-rich sequences present in its 3′-untranslated region (3′-UTR) and the ERK pathway. We present several methods that have been used to demonstrate that ERKs increase VEGF mRNA half-life. This mRNA-stabilising effect is partly due to reduction of the mRNA destabilising effects of Tristetraprolin (TTP), an AU-Rich binding protein which binds to VEGF-A mRNA 3′-UTR.

Key words: Angiogenesis, VEGF, mRNA stability, mRNA-binding protein, Tristetraprolin

1. Introduction

Blood vessel growth and maturation are highly complex and coordinated processes involve numerous growth factors and related transduction pathways. Among them, vascular endothelial growth factor (VEGF), a potent and specific mitogen/survival factor for vascular endothelial cells, represents a crucial rate-limiting step in both physiological and pathological angiogenesis (1, 2). Its expression is induced in cancer cells as a result of hypoxia and multiple genetic alterations, including p53 and PTEN loss-of-function, RAS and SRC gain-of-function, and autocrine tyrosine kinase signalling pathways (3–8).

Rony Seger (ed.), *MAP Kinase Signaling Protocols: Second Edition*, Methods in Molecular Biology, vol. 661,
DOI 10.1007/978-1-60761-795-2_28,

Several studies have provided evidence that the ERK1/2 pathway is actively involved in angiogenesis mediated by VEGF (9–11). The link between signalling pathways and VEGF production was established following a study examining VEGF mRNA expression in cells transformed with the constitutively active form of ERK1/2 (9). Moreover, the conditional expression of MAP Kinase phosphatase 3/Dual Specificity phosphatase 6 (MKP-3/DUSP-6) in Ras-transformed cells blocks VEGF production and prevents the development of vascularised tumours in nude mice (12). Pathological situations are characterised by an increase in VEGF mRNA and protein levels, for example (a) Chronic myeloid leukaemia (CML) can be characterised by a chromosomal translocation between the *Bcr* and *Abl* genes which involves an increase in ERK1/2 activity (13). The inhibition of this signalling pathway by imatinib mesylate (Gleevec) standardises the rate of VEGF production in cellular models of CML and among patients in cytogenetic remission (14). (b) *Helicobacter pylori* infection involves an increase in ERK1/2 activity, VEGF mRNA and protein levels (15). The increased VEGF expression is related to an increase in gene transcription, mRNA stability and translation. A combination of these phenomena is also possible. The link between ERK1/2 signalling and VEGF transcription has been partially determined. Milanini et al. demonstrated that ERK1/2 up-regulate the transcription of the VEGF gene, an action mediated via the ERK1/2-dependent phosphorylation of Sp1 and Sp3 transcription factors. Phosphorylation sites have been identified (16, 17). The Sp transcription factors were also described as markers of tumour aggressiveness. This characteristic is probably related to their capacity to induce excessive angiogenesis throughout tumour development (18–20). Transcription of the VEGF gene is also increased following hypoxia via Hypoxia Inducible Factor (HIF-1). However, both growth factors and hypoxia induce the expression of HIF-1. This phenomenon is partly due to direct phosphorylation of HIF-1α by ERK1/2 allowing its accumulation in the nucleus (21) and an increase in its transactivation activity (22). A number of genes are post-transcriptionally modulated by ERK1/2 (23). Pharmacological inhibitors of these MAPK modules are of high relevance for future strategies aimed at pharmacological modulation of mRNA stability.

The RAS/RAF/MEK/ERK pathway activated in response to oncogenes or growth factor stimulation, contribute to VEGF expression by activating transcription factors or inactivating proteins implicated in degradation of its mRNA. A few proteins were identified as regulators of VEGF mRNA stability: HuR (24), hnRNPL (25), PAIP2 (26), Tis11b (27), and tristetraprolin (TTP) (28). Only TTP (29) and HuR proteins (30, 31) were described as phosphoproteins. This phosphorylation event modifies their activity.

In this context, we have previously shown that ERK1/2 are involved in the regulation of TTP and the consequent degradation of VEGF mRNA (28). TTP is also able to induce a reversion of the tumour phenotype of Ras-transformed cells by a defect in VEGF production and angiogenesis (28, 32). Furthermore, we have established a link between ERK1/2 activity, the angiogenic switch and VEGF mRNA stability (28).

This chapter outlines experimental protocols designed to investigate the role played by ERKs on VEGF mRNA stability. The experimental procedures also described how TTP, an RNA-binding protein, is implicated in the control of VEGF mRNA stability. TTP interacts with VEGF mRNA 3 -UTR in vitro and in vivo. ERKs induce TTP phosphorylation and stabilisation. TTP can induce VEGF mRNA degradation but the "degradative" potential of TTP is decreased by ERKs.

2. Materials

2.1. Plasmids

Plasmids containing the luciferase cDNA cloned upstream of the rat *VEGF* 3′-UTR and downstream of the thymidine kinase promoter (pLuc-FL and truncations) (27) were derived from the pSp64 plasmids described by Levy et al. (33). pGL-TK encoding luciferase was obtained from Promega (Charbonnières-les-Bains, France).

2.2. Cell Culture and Media

1. For the tetracycline-inducible "Tet on" system, we used an expression vector encoding a tetracycline-inducible myc-tagged TTP (cloned into the pCDNA4/TO/myc-His A vector, Invitrogen, Cergy Pontoise, France). This plasmid was generated by inserting a 1 kb DNA fragment corresponding to the coding region of the TTP cDNA (32) within *Eco*RI and *Age*I sites of the pCDNA4/TO.
2. S19-R443 cells (see Note 1) expressing a tetracycline-inducible vector were cultured in Dulbecco's modified Eagle's medium (DMEM) (Invitrogen, Cergy Pontoise, France) without phenol red, containing 7.5% foetal calf serum (FCS) (Dutscher, Brumath, France), penicillin (50 U/ml), streptomycin sulphate (50 μg/ml), and G418 (400 μg/ml). (All antibiotics were from Invitrogen, Cergy Pontoise, France). This medium was supplemented with blasticidin (7.5 μg/ml for selection of the tetracycline repressor) and zeocin (500 μg/ml, for selection of the gene of interest). Growth-arrested cells were obtained by total deprivation of serum for 24 h. Induction of the transgene was obtained by stimulating the cells for 24 h with 1 μg/ml tetracycline (Tet on system).
3. *Raf:ER* cells (see Note 2) were cultured in the same media as S19-R443 cells (see above).

4. 1× phosphate-buffered saline (PBS). The solution should be autoclaved.
5. Detaching solution, 10× trypsin–EDTA: 0.5% trypsin and 5.3 mM EDTA (Invitrogen), diluted in PBS to obtain 1× solution.
6. 1% (v/v) formaldehyde in PBS: dilute fresh from 37% w/w stock solution.
7. 100- and 150-mm tissue culture dishes.
8. RIPA buffer : 50 mM Tris–HCl, pH 7.5, 1% Nonidet P-40 (NP-40), 0.5% sodium deoxycholate, 0.05% SDS, 1 mM EDTA, 150 mM NaCl.
9. Lysis buffer for phosphatase alkaline treatment: 1% Triton X-100, 50 mM Tris pH:8.5, 100 mM NaCl, 0.5 mM EDTA.
10. Blocking buffer: 5% (W/V) dry milk in PBS 1×.
11. Lysis buffer for luciferase assay: 25 mM Tris-phosphate (pH 7.8), 2 mM DTT, 2 mM 1, 2-diaminocyclohexane-*N,N,N′,N′*-tetraacetic acid, 10% glycerol, and 1% Triton X-100).
12. Luciferase assay buffer: 20 mM Tricine, 1.07 mM $(MgCO_3)$ $Mg(OH)_2$-$5H_2O$ (Sigma, L'Isle d'Abeau, Chesne, France), 2.67 mM $MgSO_4$, 3.1 mM EDTA, 33.3 mM DTT, 270 μM coenzyme A (Sigma), 470 μM beetle luciferin (Promega, Charbonnières-les-Bains, France), and 530 μM ATP.

2.3. Reverse Transcription, Polymerase Chain Reaction and Real-Time PCR

1. Trizol reagent (Invitrogen, Cergy Pontoise, France).
2. 10× and 1× morpholinopropan-sulfuric acid (MOPS) running buffer: 10× MOPS: 0.2 M MOPS, 50 mM sodium acetate, 1 mM ethylenediamine tetraacetic acid (EDTA), pH 7.0.
3. 12.3 M (37%) formaldehyde, pH > 4.0.
4. Formaldehyde-loading buffer: 50 μl formamide; 17.5 μl formaldehyde; 10 μl 10× MOPS; 22.5 μl H_2O).
5. Omniscript Reverse Transcriptase (RT) Kit (Qiagen, Courtaboeuf, France).
6. Oligo $(dT)_{12-18}$ (Promega, Charbonnières-les-Bains, France).
7. RNase inhibitor (Promega, Charbonnières-les-Bains, France).
8. Oligonucleotides (5′–3′) were from Eurogentec (Liège, Belgium): VEGF, 5′-ATGGCAGAAGGAGGGCAGCAT-3′ and 5′-TTGGTGAGGTTTGATCCGCATCAT-3′; and 36B4, 5′-GCCAACCGCGAGAAGATGACCCAG-3′ and 5′-CTCGAAGTCCAGGGCGACGTAGC-3′.
9. *Taq* polymerase chain reaction (PCR) master mix kit (Qiagen, Courtaboeuf, France).
10. RT mixture: 2 μl of 10× buffer RT; 2 μl dNTP mix (5 mM of each dNTP); 0.8 μl oligo-dT primer; 1 μl RNase inhibitor

(10 U/μl); 1 μl Omniscript Reverse Transcriptase (1 μl); 2 μl template RNA and RNase-free water (QSP 20 μl).

11. Probe-based PCR reaction master mix: TaqMan Universal PCR master mix (2×; Eurogentec, Liège, Belgium).
12. TaqMan probes and PCR primers (Applied Biosystems, Foster City, CA).
13. Real-time PCR plate and cover: 96-well Clear Optical Reaction Plates (Applied Biosystems, Foster City, CA) and Optical Adhesive Covers (Applied Biosystems, Foster City, CA).
14. Real-time PCR machine: ABI PRISM 7300 Sequence Detection System (Applied Biosystems, Foster City, CA).
15. Extracting buffer: 50 mM Tris–HCl, pH 7.0, 5 mM EDTA, 10 mM dithiothreitol (DTT), and 1% SDS.
16. Additional reagents and equipment for agarose and acrylamide gels electrophoresis.

2.4. Electrophoretic Mobility Shift Assay

1. In vitro transcription kit: Riboprobe combination system SP6/T7 Kit (Promega, Charbonnières-les-Bains, France).
2. [α-^{32}P] rNTP: 3,000 Ci/mmol (Amersham Biosciences, Buckinghamshire, UK).
3. Transcription buffer: 20 μl 5× transcription buffer, 10 μl 100 mM DTT, 2.5 μl RNasin (100 U), 20 μl 10 mM rCTP and rGTP, 12 μl 100 μM rATP, 12 μl 100 μM rUTP, 5 μl [γ-^{32}P] rUTP (50 μCi), 5 μl [γ-^{32}P] rATP (50 μCi), 2.7 μl SP6 RNA polymerase enzyme mix and RNAse-free water (QSP 100 μl).
4. Mini Quick Spin RNA columns (Roche, Mannheim, Germany).
5. Binding buffer : 10 mM HEPES pH 7.6, 5 mM $MgCl_2$, 50 mM KCl, 0.5 mM EGTA, 0.5 mM DTT, 10% Glycerol, 0.1 mg/ml tRNA, 5 mg/ml heparin.
6. Native 5% polyacrylamide gel: For 30 ml, Prepare 5 ml of 30% acrylamide/bisacrylamide solution (29:1), 25 ml of 0.5× TBE (30 mM Tris, 30 mM boric acid, 0.06 mM EDTA), 30 μl TEMED, and 0.3 ml APS (10%).

3. Methods

3.1. Measurement of mRNA Stability

The relationship between increased VEGF mRNA levels and the ERK signalling pathway was first observed in Chinese hamster fibroblasts in a quiescent state or stimulated by serum. Increased VEGF mRNA levels were also observed in cells transformed by constitutively active members of the Ras/Raf/MEK/ERK pathway. The relationship between the ERK pathway and VEGF mRNA was also addressed in Chinese Hamster lung fibroblasts expressing a fusion protein called Raf:ER, composed of the catalytic domain of c-Raf and the hormone-binding domain of the oestrogen receptor. In these cells, a strong, sustained and specific activation of ERK results from tamoxifen stimulation. Hence, a dose- and time-dependent increase in VEGF mRNA has been observed by Northern blot analysis and qPCR (9).

The regulation of VEGF mRNA stability by TTP occurs through an ERK1/2-signalling-dependent mechanism. The role of the ERK1/2 pathway in this mechanism was investigated using a cell line derived from CCL39 fibroblasts (Raf-1-ER) in which rapid and exclusive activation of ERK1/2 can be achieved in response to tamoxifen (*34*).

3.1.1. 5, 6-Dichlorobenzimidazole Riboside Pulse Chase Experiments

1. Raf1-ER cells expressing myc-tagged TTP or with a non-inducible myc-tagged TTP were first serum starved, treated with/without 1 μg/ml tetracycline for 24 h, and then treated with/without 1 μM tamoxifen for 3 h.
2. After the required time, cells were treated or not with MEK inhibitor U0126 (1 μM) and 25 μg/ml of 5, 6-dichlorobenzimidazole riboside (DRB) for 0, 1, 2, or 4 h (see Note 3).
3. Total RNA and cDNA were prepared as below and the relative expression level of transcripts was quantified by real-time RT-PCR using the Taqman PCR master mix (Eurogentec, Liège, Belgium) on an ABI Prism 7300 Sequence Detection System (Applied Biosystems, Foster City, CA) according to the manufacturer's instructions.

3.1.2. Real-Time Quantitative PCR

1. Extracted cDNA (1 μl; fivefold dilution) was used as template for PCR amplification according to the manufacturer's instructions (Applied Biosystems, Foster City, CA).
2. The qPCR reaction mix (25 μl final volume) (Eurogentec, Liège, Belgium) contained the following volumes and concentrations of reactants: 1.25 μl of sense, antisense primers and probe (10 μM), 2.5 μl 10× reaction buffer, 1 μl dNTPs (5 mM), 0.125 μl Hot GoldStar enzyme (0.025 U/μl), and 2.5 μl $MgCl_2$ (50 mM).
3. Master mix and cDNA were added to a 96-well reaction plate, which was sealed, centrifuged, and placed in the real-time PCR machine, ABI Prism 7300 Sequence Detection System (Applied Biosystems, Foster City, CA).
4. The relative amount of the gene of interest (VEGF mRNA) to the internal control (36B4), and the fold stimulation, were calculated using the equation $2^{-\Delta\Delta C^T}$ (35) (see Note 4). An example of the results is shown in Fig. 1.

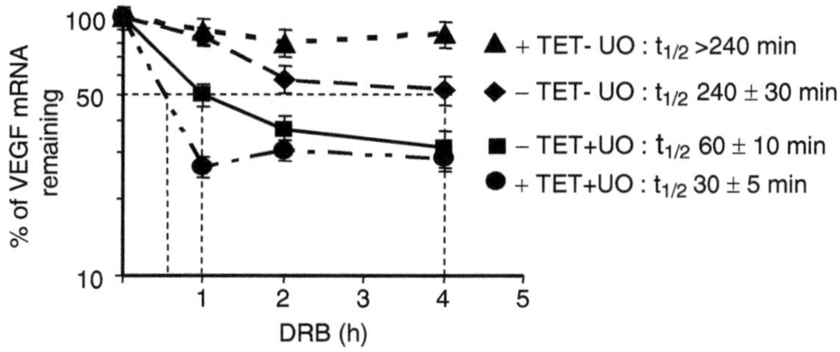

Fig. 1. TTP over-expression decreases VEGF mRNA half-life in vivo. Raf1-ER /TTP cells were serum-starved and stimulated (+) or not (–) with tetracycline (1 µg/ml) for 24 h. After 3 h of stimulation with tamoxifen, cells were incubated in the absence or presence of U0126 (10 µM) for one supplemental hour then in the presence of DRB (25 µg/ml) for the indicated times. During the DRB chase, cells were maintained or not in the presence of U0126. The amounts of VEGF mRNA remaining were quantified by real-time PCR. The values are normalised to 36B4 and the values at time 0 were taken as 100%. VEGF mRNA half-lives were deduced from the curves ($n=3$, $p<0.05$). *Comments:* In control cells, when ERKs were activated by tamoxifen treatment (–TET, –U0, maximal ERKs activity), the half-life of VEGF mRNA was 4 h ($t_{1/2}=240$ min ± 20 min; $p<0.05$). Inhibition of ERKs by U0126 (–TET, +U0, very low ERKs activity) resulted in approximately fourfold decrease of VEGF mRNA half-life ($t_{1/2}=60$ min ± 7 min; $p<0.05$). When TTP was over-expressed and ERKs were activated (+TET, –U0, maximal ERKs activity), VEGF mRNA half-life is almost the same than in absence of TTP and even superior ($t_{1/2}>240$ min; $p<0.05$). However, when TTP is over-expressed and ERKs were inactivated by U0126 (+TET, +U0, very low ERK activity), the half-life of VEGF mRNA is reduced by over eightfold ($t_{1/2}=30$ min ± 5 min; $p<0.05$).

3.2. Measurement of Protein Phosphorylation and Stability

The link between ERK and TTP protein phosphorylation was highlighted by the modification of the electrophoretic mobility and alkaline phosphatase treatment. Protein half-life was evaluated by cycloheximide pulse chase experiments followed by immunoblot analysis. An example of such experiments is shown in Figs. 2 and 3.

3.2.1. Cycloheximide Pulse Chase

1. Rafl-ER cells expressing myc-tagged TTP or with a non-inducible myc-tagged TTP were treated as described in Subheading 3.1.1.
2. After the required time, treat cells with cycloheximide (10 µg/ml) to block protein neosynthesis in the presence or absence of MEK inhibitor U0126 (10 µM).
3. Wash cells with ice-cold PBS and lyse in laemmli buffer.
4. Clarify the lysates by centrifugation at 11,228 *g* for 5 min.
5. Determine the protein concentration by BCA kit.

3.2.2. Alkaline Phosphatase Treatment

1. For experiments with Calf Intestine Phosphatase (CIP) (New England Biolabs, Beverly, Mass), cells were treated as described in Subheading 3.1.1 and lysed in lysis buffer (see Subheading 2.1).

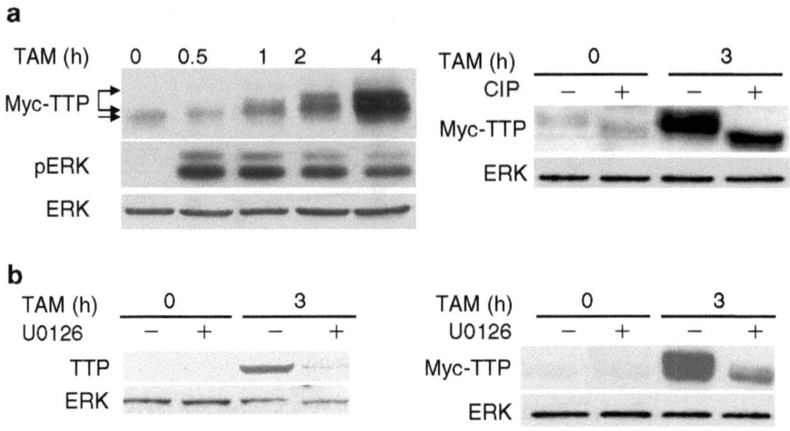

Fig. 2. Accumulation of phosphorylated forms of TTP upon ERK stimulation. Raf1-ER/TTP cells were serum-starved and stimulated with tetracycline (1 μg/ml) for 24 h prior to stimulation or not with tamoxifen for indicated times. (**a**) *Left panel:* time course of tamoxifen stimulation; *right panel:* cell extracts were treated or not with Calf Intestine Phosphatase (CIP). Protein extracts (30 μg) were then analysed by Western blotting using anti-myc, pERK, and ERK antibodies. This experiment is representative of two independent experiments. (**b**) Raf1-ER cells (*left panel*) or Raf1-ER/TTP cells (*right panel*) were serum-starved, stimulated with tamoxifen (1 μM) and incubated in the absence or presence of U0126 (10 μM) for one supplemental hour. Protein extracts (30 μg) were analysed by Western blotting using anti-TTP, myc, and ERK antibodies. This experiment is representative of two independent experiments. *Comments:* (**a**) *Left* shows that tamoxifen stimulation results in a shift in electrophoretic mobility of TTP. Phosphatase treatment induced a disappearance of the retarded bands (**a**, *right*), demonstrating that tamoxifen stimulated ERKs induced TTP phosphorylation. Upon tamoxifen treatment, TTP amounts are increased compared with growth factor-deprived cells. This induction is specific since the MEK inhibitor U0126 blocks the accumulation of both the endogenous and exogenous proteins.

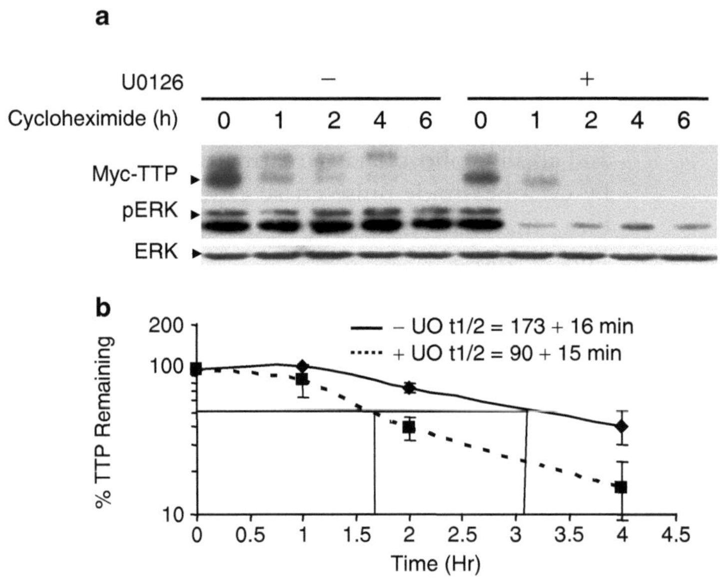

Fig. 3. TTP protein stability is regulated by the ERK pathway. Raf1-ER/TTP cells were serum-starved and stimulated (+) or not (–) with tetracycline (1 μg/ml) for 24 h prior to stimulation for 3 h with tamoxifen (1 μM). Cells were incubated in the presence of cycloheximide (10 μg/ml) for the indicated times. During the cycloheximide pulse chase, cells were maintained or not in the presence of U0126. (**a**) Protein extracts (25 μg) were analysed by Western blotting using anti-myc and pERK antibodies. ERK is shown as loading control. This experiment is representative of three independent experiments. (**b**) Quantification of three independent experiments. Arbitrary units correspond to a ratio between TTP and ERK signals. The values at time 0 were taken as 100%. Myc-TTP half-life in the presence or absence of U0126 was deduced from the logarithmic curves and is indicated on the graph (one way ANOVA, $p<0.05$). *Comments:* Cycloheximide pulse chase experiments show that phosphorylation of TTP by ERKs results in increased stability of the protein.

2. Add CIP (35 U) to the lysate and incubate for 1 h at 37°C.
3. Stop the reaction by adding Laemmli sample buffer.

3.2.3. Immunobloting

1. Boil equal amounts of proteins for 5 min at 95°C before loading onto a 10% SDS polyacrylamide gel.
2. Transfer the proteins to polyvinylidene difluoride membrane (Immobilon) following electrophoresis.
3. Block the membranes in blocking buffer for 1 h.
4. Probe the blots overnight with the primary antibody diluted in blocking buffer (1:5).
5. Detect the antigen using the HRP-conjugated secondary antibody and visualise the HRP using ECL (Pierce Biotechnology, Rockford, USA). An example of the results is shown in Figs. 2 and 3.

3.3. RNA–Protein Interactions In Vitro

3.3.1. Preparation of Radiolabelled RNA Transcripts

1. pSp64 plasmids containing the full length or different parts of the VEGF mRNA 3′-UTR were already described (33). They were linearised with the restriction enzyme *Xmn*I.
2. The digestion products were run on a 1% agarose gel, and linear plasmid DNA was purified using a gel extraction kit (Qiagen, Courtaboeuf, France) according to the manufacturer's protocol.
3. In vitro transcription is performed using Riboprobe combination system SP6/T7 kit (Promega, Charbonnières-les-Bains, France). About 2.4 μg of linearised template DNA was transcribed in Transcription buffer described in Subheading 2 (see Note 5).
4. The mixture was incubated for 2 h at 37°C.
5. Add 2.4 μl of RQ1 RNase-free DNase and incubate 15 min at 37°C.
6. To remove unincorporated nucleotides from labelled RNA, apply the mix to the G-50 Sephadex Columns (Roche, Penzberg, Germany) and spin at 371 *g*, 5 min.
7. Extract the eluate with 1 volume phenol–chloroform and chloroform, respectively.
8. Centrifuge the mix for 5 min at 15,682 *g*.
9. Precipitate the RNA from the aqueous phase with 0.5 volume of 7.5 ammonium acetate and 1 volume of isopropanol for 3 h at −80°C.
10. Pellet the RNA by centrifugation at 15,682 *g* for 5 min.
11. Wash the RNA pellet with 1 ml of 70% ethanol and allow it to air dry briefly.
12. Dissolve the RNA pellet in 20 μl nuclease-free water.
13. Aliquots were placed at −80°C until use.
14. The quality and quantity of transcripts were verified by agarose gel electrophoresis.

3.3.2. Purification of GST-Fusion Protein

The glutathione *S*-transferase (GST)-fusion protein was expressed from pGEX recombinant plasmid as previously described (29) using glutathione Sepharose 4B (BD Biosciences, Erembodegem, Belgium).

3.3.3. RNA Electrophoretic Mobility Shift Assay

1. Radiolabelled RNA transcripts (200,000 cpm/reaction) were mixed with 50 nM GST fusion proteins or GST alone, in binding buffer.
2. The reaction mixture was incubated for 30 min at 30°C.
3. 100 U of ribonuclease T1 (Roche, Mannheim, Germany) were added to the mix.
4. The mixture was incubated for 15 min at room temperature.
5. When specific or non-specific competitors were used, they were incubated for 15 min at 30°C with the proteins in binding buffer before the addition of the radiolabelled transcripts.
6. The RNA electrophoretic mobility shift assay (REMSA) samples were run on 5% native polyacrylamide gel (0.5× TBE) for 1 h 30 min at 10 mA at room temperature (see Note 6).
7. At the end of the run, separate the gel plates, adhere wet gel to 3MM Whatman paper and cover with Saran wrap.
8. Place on slab gel drier at 80°C for 1 h with vacuum.
9. Autoradiograph dried gels with intensifying screens as required (see Note 7).
10. The film was developed according to manufacturer's protocols. An example of the results obtained is shown in Fig. 4.

3.4. RNA–Protein Interaction In Cellulo

This method is adapted from chromatin IP and is very useful to study RNA–protein interactions *in cellulo*. It consists of immuno-precipitation of RNA–protein complexes using a specific antibody against a target protein. RNA is then isolated from the immuno-precipitates and analysed by RT-PCR using specific primers. We have used this approach to demonstrate that VEGF mRNA is specifically associated with TTP in vivo but ERKs do not influence its binding capacity.

3.4.1. Ribonucleoprotein Immuno-precipitation

1. Cells (10^8 cells/150-mm Petri dishes) were treated as above. RNP complexes were immuno-precipitated as previously described (37).
2. In each condition Raf-ER/TTP cells are first trypsinised with trypsin/EDTA, washed twice with 10 ml PBS, and re-suspended in 30 ml of PBS (see Note 8). Cell suspensions were then incubated for 10 min at room temperature in 1% formaldehyde (see Note 9).
3. The cross-linking reaction was stopped by the addition of glycine (pH 7.0) to a final concentration of 0.25 M, followed by incubation at room temperature for 5 min.

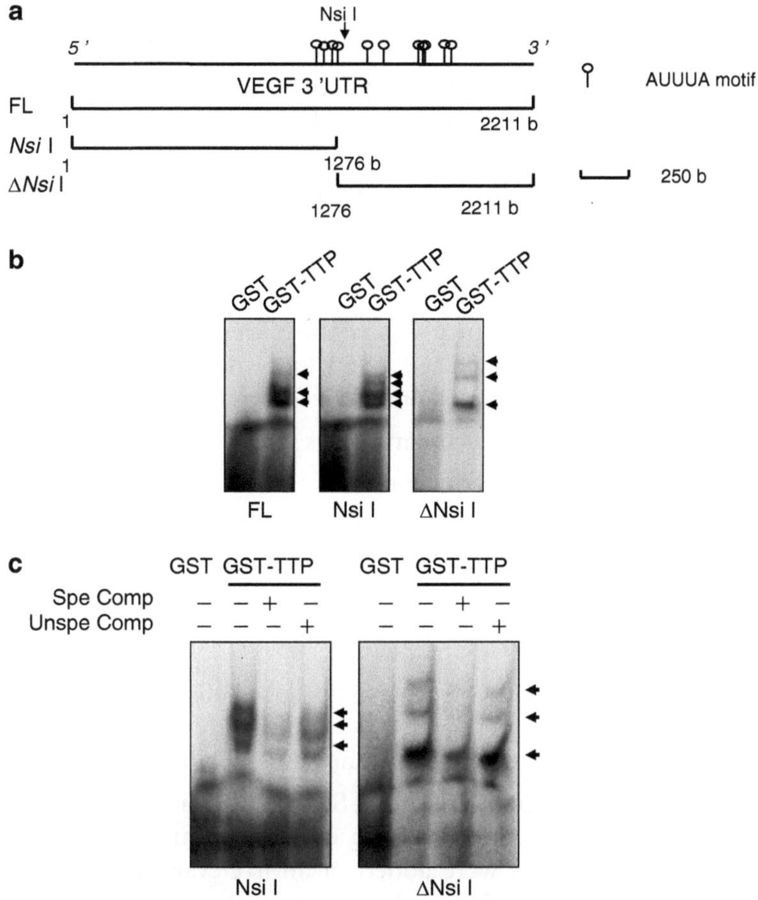

Fig. 4. TTP interacts with the VEGF mRNA 3′-UTR in vitro. (**a**) Schematic map of the VEGF 3′-UTR illustrating the templates used for generation of riboprobes for REMSA. (**b**) REMSAs were performed by incubating the radio-labelled VEGF 3′-UTR transcripts (full length, NsiI and ΔNsiI) with purified GST–TTP or GST. The *arrows point* to RNA–protein complexes. (**c**) Competition with specific competitor (100 molar excess) corresponding to un-labelled NsiI and ΔNsiI transcripts, respectively, was also performed in the presence of GST–TTP. The *arrows point* to RNA–protein complexes. *Comments:* RNA electrophoretic mobility shift assays showed that GST–TTP protein interacts specifically with full length, NsiI and ΔNsiI regions. No shifted band was observed with GST alone, demonstrating the specificity of the interaction. GST–TTP interaction can be inhibited by competition with 100× M excess of unlabeled specific probe. The same molar excess of unlabeled non-specific probe poorly affects TTP interaction demonstrating the specificity of TTP RNA binding.

4. Cells were harvested by centrifugation at 2,320 *g* for 5 min followed by two washes with ice-cold PBS (see Note 10).
5. Fixed cells were re-suspended in 3 ml of RIPA buffer containing protease inhibitors (complete, mini, EDTA-free protease inhibitor cocktail tablet, one tablet for 10 ml lysis buffer (Roche, Penzberg, Germany).
6. Cross-linked complexes were solubilised by five rounds of sonication (20 s each, in a Microson XL2007 ultrasonic homogenizer with a microprobe at an amplitude setting of 7 (output, 8–9 W) (Bioblock scientist, Illkirch, France) (see Note 11).

7. The sample (10 μl) was then examined under a microscope. It should be devoid of any intact cells.
8. Insoluble material was removed by micro-centrifugation at 18,188 *g* for 10 min at 4°C.

3.4.2. Pre-clearing Lysates

1. Solubilised cell lysate (3 ml/condition) was mixed with protein G-Sepharose beads (100 μl) (Pharmacia, Piscataway, NJ) along with non-specific competitor tRNA (final concentration of 100 μg/ml).
2. The mixture was rotated for 1 h 30 min at 4°C, and then centrifuged at 1,485 *g* for 5 min. The pre-cleared supernatant was removed and used for immuno-precipitation.
3. 20 μl of each sample was saved for a protein quantification assay.
4. Protein concentrations (BCA or other method of choice) were quantified for each sample according to the manufacturer's instructions.
5. An aliquot of this supernatant was saved for RNA extraction (input).

3.4.3. Immuno-precipitation of Cross-linked RNP Complexes

1. Protein G-Sepharose beads (100 μl/experimental data point) were incubated for 15 min with 0.5 μl of RNasin (40 U/μl; Promega, Charbonnières-les-Bains, France).
2. 100 μl of protein G-Sepharose beads were placed in an eppendorf tube and 5 μg of specific antibody (anti-myc antibody, 9E10) were added or an irrelevant antibody (anti-Pan-Ras) (VWR, Calbiochem, Fontenay-sous-Bois, France).
3. The mixture was rotated for 2 h at 4°C, and then centrifuged at 2,320 *g* for 15 s. After removal of the supernatant, the pellet was washed extensively with RIPA buffer containing protease inhibitors.
4. Then resuspended in 100 μl of cold RIPA buffer.
5. 2 mg of pre-cleared lysate was placed into an eppendorf tube and diluted with complete RIPA to achieve equal volumes in each eppendorf.
6. The lysates were mixed with the antibody-coated beads and rotated for 2 h at room temperature.
7. Samples were centrifuged for 15 s at 9,279 *g* and the supernatant was saved for RNA extraction.
8. The beads were washed six times with RIPA (500 μl per wash), then collected by centrifuging at 13,362 *g* for 30 s, and resuspending in 200 μl of extraction buffer.

3.4.4. Reversal of Cross-links and Trizol RNA Extraction

1. Resuspended beads were incubated at 70°C for 45 min to reverse the cross-links.

2. The RNA was extracted from these samples using Trizol reagent according to the manufacturer's protocol (Invitrogen, Cergy Pontoise, France).
3. 200 μl of sample was mixed with 800 μl of Trizol reagent. Then 200 μl of chloroform was added and the mixture was incubated on ice for 5 min.
4. After centrifugation at 15,682 *g* for 15 min at 4°C, the aqueous phase (colourless) was transferred to a fresh tube and precipitated with 1 ml of isopropanol (see Note 12).
5. RNA precipitates were incubated for 10 min at room temperature and centrifuged at 15,682 *g* at 4°C.
6. The supernatant was discarded and the pellet was washed with 1 ml 75% ethanol. Then centrifuged at 5,939 *g* for 5 min at 4°C.
7. The RNA pellet was air-dried (see Note 13).
8. RNA was resuspended in an equal volume of RNase-free water (see Note 14).

3.4.5. Analysis of Immuno-precipitated RNA by RT-PCR

1. Two microlitres of RNA (1/10th of the total sample) purified from the previous step were used as a template to synthesise cDNA using the Superscript First-Strand Synthesis System (Qiagen, Courtaboeuf, France), with oligo(dT) to prime first-strand synthesis according to the manufacturer's protocol.
2. Prepare the RT mixture with 2 μl template RNA.
3. The RT mixture was then incubated for 60 min at 37°C and stored on ice before proceeding with the PCR.
4. PCR was performed in a 30 μl volume containing 15 μl *Taq* master mix from (Qiagen, Courtaboeuf, France), 1 μl of each oligonucleotide (from a 10 μM stock), 11 μl sterile water and 2 μl of RT enzyme.
5. The reaction mixtures were subjected to 20, 30, and 40 cycles of amplification. The conditions depended on the oligonucleotide Tm and the length of the fragment to be amplified (see Note 15).
6. PCR products (2 μl) were analysed on 2% agarose gels and visualised by ethidium bromide staining. An example of the results obtained is shown in Fig. 5.

3.5. Transient Transfections and Luciferase Assays

The use of a luciferase reporter vector allowed us to quantify the influence of proteins that modulate gene expression. In order to assess the direct role of TTP in VEGF 3′-UTR RNA decay, we used constructs in which different domains of VEGF-A mRNA 3′-UTR were cloned downstream of the luciferase reporter gene under the control of the thymidine kinase promoter.

1. 293 Raf1-ER cells (36) seeded in 12-well dishes (10^5 cells/well) were transiently transfected by $CaPO_4$ precipitation.

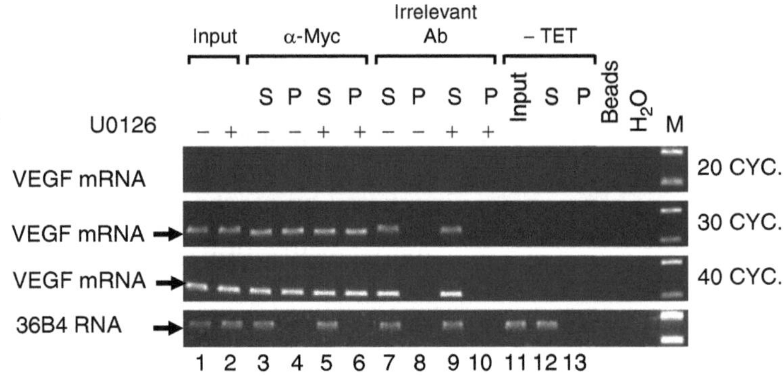

Fig. 5. TTP interacts with VEGF mRNA in vivo. Tamoxifen-treated Raf1-ER/TTP cell extracts were immuno-precipitated with no antibody, with the anti-myc antibody, or with an irrelevant antibody (anti-ras antibody) in the presence or absence of MEK inhibitor U0126 (10 μM). Equal aliquots of purified total RNA isolated from the immuno-precipitates (P), and from the supernatants (S), were assayed by reverse transcriptase-PCR to detect the VEGF (*upper panel*) and 36B4 (*lower panel*) transcripts. *Comments:* The figure shows that immuno-precipitation of TTP results in the co-precipitation of VEGF mRNA (*lane 4*, *top*). This co-precipitation is specific, because immuno-precipitation with an irrelevant antibody or using cells in which TTP was not induced does not co-precipitate VEGF mRNA (*lanes 8*, *10*, *11*, and *13*, *top*). This in vivo interaction was not influenced by the ERK pathway as immuno-precipitation with specific antibody after U0126 treatment does not modify the amount of immuno-precipitated VEGF mRNA level (*lane 6*, *top*). VEGF mRNA was also detected in total RNA prepared from Raf1-ER/TTP cell extracts used for immuno-precipitations (input, *lanes 1* and *2*). Immuno-precipitation of TTP does not lead to co-precipitation of a non-specific transcript such as 36B4 RNA (*lanes 4* and *6*, *bottom*), which was used as a negative control. These results evidenced that VEGF mRNA is specifically associated with TTP in vivo but ERKs do not influence its binding capacity.

In our experiments, we used 300 ng/well of a reporter plasmid plus 50, 100, and 150 ng of a vector encoding for TTP and the empty vector to normalise the quantity of DNA transfected to a total of 600 ng/well.

2. Eight hours after transfection, cells were serum-starved for 14 h, followed by stimulation with/without 1 μM tamoxifen for the activation of the Raf1-ER chimaera for 5 h.
3. Cells were washed twice with cold PBS.
4. Cells were lysed in lysis buffer for 15 min at room temperature.
5. The lysate was cleared by centrifugation for 10 min at 13,362 *g*.
6. 20 μl of cell extract was mixed with 20 μl of luciferase assay reagent.
7. Results were quantified with a MicroBeta TRILUX luminescence counter (Beckman, Paris, France) and the luciferase activity was normalised using total protein extracts. Protein concentration was measured using the Bradford protein assay kit (BIO-RAD, Marnes-la-Coquette, France) with bovine serum albumin as a standard (see Note 16). An example of the results obtained is shown in Fig. 6.

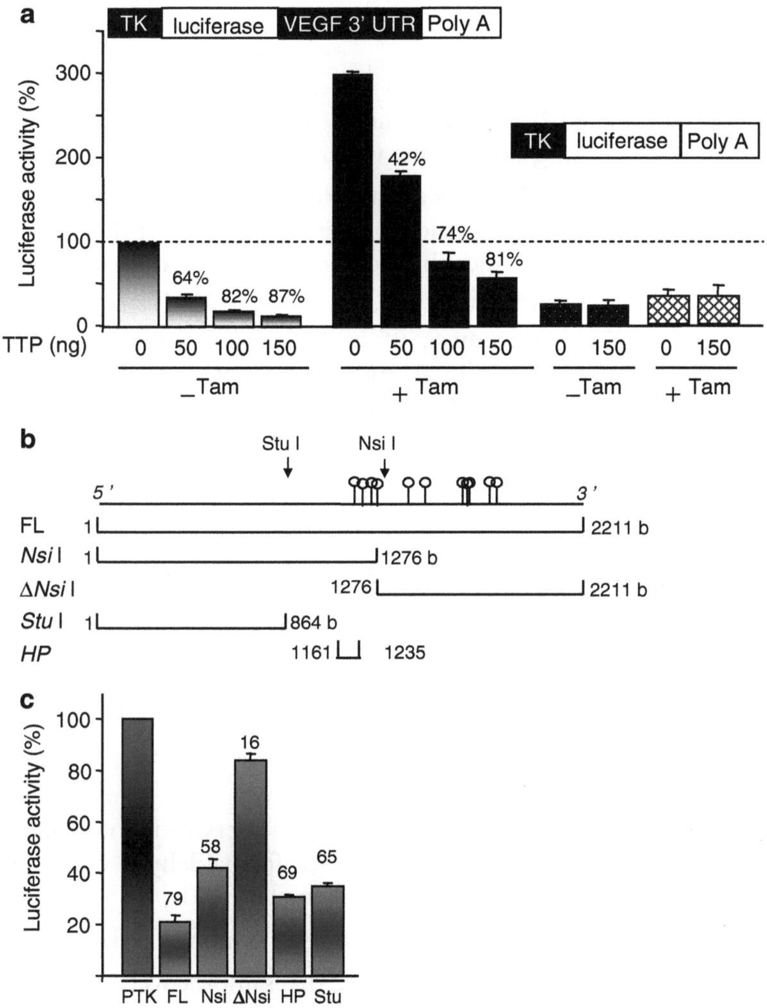

Fig. 6. (**a**) TTP induces an inhibitory effect on VEGF mRNA-3′UTR-luciferase reporter constructs: attenuation of this effect by ERKs. 293 Raf1-ER cells were co-transfected with a VEGF mRNA-3′UTR-luciferase reporter construct and different amounts of TTP expression plasmids in the presence or absence of tamoxifen (1 μM). Luciferase assays were conducted 24 h after transfection. Relative luciferase activity was normalised to total protein amount. The luciferase counts obtained with the construct luciferase/VEGF 3′UTR in the absence of TTP was taken as the value of reference (100%). The percentages of inhibition exerted by TTP are also indicated on the figure. Results are reported as the mean ± SE of three independent experiments performed in triplicate. (**b**) Mapping of different sequences in VEGF mRNA 3′-UTR used for TTP inhibition of reporter gene activity. (**c**) Transfections were performed as described in (**a**) in the presence or absence of 100 ng of TTP. The luciferase values obtained for each construct in the absence of TTP were taken as the value of reference (100%). Plotted are the percentages of remaining luciferase activity in the presence of TTP. Results are reported as the mean ± SE of three independent experiments performed in triplicate. *Comments:* TTP induced a dose-dependent inhibitory effect on luciferase activity on the full-length 3′-UTR, reaching 80% inhibition in serum deprived cells. Tamoxifen stimulation results in an increase of luciferase activity demonstrating that ERKs are implicated in VEGF mRNA stability through its 3′-UTR. In the presence of tamoxifen, TTP is still able to reduce luciferase activity but to a lesser extent (**a**). Thus, ERKs partially prevents TTP-dependent degradation of VEGF 3′-UTR reporter mRNA. (**c**) Indicated that, whereas TTP had no effect on luciferase activity mediated by a control vector, it principally inhibits the activity of the VEGF 3′-UTR throughout the NsiI region. The Luc-HP construct, which contains five AU consensus domains, similarly responded to TTP, although slightly less than the full-length construct. Deletion of a large fragment of the VEGF 3′-UTR containing all ARE motifs (*StuI*), did not abrogate the inhibitory effect of TTP on reporter gene activity. This suggests the existence of an indirect mode of regulation that does not depend on TTP binding.

4. Notes

1. S19-R443 cells are a derivative of CCL39 Chinese hamster lung fibroblasts, which stably express a tetracycline repressor.
2. *Raf:ER* cells are a derivative of CCL39 fibroblasts that stably express a fusion protein comprised of the catalytic domain of Raf-1 and the hormone-binding domain of the oestrogen receptor (34).
3. Measurement of mRNA decay following inhibition of the transcriptional machinery with DRB permits changes in mRNA half-life to be determined, indicating whether mRNA stability is altered.
4. Each individual induction is the mean of two amplifications. Each difference in gene expression was estimated within the limits of a 95% confidence interval (95% CI).
5. All reagents should be kept on ice before incubation.
6. Before loading samples, the gel should be pre-run for 1 h at 10 mA.
7. The exposure time should be adjusted to obtain desired signal.
8. Two 150-mm Petri dishes are required for each experimental data point.
9. It was verified that the buffers did not contain primary or secondary amines, since formaldehyde could crosslink these amines to proteins or nucleic acids.
10. At this stage, the cell pellet can be stored at –80°C until used.
11. Between each cycle, the samples were kept in an ice-water bath for at least 2 min.
12. At this stage, the isopropanol mixture can be stored at –20°C until used.
13. Over-dried RNA is difficult to resuspend.
14. At this point, RNA can be either conserved at –20°C or incubated for 5 min at 65°C and set on ice.
15. It is important to set up PCR conditions in such a way that the reaction is not performed in saturating conditions but in the exponential phase of PCR.
16. We have not included a "normalisation" plasmid in our transient transfection experiments. This is due to the fact that the different classical control plasmids used for normalising luciferase assays experiments were affected by the different conditions tested. The βGal, or Renilla luciferase-expressing plasmids and more specifically pRL-TK, were responsive to

the ERK pathway. It is important to stress that different publications have reported normalisation problems linked to the use of pRL-TK (38–42). In order to avoid a misinterpretation of the data we have preferred to repeat the experiment numerous times with different preparations of plasmids and normalised the luciferase counts to the amount of proteins.

Acknowledgements

This work was supported by the French association for cancer research (ARC, 4932), by the Association for International Cancer Research (AICR) and the French National Institute of Cancer (INCa).

References

1. Folkman, J., and Shing, Y. (1992) Angiogenesis. *J Biol Chem* **267**, 10931–4.
2. Folkman, J., and D'Amore, P. A. (1996) Blood vessel formation: what is its molecular basis? *Cell* **87**, 1153–5.
3. Mukhopadhyay, D., Tsiokas, L., Zhou, X. M., Foster, D., Brugge, J. S., and Sukhatme, V. P. (1995) Hypoxic induction of human vascular endothelial growth factor expression through c-Src activation. *Nature* **375**, 577–581.
4. Arbiser, J. L., Moses, M. A., Fernandez, C. A., Ghiso, N., Cao, Y., Klauber, N., Frank, D., Brownlee, M., Flynn, E., Parangi, S., Byers, H. R., and Folkman, J. (1997) Oncogenic H-ras stimulates tumor angiogenesis by two distinct pathways. *Proc Natl Acad Sci USA* **94**, 861–6.
5. Petit, A. M., Rak, J., Hung, M. C., Rockwell, P., Goldstein, N., Fendly, B., and Kerbel, R. S. (1997) Neutralizing antibodies against epidermal growth factor and ErbB-2/neu receptor tyrosine kinases down-regulate vascular endothelial growth factor production by tumor cells in vitro and in vivo: angiogenic implications for signal transduction therapy of solid tumors. *Am J Pathol* **151**, 1523–30.
6. Akagi, Y., Liu, W., Zebrowski, B., Xie, K., and Ellis, L. M. (1998) Regulation of vascular endothelial growth factor expression in human colon cancer by insulin-like growth factor-I. *Cancer Res* **58**, 4008–14.
7. Ellis, L. M., Staley, C. A., Liu, W., Fleming, R. Y., Parikh, N. U., Bucana, C. D., and Gallick, G. E. (1998) Down-regulation of vascular endothelial growth factor in a human colon carcinoma cell line transfected with an antisense expression vector specific for c-src. *J Biol Chem* **273**, 1052–7.
8. Yen, L., You, X. L., Al Moustafa, A. E., Batist, G., Hynes, N. E., Mader, S., Meloche, S., and Alaoui-Jamali, M. A. (2000) Heregulin selectively upregulates vascular endothelial growth factor secretion in cancer cells and stimulates angiogenesis. *Oncogene* **19**, 3460–9.
9. Milanini, J., Vinals, F., Pouysségur, J., and Pagès, G. (1998) p42/p44 MAP Kinase module plays a key role in the transcriptional regulation of vascular endothelial growth factor gene in fibroblasts. *J Biol Chem* **273**, 18165–72.
10. Rak, J., Mitsuhashi, Y., Sheehan, C., Tamir, A., Vioria-Petit, A., Filmus, J., Mansour, S. J., Ahn, N. G., and Kerbel, R. S. (2000) Oncogenes and tumor angiogenesis: differential modes of vascular endothelial growth factor up-regulation in ras transformed epithelial cells and fibroblasts. *Cancer Res* **60**, 490–498.
11. Pagès, G., Milanini, J., Richard, D. E., Berra, E., Gothie, E., Vinals, F., and Pouysségur, J. (2000) Signaling angiogenesis via p42/p44 MAP kinase cascade. *Ann N Y Acad Sci* **902**, 187–200.
12. Marchetti, S., Gimond, C., Roux, D., Gothie, E., Pouysségur, J., and Pagès, G. (2004) Inducible expression of a MAP kinase phosphatase-3-GFP chimera specifically blunts fibroblast growth and ras-dependent tumor formation in nude mice. *J Cell Physiol* **199**, 441–50.

13. Aguayo, A., Kantarjian, H., Manshouri, T., Gidel, C., Estey, E., Thomas, D., Koller, C., Estrov, Z., O'Brien, S., Keating, M., Freireich, E., and Albitar, M. (2000) Angiogenesis in acute and chronic leukemias and myelodysplastic syndromes. *Blood* **96**, 2240–5.
14. Legros, L., Bourcier, C., Jacquel, A., Mahon, F. X., Cassuto, J. P., Auberger, P., and Pagès, G. (2004) Imatinib mesylate (STI571) decreases the vascular endothelial growth factor plasma concentration in patients with chronic myeloid leukemia. *Blood* **104**, 495–501.
15. Strowski, M. Z., Cramer, T., Schafer, G., Juttner, S., Walduck, A., Schipani, E., Kemmner, W., Wessler, S., Wunder, C., Weber, M., Meyer, T. F., Wiedenmann, B., Jons, T., Naumann, M., and Hocker, M. (2004) *Helicobacter pylori* stimulates host vascular endothelial growth factor-A (vegf-A) gene expression via MEK/ERK-dependent activation of Sp1 and Sp3. *FASEB J* **18**, 218–20.
16. Milanini-Mongiat, J., Pouyssegur, J., and Pagès, G. (2002) Identification of two Sp1 phosphorylation sites for p42/p44 mitogen-activated protein kinases: their implication in vascular endothelial growth factor gene transcription. *J Biol Chem* **277**, 20631–9.
17. Pagès, G. (2007) Sp3-Mediated VEGF regulation is dependent on phosphorylation by extra-cellular signals regulated kinases (Erk). *J Cell Physiol* **213**, 454–63.
18. Abdelrahim, M., Smith, R., 3rd, Burghardt, R., and Safe, S. (2004) Role of Sp proteins in regulation of vascular endothelial growth factor expression and proliferation of pancreatic cancer cells. *Cancer Res* **64**, 6740–9.
19. Abdelrahim, M., Baker, C. H., Abbruzzese, J. L., and Safe, S. (2006) Tolfenamic acid and pancreatic cancer growth, angiogenesis, and Sp protein degradation. *J Natl Cancer Inst* **98**, 855–68.
20. Essafi-Benkhadir, K., Grosso, S., Puissant, A., Robert, G., Essafi, M., Deckert, M., Chamorey, E., Dassonville, O., Milano, G., Auberger, P., and Pagès, G. (2009) Dual role of Sp3 transcription factor as an inducer of apoptosis and a marker of tumour aggressiveness. *PLoS One* **4**, e4478.
21. Mylonis, I., Chachami, G., Samiotaki, M., Panayotou, G., Paraskeva, E., Kalousi, A., Georgatsou, E., Bonanou, S., and Simos, G. (2006) Identification of MAPK phosphorylation sites and their role in the localization and activity of hypoxia-inducible factor-1alpha. *J Biol Chem* **281**, 33095–106.
22. Richard, D. E., Berra, E., Gothie, E., Roux, D., and Pouysségur, J. (1999) p42/p44 mitogen-activated protein kinases phosphorylate hypoxia-inducible factor 1alpha (HIF-1alpha) and enhance the transcriptional activity of HIF-1. *J Biol Chem* **274**, 32631–7.
23. Andoh, A., Shimada, M., Bamba, S., Okuno, T., Araki, Y., Fujiyama, Y., and Bamba, T. (2002) Extracellular signal-regulated kinases 1 and 2 participate in interleukin-17 plus tumor necrosis factor-alpha-induced stabilization of interleukin-6 mRNA in human pancreatic myofibroblasts. *Biochim Biophys Acta* **1591**, 69–74.
24. Levy, N. S., Chung, S., Furneaux, H., and Levy, A. P. (1998) Hypoxic stabilization of vascular endothelial growth factor mRNA by the RNA-binding protein HuR. *J Biol Chem* **273**, 6417–23.
25. Shih, S. C., and Claffey, K. P. (1999) Regulation of human vascular endothelial growth factor mRNA stability in hypoxia by heterogeneous nuclear ribonucleoprotein L. *J Biol Chem* **274**, 1359–65.
26. Onesto, C., Berra, E., Grépin, R., and Pagès, G. (2004) Poly(A)-binding protein-interacting protein 2, a strong regulator of vascular endothelial growth factor mRNA. *J Biol Chem* **279**, 34217–26.
27. Ciais, D., Cherradi, N., Bailly, S., Grenier, E., Berra, E., Pouysségur, J., Lamarre, J., and Feige, J. J. (2004) Destabilization of vascular endothelial growth factor mRNA by the zinc-finger protein TIS11b. *Oncogene* **23**, 8673–80.
28. Essafi-Benkhadir, K., Onesto, C., Stebe, E., Moroni, C., and Pagès, G. (2007) Tristetraprolin inhibits Ras-dependent tumor vascularization by inducing vascular endothelial growth factor mRNA degradation. *Mol Biol Cell* **18**, 4648–58.
29. Cao, H., Deterding, L. J., Venable, J. D., Kennington, E. A., Yates, J. R., 3rd, Tomer, K. B., and Blackshear, P. J. (2006) Identification of the anti-inflammatory protein tristetraprolin as a hyperphosphorylated protein by mass spectrometry and site-directed mutagenesis. *Biochem J* **394**, 285–97.
30. Doller, A., Huwiler, A., Muller, R., Radeke, H. H., Pfeilschifter, J., and Eberhardt, W. (2007) Protein kinase C alpha-dependent phosphorylation of the mRNA-stabilizing factor HuR: implications for posttranscriptional regulation of cyclooxygenase-2. *Mol Biol Cell* **18**, 2137–48.
31. Doller, A., Akool el, S., Huwiler, A., Muller, R., Radeke, H. H., Pfeilschifter, J., and Eberhardt, W. (2008) Posttranslational modification of the AU-rich element binding protein HuR by protein kinase Cdelta elicits angiotensin II-induced stabilization and nuclear export of cyclooxygenase 2 mRNA. *Mol Cell Biol* **28**, 2608–25.

32. Stoecklin, G., Gross, B., Ming, X. F., and Moroni, C. (2003) A novel mechanism of tumor suppression by destabilizing AU-rich growth factor mRNA. *Oncogene* **22**, 3554–61.
33. Levy, A. P., Levy, N. S., and Goldberg, M. A. (1996) Post-transcriptional regulation of vascular endothelial growth factor by hypoxia. *J Biol Chem* **271**, 2746–53.
34. Lenormand, P., McMahon, M., and Pouysségur, J. (1996) Oncogenic Raf-1 activates p70 S6 kinase via a mitogen-activated protein kinase-independent pathway. *J Biol Chem* **271**, 15762–8.
35. Schmittgen, T. D., and Livak, K. J. (2008) Analyzing real-time PCR data by the comparative C(T) method. *Nat Protoc* **3**, 1101–8.
36. Cagnol, S., Van Obberghen-Schilling, E., and Chambard, J. C. (2006) Prolonged activation of ERK1,2 induces FADD-independent caspase 8 activation and cell death. *Apoptosis* **11**, 337–46.
37. Niranjanakumari, S., Lasda, E., Brazas, R., and Garcia-Blanco, M. A. (2002) Reversible cross-linking combined with immunoprecipitation to study RNA-protein interactions in vivo. *Methods* **26**, 182–90.
38. Thavathiru, E., and Das, G. M. (2001) Activation of pRL-TK by 12S E1A oncoprotein: drawbacks of using an internal reference reporter in transcription assays. *Biotechniques* **31**, 528–30, 532.
39. Matuszyk, J., Ziolo, E., Cebrat, M., Kochel, I., and Strzadala, L. (2002) Nurr1 affects pRL-TK but not phRG-B internal control plasmid in genetic reporter system. *Biochem Biophys Res Commun* **294**, 1036–9.
40. Sims, R. J., 3rd, Liss, A. S., and Gottlieb, P. D. (2003) Normalization of luciferase reporter assays under conditions that alter internal controls. *Biotechniques* **34**, 938–40.
41. Mulholland, D. J., Cox, M., Read, J., Rennie, P., and Nelson, C. (2004) Androgen responsiveness of Renilla luciferase reporter vectors is promoter, transgene, and cell line dependent. *Prostate* **59**, 115–9.
42. Ho, C. K., and Strauss, J. F., 3rd. (2004) Activation of the control reporter plasmids pRL-TK and pRL-SV40 by multiple GATA transcription factors can lead to aberrant normalization of transfection efficiency. *BMC Biotechnol* **4**, 10.

Chapter 29

Studies on MAP Kinase Signaling in the Immune System

Hongbo Chi and Richard A. Flavell

Abstract

The primary function of the immune system is to protect the organism from invading pathogens. In response to pathogen invasion, multiple signaling pathways are activated in immune cells, leading to diverse immune defense mechanisms. Chief among these pathways is the activation of MAPKs, which are crucial for transcriptional and nontranscriptional responses of the immune system. Here we describe protocols to study the roles of MAPKs in T lymphocytes, a cell type central for immune regulation. Specifically, we describe flow cytometry-based assays to analyze the roles of MAPKs in the development, homeostasis, proliferation, and apoptosis of murine T cells. We also describe methods to examine the activation of MAPKs in T cells.

Key words: MAPK, T cells, Flow cytometry, Phosphorylation, Immune regulation

1. Introduction

The primary function of the immune system is to protect the organism from invading pathogens. To perform this function, the mammalian immune system has developed two components: innate immunity and adaptive immunity. Both arms of immunity recognize invading pathogens as nonself, although they utilize different receptor systems. Innate immune cells, including macrophages and dendritic cells, utilize pattern-recognition receptors (PRRs) such as Toll-like receptors (TLRs) to recognize pathogens rapidly, which provide the first line of defense against invading pathogens (1). In adaptive immunity, T and B lymphocytes recognize nonself through antigen-specific receptors, such as T-cell receptors and immunoglobulins. Adaptive immune responses are slow to develop, and therefore generally mediate protection only after several days or more postinfection.

Rony Seger (ed.), *MAP Kinase Signaling Protocols: Second Edition*, Methods in Molecular Biology, vol. 661,
DOI 10.1007/978-1-60761-795-2_29, © Springer Science+Business Media, LLC 2010

While pathogen recognition begins at the receptor level, it is the signaling components downstream of each receptor and the way they interact with each other that ultimately determine the specific transcriptional response and immunological outcome. Among the central pathways activated in immune cells are the MAP kinases (MAPKs), a family of serine/threonine kinases that include the extracellular signal-regulated kinases (ERK1/2), c-Jun NH_2-terminal kinases (JNK-1, JNK-2, and JNK-3), p38 (p38α, p38β, p38γ and p38δ), and ERK5. MAPKs contain the signature sequence –TXY–, where T and Y are threonine and tyrosine, and X is glutamate, proline or glycine, in ERK, JNK, or p38, respectively (2). Phosphorylation of both the threonine and tyrosine within this signature sequence is required for MAPK activation. Phosphorylation of MAPKs is achieved via a signaling cascade involving a MAPK kinase (MAPKK or MAP2K) that is responsible for phosphorylation of the appropriate MAPK, and a MAPK kinase kinase (MAPKKK or MAP3K) that phosphorylates and activates MAPKK (2).

Recent gene knockout studies have revealed that MAPKs play an important function in both the innate and adaptive immune systems. In particular, T lymphocytes (T cells), a cell type central for adaptive immunity, utilize the MAPKs for proper regulation of their development, activation and function. For example, deficiency in TAK1, a MAP3K for JNK and p38, results in impaired development, survival and homeostasis of T cells under steady-state levels, as well as diminished responses of mature T cells to antigen and cytokine stimulation (3). Combined deletion of both ERK1 and ERK2 causes defects in multiple steps of T-cell development in the thymus (4). In contrast, deficiency in JNK1 or JNK2 leads to impaired T-cell activation and differentiation (5). Given that MAPK signaling is best characterized in T cells, we center our discussions on the roles of MAPKs in the development, homeostasis, proliferation, and apoptosis of T cells. Notably, the use of flow cytometry has greatly advanced our understanding of MAPK signaling in T cells over the last decade. Therefore, we mainly focus on flow cytometry-based approaches to address the physiological and biochemical processes in T cells. First, we describe methods to isolate and purify T cells from murine lymphoid organs. We then describe assays to analyze the roles of MAPK signaling in the development, homeostasis, activation, and apoptosis of T cells. Finally, we present flow cytometry-based protocols to study MAPK activation in T cells. As compared with more conventional approaches such as Western blotting, the flow cytometry-based system is rapid, sensitive, and provides single-cell resolution. By combining surface marker expression analysis, it enables the evaluation of mechanistic and kinetic information of subsets-specific signaling.

Notably, although we describe the isolation and analysis of murine T cells, similar approaches can be implemented to study human lymphocytes.

2. Materials

2.1. Isolation of Lymphocytes and Purification of Peripheral CD4 T Cells

1. HBSS/1% FBS: Hank's balanced salt solution (HBSS, Invitrogen) supplemented with 1% fetal bovine serum (FBS).
2. MACS buffer: 1× PBS supplemented with 0.5% Bovine serum albumin (BSA) and 2 mM EDTA. Filter to sterilize and store at 4°C.
3. ACK lysing buffer (Invitrogen).
4. Cell strainer (BD falcon).
5. Mouse CD4 (L3T4) microbeads (Miltenyi Biotec).
6. MACS MS columns (Miltenyi Biotec; capacity: 10^7 positive cells).

2.2. Analysis of MAPK Functions in T-Cell Development and Homeostasis by Flow Cytometry

1. FACS-staining buffer: 1× PBS, 1% BSA, and 0.1% sodium azide.
2. Antibodies suitable for flow cytometry (BD Biosciences or ebiosciences).

2.3. Analysis of MAPK Functions in T-Cell Proliferation by CFSE Dilution Assays

1. T-cell culture medium: EHAA Clicks Medium (Irvine Scientific) supplemented with 10% fetal bovine serum (FBS) and antibiotics (Penicillin /streptomycin).
2. Anti-CD3 (2C11) (BD Biosciences).
3. Anti-CD28 (37.51) (BD Biosciences).
4. CFSE (5-(and-6)-carboxyfluorescein succinimidyl ester) (Invitrogen).

2.4. Analysis of MAPK Functions in T-Cell Apoptosis by Annexin V Staining

1. 10× Annexin V Binding Buffer (BD Biosciences). Dilute to 1× prior to use.
2. FITC-Annexin V (BD Biosciences) or PE-Annexin V (BD Biosciences).
3. 7-AAD (Sigma).

2.5. Analysis of MAPK Activation by Flow Cytometry of Phosphorylated Proteins

1. Mouse CD3/CD28 T-cell expander (Invitrogen).
2. Phosflow Lyse/Fix Buffer I (BD Biosciences).
3. Phosflow Perm Buffer used III (BD Biosciences).

3. Methods

3.1. Isolation of Lymphocytes and Purification of Peripheral CD4 T Cells

1. Harvest the spleen and peripheral lymph nodes from a mouse and place in HBSS/1% FBS.
2. Place the lymphoid organs in a cell strainer sitting above a 50 ml tube, and smash the organs using a syringe piston. Wash the lymphoid organs with HBSS/1% FBS and continue the smashing if necessary. Pellet the cells by centrifugation (300 × *g*) for 5–10 min, and remove supernatant.
3. To lyse red blood cells in the murine spleen, add 2.0 ml of ACK lysing solution to the spleen cell pellet from one mouse, resuspend thoroughly, and incubate at room temperature for 2–5 min until the solution becomes clear. Add 5 ml of HBSS/1% FBS to stop the reaction, and pellet the cells by centrifugation (see Note 1).
4. Wash the cell pellet once with HBSS/1% FBS. Save an aliquot of cells for FACS analysis of the development and homeostasis of the immune system (see Subheading 3.2). Combine the remaining cell suspensions from the spleen and peripheral lymph nodes, and pellet the cells by centrifugation (see Note 2).
5. Prepare CD4 microbeads in MACS buffer: for each mouse, mix 10–20 μl of CD4 microbeads with 150 μl binding buffer (see Note 3).
6. Add the diluted microbeads to the cell pellet, resuspend thoroughly, and incubate at 4–8°C for 20 min with occasional mixing.
7. Add 10 ml of MACS buffer to the cell suspension and pellet the cells by centrifugation. In the mean time, set up the MACS MS column in the magnetic field of a suitable MACS Separator and wash the column with 0.5 ml of MACS buffer.
8. Resuspend the cell pellet in 1 ml of MACS buffer, and apply the cell suspension onto the MACS MS column. Allow the cells to pass through.
9. Wash the column three times with 0.5 ml of MACS binding buffer.
10. Take the column off the magnetic field, and elute the cells using 1 ml of MACS buffer.
11. Add 10 ml HBSS/1% FBS to the cell suspension and pellet the cells by centrifugation. Resuspend the cell pellet in the appropriate medium for the next step.

3.2. Analysis of MAPK Functions in T-Cell Development and Homeostasis by Flow Cytometry

1. Harvest cells from various lymphoid organs (including spleen, peripheral lymph nodes, and thymus), and prepare single-cell suspensions in HBSS/1% FBS. Red blood cells from murine spleen cell suspension should be lysed as described above (see Subheading 3.1).
2. Take approx 1×10^6 cells for FACS staining. Save some cells for staining controls, which should include nonstained and single-color-stained samples. Pellet the cells by centrifugation ($300 \times g$) for 5–10 min, and remove supernatant.
3. Dilute primary mAbs to predetermined optimal concentrations in FACS-staining buffer and use 20–50 μl for each sample or staining control (see Note 4).
4. Incubate at 4–8°C for 20–40 min in the dark.
5. Add 1.5 ml of FACS-staining buffer to the cell suspension, and pellet the cells by centrifugation.
6. If a second-step reagent is needed, resuspend the cell pellet in 20–50 μl of appropriately diluted secondary reagent (e.g., fluorochrome-conjugated streptavidin). Incubate at 4–8°C for 20 min in the dark, and wash the cells as in step 5.
7. Resuspend the cell pellet in 300–400 μl FACS-staining buffer and transfer to FACS tubes.
8. Acquire sample data on flow cytometer as soon as possible after staining (see Note 5).
9. Analyze the FACS data using the software Flow Jo (Tree Star). Figure 1 is an example of the expression of various markers in T cells.

3.3. Analysis of MAPK Functions in T-Cell Proliferation by CFSE Dilution Assays

1. Prepare anti-CD3 or anti-CD3/CD28-coated plates. For a 24-well plate, add 400 μl of PBS that contains 5 μg/ml anti-CD3 or 5 μg/ml anti-CD3 plus 5 μg/ml anti-CD28 to each well, and incubate at 37°C for 2 h or longer (see Note 6).

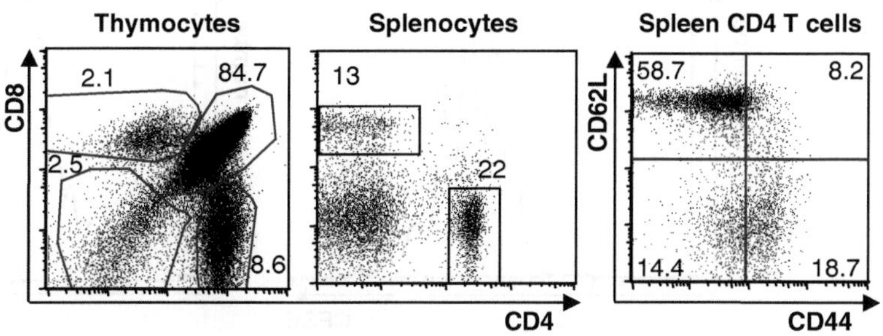

Fig. 1. Development and homeostasis of T cells. Single-cell suspensions from murine thymus and spleen were stained with anti-CD4 and anti-CD8 antibodies, and analyzed by flow cytometry (*left* and *middle panels*). Spleen cells were additionally stained with CD62L and CD44, whose expression on gated CD4+ cells was shown on the *right panel*. Numbers adjacent to outlined areas indicate percent cells in each gate.

2. Purify peripheral CD4 T cells from mouse spleen and peripheral lymph nodes as described above (see Subheading 3.1). Resuspend cells in HBSS/10% FBS to 10^6–10^7/ml.
3. Dilute CFSE stock solution to 6–10 μM in PBS, and mix with the cells at 1:1. The final concentration of CFSE is 3–5 μM, which labels T cells efficiently in the presence of 5% FBS without significant toxic effects.
4. Incubate the cells at 37°C 25 min in the dark with occasional mixing.
5. Wash the cells with HBSS/10% FBS three times (see Note 7).
6. Resuspend CFSE-labeled cells in T-cell culture medium at $1–2 \times 10^6$/ml.
7. Remove the antibody solution from the 24-well plate, and wash with PBS solution twice. Add 1 ml of cells to each well ($1–2 \times 10^6$/well).
8. Incubate the cells at 37°C (5% CO_2) for 72 h (see Note 8).
9. Harvest the cells, and perform additional staining if necessary.
10. Acquire sample data on flow cytometer to analyze the dilution of CFSE, which is detected in the FITC channel. Figure 2a is an example of dilution of CFSE in activated T cells.

3.4. Analysis of MAPK Functions in T-Cell Apoptosis by Annexin V Staining

1. Activate T cells as described above (see Subheading 3.3), with or without CFSE labeling.
2. Harvest the cells between 24 and 72 h following the start of the culture. Wash cells with 1× Annexin V binding buffer (see Note 9).

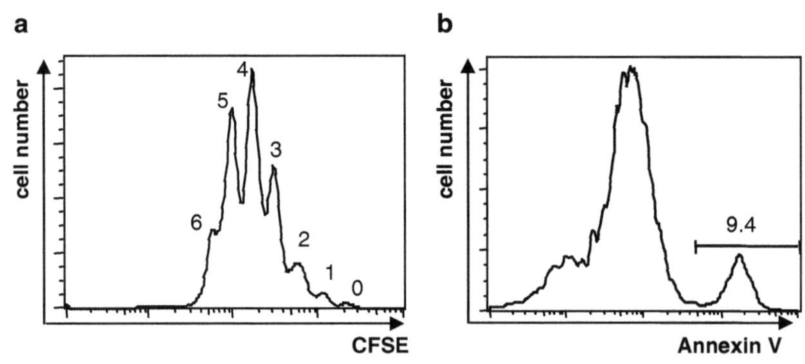

Fig. 2. Proliferation and apoptosis of activated T cells. (**a**) Purified CD4 T cells were labeled with CFSE, and activated with anti-CD3/CD28 for 3 days. Dilution of CFSE was analyzed by flow cytometry. Numbers above each peak indicate the numbers of cell cycles progressed. (**b**) Purified CD4 T cells were activated with anti-CD3/CD28 for 1 day, and stained with Annexin V and 7-AAD. The expression of Annexin V on gated 7-AAD$^-$ population was showed. The number indicates the percentage of Annexin V$^+$ early apoptotic cells.

3. Make staining master mix: 1:20 dilution of FITC or PE-Annexin V and 1:100 dilution of 7-AAD (1 μg/ml) in 1× Annexin V binding buffer. Resuspend the cells in 20–50 μl of the staining mix.
4. Incubate at RT for 15 min in the dark.
5. Wash the cells with 1× Annexin V binding buffer, resuspend in the same buffer and leave on ice.
6. Acquire sample data on flow cytometer as soon as possible after staining (within 1 h). Live cells should be gated as the $FSC^{hi}7AAD^{-}$ population, and the early apoptotic cells are indicated by the Annexin V^{+} population. Figure 2b is an example of the detection of apoptotic cells by Annexin V staining in activated T cells.

3.5. Analysis of MAPK Activation by FACS Analysis of Phosphorylated Proteins

1. Purify peripheral CD4 T cells from mouse spleen and peripheral lymph nodes as described above (see Subheading 3.1). Resuspend purified T cells in 100 μl T cell culture medium.
2. Add the CD3/CD28 beads at 1 bead: 1 cell ratio (see Note 10).
3. Incubate the cells at 37°C for 5 min to 1 h.
4. Fix the cells immediately in order to maintain phosphorylation state by mixing one volume of single-cell suspension with 20 volumes of prewarmed 1× Phosflow Lyse/Fix Buffer.
5. Incubate the cells at 37°C for 10 min. Pellet by centrifugation (300 μg) for 5–10 min and remove supernatant.
6. Vortex or mix to disrupt the pellet. Permeabilize the cells by adding Phosflow Perm Buffer III ($1–10 \times 10^6$ cells/ml) and incubating for 30 min on ice (see Note 11).
7. Wash cells twice with PBS/1% BSA twice.
8. Add p-ERK antibody (diluted 1:10 in PBS/1% BSA) at 50 μl/1×10^6 cells and incubate for 30 min at room temperature.
9. Wash the cells with PBS/1% BSA, and resuspend in the same buffer prior to flow cytometric analysis. Figure 3 is an example of the detection of ERK activation in T cells stimulated with anti-CD3 plus anti-CD28.

4. Notes

1. For murine lymph node cells, lysis of red blood cells is not necessary. However, for murine thymus, lysis of red blood cells should be avoided because thymocytes are sensitive to such treatment.

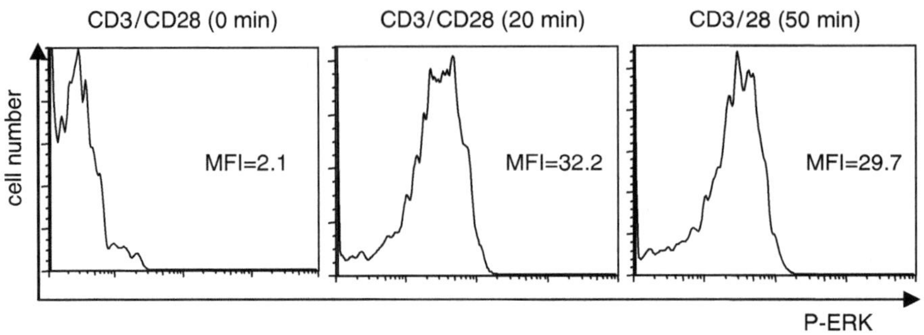

Fig. 3. Activation of ERK in stimulated T cells. Purified CD4 T cells were left unstimulated, or stimulated with anti-CD3/CD28 for 20 and 50 min, followed by fixation and permeabilization using Phosflow Perm Buffer III. The samples were stained with anti-phospho-ERK antibody and analyzed by flow cytometry.

2. The percentages of CD4 T cells are different between the murine spleen and lymph nodes, and thus development and homeostasis of these lymphoid organs should be individually analyzed. However, for functional and biochemical studies, we routinely pool these organs to purify CD4 T cells in order to obtain sufficient cell numbers.
3. The method described here refers to the isolation of CD4 T cells from one mouse. If there is a need to pool cells from different mice, the amount of microbeads should be increased proportionally. In addition, the LS column (Miltenyi Biotec; capacity: 10^8 positive cells) should be used to isolate CD4 T cells from more than one mouse.
4. In general, FACS antibodies are used at 1:100–1:1,000 dilutions, and 10× working solutions (i.e., 1:10–1:100 dilutions) can be prepared in FACS-staining buffer and stored for several months at 4°C. For single-color-stained controls, resuspend cells in 18 μl of FACS-staining buffer and add 2 μl of 10× working solutions for each antibody. For experimental samples, make a master mix by combining the antibodies with different fluorochromes, and add 20–50 μl to each sample. Also, to reduce FcgII/IIIR-mediated antibody binding that could contribute to a high background, anti-mouse CD32/CD16 (BD Biosciences) can be used prior to the addition of antibody mix.
5. We routinely use FACSCalibur (Becton Dickinson) for four-color flow cytometric analysis, and LSRII (Becton Dickinson) for multicolor flow cytometric staining. Run nonstained and single-color-stained samples first to set the voltage and compensation by following the appropriate instructions of the software (CELLQuest for FACSCalibur and FACSDiva for LSRII).

6. The method described here refers to the use of plate-bound anti-CD3 antibody to activate T cells. Alternatively, soluble anti-CD3 antibody can be added directly to the cell suspension, together with irradiated splenocytes as antigen-presenting cells (APC). To prepare APC, irradiate single-cell suspension of splenocytes from wild-type mice at 20–30 Gy (2,000–3,000 rads).
7. The presence of a high concentration of serum is required to bind and eliminate the excess CFSE that does not label the cells.
8. It is critical to choose the correct time points to analyze the dilution of CFSE in labeled cells, because either limited or excessive dilution of CFSE may not be informative. We have found that analysis of T cells activated for 72 h gives a reasonable starting time for further optimization.
9. In apoptotic cells, the membrane phospholipid phosphatidylserine (PS) is translocated from the inner to the outer leaflet of the plasma membrane, thereby exposing PS to the external cellular environment. Annexin V is a Ca^{2+}-dependent phospholipid-binding protein that has a high affinity for PS, and binds to cells with exposed PS. Therefore, the use of a Ca^{2+}-containing buffer throughout the staining process is crucial.
10. For the analysis of MAPK activation in response to anti-CD3 antibody alone, cells can be first stained on ice with anti-CD3-biotin (final concentration is 20 μg/ml; BD Biosciences), washed, and then stained on ice with streptavidin (final concentration is 10 μg/ml; Sigma). Afterwards, incubate the stained cells at 37°C to start the reaction.
11. Depending on the phosphospecific antibodies and conjugates, different permeabilization methods may be used. BD Biosciences have a number of different choices available: Perm/Wash I, Perm Buffer II, or Perm Buffer III.

Acknowledgments

We thank the Chi and Flavell laboratories for discussions. This work was supported by US National Institutes of Health R01 NS064599 and Cancer Center Support Grant CA021765, National Multiple Sclerosis grant RG4180-A-1, the Hartwell Foundation Individual Biomedical Research Award, Cancer Research Institute Investigator Award, and the American Lebanese Syrian Associated Charities (to H.C.). R.A.F. is an investigator of Howard Hughes Medical Institute.

References

1. Akira, S., Uematsu, S., and Takeuchi, O. (2006) Pathogen recognition and innate immunity, *Cell* **124**, 783–801.
2. Kyriakis, J. M., and Avruch, J. (2001) Mammalian mitogen-activated protein kinase signal transduction pathways activated by stress and inflammation, *Physiol Rev* **81**, 807–869.
3. Wan, Y. Y., Chi, H., Xie, M., Schneider, M. D., and Flavell, R. A. (2006) The kinase TAK1 integrates antigen and cytokine receptor signaling for T cell development, survival and function, *Nat Immunol* 7, 851–858.
4. Fischer, A. M., Katayama, C. D., Pages, G., Pouyssegur, J., and Hedrick, S. M. (2005) The role of erk1 and erk2 in multiple stages of T cell development, *Immunity* **23**, 431–443.
5. Dong, C., Davis, R. J., and Flavell, R. A. (2002) MAP kinases in immune response, *Annu Rev Immunol* **20**, 55–72.

Chapter 30

Methods to Study MAP Kinase Signalling in the Central Nervous System

Bettina Wagner and Maria Sibilia

Abstract

The mitogen-activated protein kinase (MAPK) family of intracellular signal transducers includes ERK1/2, ERK5, JNK/SAPK, and p38 and has been shown to control survival, proliferation and differentiation of cells composing the central and peripheral nervous system. Some MAPKs preferably induce the differentiation of neural precursor cells into the neuronal lineage, whereas others into the glial lineages, which comprises astrocytes, oligodendrocytes, and Schwann cells. MAPKs and their upstream signalling receptors play also an important role in the development of neurodegenerative diseases due to their capacity to control neural cell apoptosis. It is therefore of vital importance to better define the processes controlled by MAPKs to design therapies aimed at preventing neurodegenerative disorders in the future. The methods described in this chapter about how to culture and analyse primary astrocytes and neurons in culture have allowed us to improve our understanding on the role of the EGFR and its downstream signalling pathways in neural cell development and neurodegeneration.

Key words: Brain, Astrocytes, Neurons, Neuronal–astrocyte co-cultures, EGFR, MAPK, Immunofluorescence staining, Apoptosis, Protein lysates

1. Introduction

Intracellular signalling cascades like mitogen-activated protein kinases (MAPKs) and PI3K (phosphoinositide 3-kinase)/Akt (protein kinase B) play multiple roles in central nervous system (CNS) development, function and disease (1). MAPK actively mediate processes such as learning and memory, pain sensitivity, and neuronal survival (2–4). In general, it is believed that members of the ERK family of MAPKs mediate growth and survival responses, while members of the p38 and JNK families are promoters of cell death (1, 5, 6). However, recent reports have also

Rony Seger (ed.), *MAP Kinase Signaling Protocols: Second Edition*, Methods in Molecular Biology, vol. 661,
DOI 10.1007/978-1-60761-795-2_30, © Springer Science+Business Media, LLC 2010

suggested that ERK activation can lead to neuronal cell death in various neurodegeneration models highlighting the complex role of these proteins in the CNS (7). For each of these families several members are expressed in the developing and the mature brain, mediating a diverse repertoire of cellular functions (8).

Initial in vitro experiments indentified the ERKs (ERK1 and ERK2) as key factors for neurotrophin, growth factor, and glutamate signalling (9, 10). These findings are substantiated knock-out experiments showing that mice lacking ERK2 in neural precursors exhibit profound learning deficits (11). Functional evidence on the role of the stress kinases in neuronal cell death was initially derived from cell culture experiments as well. Activation of JNK kinases in neuronal cells occurs following various stress stimuli resulting in increased apoptosis (12). These results are corroborated by the analysis of mice lacking JNK3, a member of the JNK family which is only expressed in the brain. These mice are protected from excitotoxicity and ischemia induced cell death, emphasizing the role of JNK3 in neuronal survival (13, 14). Similarly, mice in which the endogenous Jun AP1 transcription factor was replaced by a mutant Jun version, in which the JNK phosphorylation sites serines 63 and 73 were mutated to alanines, were also resistant to epileptic seizures and neuronal apoptosis induced by the excitatory amino acid kainite demonstrating that Jun is the main substrates of JNK3 during neuronal apoptosis (15).

Neurodegenerative diseases can affect neuronal and glial cells in the nervous system and usually lead to cell death by apoptosis. The survival and differentiation of neurons depend on a variety of growth and neurotrophic factors which are only partly known to date. All MAPK have been shown to be involved in the survival, proliferation, and differentiation of nervous cells including neural stem cells (1, 16). The epidermal growth factor receptor (EGFR) can activate a number of downstream pathways, including MAPK and PI3K/Akt that regulate the migration, proliferation, and survival of neural stem cells (1, 17, 18). Mobilization of neural progenitor cells and stimulation of their differentiation into neurons or glial cells might offer new therapeutic options in treating neurodegeneration (16). However, up to now there are no therapeutic agents known to promote neurogenesis following brain injury. In a recent study, it was shown that agents known to stimulate PI3K/Akt and ERK were able to enhance neurogenesis induced by brain ischemia and might therefore provide an interesting therapeutic option in the future (19).

Although many roles of MAPK signalling in neuronal systems have been described, our understanding of their individual functions is still limited. Some of them can only be answered by in vivo experiments, but many questions can be resolved more easily using neuronal cultures. These culture systems have allowed us to dissect the role of EGFR and its downstream pathways in the development of the postnatal neurodegeneration observed in the frontal cortex and olfactory bulb of EGFR-deficient mice, which affects neurons and glial cells (20–22). Cerebral cortices of EGFR-deficient mice contain lower numbers of astrocytes and their expansion in vitro is severely compromised (21–23). Interestingly, in the absence of EGFR only cortical astrocytes were affected whereas astrocytes from other brain regions, e.g. midbrain were totally normal. Therefore, EGFR signalling seems to play a key role in controlling cortical neurodegeneration by regulating the survival of cortical astrocyte thus providing a mechanism for the region-specific neurodegeneration (22). Interestingly, MAPK activation does not seem to be involved downstream of the EGFR as ERK1/2 activation was not affected in EGFR-deficient cortical astrocytes as shown by immunofluorescence staining and Western blot analysis (data not shown). In contrast, in the absence of EGFR cortical astrocytes displayed increased apoptosis mediated by an Akt-caspase-dependent mechanism. Consequently, EGFR-deficient cortical astrocytes are not able to keep neurons alive in culture, whereas midbrain astrocytes are competent to support neuronal survival independently of EGFR expression (22).

In this chapter, we want to describe the methods for culturing cortical astrocytes and neurons which are currently used in our laboratory (22, 24). Neurons are usually prepared from embryonic mouse brain between day 12.5 and day 17.5 of gestation during the peak of neurogenesis. Neuronal cultures obtained during this period contain only very few glial cells. After day 17.5 gliogenesis already proceeds and purity of the cultures becomes an issue. We use two types of culture systems: one where neurons are cultured alone using a defined medium (referred as neuronal monocultures), while in the other type of system neurons are co-cultured with astrocytes (referred as co-cultures). These two culture systems allow answering different biological questions.

For the neuronal monocultures, neurons are seeded on poly-L-lysine-coated dishes and cultured in a serum-free defined neuronal medium. Depending on the seeding density these cultures can be kept for up to 2 weeks and are usually analyzed by immunofluorescence (Fig. 1a), Western Blot, Real-Time PCR, or RNase protection assay. The advantage of this culture system is that there is no or only very little contamination by glial cells, and therefore the results obtained with these assays reflect the cellular and molecular conditions of the neuronal population alone.

For the astrocyte–neuronal co-culture, astrocytes are seeded directly on a tissue culture dish while neurons are cultured on coverslips that have been previously treated with poly-L-lysine and endowed with wax dots necessary for the proper set-up of the co-cultures in the astrocyte-containing dishes (Fig. 2). The neurons are seeded on coverslips in medium-containing horse serum (Fig. 1b). After the neurons have attached to the surface, the

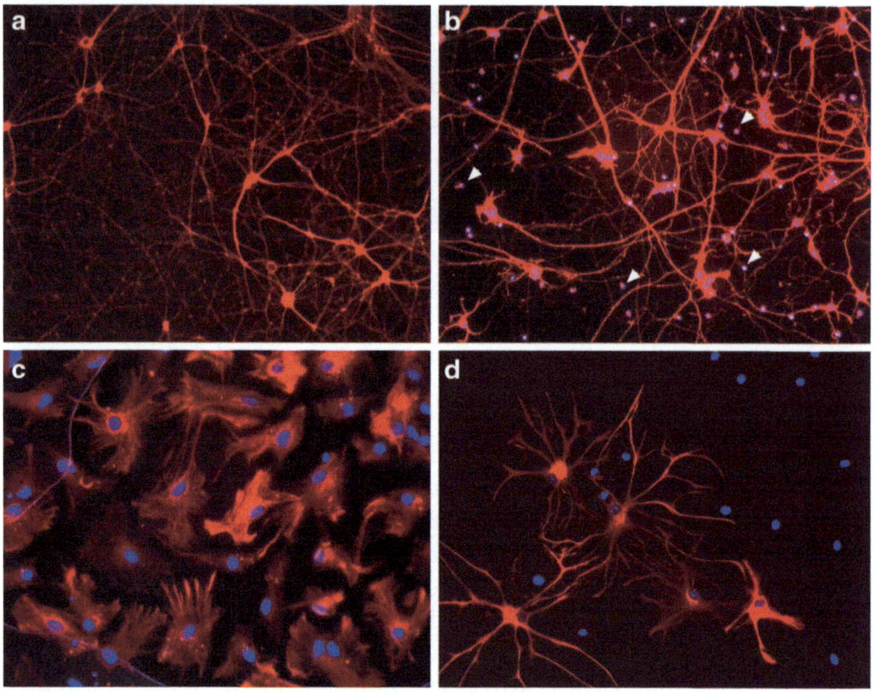

Fig. 1. Morphological appearance of neurons and astrocytes cultured by the various methods described in this chapter. (**a**) Appearance of cortical neurons cultured as neuronal monocultures and stained for the neuronal marker GAP-43 (*red*). (**b**) Appearance of cortical neurons grown on coverslips in astrocyte–neuronal co-cultures. Immunofluorescence and TUNEL staining was performed 12 days after co-culture using GAP-43 (*red*) as a neuronal marker and DAPI (*blue*) as a nuclear counterstain and TUNEL (*green*) for apoptotic neurons. *Arrowheads* point to apoptotic nuclei. (**c**, **d**) Immunofluorescence staining for glial fibrillary acidic protein (GFAP) (*red*) of cultured cortical astrocytes at two different magnifications. DAPI (*blue*) was used as a nuclear counterstain. Expression and activation of MAPK was also monitored by immunofluorescence staining in all these cultures (data not shown).

coverslips are transferred to dishes containing astrocytes whereby the paraffin dots present on the coverslips prevent direct contact between neurons and astrocytes (Fig. 2). In this system, the astrocytes provide the growth factors necessary for neuronal survival. Therefore, this culture method is best suited to study neuronal–glial interactions in vitro. The best period for preparation of astrocytes is the first 2 days after birth, when gliogenesis peaks. It is also possible to prepare astrocytes from late embryonic stages, but the astrocyte yield is less and the contamination with neuronal–glial precursors higher. After preparation astrocytes are seeded into tissue culture flasks. When the astrocyte cultures reach confluency they are trypsinised and seeded at the appropriate cell density in the dishes in which the co-cultures will be established (Fig. 2). We usually keep the co-cultures for about 1–2 weeks, and mainly analyze them by immunofluorescence (Fig. 1c, d), although analysis by quantitative RT-PCR would also be possible. This culture system is more difficult to set-up and maintain, but it allows

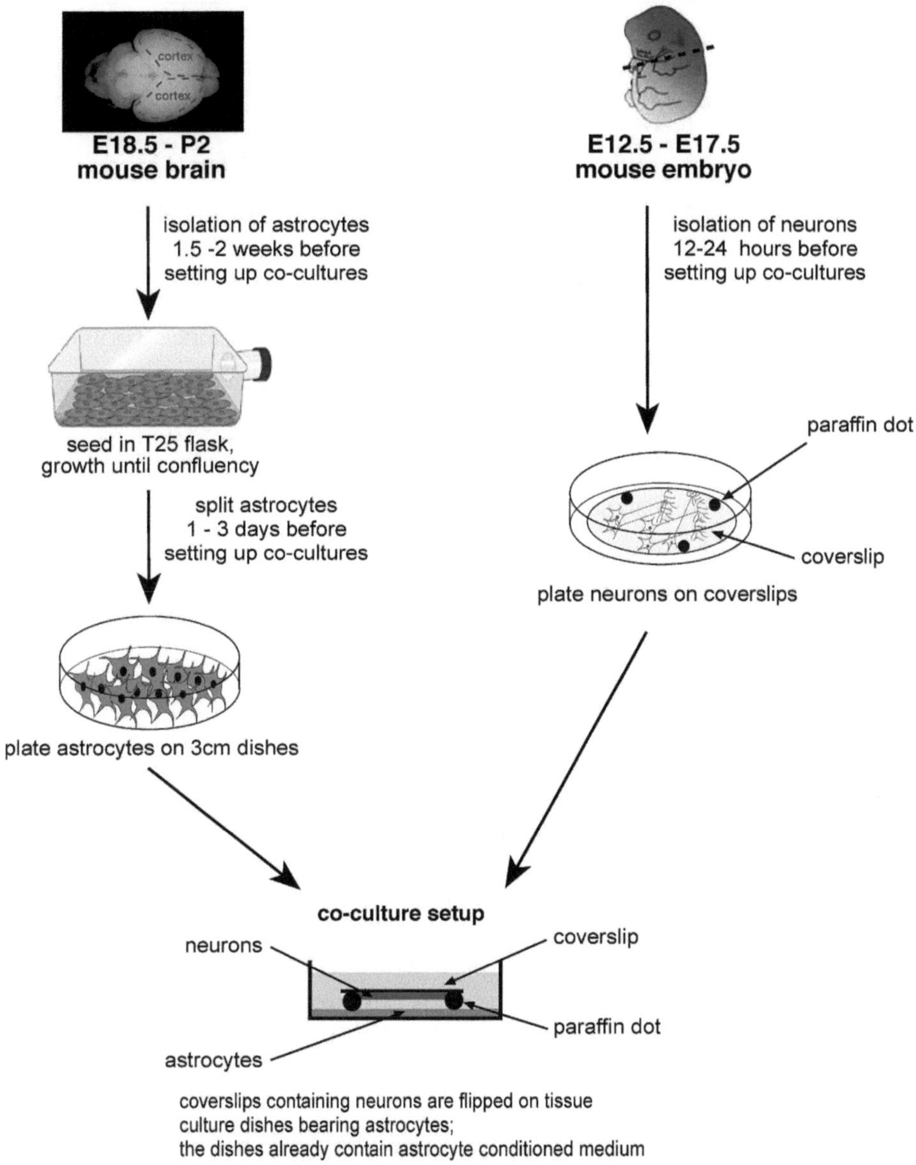

Fig. 2. Schematic representation of how to set up an astrocyte–neuronal co-culture. A summary of the procedure and timeline how and when to prepare astrocytes and neurons for co-culture is shown on the *left* and *right side* of the scheme, respectively.

studying the effects of exogenously added or astrocyte-produced factors on neuronal survival, proliferation, or differentiation (22).

Astrocyte and neuronal monocultures can be used to investigate intracellular signalling pathways such as MAPK by Western blot or immunofluorescence. They are also suited to investigate the impact of exogenously added factors and drugs on MAPK activation. In neuronal–astrocyte co-cultures, activation of intracellular signalling pathways can only be studied by immunofluorescence.

2. Materials

2.1. Instruments

1. Dissecting Microscope.
2. 11.5 cm Tough Cut Scissors, sharp edge (Fine Science Tools).
3. 10.5 cm Fine Scissors, straight (Fine Science Tools).
4. Dumont #5 forceps, standard tip (Fine Science Tools).
5. Dumont #5 forceps, biologie tip (Fine Science Tools).
6. Dumont #5 forceps, fine tip (Fine Science Tools).
7. 10 cm forceps, straight tip (Fine Science Tools).
8. 10 cm forceps, curved tip (Fine Science Tools).
9. Bunsen Burner.
10. Pasteur Pipettes, Stuffed.
11. Pasteur Pipettes.
12. Sterile tweezers.
13. Neubauer chamber.

2.2. Cell Culture Reagents

1. 65% HNO_3.
2. Distilled water, cell culture grade.
3. Phosphate-buffered saline (PBS).
4. Trypan Blue 0.4% solution.

2.3. Cell Culture Media

1. Neuronal plating medium: 1× MEM (Invitrogen), 10% Horse Serum (Invitrogen, #16050-122), 2 mM glutamine, 0.6% glucose solution, 0.22% $NaHCO_3$ solution, 10× MEM non-essential amino acids (Invitrogen), 5× MEM essential amino acids (Invitrogen, #11130-036).
2. Neurobasal medium: Neurobasal (Invitrogen) supplemented with 1× B27-supplement (Invitrogen) and 2 mM Glutamine.
3. N2-medium: 1× MEM (Invitrogen), 1× Pyruvate solution, 2 mM glutamine, 0.6% glucose, 0.22% $NaHCO_3$, 1× N2 supplement (Invitrogen #17502-048), 0.1% ovalbumin.
4. Astrocyte growth medium: DMEM high glucose (Invitrogen), supplemented with 10% FCS, 2 mM Glutamine, 1× Penicillin/Streptomycin.

2.4. Cell Culture Solutions

If not indicated otherwise, cell culture solutions are kept at 4°C for up to 2 weeks.

1. Hepes pH 7.3: dissolve Hepes (Sigma, #H6147, mouse embryo tested) in cell culture water to a concentration of 1 M, set pH to 7.3 with NaOH and sterilize by filtration through a 0.22 μM filter unit.

2. Brain-HBSS: Hank's balanced salt solution (HBSS) (Invitrogen, #14025-050) supplemented with 1× penicillin/streptomycin and 7 mM Hepes pH 7.3.
3. Brain-Trypsin: 0.05% Trypsin-EDTA (Invitrogen) supplemented with 1× penicillin/streptomycin and 7 mM Hepes pH 7.3; after addition of the supplements, the solution is aliquoted and can be stored at –20°C for at least 6 months.
4. 1% Ovalbumin solution: dissolve 1 g Ovalbumin (Sigma) in 100 ml N2, sterilize by consecutive filtration through a 0.45 μM and a 0.22 μM filter.
5. Borate Buffer: Prepare Borate Buffer by dissolving 1.9 g Borax (Sigma) and 3.2 g Boric Acid (Sigma) in 400 ml cell culture water.
6. Poly-L-lysine solution: Prepare poly-L-lysine solution by dissolving the appropriate amount of poly-L-lysine (Sigma) in Borate Buffer to give a solution containing 0.5 mg/ml poly-L-lysine. Sterile filter the solution using a 0.22 μM filter.
7. Pyruvate solution (100×): Dissolve 1.1 g pyruvate (Sigma) in cell culture water to a final volume of 100 ml and sterilize by filtration through a 0.22 μM filter.
8. Glucose solution: dissolve 20 g glucose (Sigma) in cell culture water to a final volume of 100 ml and sterilize by filtration through a 0.22 μM filter.
9. $NaHCO_3$ solution: dissolve 5.5 g (Sigma) in cell culture water to a final volume of 100 ml and sterilize by filtration through a 0.22 μM filter.

2.5. Cell Culture Materials

Nunclon 3-cm dishes.
Nunclon 6-cm dishes.
Tissue culture flasks.
Coverslips, No 1 (VWR).
150 ml Tube Top Vacuum Filter System, 0.22 μM.
150 ml Tube Top Vacuum Filter System, 0.45 μM.

2.6. Other Reagents and Solutions

Paraformaldehyde (PFA).
Bovine Serum Albumin (Sigma).
Paraffin wax.
In Situ Cell Death Detection Kit (Roche).
Permeabilisation Solution: 0.1% Sodiumcitrate, 0.1% Triton X-100.
Protein Lysis Buffer: 50 mM Hepes pH 7.3, 150 mM NaCl, 10% Glycerol, 1.5 mM $MgCl_2$, 1% Triton X-100, 1 mM EGTA pH 8, 10 mM $Na_2P_2O_7$, supplemented with 0.001% Aprotinin, 0.001% Leupeptin, 25 mM NaF, 1 mM $NaVO_3$, 1 mM PMSF, 20 mM PNPP.

3. Methods

3.1. Dissection of Brain from Mouse Embryos

3.1.1. Isolation of Embryos

1. Sacrifice pregnant mice by cervical dislocation.
2. Place the mouse on a piece of cork and fix the animal with the ventral side facing upwards. Clean the abdomen by spraying it with 70% Ethanol.
3. Make an incision through the skin reaching from the pelvis to the thorax with tough cut scissors.
4. Gently lift the abdominal membrane with straight-tipped forceps and make an incision along the linea alba. The incision should be large enough to remove the uteri (see Note 1).
5. Hold a uterus with straight-tipped forceps and cut it off just below the ovary. Remove the uterus from the abdominal cavity and place it in a 6-cm dish containing pre-warmed Brain-HBSS.
6. Hold the uterus at one end with straight-tipped forceps and gently cut it open with fine scissors.
7. Take out the embryos which are still enclosed in the amnion and transfer them to a new dish containing pre-warmed Brain-HBSS (see Note 2).
8. Lift the amnion with forceps, gently tear it open and separate the embryo from the placenta.

3.1.2. Dissection of the Brain

1. Gently fix the embryo with forceps around the abdomen and swiftly decapitate the embryo with scissors (Fig. 2) (see Note 3).
2. Transfer the head to a new dish containing pre-warmed brain-HBSS (see Note 4) and place the dish under a dissecting microscope.
3. Fix the head by holding it with round end forceps just above the nose and cut the skin along the middle starting from caudal. Pull the skin down on both sides of the head. Insert #5 Biologie forceps through the skull between the eyes and gently move it towards the neck. The skull is now divided into two parts and can easily be pulled off on both sides of the head.
4. Slide a straight-tipped forceps below the forebrain and the push the brain gently out of the remaining skull.
5. Transfer the brain to a new dish containing pre-warmed brain-HBSS with the dorsal side facing upwards.
6. Hold the brain gently with forceps and dissect both cortices from the diencephalon. First separate the hemispheres with #5 fine forceps, then insert the forceps between the cortex and the diencephalon and gently push sidewards.

7. Place the cortex on its lateral side and remove the olfactory bulbs, remaining traces of subcortical tissues and the hippocampus (see Note 5).
8. Turn the cortex to its internal side and remove the meninges by pulling them off with #5 fine forceps starting at the most caudal part of the cortex (see Note 6).

3.2. Isolation of Cortical Neurons

1. Transfer the meninges-free cortices to a falcon tube containing pre-warmed Brain-HBSS (see Note 7).
2. Wait until cortices have settled at the bottom of the falcon tube then aspirate off the Brain-HBSS.
3. Add 5 ml pre-warmed Brain-Trypsin and incubate at 37°C in a waterbath with slight agitation for 15 min.
4. Aspirate off the Brain-Trypsin and add 5 ml pre-warmed Brain-HBSS (see Note 8).
5. Incubate the tubes for 5 min at room temperature and then aspirate off the Brain-HBSS.
6. Repeat the washing procedure twice.
7. During the washing procedure fire polish-stuffed Pasteur pipettes so that the port diameter is reduced to about 2 mm.
8. After washing is completed add 2.5 ml Brain-HBSS to each falcon tube.
9. Pipette the cortices up and down in a stuffed Pasteur pipette twice to dissociate the brain.
10. Take a fire polished stuffed Pasteur pipette and dissociate the brains completely by pipetting. The solution should become cloudy.
11. Count the cells in a Haemocytometer using Trypan blue exclusion. The number of blue cells should not exceed 15%.
12. Dilute the cells to 1×10^6cells/ml in pre-warmed Brain-HBSS.
13. Neurons can be plated as monocultures (see Subheading 3.3) or co-cultures (see Subheading 3.4).

3.3. Setting Up Astrocyte-Free Neuronal Monocultures

1. Add poly-L-lysine solution to cell culture dishes (approximately 0.3 ml for a 3-cm dish) and distribute the solution by gently shaking the dish until the growth area is completely covered (see Note 9).
2. Incubate the dishes in the dark over night at room temperature (see Note 10).
3. Aspirate the poly-L-lysine.
4. Wash dishes thoroughly 3× with cell culture water (3–4 ml/dish) (see Note 11).
5. Add the appropriate amount of Neurobasal medium (1 ml per four-well dish, 4 ml per 3-cm dish, 6 ml per 6-cm dish).

6. Incubate the dish in a cell culture incubator at 37°C, 5% CO_2 for at least 6 h before seeding the cells in order to ensure proper equilibration of the medium.
7. Add the appropriate amount of cell suspension (1×10^5 to 5×10^5 cells per 3-cm dish, 5×10^5 to 2×10^6 per 6-cm dish) to the dishes (see Note 12), and return them to the incubator as fast as possible (see Note 13).
8. Depending on seeding density and experiment design these cultures can be analyzed at any time between day 2 and day 14 of the culture. If the cultures are kept for more than a week, a medium change becomes necessary (see Note 14).

3.4. Seeding Neurons for Co-cultures

For co-cultures we recommend to seed the astrocytes on culture dishes and the neurons on coverslips (see also Fig. 2).

1. Incubate coverslips in Nitric Acid in a glass beaker over night.
2. Wash the coverslips twice thoroughly with cell culture water and dry them by heat sterilization.
3. Heat paraffin was until it becomes liquid.
4. Transfer several coverslips to a bigger dish using disposable sterile plastic forceps and put three wax dots on each of the coverslips (see Note 15).
5. Coat the coverslips with an appropriate amount of poly-L-lysine solution and proceed with coating and washing as described in Subheading 3.3, steps 1–4.
6. After washing is completed transfer each coverslip to a fresh 3-cm dish using disposable sterile plastic forceps and add 4 ml of Neuronal Plating Medium.
7. Incubate the dish in a cell culture incubator at 37°C, 5% CO_2 for at least 6 h before seeding the cells in order to ensure proper equilibration of the medium.
8. Add the appropriate amount of cell suspension (1×10^5 cells per 3-cm dish) to the dishes, and return them to the incubator as fast as possible.
9. Co-cultures should be set up between 12 and 24 h after plating.

3.5. Brain Dissection from Newborn Mice

To obtain sufficient numbers of astrocytes it is recommended to prepare them from mice between E18.5 and postnatal day 2 (Fig. 2). The preparation procedure is identical to the one described in Subheading 3.2 with the following exceptions.

1. For decapitation gently hold the mouse between your index finger and your thumb and cut off the head with a single cut.
2. Dissection of the brain from the skull may be carried out without a microscope. Only the isolation of the cortices needs to be performed under a microscope.

3.6. Isolation of Cortical Astrocytes

Isolation of cortical astrocytes is performed as described in Subheading 3.2 for cortical neurons.

At the end of the dissociation procedure cell counting is rather difficult since there are many dead cells in the solution so we usually skip it and instead seed the cells prepared from two cortices in one T25 flask. Astrocytes are seeded into tissue culture flasks in astrocyte growth medium. Medium changes are usually performed once a week (see Note 16). The astrocyte cultures reach confluence after 1.5–2 weeks and can then be replated to 3-cm dishes for co-culture (see also Fig. 2).

3.7. Co-culture System Between Neurons and Astrocytes

1. For replating confluent astrocytes are washed once with PBS.
2. Trypsinisation is performed with 1 ml Brain-Trypsin per T25 flask for 3 min at 37°C.
3. Cells are resuspended in 6 ml astrocyte growth medium and the cell number is determined.
4. The astrocytes are plated at 30,000 cells/3-cm dish in astrocyte growth medium.
5. 1–3 days later the medium is changed to N2-medium.
6. 12–24 h after changing to N2-medium coverslips containing neurons can be added to the culture (see Note 13).
7. Co-cultures are usually kept for a week without medium changes (see Note 17).

3.8. Immunofluorescence Staining of Cultured Brain Cells

Immunofluorescence staining is an important tool in the analysis of cultured brain cells. Generally, it is advisable to analyze the purity of the neuronal/glial cultures by staining them for their respective markers. Additionally important cellular processes such as proliferation, cell death, or differentiation can be analysed along with the activation of intracellular signalling pathways such as MAPK.

For immunofluorescence it is recommended to seed the cells on coverslips, since we often observed background problems when immunofluorescence was performed on cells seeded on plastic dishes.

3.8.1. Immunofluorescence Staining for Neuronal/Glial Marker Proteins

1. Aspirate off medium and wash the cells twice with PBS.
2. Fix the cells with 4% paraformaldehyde (PFA) in PBS for 1 h.
3. Wash the cells 3× with PBS.
4. Add Permeabilisation Solution and incubate for 2 min on ice.
5. Wash the cells 2× with PBS.
6. Block with 0.2% BSA in PBS for 1 h.
7. Incubate the cells with the primary antibody in 0.2% BSA/PBS for 1 h at room temperature.

8. Wash the cells 3× with PBS.
9. Incubate the cells with the secondary antibody diluted in 0.2% BSA/PBS for 30 min at room temperature.
10. Wash the cells 3× with PBS.
11. Mount the coverslips on slides using mounting-medium containing DAPI.

3.8.2. Detection of Apoptotic Cells Using TUNEL Staining

For the detection of apoptotic cells we used the In Situ Cell Death Detection Kit, Fluorescein from Roche.

1. Staining for apoptotic cells is performed as described in Subheading 3.8.1 with the following additions:
2. Following permeabilization the cells are washed twice with PBS.
3. The coverslips are incubated with TUNEL reaction mixture for 1 h at 37°C in a humidified chamber in the dark.
4. Wash the coverslips 3× with PBS.
5. Proceed with blocking as described in Subheading 3.8.1. Perform all future incubations in the dark to prevent the fluorescin from fading.

3.9. Preparation of Protein Lysates from Cultured Brain Cells for Western Blot Analysis

1. Starve cultured cells (about 80% confluent) overnight in serum/growth factor-free medium.
2. Stimulate the cells with your growth factor of choice (e.g. 5 ng/ml EGF for 5 min).
3. Transfer the culture dishes to an ice-bucket.
4. Remove the medium and wash twice with ice-cold PBS.
5. Add the appropriate amount of protein lysis buffer (50–100 μl per 3-cm dish) to the dish and scrape off the cells with a cell scraper.
6. Transfer the cell lysate to a reaction tube and incubate on ice for 30 min.
7. Centrifuge the lysate for 10 min in an Eppendorf centrifuge at full speed.
8. Transfer the cleared lysate to a fresh tube and determine the protein concentration.
9. Shock freeze the lysate by immersing it in liquid nitrogen and store the lysate at –80°C until it is used for Western Blot analysis.

3.10. RNA Preparation from Cultured Brain Cells for Gene Expression Analysis

1. For expression analysis RNA is prepared using either the Absolutely RNA Miniprep or the Absolutely RNA Micro Kit.
2. Transfer the culture dishes to a bucket containing ice.
3. Remove the medium and wash twice with ice-cold PBS.

4. Add the appropriate amount of lysis buffer (50–100 μl for a 3-cm dish) to the dish and scrape off the cells with a cell scraper.
5. Collect the cell lysate in an RNase free Eppendorf tube.
6. At this point lysates can be shockfrozen and stored at −80°C.
7. After thawing proceed with RNA preparation as described in the manufacturer's protocol.
8. The RNA can be easily used for Real-Time PCR or RNase Protection Assay.

4. Notes

1. All instruments should be cleaned with 70% ethanol before dissection and between dissection steps. Before use they should be briefly submerged in Brain-HBSS to remove superfluous ethanol.
2. Dissection of embryos is a very time-consuming process. Since neuronal survival is improved by a swift dissection procedure we recommend not doing too many embryos at the same time.
3. If you use embryos from earlier stages of pregnancy they are usually immobile. However, if you use embryos from later stages of development they already respond to tactile stimuli and may squirm if you try to fix them with the tweezers. Try to work as swiftly as possible to avoid unnecessary suffering. After opening the amnion proceed immediately to decapitation.
4. During the different stages of dissection blood and tissue debris cloud the dissociation solution. Changing dishes frequently ensures a clean working environment, which makes preparation a lot easier.
5. The hippocampi can be collected in a separate tube and can be used for seeding hippocampal neuronal cultures. However the cellular yield is much lower compared to cortical cultures.
6. Removal of the meninges is an essential step during the dissection procedure. Any traces that are left will greatly impair the washing procedure after trypsinisation.
7. Limit the number of cortices to a maximum of six per falcon tube – if more cortices are used dissociation becomes more difficult and often results in reduced neuronal survival.
8. For this step the previous removal of meninges is essential. Cortices retaining pieces of meninges will not sink to the bottom of the tube during washing and may be aspirated off together with the brain-HBSS. Diligent removal of the meninges is therefore a prerequisite for a proper neuronal yield.

9. Poly-L-lysin2 is very hygroscopic, it easily clumps once the bottle is opened and weighing in becomes a very tedious procedure. Therefore, we usually dissolve the contents of a whole bottle in borate buffer, aliquot the solution and keep the aliquots at −20°C protected from light for up to 2 months.
10. In the literature incubation periods for coverslips/dishes with poly-L-lysine vary from 10 min up to 3 days. We recommend not keep the coverslips/dishes in poly-L-lysine for too long.
11. Washing needs to be performed thoroughly. Residual traces of poly-L-lysine in the medium may impair neuronal survival.
12. The amount of cells per dish can vary depending on your purpose. If too few cells are seeded, neurons rarely do survive. If too many cells are seeded, nutrients are used up very quickly and the medium needs to be changed frequently. Neurons are very sensitive to medium changes, and too frequent medium changes can harm your culture.
13. For setting up cultures take a few dishes from the incubator, seed the neurons and swiftly return the dishes to the incubator. Neurons are very sensitive to temperature and pH changes, and keeping them outside for too long will definitely diminish their survival. This precaution also applies for setting up co-cultures.
14. Replace one half of the culture medium with fresh Neurobasal medium. Before adding the fresh medium, equilibrate it in a tissue culture incubator for a few hours.
15. If the wax is too hot the dots will melt away, if the wax is too cold the dots do not adhere properly to the coverslip.
16. During the first few days after seeding astrocyte cultures do not look good, with lots of debris in the medium. At this stage medium changes to remove the debris do not positively affect the development of the culture. After a few days, patches of astrocytes are visible in the flasks and the flasks usually become confluent.
17. Although the astrocytes grow slowly in N2 medium, after some time they use up too many nutrients and the neurons can no longer survive. Astrocyte growth can be prevented by adding Cytosine Arabinoside (Ara C), but we usually circumvent this problem by keeping the cultures for a limited period.

Acknowledgments

This work was supported by grants from the Austrian Science Fund (FWF grants P18421, P18782, F23-03 and W1212), the European Community (LSHC-CT-2006-03773) and the Austrian Genome Program GENAU (Austromouse) (GZ 200.147/1-VI/1/2006).

References

1. Frebel, K., and Wiese, S. (2006) Signalling molecules essential for neuronal survival and differentiation, *Biochem Soc Trans* 34, 1287–1290.
2. Ji, R. R., Baba, H., Brenner, G. J., and Woolf, C. J. (1999) Nociceptive-specific activation of ERK in spinal neurons contributes to pain hypersensitivity, *Nat Neurosci* 2, 1114–1119.
3. Thomas, G. M., and Huganir, R. L. (2004) MAPK cascade signalling and synaptic plasticity, *Nat Rev* 5, 173–183.
4. Xia, Z., Dickens, M., Raingeaud, J., Davis, R. J., and Greenberg, M. E. (1995) Opposing effects of ERK and JNK-p38 MAP kinases on apoptosis, *Science* 270, 1326–1331.
5. Bogoyevitch, M. A. (2006) The isoform-specific functions of the c-Jun N-terminal Kinases (JNKs): differences revealed by gene targeting, *Bioessays* 28, 923–934.
6. Takeda, K., and Ichijo, H. (2002) Neuronal p38 MAPK signalling: an emerging regulator of cell fate and function in the nervous system, *Genes Cells* 7, 1099–1111.
7. Cheung, E. C., and Slack, R. S. (2004) Emerging role for ERK as a key regulator of neuronal apoptosis, *Sci STKE* 2004, PE45.
8. Wang, X., Destrument, A., and Tournier, C. (2007) Physiological roles of MKK4 and MKK7: insights from animal models, *Biochim Biophys Acta* 1773, 1349–1357.
9. Kaplan, D. R., and Miller, F. D. (2000) Neurotrophin signal transduction in the nervous system, *Curr Opin Neurobiol* 10, 381–391.
10. Wang, J. Q., Tang, Q., Parelkar, N. K., Liu, Z., Samdani, S., Choe, E. S., Yang, L., and Mao, L. (2004) Glutamate signaling to Ras-MAPK in striatal neurons: mechanisms for inducible gene expression and plasticity, *Mol Neurobiol* 29, 1–14.
11. Samuels, I. S., Karlo, J. C., Faruzzi, A. N., Pickering, K., Herrup, K., Sweatt, J. D., Saitta, S. C., and Landreth, G. E. (2008) Deletion of ERK2 mitogen-activated protein kinase identifies its key roles in cortical neurogenesis and cognitive function, *J Neurosci* 28, 6983–6995.
12. Harper, S. J., and LoGrasso, P. (2001) Signalling for survival and death in neurones: the role of stress-activated kinases, JNK and p38, *Cell Signal* 13, 299–310.
13. Kuan, C. Y., Whitmarsh, A. J., Yang, D. D., Liao, G., Schloemer, A. J., Dong, C., Bao, J., Banasiak, K. J., Haddad, G. G., Flavell, R. A., Davis, R. J., and Rakic, P. (2003) A critical role of neural-specific JNK3 for ischemic apoptosis, *Proc Natl Acad Sci U S A* 100, 15184–15189.
14. Yang, D. D., Kuan, C. Y., Whitmarsh, A. J., Rincon, M., Zheng, T. S., Davis, R. J., Rakic, P., and Flavell, R. A. (1997) Absence of excitotoxicity-induced apoptosis in the hippocampus of mice lacking the Jnk3 gene, *Nature* 389, 865–870.
15. Behrens, A., Sibilia, M., and Wagner, E. F. (1999) Amino-terminal phosphorylation of c-Jun regulates stress-induced apoptosis and cellular proliferation, *Nat Genet* 21, 326–329.
16. Miloso, M., Scuteri, A., Foudah, D., and Tredici, G. (2008) MAPKs as mediators of cell fate determination: an approach to neurodegenerative diseases, *Curr Med Chem* 15, 538–548.
17. Deleyrolle, L. P., and Reynolds, B. A. (2009) Isolation, expansion, and differentiation of adult Mammalian neural stem and progenitor cells using the neurosphere assay, *Methods Mol Biol* 549, 91–101.
18. Tropepe, V., Sibilia, M., Ciruna, B. G., Rossant, J., Wagner, E. F., and van der Kooy, D. (1999) Distinct neural stem cells proliferate in response to EGF and FGF in the developing mouse telencephalon, *Dev Biol* 208, 166–188.
19. Shioda, N., Han, F., and Fukunaga, K. (2009) Role of Akt and ERK signaling in the neurogenesis following brain ischemia, *Int Rev Neurobiol* 85, 375–387.
20. Sibilia, M., Kroismayr, R., Lichtenberger, B. M., Natarajan, A., Hecking, M., and Holcmann, M. (2007) The epidermal growth factor receptor: from development to tumorigenesis, *Differentiation* 75, 770–787.
21. Sibilia, M., Steinbach, J. P., Stingl, L., Aguzzi, A., and Wagner, E. F. (1998) A strain-independent postnatal neurodegeneration in mice lacking the EGF receptor, *EMBO J* 17, 719–731.
22. Wagner, B., Natarajan, A., Grunaug, S., Kroismayr, R., Wagner, E. F., and Sibilia, M. (2006) Neuronal survival depends on EGFR signaling in cortical but not midbrain astrocytes, *EMBO J* 25, 752–762.
23. Kornblum, H. I., Hussain, R., Wiesen, J., Miettinen, P., Zurcher, S. D., Chow, K., Derynck, R., and Werb, Z. (1998) Abnormal astrocyte development and neuronal death in mice lacking the epidermal growth factor receptor, *J Neurosci Res* 53, 697–717.
24. Celis, J. E. (1998) *Cell biology: a laboratory handbook*, 2nd ed., Academic Press, San Diego.

Chapter 31

MAP Kinase Regulation of the Mitotic Spindle Checkpoint

Eva M. Eves and Marsha Rich Rosner

Abstract

Maintaining the integrity of the cell cycle is critical for ensuring that cells only undergo DNA replication and proliferation under controlled conditions in response to discrete stimuli. One mechanism by which the fidelity of this process is guaranteed is through the activation of cell cycle checkpoints. The mitotic spindle checkpoint, which is regulated by Aurora B kinase, ensures proper kinetochore attachment to chromosomes leading to equal distribution of chromosomes to daughter cells. We demonstrated that the mitogen-activated protein kinase (MAPK) cascade regulates mitotic progression and the spindle checkpoint. As demonstrated by immunofluorescence at kinetochores, depletion of Raf Kinase Inhibitory Protein (RKIP), an inhibitor of Raf/MEK/ERK signaling, causes an increase in MAPK activity that inhibits Aurora B kinase activity. By monitoring mitotic index and transit time from nuclear envelope breakdown to anaphase, we demonstrated that RKIP depletion leads to a defective spindle checkpoint and genomic instability, particularly in response to drugs that disrupt microtubule function.

Key words: Cell cycle, Mitosis, Spindle checkpoint, MAP kinase, ERK, Immunofluorescence

1. Introduction

The spindle checkpoint is activated during mitosis by defects in spindle tension or attachment, causing a delay in chromosome segregation until the problem is corrected by Aurora B kinase. A conserved kinase, Aurora B kinase is associated with chromosomal kinetochores in complex with other "chromosomal passenger" proteins including Borealin/Dasra B, Survivin, and INCENP (1). Increased mitogen-activated protein kinase (MAPK) activity resulting from loss or depletion of Raf Kinase Inhibitory Protein (RKIP) causes a reduction in the mitotic index due to a decrease in mitotic traversal time from nuclear envelope breakdown to anaphase (2). When cells are exposed to drugs that reduce spindle tension such as Taxol, the spindle checkpoint is normally activated.

Rony Seger (ed.), *MAP Kinase Signaling Protocols: Second Edition*, Methods in Molecular Biology, vol. 661, DOI 10.1007/978-1-60761-795-2_31, © Springer Science+Business Media, LLC 2010

However, RKIP depletion and enhanced MAPK activity causes override of the spindle checkpoint and increased chromosomal defects following Taxol treatment.

Expression, activation, or localization of MAPK (ERK1,2) or its regulators can be monitored across mitosis by immunofluorescence. This methodology, as opposed to methods such as flow analysis, enables one to follow MAPK signaling in the context of specific stages of mitosis. Similarly, fluorescent DNA staining in fixed cells to show chromatin morphology and time-lapse microscopy of cells traversing mitosis reveal the effects of changes in ERK activity on mitotic progression and the spindle checkpoint. Finally, synchronization and/or arrest with drugs that disrupt the spindle enable biochemical analysis of mitotic cell populations in addition to single cells.

2. Materials[1]

2.1. Cell Culture, Synchronization, and Mitotic Arrest

1. Dulbecco's Modified Eagle's Medium with high glucose (Gibco/Invitrogen, Bethesda, MD) (DMEM) supplemented with 10% fetal bovine serum, 50 u/ml penicillin, and 50 μg/ml streptomycin (100× stock, Gibco/Invitrogen, Bethesda, MD) and selection antibiotics appropriate for any ectopically expressed markers.
2. Thymidine is dissolved in DMEM at 100 mM and sterilized by passage through a 0.22 μm filter.
3. Taxol (paclitaxel) (cat. no. 1097, Tocris Bioscience, Ellisville, MD) stock, 5 mM in DMSO.
4. Nocodazole stock, 20 mM in DMSO.
5. 0.5% trypsin (BD Diagnostic Systems, Sparks, MD), 5 mM EDTA solution in PBS (see Subheading 2.2 below).
6. 15 mm round #1 glass coverslips.
7. Small beaker of acetone.
8. Small beaker of 100% ethanol.
9. Bunsen or alcohol burner.
10. Forceps.

2.2. Cell Fixation and Immunofluorescence

1. Freshly made 4% paraformaldehyde (cat. no. PX0055-3, EMD Chemicals, Gibbstown, NJ) in 0.1 M Na phosphate buffer, pH = 7.4. Store at 4°C for no more than 3 days.
2. Dulbecco's Ca^{++}-free, Mg^{++}-free phosphate-buffered saline (PBS).
3. 2% sodium azide in H_2O (Note 1) (100× stock).
4. 5.0% IgG-free bovine serum albumin (BSA) (cat. no. 001-000-161, Jackson ImmunoResearch, West Grove, PA) in PBS with 0.02% sodium azide.

[1] All chemicals are from Sigma (St. Louis, MO) unless otherwise indicated.

5. Primary and fluorochrome-conjugated secondary antibodies appropriately diluted in 1.0% BSA in PBS plus 0.02% azide.
   (a) Primary Antibodies: We have used the following antibodies. Many others are available.
   (b) Mitotic cells:
      - Anti-cyclin B1 (cat. no. sc-245, Santa Cruz Biotechnology, Santa Cruz, CA).
      - Anti-phosphoHistone H3 (cat. no. 08-570, Upstate Biotechnology/Millipore, Lake Placid, NY).
      - Anti-phosphoS153 RKIP (custom, affinity purified with phosphopeptide).
      - Anti-phospho CENP-A (cat. no. 05-792, 07-232, Upstate Biotechnology/Millipore, Lake Placid, NY).
   (c) Centrosomes:
      - Anti-Nek2 (cat. no. 610593, BD Biosciences, San Jose, CA).
   (d) Centromeres/Kinetochores:
      - Anti-CENP-A (cat. no. D115-3, MBL International, Woburn, MA).
      - Anti-phosphoCENP-A (see above).
6. 1.0 mg/ml Höechst 33342 (#H1399, Molecular Probes/Invitrogen, Eugene, OR) in H_2O, 1,000× stock. Store, protected from light, at 4°C.
7. Vectashield Hard Set mounting medium (cat. no. H1400, Vector Laboratories, Burlingame, CA).
8. 1 in. × 3 in. precleaned glass slides.

2.3. Cell Lysis for Immunoblotting

1. Lysis buffer: 20 mM Hepes, pH = 8.0, 0.1% SDS, 1.0% Triton X-100, 0.5% DOC, 5 mM EDTA pH = 8.0, 50 mM NaCl. Add fresh before using, 25 mM sodium pyrophosphate, 5 mM NaF, 50 μM Na_3VO_4, 10 mM phosphatase substrate (Sigma, P4744), 20 mM β-glycerophosphate, 0.5% Protease Inhibitor Cocktail III (cat. no. 539134, Calbiochem, La Jolla, CA).
2. Cell scrapers.
3. 1 ml syringes with 26-gauge needles.

3. Methods

In our studies, we determined whether depletion or augmentation of ERK signaling has an effect on the progression of mitosis or on a particular stage of mitosis. When analyzing RKIP-depleted cells, a reduction in the frequency of metaphase cells was the initial clue that the spindle checkpoint was compromised. In most eukaryotic cells, the stages of mitosis, i.e., prophase, pro-metaphase, metaphase, anaphase, telophase, and cytokinesis, are defined by chromosomal morphology and the dissolution and reformation of the nuclear membrane. Chromatin or chromosomes in individual cells can be visualized with a DNA-specific fluorescent dye. Since most mammalian cells are permeable to Höechst 33342, the dye can be used on viable cells (Note 7). Höechst 33342, Höechst 33258, and DAPI (4′,6′-diamidino-2-phenylindole) all work well to stain the chromatin of fixed/permeabilized cells that can be immunostained for other markers. Mitotic chromatin can also be visualized without fluorescence optics as its density makes it readily visible in most cells using phase microscopy. Time-lapse photomicroscopy using phase optics can be used to detect differences in the timing of mitotic progression.

The subcellular localization during mitosis of a molecule of interest can help to elucidate its function. Using this approach, one can monitor expression of specific proteins during mitosis by immunofluorescence. For example, in Fig. 31.1, the increase in phosphoS153 Raf Kinase Inhibitory Protein (pRKIP), a modified RKIP that does not bind Raf-1, can be readily detected throughout mitosis although it is not detectable in interphase cells. Since mitosis is a highly dynamic process with differential cellular localization of molecules dependent upon mitotic stage (1), chromatin staining to define the mitotic stage as well as immunostaining of landmarks such as centrosomes or kinetochores is critical. To visualize centrosomes and kinetochores, we have used antibodies to Nek2 (3) and CENP-A (4), respectively.

The spindle checkpoint is activated by drugs that disrupt mitosis such as Taxol, which stabilizes microtubules, or nocodazole, which prevents microtubule polymerization. However, many cells will de-condense chromatin with time and enter G1 in spite of the presence of such "spindle poisons" (5). Changes in the timing or rate of escape can indicate relaxation or strengthening of the spindle checkpoint in response to depletion or augmentation of ERK signaling. In the methods detailed below we have used HeLa cells. However, other cells could also be tested for synchronization and response to spindle poisons.

All of the methods described above can be carried out on normally proliferating cell cultures. However, the incidence of detectable mitotic cells at any one time is usually a few percent or less (2). Cell synchronization, while never perfect, not only allows the visualization of more mitotic cells but makes it possible to collect sufficient numbers of mitotic cells for biochemical analyses such as immunoblotting (IB), immunoprecipitation (IP), or enzyme activity assays. In addition, synchronization is vital for

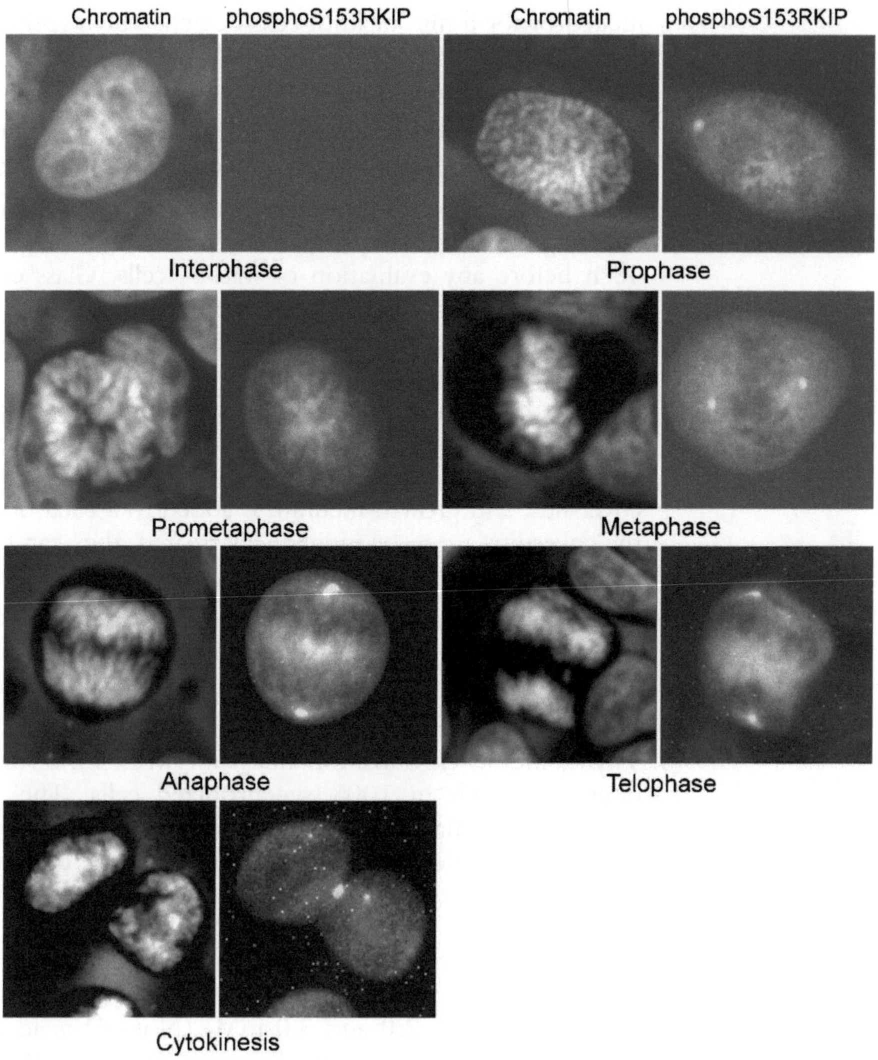

Fig. 1. Subcellular localization of phosphoS153 RKIP during different stages of mitosis. Proliferating HeLa cells were fixed and immunostained for phosphoS153 RKIP using anti-phosphoS153 RKIP antibody. Chromatin was visualized with Höechst 33342.

time-lapse photomicroscopy of mitosis since the likelihood of mitotic cells in the chosen field(s) is greatly enhanced. To synchronize HeLa cells, we used a double-thymidine block (6).

3.1. Cell Culture, Thymidine Synchronization, and Mitotic Arrest

1. HeLa cells are cultured at 37°C in a 5% CO_2 incubator in DMEM with 10% FBS and penicillin/streptomycin (growth medium).
2. The growth medium of exponentially growing HeLa cells is removed by aspiration and trypsin/EDTA at 37°C is added (5 ml per 100 mm tissue culture dish, and proportionally by area for other size dishes). The plates are returned to the

incubator for 5 min or until cells detach. The trypsin/EDTA and cells are added to 2 ml of growth medium and centrifuged at 300 × g for 3 min at room temperature.

3. The cell pellet is resuspended in growth medium and the cells are counted. Cells are diluted as needed for further use.
4. If the cells are to be immunostained, they are plated on glass coverslips at least 24 h prior to any treatment and, preferably, 48 h before any evaluation of mitotic cells. Glass coverslips (15 mm round) are cleaned and sterilized by dipping sequentially in acetone and ethanol. The coverslips are then flamed and placed in 60 mm Petri dishes (Note 2). 30,000 cells in 0.3 ml growth medium are carefully added to the top of each coverslip, and dishes are placed in an incubator. After 2 h or more, cells are checked to be sure that they have adhered to the coverslips, and growth medium is added to a total of 4 ml. If the coverslips are to be individually treated, they can be transferred cell-side up to 12-well tissue culture plates the next day.
5. Cells for chromatin staining only (live or fixed), for biochemical analyses, or for synchronization should be plated on tissue culture plastic with densities and dish or well sizes appropriate to the assay to be performed.
6. Thymidine synchronization does not work on all cells and does not result in 100% synchronized cells. The method described here has been optimized for the HeLa cells we used in our assays (Note 3). Pilot studies should be done to optimize times and dosage for each cell line. Three days prior to analysis, cells to be synchronized should be plated at 4×10^5 cells per 100 mm tissue culture dish (and proportionally by area to other size dishes). The following day, 2.5 mM thymidine (or between 2.0 and 3.0 mM) (Note 4) is added, and cells are returned to the incubator for 16–18 h. The thymidine-containing medium is removed, and cells are rinsed three times with DMEM (no supplements). This rinsing should be vigorous in order to remove rounded up mitotic cells that were not blocked by the thymidine. Growth medium is added, and cells are returned to the incubator for 6–8 h. Again, 2.5 mM thymidine is added to the cells for 16–18 h, cells are rinsed three times with DMEM, and growth medium is added back to the cells. Rounded up mitotic cells are evident about 7 h after release from thymidine, and cell rounding is maximal at 9–10 h after release. Cultures can be fixed (see below) at various times postrelease to evaluate mitotic progression. Further enrichment for mitotic cells for biochemical analyses can be attained by "shake-off" (see Subheading 3.3 below).
7. Alternatively, cells in culture can be followed by time-lapse photomicroscopy to detect differences in the timing of mitotic

progression. Time-lapse photomicroscopy was performed with a 20× objective on a Zeiss Axiovert 100TV microscope. The microscope was enclosed in an environmental box (37°C, 5% CO_2). Images were acquired at 3 min intervals by a Princeton Instruments MicroMAX 1300Y cooled CCD camera with SlideBook software (Intelligent Imaging Innovations).

8. Taxol arrests cells in mitosis by stabilizing the spindle microtubules. Cell sensitivity to Taxol varies, so dosage and time of exposure should be determined empirically for each cell line. In our studies, we used conditions that produced a two- to threefold increase in detectable mitotic cells in nonsynchronized cultures of wild-type cells. Those conditions are 1 nM Taxol for 5 h for HeLa cells and 10 nM for 7 h for a rat hippocampal line (H19-7) (Note 5). Most established cell lines will eventually escape from these levels of Taxol. Higher doses of Taxol (10 μM for 12 h for HeLa) will produce a more stringent arrest. However, cells in which the spindle checkpoint is compromised will de-condense their chromatin into micronuclei that are clearly distinguishable from the chromatin in arrested cells (see Fig. 4a from (2)) (Note 6).
9. Nocodazole can also be used to induce the spindle checkpoint in cells. Dosage and timing should be determined empirically for each cell line. For RKIP-depleted HeLa cells, longer incubations in nocodazole, compared to Taxol, were required to reveal the compromised spindle checkpoint.

3.2. Cell Fixation and Immunostaining

1. In mammalian cell cultures, mitotic cells are usually rounded up and poorly adherent. To fix cells, the growth medium must be gently aspirated from the edge of the dish or well, and the cells should not be rinsed. Cold paraformaldehyde fixative (4°C) should be gently added and the cells fixed for 10 min at room temperature (RT) (1 ml per well of a 12-well plate, 2 ml for a 35 mm dish, etc).
2. For immunostaining, the fixative is aspirated and the cells are rinsed once with RT PBS. The same volume as fixative of RT 0.1% Triton X-100 is then added, and cells are incubated for 2 min at RT to permeabilize them. Following this step, one should proceed to step 4 below.
3. For chromatin morphology only, i.e., to count mitotic cells or identify stages of mitosis, the fixative is aspirated, cells are rinsed with RT PBS, and 1.0 μg/ml Höechst 33342 in PBS is added. Cells are incubated at RT until the nuclei are completely stained (this step should be monitored on a fluorescent microscope at 10 min intervals). Finally, cells are rinsed with RT PBS, and stored for counting in PBS containing 0.02% azide at 4°C.

4. Triton X-100 is aspirated off the cells, 5% BSA is added, and cells are blocked overnight at 4°C.
5. The block is aspirated, and primary antibody is added at the appropriate dilution in 1% BSA. For 15 mm glass coverslips, 50–60 μl of the primary antibody can be spotted onto Parafilm, and the coverslips inverted onto it. Incubation should continue for 2 h at RT in a moist environment.
6. The coverslips are returned to 12-well plates and washed three times at RT with PBS for 5 min with gentle rocking.
7. About 200 μl fluorochrome-conjugated secondary antibody at the appropriate dilution in 1% BSA is added to each well and incubated, protected from light, at RT for 1.5 h with gentle rocking. If 200 μl does not fully cover the bottom of the well, more should be used.
8. Cells are washed three times at RT with PBS for 5 min with gentle rocking. Höechst 33342 (1 μg/ml) is added to the final wash, and incubation is continued until the nuclei are completely stained. This step can be monitored on a fluorescent microscope at 10-min intervals. Cells are then rinsed with RT PBS. Fixed cells are stored, protected from light, at 4°C in PBS containing 0.02% azide until mounting (within 24 h).
9. 15 μl Vectashield HardSet is dropped onto a microscope slide and a coverslip is inverted onto the drop. The slides are allowed to dry fully, protected from light.

3.3. Mitotic Cell Collection and Lysis for Immunoblot or Other Biochemical Analysis

For thymidine-synchronized cultures (see Subheading 3.1 above), the time after release from thymidine that the maximum number of cells is at the mitotic stage of choice should be determined by pilot experiments. For example, in our wild-type HeLa cells, metaphase cells were most numerous at ~9 h postrelease.

1. Growth medium is removed from the cells by gentle aspiration. Lysis buffer (at 4°C) is added to cover the surface, and cells are incubated on ice for 15 min. Cells are scraped and passed through a 26-gauge needle.
2. To enrich for rounded up mitotic cells, the flasks are vigorously shaken or the plates tapped to loosen the rounded up cells which are then collected from the medium by gentle centrifugation (300 × *g*). The pellet is resuspended in 4°C lysis buffer, incubated for 15 min at 4°C, and then passed through a 26-gauge needle.
3. The lysates are centrifuged at 16,000 × *g* at 4°C for 15 min.
4. The cleared lysates are analyzed with a biochemical assay of choice.

4. Notes

1. All water used is 18.2 MΩ.
2. Do not use tissue culture plastic dishes when plating cells on coverslips.
3. In our studies of thymidine-synchronized cells, we used populations of chromosomally heterogeneous HeLa cells. It is possible that clonal isolates of near-diploid cells would exhibit tighter synchronization.
4. The optimal concentration of thymidine to block mitosis is in the low millimolar range, 2.5 mM for our HeLa cells. The optimal concentrations for other cells should be determined empirically. If the concentration is too low, fewer cells will be blocked. If the concentration is too high, cells will not progress after the thymidine is removed.
5. Cell line sensitivity to the described spindle perturbing treatments can vary enormously. Appropriate dose–response and time course pilot experiments should be done for each cell line.
6. Spindle poisons (Taxol and nocodazole) perturb the spindle structure and the morphology of mitotic chromatin so that although a cell can be identified as "mitotic," the stages of mitosis cannot be discriminated with certainty.
7. Höechst 33342 intercalates into the DNA so the effects of long exposure on viable cells can be presumed to be mutagenic, and viable cells stained with Höechst 33342 should not be maintained for further use.

References

1. Ruchaud S, Carmena M, Earnshaw WC. Chromosomal passengers: conducting cell division. Nat Rev Mol Cell Biol 2007;8:798–812.
2. Eves EM, Shapiro P, Naik K, Klein UR, Trakul N, Rosner MR. Raf kinase inhibitory protein regulates aurora B kinase and the spindle checkpoint. Mol Cell 2006;23:561–74.
3. Fry AM, Meraldi P, Nigg EA. A centrosomal function for the human Nek2 protein kinase, a member of the NIMA family of cell cycle regulators. Embo J 1998;17:470–81.
4. Kunitoku N, Sasayama T, Marumoto T, et al. CENP-A phosphorylation by Aurora-A in prophase is required for enrichment of Aurora-B at inner centromeres and for kinetochore function. Dev Cell 2003;5:853–64.
5. Brito DA, Rieder CL. Mitotic checkpoint slippage in humans occurs via cyclin B destruction in the presence of an active checkpoint. Curr Biol 2006;16:1194–200.
6. Bootsma D, Budke L, Vos O. Studies on synchronous division of tissue culture cells initiated by excess thymidine. Exp Cell Res 1964;33:301–9.

Chapter 32

Using High-Content Microscopy to Study Gonadotrophin-Releasing Hormone Regulation of ERK

Christopher J. Caunt, Stephen P. Armstrong, and Craig A. McArdle

Abstract

Gonadotrophin-releasing hormone (GnRH) is a hypothalamic peptide that acts via $G_{q/11}$-coupled 7TM receptors on pituitary gonadotrophs and mediates the central control of reproduction. Recent evidence also indicates that GnRH can affect numerous tissues, but the molecular mechanisms of GnRH receptor stimulation are cell type-specific. Extracellular signal-regulated kinase (ERK) 1 and 2 are key regulators of GnRH function in several cell types, but they also integrate signals from a wide variety of other stimuli. This leads to the obvious question of how specific cellular responses to ERK activation occur, and it is now clear that this is, in part, achieved through strict spatiotemporal control of ERK activity. This means that, in order to infer the function of ERK regulation accurately, multiple readouts for ERK activity, localisation and downstream consequences (e.g. transcriptional activation or cell growth) must be compared simultaneously. Here, we describe some of our findings in the investigation of GnRH signalling to ERK, with particular emphasis on novel, high-content microscopy methods for studying ERK regulation.

Key words: ERK, GnRH, siRNA, Phorbol ester, EGF, MKP, DUSP, High content analysis

1. Introduction

ERK1 and ERK2 are ubiquitously expressed, prototypic members of the mitogen-activated protein kinase family. The majority of extracellular signalling molecules activate ERK1/2, which in turn control cell differentiation, survival, division, and metabolism (1–5). The deregulation of ERK signalling is associated with a host of pathological conditions (notably in several tumour types and cardiac pathologies), posing two obvious questions: how are particular responses to ERK activation achieved, and how can ERKs be therapeutically targeted with any degree of specificity? It is now clear that the specificity of ERK action is, to a large extent,

Rony Seger (ed.), *MAP Kinase Signaling Protocols: Second Edition*, Methods in Molecular Biology, vol. 661,
DOI 10.1007/978-1-60761-795-2_32,

governed by the kinetics and compartmentalisation of the ERK response (1–5). In resting cells, the majority of inactive ERK is bound to its upstream activator, MEK (MAPK/ERK kinase) in the cytoplasm, and the rate of exchange between nuclear and cytosolic compartments is low (6–9). Dual phosphorylation of ERK1/2 within the TEY motif in the activation loop causes their release from MEK, and an increase in shuttling to and from the nucleus (6, 7). Activated ERK1/2 can then phosphorylate a growing list of substrate proteins throughout the cell, unless restricted to particular compartments by scaffolding proteins, or dephosphorylated (2, 6). The repertoire of upstream activators, substrates and binding partners of ERK are likely to underlie cell and tissue-specific differences in ERK activity and its consequences.

The issues described above are exemplified by studies on gonadotrophin-releasing hormone receptor (GnRHR) signalling to ERK. GnRH is secreted by the hypothalamus in a pulsatile fashion to act on 7 transmembrane GnRHRs in the pituitary (10). These activated receptors act via classical $G_{q/11}$-coupled second messenger cascades, causing increases in intracellular Ca^{2+} and activation of protein kinase C (PKC) isoforms (11, 12). This causes the synthesis and secretion of luteinising hormone (LH) and/or follicle stimulating hormone (FSH) (10). The pulse frequency of GnRH release from the hypothalamus governs hormone release and transcriptional programmes induced by GnRHR signalling (13–16). GnRH therefore mediates central control of reproduction and is essential for mammalian reproduction to occur. ERK signalling mediates GnRH effects on LH and FSH transcription in several in vitro cell models (i.e. transcription of the common α-subunits, as well as the distinct β-subunits), and recent work has shown that pituitary-specific knock-down of ERK1/2 causes infertility in female mice (17). This is caused by a reduction in ERK-dependent induction of the immediate early gene product, Egr-1, which is in turn required for LH synthesis (17). A wealth of additional data shows that the mechanisms of ERK activation by GnRH are dependent on cellular context. For example, ERK activation in gonadotroph-lineage cells appears to be almost solely through PKC-dependent activation of Raf (18), while in GT1–7 neurons, it occurs exclusively through PKC-dependent transactivation of epidermal growth factor (EGF) receptors (19, 20). Other mechanisms described include signalling to ERK via focal adhesion complexes in HEK293 cells (21), and via G_i-coupled routes in prostate cancer cells (22). This clearly affects the localisation and cellular consequences of ERK signalling. For example, GnRH-mediated ERK activation in focal adhesions mediates cytoskeletal rearrangement (21), while PKC-dependent activation causes ERK translocation to the nucleus and transcriptional activation (18).

The fact that the spatiotemporal characteristics of ERK regulation are so important in its role in physiological control of cellular systems means that laboratory study of ERK behaviour can be complex. The main methods for doing this, along with specific limitations of each, are detailed below.

1. *Western blotting*. The availability of antibodies that are highly specific for ERK1/2 (irrespective of phosphorylation state) as well as the TEY dual-phosphorylated, active form of ERK1/2 (ppERK1/2) enables the study of changes in ERK expression level and phosphorylation state. The most common method of doing this is by western blotting of whole cell lysates, allowing the comparison of multiple samples separated on gels. The main drawback to this method is obviously that changes in localisation cannot be observed, and the number of samples one can compare on a gel is often a limiting factor. Alternatively, fractionated cell preparations can be used to determine subcellular localisation of ERK. These techniques can include isolation of nuclear and cytoplasmic compartments by partial lysis or the use of sucrose density gradients to separate organelles. These methods can be much more informative, but are also much more labour-intensive and technically demanding.
2. *Immunofluorescence microscopy*. The quality of antibodies available to ERK1/2 and ppERK1/2 also enables immunofluorescence staining of cells and tissues to determine localisation. When used in combination with fluorescence or confocal microscopy, a direct measurement of ERK1/2 and ppERK1/2 can be achieved by simply comparing the intensity of ERK staining in different cellular compartments. That is, the intensity of fluorescent signal is directly proportional to the amount of ERK1/2 or ppERK1/2 in a given region of the cell. The use of different fluorophores to stain for ERK1/2 and ppERK1/2 in the same cells allows comparison of localisation and activity in a range of samples in an internally controlled manner. One drawback to this method, however, is that both image capture and analysis are time consuming. For example, even a relatively modest experiment involving the analysis of ~30 cells per field in two fluorophores can take hours when considering all the variables in experimental design (e.g. random field selection of more than one area, replicates per condition and the necessary number of timepoints). Some of these issues can be overcome by using fluorescence-activated cell sorting (FACS) of stained cells (which has been used to define "graded" versus "switch-like" behaviour of ERK in cells (23)), but this involves trypsinisation of adherent cultures prior to staining, and most FACS systems cannot be used to visualise subcellular localisation of signals. The other major issues with

immunofluorescence are that cells need to be fixed for staining, increasing the number of samples necessary and that dynamic information about ERK regulation can be missed.

3. *Live cell study of ERK labelled with fluorescent proteins.* Several groups have successfully undertaken studies of ERK regulation in which expression vectors are used to introduce ERK1 or ERK2 fused to fluorescent proteins (on the N- or C-terminus (24, 25)) and subsequently used to track ERK behaviour in live cells. This can be done using a range of approaches, the simplest being to use time-lapse image capture of cells to monitor ERK movement (24), but more complex approaches using fluorescence recovery after photobleaching (FRAP) to define rates of diffusion or traffic (6, 7) or fluorescent resonance energy transfer (FRET) (26, 27) to measure dynamic association with binding partners or activation can also be employed. These have most commonly taken place using green fluorescent protein (GFP) or yellow fluorescent protein (YFP) conjugates, but notable studies on dynamic ERK2 shuttling using methods similar to FRAP have also used fusions with the photo-activatable protein, Dronpa (28). The chief drawback here is that unless endogenous ERK is removed from cells, the ERK fusion protein will be expressed beyond physiological levels. This, in turn, means there will be a non-physiological stoichiometry of ERK to scaffolds and other regulators that shape ERK responses, and the normal regulation of ERK location could easily be masked because the normally available binding sites for interaction may be swamped by excess levels of fusion protein. Co-expression with candidate scaffold proteins, such as MEK or β-arrestin, can reveal much about the nature of their interaction with ERK as this will undoubtedly diminish effects of the multiple other proteins acting on ERK in physiological scenarios (9, 29, 30). Other groups have circumvented the over-expression problem by selecting cells expressing very low levels of ERK2-GFP for study (6, 7), or, as we have done, by using short inhibitory RNA (siRNA) to remove endogenous ERKs prior to reconstitution with fusion constructs (see Note 2 and refs. 31–33).

4. *Studying ERK regulation using high-content microscopy and analysis.* Recently, we have developed methods using automated fluorescence microscopy (high-content analysis) to explore ERK signalling based on immunostaining of endogenous ERK1/2 and a system where siRNAs are used to knock-down endogenous ERK1/2 and recombinant adenovirus is used to add-back ERK2-GFP reporters (Fig. 1). These systems can also be paralleled by staining for ppERK1/2, providing an efficient system for simultaneously monitoring ERK activity and compartmentalisation. It also enables genetic manipulation of ERK signalling (add-back of mutated

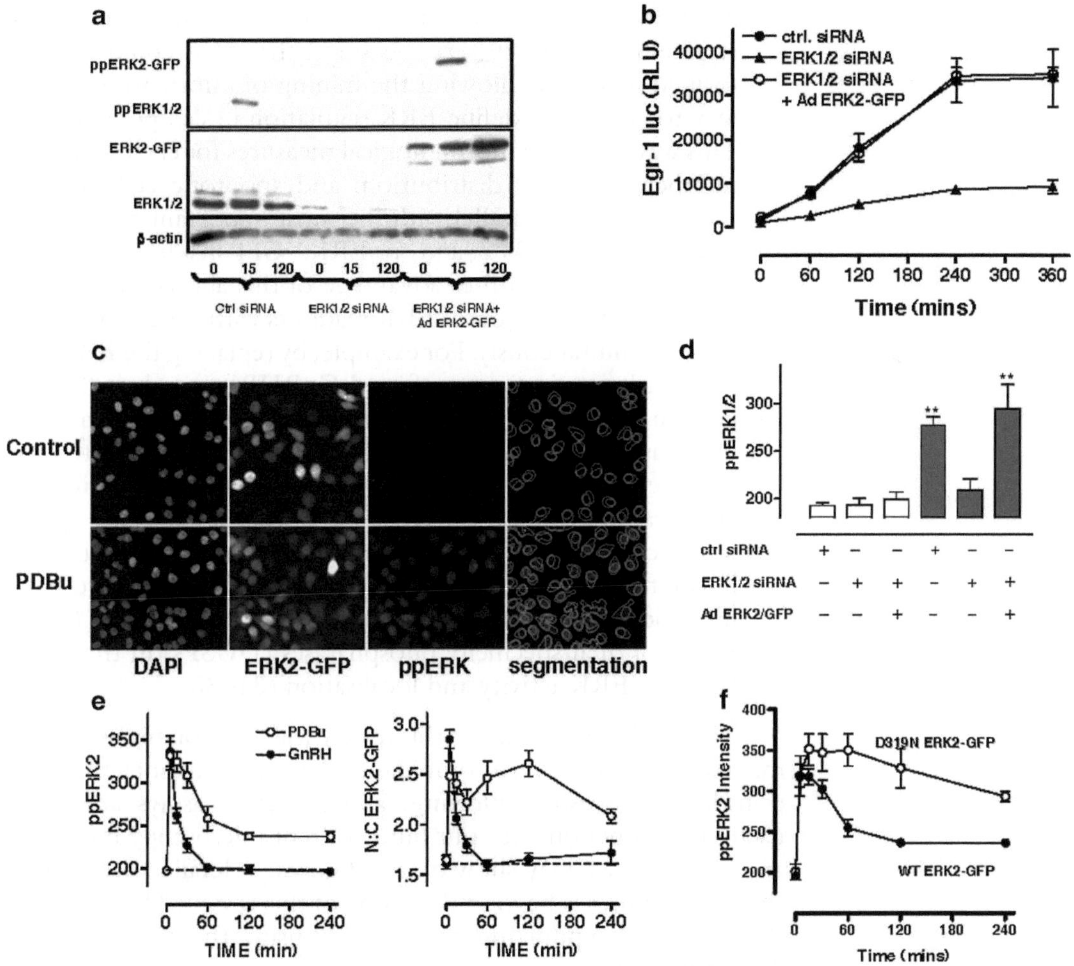

Fig. 1. A knock-down and add-back model for ERK signalling. We have developed a high throughput imaging model for ERK signalling based on reverse transfection with siRNAs targeting non-coding regions of ERKs 1 and 2, and transduction with adenovirus expressing ERK2-GFP to express the reporter at a physiological (for endogenous ERK) level. Cells can then be stained for ppERK (active ERK) facilitating automated imaging to monitor ERK compartmentalisation and activity in parallel. *Panel* (**a**) shows western blots for ERK and ppERK in cells transfected with control or ERK1/2 siRNA and transduced with Ad ERK2-GFP, before stimulation for 0, 5 or 120 min with PDBu. *Panel* (**b**) shows use of an ERK-dependent transcription reporter (Egr-1 luciferase) to document functional effects of the ERK knock-down and add-back in cells stimulated from 0 min with PDBu. *Panel* (**c**) shows images acquired with the IN Cell 1000 for the nuclear stain (DAPI), ERK2-GFP and ppERK in control and PDBu-stimulated (15 min) cells, as well as the automated image segmentation used to define nuclear and cellular perimeters. This enables quantification of whole cell ppERK as a measure for ERK activation and the nuclear:cytoplasmic (N:C) ERK2-GFP ratio as a measure of ERK localisation (*Panels* (**d**) and (**e**)). *Panel* (**e**) shows imaging data from cells stimulated with PDBu which causes sustained ERK activation and translocation, or with GnRH which causes transient ERK activation and translocation. These data mirror western blotting results although throughput is much greater with the imaging assays. An important feature of this model is that it facilitates add-back of ERK2-GFP mutants, as illustrated in *Panel* (**f**) where add-back of D319N ERK2-GFP prolongs PDBu-stimulated ERK activation. This mutation prevents D-domain-dependent binding of ERK to MKPs, supporting the possibility that they are negative regulators of the PDBu effect. These data were generated in HeLa cells (see refs. 31–33 for details).

ERK2-GFP see Fig. 1 and Notes 1–4); it can be linked with biochemical assays and used with downstream readouts of ERK activation (e.g. transcriptional activation) and also to monitor ERK movement in live cells. Using these techniques, many more cells (thousands per condition) and parameters (typically 45 per cell) are measured than that could be performed manually, allowing the mining of information necessary to accurately define ERK regulation in cell populations. This way, multiple morphological measures for each stain, cell number, cell-cycle distribution, and apoptotic cell number are measured in parallel with the most important measures – ppERK1/2, ERK1/2, and/or ERK2-GFP inside and outside of the nucleus. Another advantage of this approach is that it allows the comparison of ERK readouts with other important factors simultaneously. For example, by replacing the ERK1/2 stain with a stain for transfected GnRHR, a single-cell comparison of ppERK1/2 response to receptor number can easily be measured. This increased throughput is ideally suited for large-scale studies involving, for example, the testing of chemical libraries for effectiveness in cell-based assays or for large scale gain or loss of function screens. This is suitably illustrated by some of our recent studies, in which we have identified novel functions of the MAPK phosphatase (MKP) family of dual-specificity phosphatases (DUSPs) in the regulation of ERK activity and localisation (31–33).

The methods outlined in this review will focus on high-content microscopy methods. Recombinant adenovirus manufacture and transduction, western blotting, and luciferase assays are often used in conjunction with our high-content microscopy methods, and although we may show data in figures, a detailed account of these techniques is beyond the scope of this review. However, we have included supporting data gleaned using these methods for illustrative purposes.

2. Materials

2.1. Cell Culture and siRNA Transfection

1. HeLa and MCF7 cells (European Collection of Cell Cultures).
2. Dulbecco's Modified Eagle's Medium (DMEM) (Invitrogen, Paisley, UK).
3. Opti-MEM minimum essential medium (Invitrogen).
4. Clear bottomed, black-walled 96-well plates for imaging (Corning Costar plate 3904).
5. Fetal calf serum (FCS) (First Link, Brierly Hill, UK).
6. Trypsin (0.25%) and ethylenediamine tetra-acetic acid (EDTA) mix (Invitrogen) dissolved in phosphate buffered saline (PBS) (Sigma, Poole, UK).
7. PBS (Sigma): 137 mM NaCl, 2.7 mM KCl, 4.3 mM Na_2HPO_4, and 1.47 mM KH_2PO_4, pH 7.4.

8. siGENOME, non-targeting siRNA sequence no. 2, and DUSP targeting SMARTpools (Dharmacon, Cramlington, UK).
9. ERK1/2 siRNA sequences (Qiagen, Crawley, UK):
   ERK1;
   5′-CGUCUAAUAUAUAAAUAUAdTdT-3′ (sense)
   5′-UAUAUUUAUAUAUUAGACGdGdG-3′ (antisense)
   5′-CCCUGACCCGUCUAAUAUAdTdT-3′ (sense)
   5′-UAUAUUAGACGGGUCAGGGdAdG-3 (antisense).
   ERK2;
   5′-CACUUGUCAAGAAGCGUUAdTdT-3′ (sense)
   5′-UAACGCUUCUUGACAAGUGdTdT-3′ (antisense)
   5′-CAUGGUAGUCACUAACAUAdTdT-3′ (sense)
   5′-UAUGUUAGUGACUACCAUGdAdT-3′ (antisense).
10. Lipofectamine RNAiMAX transfection reagent (Invitrogen).
11. EGF (Sigma) is stored in 10 mM acetic acid containing 0.1% bovine serum albumin (BSA) and stored in single-use aliquots at −20°C. Working stock is prepared by dilution in PBS.
12. GnRH (Sigma) is stored in 10 mM acetic acid containing 0.1% bovine serum albumin (BSA) and stored in single-use aliquots at −20°C. Working stock is prepared by dilution in PBS.
13. Phorbol 12, 13 dibutyrate (PDBu) (Sigma) is stored in dimethyl sulphoxide (DMSO) in single use aliquots at −20°C. Working stock is prepared by dilution in PBS.

2.2. Immunofluorescence Staining of ERK1/2 and ppERK1/2

1. Paraformaldehyde (PFA) (Sigma) made up as a 10% stock solution in PBS. Warm PBS to 65°C, add PFA, and stir, allowing the solution to cool to room temperature overnight in a fume cupboard. Store in aliquots at −20°C, and dilute to 4% PFA in PBS before use.
2. 100% methanol (Merck).
3. Normal goat serum (Sigma) stored in single-use aliquots at −20°C. Working stock is prepared by dilution into a 5% solution with PBS.
4. Anti ERK1/2 rabbit monoclonal antibody (clone 137F5) (Cell Signaling Technology).
5. Anti ppERK1/2 mouse monoclonal antibody (clone MAPK-YT) (Sigma).
6. Alexa 488-conjugated, highly cross adsorbed goat anti-mouse secondary antibody (Invitrogen).
7. Alexa 546-conjugated, highly cross adsorbed goat anti-mouse secondary antibody (Invitrogen).

8. Alexa 488-conjugated, highly cross adsorbed goat anti-rabbit secondary antibody (Invitrogen).
9. Alexa 546-conjugated, highly cross adsorbed goat anti-rabbit secondary antibody (Invitrogen).
10. 3 mM DAPI (4,6-diamidino-2-phenylindole) (Sigma) stock in water. Dilute 1:5,000 in PBS for working solution (600 nM).

3. Methods

The high quality and specificity of antibodies available to ERK1/2 and ppERK1/2 enables sensitive detection of both at the single cell level using immunostaining of endogenous proteins. As noted in the introduction, we routinely use a system where endogenous ERK1/2 are removed from cells and are replaced with ERK2-GFP tagged variants (using adenoviral vectors) to allow either live imaging or the comparison of ERK2-GFP mutants at endogenous levels of expression. Firstly, this means that the siRNAs used must not interfere with ERK2-GFP expression, and secondly that responses of the endogenous ERK1/2 kinases and the extent of the ERK1/2 knock-down must be documented in each experiment to ensure that the ERK2-GFP fusion accurately reflects wild-type ERK behaviour. The multiwell high-content analysis format is ideal for this as directly quantitative information from individual cells can be collected from numerous stimulus conditions simultaneously. Thus, we routinely compare the level of total (ERK1/2 or ERK2-GFP) and ppERK1/2 (or ppERK2-GFP) in single cells, as well as the localisation of total and ppERK in the cytoplasm and nucleus. These data are usually paralleled with luciferase assays for ERK-dependent transcriptional activation (e.g. of the *egr-1* promoter) to give a comprehensive view of how changes in ERK activity and compartmentalisation impact on downstream consequences.

3.1. "Reverse" Transfection of Cells with siRNAs

The "reverse" transfection protocol merely refers to the fact that transfection mixtures are spotted onto plates prior to cells and media. We routinely use this method for siRNA transfection, partly because it saves time (cell plating and transfection occurs simultaneously) and also because we have found it to be more efficacious than conventional "forward" transfection.

1. Amounts detailed here are for the transfection of a single well of a 96-well plate at a concentration of 1 nM siRNA. Assays are then scaled up accordingly. Mix either 0.12 pmol. control siRNAs (either scrambled siRNAs that do not target mRNA sequences or siRNAs known to target non-human sequences,

such as luciferase or GFP) or ERK1/2 siRNAs (all four ERK1/2 duplexes mixed in equal proportions) or 10 nM SMARTpool siRNAs (mixture of four duplexes) with a total volume of 20 μl Opti-MEM. Make up enough for ~10 wells extra to compensate for volumes lost during liquid handling.

2. Add 0.2 μl Lipofectamine RNAiMAX per well (or per 0.12 pmol. of siRNA) and mix gently by pipetting the mix up and down four to five times. Spot the transfection mixtures on to tissue culture-treated 96-well black-wall microscopy plates at 20 μl/well using a repeater pipette. Incubate at room temperature for 10–20 min.
3. Meanwhile, trypsinise a subconfluent flask of cells kept in antibiotic-free media (antibiotics in the media can inhibit the formation of transfection complexes). Dilute cells to a suspension containing 5×10^3 HeLa cells per 100 μl of 10% FCS containing DMEM (cell number can vary for other cell types, e.g. 7×10^3 per well for MCF7 cells).
4. Add 100 μl cell suspension/well containing siRNA transfection mixtures. Return cells to the incubator overnight.
5. Transduce cells with adenovirus expressing mouse GnRHR (normally 3×10^6 plaque forming units (pfu) per ml for expression of receptors in the physiological range in HeLa cells). If necessary, transduce cells with an appropriate adenoviral titre of wild-type or mutated ERK2-GFP vector (normally 2×10^6 pfu/ml for reconstitution of endogenous ERK levels in HeLa cells). To do this, remove media from cells and replace with normal media (10% FCS DMEM) containing adenovirus at 100 μl/well. Return cells to the incubator for 4 h.
6. Remove media from all cells, wash once with PBS, and replace with 90 μl/well DMEM containing 0.1% FCS to serum-starve the cells overnight. In HeLa cells (and many other cell types), this is necessary for cells to adopt the normal "resting" ERK localisation in the cytoplasm rather than being in equilibrium between the nucleus and cytosol in cells chronically stimulated with serum.

3.2. Fixation and Staining of Cells for ERK1/2 and ppERK1/2

1. 16 h after media change, add agonists to cells for the desired period. We normally dilute stocks of agonist on the day of use in PBS to 10× the desired concentration, allowing the rapid addition of 10 μl of the 10× agonist to the 90 μl of media covering cells to stimulate cells without necessitating a further media change.
2. Wash cells in 150 μl/well cold (<4°C) PBS and place cells on ice. Dilute 10% PFA to 4% in PBS before use. Remove PBS from cells and fix by adding 50 μl/well of 4% PFA in PBS and

incubating on a rocking platform for 10 min at room temperature.

3. Remove PFA and permeabilise cells by adding 50 μl/well of −20°C methanol and incubating for 5 min in the freezer.
4. Remove methanol and wash cells with 150 μl/well PBS at room temperature.
5. Add 30 μl/well 5% normal goat serum in PBS and incubate plates at room temperature on a rocking platform for 2 h to block cells for immunostaining.
6. Wash cells once with PBS.
7. Add 1:200 dilution of primary mouse anti-ppERK1/2 antibody in 1% normal goat serum/PBS at 30 μl/well, with or without 1:100 dilution of primary rabbit anti-ERK1/2. Incubate overnight on a rocking platform at 4°C.
8. Wash cells three times with PBS.
9. Remove PBS and add 1:200 dilution of either or both Alexa 546 or 488 labelled goat anti-mouse or anti-rabbit secondary antibody in 1% normal goat serum/PBS at 30 μl/well. Incubate for 90 min on a rocking platform at room temperature.
10. Wash cells three times with PBS.
11. Remove PBS and add 100 μl/well of 600 nM DAPI in PBS. Leave for at least 20 min at room temperature before acquiring and analysing images, or store at 4°C until ready to process.

3.3. High Content Microscopy and Analysis

Protocols outlined here specifically pertain to the use of the IN Cell Analyzer 1000 microscope and associated IN Cell Investigator software, although the assay should be adaptable to most high-content imaging and analysis platforms.

1. Plates must firstly be equilibrated to room temperature prior to processing, to prevent condensation forming on the bottom of the plate (which can impair image quality). Clean the underside of the plate surface with a lens cleaning tissue.
2. Ensure that the 10× objective is inserted and that trichroic mirror number 60012 is in the light path on the carousel under the objective. Dock the plate into the microscope, ensuring that well A1 is in the top left-hand corner.
3. Set up an acquisition protocol to capture three fluorescent images per well, and check that the plate dimensions programmed are correct for Costar plate type 3904.
4. In the "image acquisition" window, set excitation (ex) filters to 360 nm (for DAPI), 475 nm (for GFP or Alexa 488), and 535 nm (for Alexa 546) and emission (em) filters at 460 nm,

535 nm, and 620 nm, respectively. Set exposure times at 600 ms for DAPI (360 nm ex, 460 nm em), 2 s for GFP/ Alexa 488 (475 nm ex, 535 nm em), and 2 s for Alexa 546 (535 nm ex, 620 nm em). Try "autofocus" option on a well where you expect cells to be and you also expect a strong signal from all three fluorophores, i.e. choose a well in which plenty of cells are plated and you expect ERK to be maximally expressed and phosphorylated. Click the "auto-offset" button and set the parameters to search in 5 μm steps. The microscope will now automatically find the optimal distance above the plastic-PBS interface of the sample plate at which to capture images to give the best image quality.

5. Scroll on to the "acquisition options" window and set to capture two random fields from near the centre of each well from each fluorophore.
6. Once the acquisition protocol is set up, do not change it between assays, but redefine the "auto-offset" at the start of each session. Acquire images from sample wells as appropriate.
7. Once acquired images are saved on to the server, proceed to off-line image analysis, using the "multi-target analysis" module in IN Cell Investigator software. In the table requesting "define input images", scroll down and check the box next to "Reference 1". You should now have ticks next to the labels marked "Nuclei", "Cells", and "Reference 1".
8. Define input images for the nuclei (blue, wavelength 1), ERK2-GFP/total ERK1/2 (green, "cells", wavelength 2), and for ppERK1/2 (red, "Reference 1", wavelength 3). This now means the software will define nuclear and cytosolic regions using DAPI and total ERK1/2 (or ERK2-GFP) stains, and use this definition as a "mask" to record data from the ppERK1/2 readout in the third wavelength.
9. In the "segmentation" window use "Top-hat" method to segment nuclei. Type in a minimal typical area of a nucleus in the box (usually around 50 μm^2). Set the sensitivity to around 65, and then click on the "filter" button. Check the box where it says: "filter objects according to size", and set the program to "keep objects below 500 μm^2"-this will stop any large objects that aren't really nuclei from being included in the analysis. Now click on the "Cells" bar in the table and select either "Multiscale top-hat" or "Region growing". Type in a characteristic area for the cells (usually 200–300 μm^2 as a starting point), with the sensitivity set to 60. Again, click the "filter" button and set this to "keep objects below 2,000 μm^2". Click on "Reference 1" bar to highlight it. Where it says "Use objects from", select "Cells" from the drop-down menu next to it.

10. It is important to rigorously define single cells for analysis in most cases, which means filters need to be introduced to ensure this is the case. Be sure to include all measures in the analysis report. In the "filters" window, use a "decision tree" to first include a filter that will exclude aberrantly defined nuclei. Use the "chord ratio" measure of nuclei (a ratio of the longest to the shortest line across the nucleus, passing through a central point) to exclude elongated shapes (these are generally shapes where the software has been unable to distinguish between several nuclei). Add a second filter to exclude abnormally bright nuclear shapes, as these are usually contaminating particles in the well and not actual cells. Similarly, add intensity filters to exclude bright shapes in both the green ("Cells") and red ("Reference 1") wavelengths as well, as dust particles can fluoresce brightly at longer wavelengths without affecting the DAPI stain. Finally, a filter for use in ERK2-GFP expressing cells can be added to exclude cells that are expressing the transgene below or beyond endogenous levels. This should be defined in control experiments where endogenous ERK levels are compared to those in ERK1/2 siRNA-transfected cells and in cells expressing Ad wild-type and mutant ERK2-GFP. Immunostaining and western blotting with anti-ERK1/2 should be used to compare ERK expression levels in lysates and individual cells, and define both Ad titres and filter ranges (see Note 3).
11. Set the program to export as MS Excel file and XML text file and to save the IN Cell Analyzer data file. Retrieve the MS Excel file from the server for data analysis.

3.4. Results and Conclusions

Spatiotemporal aspects of ERK signalling dictate the downstream consequences of ERK activation. With this in mind, we have developed efficient methods for monitoring ERK activity and compartmentalisation based on knock-down of endogenous ERKs with inhibitory RNA and add-back of wild-type or mutant ERK2-GFP reporters. Validation of these assays has revealed that the ERK2-GFP reporter faithfully mirrors the behaviour of endogenous ERKs (in terms of potency, amplitude, kinetics, and compartmentalisation of ERK responses with a range of different stimuli). Here, a key advantage of this model is that it enables reporters to be expressed at levels that are physiologically relevant for endogenous ERKs. The model also enables function to be probed by "add-back" of reporters containing mutant ERKs. The use of automated cell imaging end-points has clear advantages in terms of throughput, and this has facilitated its use in screening for DUSP effects using RNA inhibition. It can also be used in combination with other imaging readouts (e.g. cell morphology, cell cycle information from nuclear stains, and imaging of co-expressed receptors) providing a greater depth of information than with simpler "single variable" experimental readouts. It also provides the potential for single cell analysis enabling, for example, frequency distributions to be explored with a single experimental end-point (e.g. ppERK2 or N:C ERK2-GFP) or relationships between receptor expression and response to be explored on a cell-by-cell basis. These aspects of the high-content ERK imaging assays are illustrated in Figs. 1–3 and are described in more detail in refs. 31–33.

4. Notes

1. *Assay validation.* Automated imaging of ERK1/2, ppERK1/2, ERK2-GFP, or ppERK2-GFP can greatly increase throughput for interrogation of ERK signalling, but validation is necessary to ensure that the reporter faithfully mimics behaviour of the endogenous ERKs. We have found that ERK2-GFP and ppERK2-GFP responses are very similar to those seen with immunostaining of endogenous ERK1/2 and ppERK1/2 in time-course and dose–response studies with stimuli giving characteristically brief or sustained responses. Moreover, whole cell ppERK2-GFP responses measured by western blotting mirror those seen with endogenous ERKs (Fig. 1, refs. 31–33, and unpublished data). Although these protocols have been extensively validated in HeLa and MCF7 cells, further validation of this nature would be needed before work in other cell types.
2. *Expression level in the cell population.* Many groups have found that over-expression of ERKs increases their expression in the nucleus as cytoplasmic anchors are saturated. This can be prevented by over-expression of MEK which anchors ERKs in the cytoplasmic and can recapitulate the distribution of endogenous ERKs in unstimulated cells (8, 9, 25, 34). However, we considered it inappropriate to monitor spatiotemporal aspects of ERK signalling with a cytoplasmic anchor over-expressed and therefore developed the knock-down/add-back protocol described above. Here, the aim is normally to reduce endogenous ERK expression as much as possible and then to add-back ERK2-GFP to a physiological level (i.e. to a level that would be physiological for endogenous ERKs). Efficiency of the knock-down and expression levels after add-back can be assessed by western blotting or immunostaining (Fig. 1). This would normally require optimisation of transfection protocols (for knock-down with siRNA) and transduction protocols (choice of Ad ERK2-GFP titre)

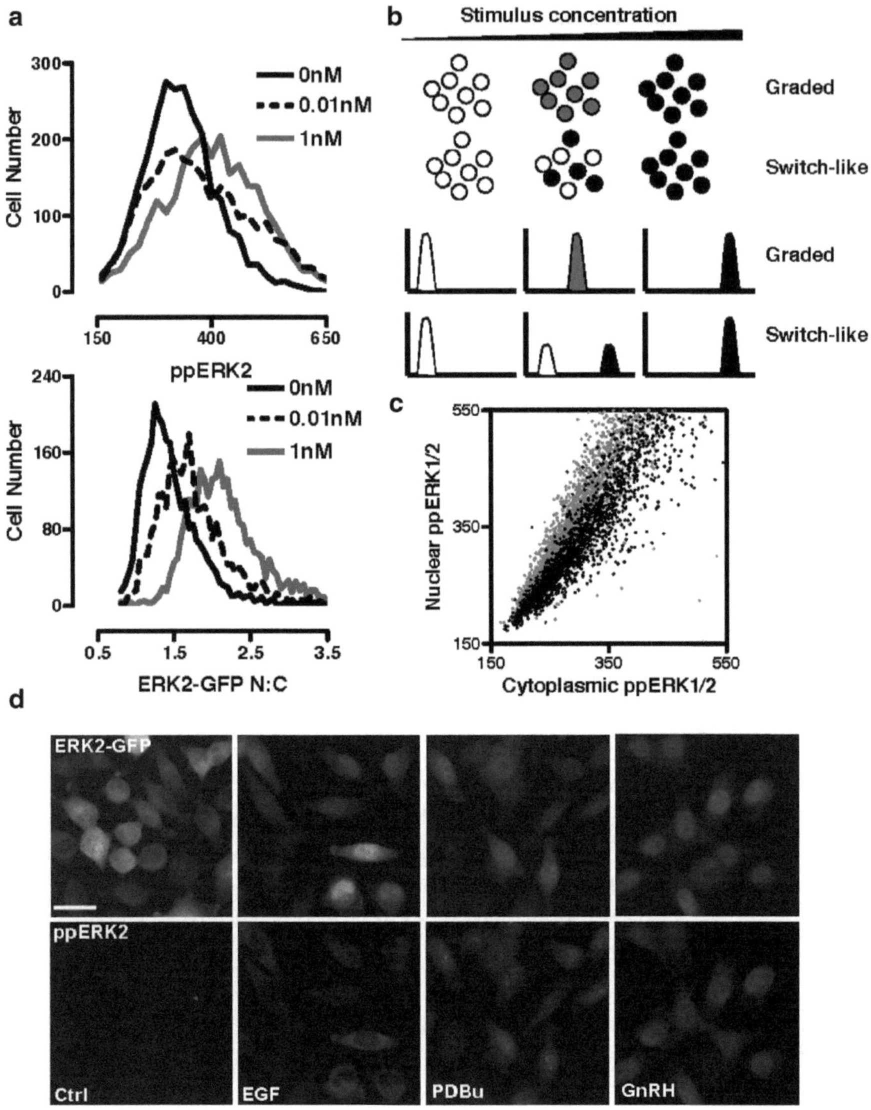

Fig. 2. Single-cell analysis of ERK phosphorylation and localisation. HeLa cells were transfected in 96-well plates with 1 nM ERK1/2 siRNAs before addition of Ad ERK2-GFP. It has been postulated that differences in ERK signal transmission can be achieved through graded or switch-like phosphorylation and/or nuclear translocation (see ref. 23), which can only be revealed by single cell analysis. *Panel* (**a**) The graphs show frequency distribution plots of binned single cell data derived from cells stimulated with 0, 0.01 or 1 nM EGF prior to ppERK staining and image analysis (as shown in Fig. 1) showing EGF-induced changes in ERK phosphorylation (ppERK2, *top panel*) and distribution (ERK2-GFP N:C ratio, *bottom panel*). In both cases, increasing stimulus causes graded responses in these cells. *Panel* (**b**) describes how graded versus "switch-like" (all or nothing) ERK signalling can occur in cells, and how these modes would differ in frequency distribution profiles. *Panel* (**c**) The scatterplot represents single cell data derived from cells stimulated for 5 min with 10 nM EGF (*black dots*) or 1 μM GnRH (*grey dots*) prior to ppERK1/2 staining and image analysis (as shown in Fig. 1), where each dot represents a single cell. The intensity of ppERK1/2 staining in cytoplasmic and nuclear regions is shown on the *x* and *y*-axis, respectively. *Panel* (**d**) Representative images of cells are shown, in which cells were stimulated for 5 min with control vehicle (Ctrl), 10 nM EGF, 1 μM PDBu or 1 μM GnRH, as indicated, prior to ppERK1/2 staining and image acquisition (as shown in Fig. 1). It is clear from the scatterplot in *Panel* (**c**) and the images in *Panel* (**d**) that the GnRH (and PDBu)-mediated ppERK2 signal is much more nuclear than the EGF-induced response, despite the fact that there is little difference in the net redistribution of ERK2-GFP in the same cells.

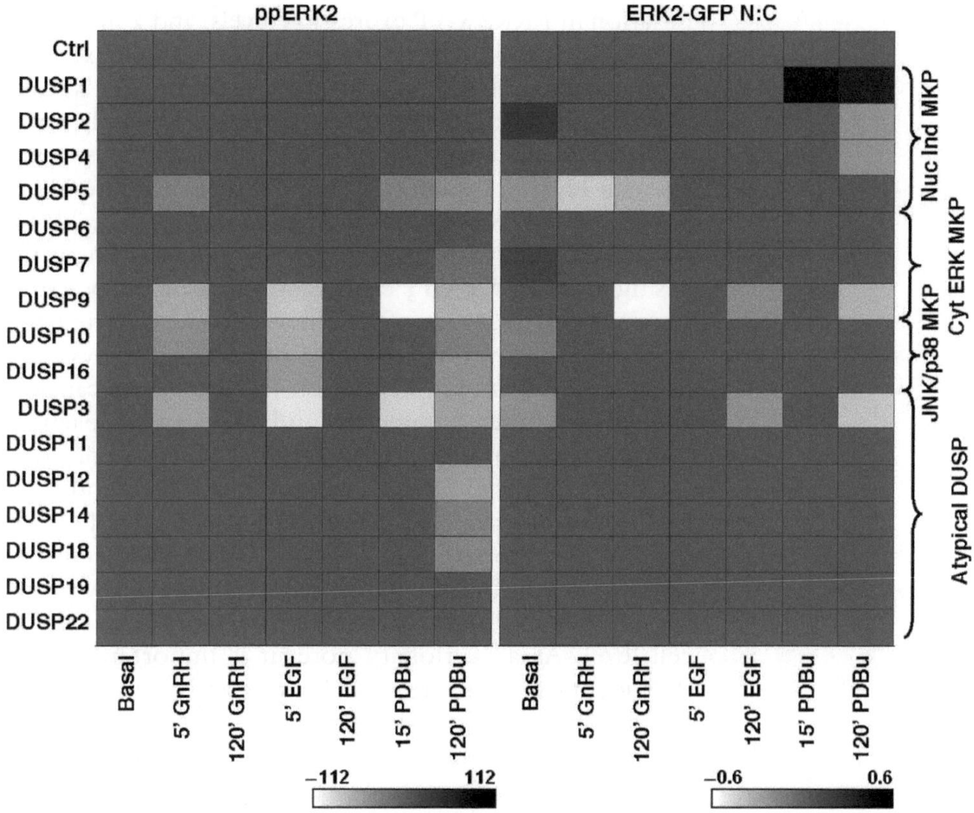

Fig. 3. Using image-based readouts to study ERK regulation by DUSPs in large scale loss of function assays. Cells were transfected in 96-well plates with 1 nM ERK1/2 siRNAs and 10 nM control (*Ctrl*) or siRNA SMARTpools targeting individual DUSPs (as indicated) before addition of Ad ERK2-GFP. Cells were stimulated with 1 μM GnRH, 10 nM EGF or 1 μM PDBu as indicated, prior to ppERK2 staining and imaging. Data are expressed in the heat map as the extent of difference above or below control values for each condition and time point for ppERK2 intensity (*left panel*) and ERK2-GFP N:C ratio (*right panel*) from four separate experiments performed in duplicate. Targets are grouped according to sequence similarity and substrate specificity. Statistical analysis was performed using one-way ANOVA and Dunnett's post hoc test, accepting $p<0.05$ as significant. Non-significant changes are shown as *grey mid-colour blocks* for both experiments. Figure adapted from refs. 31, 33.

since efficiency of transfection and transduction varies greatly between different cell types. Another cautionary note here is that protein expression levels should be monitored throughout the duration of the experimental time-course. Many vectors utilise strong viral promoters that can be regulated by growth factors and other stimuli, and their effect on exogenous ERK2 levels throughout the period of study could affect data interpretation.

3. *Expression levels in individual cells.* One of the advantages of high-content microscopy is the unbiased acquisition of data from vast numbers of individual cells which can be systematically filtered as required. Using the knock-down/add-back protocol described above, we have noted considerable cell-to-cell

variation in ERK2-GFP expression levels, and a subset of cells is typically seen in which ERK2-GFP expression exceeds that of endogenous ERK1/2. In most experiments, we aim to work at physiological levels so we routinely apply an automated data filter to exclude over-expressing cells from the analysis. Here, it is important to note that such filtering is not possible with western blotting where data are inevitably biased toward the cells with highest expression levels. The variation in single-cell ERK2-GFP expression level can also be exploited to probe the relevance of stoichiometry (i.e. filtering cells to follow ERK activation with physiological, sub-physiological, or super-physiological expression levels). Filters can also be applied on other parameters and this can be particularly valuable where multiple reporters are expressed. For example, we have used the ERK knock-down/add-back protocol to monitor responses to co-expressed epitope-tagged G-protein coupled receptors. Here, cells can be filtered according to receptor expression, enabling the relationship between receptor number and ERK activation to be monitored at the single cell level. As a cautionary note, it is important to ensure that identical filters are used for test and control groups to avoid user bias. Typically, filters are set by examination of frequency distribution plots for the entire (unfiltered) data set or are set using machine-assisted classifiers. Additionally, analysis algorithm definition and filter limits should be regularly monitored by checking cell definition and numbers in each experiment.

4. *Relevance of cell shape and widefield image acquisition.* We use an IN Cell Analyzer 1000 high-content analysis platform with a wide-field fluorescence microscope and routinely acquire images with a 10× objective for nuclear translocation assays. With this system, the depth of field is comparable to the depth of the cell (we routinely work with HeLa and MCF7 cells), but it is important to recognise that with this system the mean fluorescence intensity measured is dependent not only on the fluorophore concentration but also on the depth of the structure containing the fluorophore. For cells in which the nuclear region is deeper than the cytoplasmic region (many cells have a "fried egg"-like profile in culture) or for cells where there is considerable cytoplasm above or below the nucleus, N:C ratios based on mean fluorescence intensities will underestimate the actual N:C ratio of fluorophore concentrations. For most assays, this is not a major problem (because the underestimation is identical in test and control groups), but where more precise information is needed, a "pseudo-confocality" module can be used to obtain *z*-section data and a more accurate measure of fluorophore concentration (albeit with considerable

loss of throughput). A related issue is that treatments causing changes in cell shape could conceivably also influence the measured N:C ratio. For example, cell rounding could reduce the cross-sectional area of the cytoplasm which, without any change in fluorophore concentration, would increase the mean fluorescence intensity of the cytoplasmic segment. If this occurs, it may be necessary to calculate N:C ratios using integrated measures of nuclear and cytoplasmic fluorescence (rather than mean fluorescence intensity). We have seen little evidence of shape change with the assays described herein and have found that spatiotemporal aspects of ERK responses are generally indistinguishable using mean fluorescence and integrated fluorescence for calculation of N:C ratios, but this issue may well prove important in models where pronounced changes in cell shape occur.

References

1. Caunt, C. J., Finch, A. R., Sedgley, K. R., and McArdle, C. A. (2006) GnRH receptor signalling to ERK: kinetics and compartmentalization, *Trends Endocrinol. Metab. 17*, 308–313.
2. Caunt, C. J., Finch, A. R., Sedgley, K. R., and McArdle, C. A. (2006) Seven-transmembrane receptor signalling and ERK compartmentalization, *Trends Endocrinol. Metab. 17*, 276–283.
3. Ebisuya, M., Kondoh, K., and Nishida, E. (2005) The duration, magnitude and compartmentalization of ERK MAP kinase activity: mechanisms for providing signaling specificity, *J. Cell Sci. 118*, 2997–3002.
4. Murphy, L. O. and Blenis, J. (2006) MAPK signal specificity: the right place at the right time, *Trends Biochem. Sci. 31*, 268–275.
5. Raman, M., Chen, W., and Cobb, M. H. (2007) Differential regulation and properties of MAPKs, *Oncogene 26*, 3100–3112.
6. Costa, M., Marchi, M., Cardarelli, F., Roy, A., Beltram, F., Maffei, L., and Ratto, G. M. (2006) Dynamic regulation of ERK2 nuclear translocation and mobility in living cells, *J. Cell Sci. 119*, 4952–4963.
7. Marchi, M., D'Antoni, A., Formentini, I., Parra, R., Brambilla, R., Ratto, G. M., and Costa, M. (2008) The N-terminal domain of ERK1 accounts for the functional differences with ERK2, *PLoS One 3*, e3873.
8. Fukuda, M., Gotoh, I., Adachi, M., Gotoh, Y., and Nishida, E. (1997) A novel regulatory mechanism in the mitogen-activated protein (MAP) kinase cascade. Role of nuclear export signal of MAP kinase kinase, *J. Biol. Chem. 272*, 32642–32648.
9. Fukuda, M., Gotoh, Y., and Nishida, E. (1997) Interaction of MAP kinase with MAP kinase kinase: its possible role in the control of nucleocytoplasmic transport of MAP kinase, *EMBO J. 16*, 1901–1908.
10. Millar, R. P., Lu, Z. L., Pawson, A. J., Flanagan, C. A., Morgan, K., and Maudsley, S. R. (2004) Gonadotropin-releasing hormone receptors, *Endocr. Rev. 25*, 235–275.
11. McArdle, C. A., Franklin, J., Green, L., and Hislop, J. N. (2002) The gonadotrophin-releasing hormone receptor: signalling, cycling and desensitisation, *Arch. Physiol. Biochem. 110*, 113–122.
12. McArdle, C. A., Franklin, J., Green, L., and Hislop, J. N. (2002) Signalling, cycling and desensitisation of gonadotrophin-releasing hormone receptors, *J. Endocrinol. 173*, 1–11.
13. Belchetz, P. E., Plant, T. M., Nakai, Y., Keogh, E. J., and Knobil, E. (1978) Hypophysial responses to continuous and intermittent delivery of hypothalamic gonadotropin-releasing hormone, *Science 202*, 631–633.
14. Kanasaki, H., Bedecarrats, G. Y., Kam, K. Y., Xu, S., and Kaiser, U. B. (2005) Gonadotropin-releasing hormone pulse frequency-dependent activation of extracellular signal-regulated kinase pathways in perifused LbetaT2 cells, *Endocrinology 146*, 5503–5513.
15. Ferris, H. A. and Shupnik, M. A. (2006) Mechanisms for pulsatile regulation of the

gonadotropin subunit genes by GNRH1, *Biol. Reprod. 74*, 993–998.

16. Lawson, M. A., Tsutsumi, R., Zhang, H., Talukdar, I., Butler, B. K., Santos, S. J., Mellon, P. L., and Webster, N. J. (2007) Pulse sensitivity of the luteinizing hormone beta promoter is determined by a negative feedback loop Involving early growth response-1 and Ngfi-A binding protein 1 and 2, *Mol. Endocrinol. 21*, 1175–1191.
17. Bliss, S. P., Miller, A., Navratil, A. M., Xie, J., McDonough, S. P., Fisher, P. J., Landreth, G. E., and Roberson, M. S. (2009) ERK signaling in the pituitary is required for female but not male fertility, *Mol. Endocrinol. 23*, 1092–1101.
18. Liu, F., Austin, D. A., Mellon, P. L., Olefsky, J. M., and Webster, N. J. (2002) GnRH activates ERK1/2 leading to the induction of c-fos and LHbeta protein expression in LbetaT2 cells, *Mol. Endocrinol. 16*, 419–434.
19. Shah, B. H., Farshori, M. P., Jambusaria, A., and Catt, K. J. (2003) Roles of Src and epidermal growth factor receptor transactivation in transient and sustained ERK1/2 responses to gonadotropin-releasing hormone receptor activation, *J. Biol. Chem. 278*, 19118–19126.
20. Shah, B. H., Soh, J. W., and Catt, K. J. (2003) Dependence of gonadotropin-releasing hormone-induced neuronal MAPK signaling on epidermal growth factor receptor transactivation, *J. Biol. Chem. 278*, 2866–2875.
21. Davidson, L., Pawson, A. J., Millar, R. P., and Maudsley, S. (2004) Cytoskeletal reorganization dependence of signaling by the gonadotropin-releasing hormone receptor, *J. Biol. Chem. 279*, 1980–1993.
22. Maudsley, S., Davidson, L., Pawson, A. J., Chan, R., de Maturana, R. L., and Millar, R. P. (2004) Gonadotropin-releasing hormone (GnRH) antagonists promote proapoptotic signaling in peripheral reproductive tumor cells by activating a Galphai-coupling state of the type I GnRH receptor, *Cancer Res. 64*, 7533–7544.
23. MacKeigan, J. P., Murphy, L. O., Dimitri, C. A., and Blenis, J. (2005) Graded mitogen-activated protein kinase activity precedes switch-like c-Fos induction in mammalian cells, *Mol. Cell Biol. 25*, 4676–4682.
24. Caunt, C. J., Finch, A. R., Sedgley, K. R., Oakely, L., Luttrell, L. M., and McArdle, C. A. (2006) Arrestin-mediated ERK activation by gonadotropin-releasing hormone receptors (GnRHRs): Receptor-specific activation mechanisms and compartmentalization, *J. Biol. Chem. 281*, 2701–2710.
25. Chuderland, D., Konson, A., and Seger, R. (2008) Identification and characterization of a general nuclear translocation signal in signaling proteins, *Mol. Cell 31*, 850–861.
26. Burack, W. R. and Shaw, A. S. (2005) Live Cell Imaging of ERK and MEK: simple binding equilibrium explains the regulated nucleocytoplasmic distribution of ERK, *J. Biol. Chem. 280*, 3832–3837.
27. Fujioka, A., Terai, K., Itoh, R. E., Aoki, K., Nakamura, T., Kuroda, S., Nishida, E., and Matsuda, M. (2006) Dynamics of the Ras/ERK MAPK cascade as monitored by fluorescent probes, *J. Biol. Chem. 281*, 8917–8926.
28. Ando, R., Mizuno, H., and Miyawaki, A. (2004) Regulated fast nucleocytoplasmic shuttling observed by reversible protein highlighting, *Science 306*, 1370–1373.
29. Luttrell, L. M., Roudabush, F. L., Choy, E. W., Miller, W. E., Field, M. E., Pierce, K. L., and Lefkowitz, R. J. (2001) Activation and targeting of extracellular signal-regulated kinases by beta-arrestin scaffolds, *Proc. Natl. Acad. Sci. USA 98*, 2449–2454.
30. Luttrell, L. M., Ferguson, S. S., Daaka, Y., Miller, W. E., Maudsley, S., Della Rocca, G. J., Lin, F., Kawakatsu, H., Owada, K., Luttrell, D. K., Caron, M. G., and Lefkowitz, R. J. (1999) Beta-arrestin-dependent formation of beta2 adrenergic receptor-Src protein kinase complexes, *Science 283*, 655–661.
31. Armstrong, S. P., Caunt, C. J., and McArdle, C. A. (2009) Gonadotropin-releasing hormone and protein kinase C signaling to ERK: spatiotemporal regulation of ERK by docking domains and dual-specificity phosphatases, *Mol. Endocrinol. 23*, 510–519.
32. Caunt, C. J., Rivers, C. A., Conway-Campbell, B. L., Norman, M. R., and McArdle, C. A. (2008) Epidermal growth factor receptor and protein kinase C signaling to ERK2: spatiotemporal regulation of ERK2 by dual specificity phosphatases, *J. Biol. Chem. 283*, 6241–6252.
33. Caunt, C. J., Armstrong, S. P., Rivers, C. A., Norman, M. R., and McArdle, C. A. (2008) Spatiotemporal regulation of ERK2 by dual specificity phosphatases, *J. Biol. Chem. 283*, 26612–26623.
34. Horgan, A. M. and Stork, P. J. (2003) Examining the mechanism of Erk nuclear translocation using green fluorescent protein, *Exp. Cell Res. 285*, 208–220.

Index

Rony Seger (ed.), *MAP Kinase Signaling Protocols: Second Edition*, Methods in Molecular Biology, vol. 661, DOI 10.1007/978-1-60761-795-2, © Springer Science+Business Media, LLC 2010

K

L

M

N

O

P

T

U

V

W

X

Y

MIX
Papier aus verantwortungsvollen Quellen
Paper from responsible sources
FSC® C105338

If you have any concerns about our products,
you can contact us on
ProductSafety@springernature.com

In case Publisher is established outside the EU,
the EU authorized representative is:
Springer Nature Customer Service Center GmbH
Europaplatz 3, 69115 Heidelberg, Germany

Printed by Libri Plureos GmbH
in Hamburg, Germany